Stochastic Methods and Their Applications to Communications

Stochastic Methods and Their Applications to Communications
Stochastic Differential Equations Approach

Serguei Primak
University of Western Ontario, Canada

Valeri Kontorovich
Cinvestav-IPN, Mexico

Vladimir Lyandres
Ben-Gurion University of the Negev, Israel

John Wiley & Sons, Ltd

Other Wiley Editorial Offices

John Wiley & Sons Inc., 111 River Street, Hoboken, NJ 07030, USA

Jossey-Bass, 989 Market Street, San Francisco, CA 94103-1741, USA

Wiley-VCH Verlag GmbH, Boschstr. 12, D-69469 Weinheim, Germany

John Wiley & Sons Australia Ltd, 33 Park Road, Milton, Queensland 4064, Australia

John Wiley & Sons (Asia) Pte Ltd, 2 Clementi Loop #02-01, Jin Xing Distripark, Singapore 129809

John Wiley & Sons Canada Ltd, 22 Worcester Road, Etobicoke, Ontario, Canada M9W 1L1

Wiley also publishes its books in a variety of electronic formats. Some of the content that appears in
print may not be available in electronic books.

British Library Cataloguing in Publication Data

A catalogue record for this book is available from the British Library

0-470-84741-7

Typeset in 10/12pt Times by Thomson Press (India) Limited, New Delhi
Printed and bound in Great Britain by Antony Rowe Ltd, Chippenham, Wiltshire
This book is printed on acid-free paper responsibly manufactured from sustainable forestry
in which at least two trees are planted for each one used for paper production.

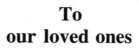

To
our loved ones

Contents

1. **Introduction** 1
   1.1 Preface 1
   1.2 Digital Communication Systems 3

2. **Random Variables and Their Description** 7
   2.1 Random Variables and Their Description 7
      2.1.1 Definitions and Method of Description 7
          2.1.1.1 Classification 7
          2.1.1.2 Cumulative Distribution Function 8
          2.1.1.3 Probability Density Function 9
          2.1.1.4 The Characteristic Function and the
                    Log-Characteristic Function 10
          2.1.1.5 Statistical Averages 11
          2.1.1.6 Moments 12
          2.1.1.7 Central Moments 12
          2.1.1.8 Other Quantities 13
          2.1.1.9 Moment and Cumulant Generating
                    Functions 14
          2.1.1.10 Cumulants 15
   2.2 Orthogonal Expansions of Probability Densities:
       Edgeworth and Laguerre Series 16
      2.2.1 The Edgeworth Series 17
      2.2.2 The Laguerre Series 20
      2.2.3 Gram–Charlier Series 22
   2.3 Transformation of Random Variables 23
      2.3.1 Transformation of a Given PDF into an Arbitrary PDF 25
      2.3.2 PDF of a Harmonic Signal with Random Phase 25
   2.4 Random Vectors and Their Description 26
      2.4.1 CDF, PDF and the Characteristic Function 26
      2.4.2 Conditional PDF 28
      2.4.3 Numerical Characteristics of a Random Vector 30
   2.5 Gaussian Random Vectors 32
   2.6 Transformation of Random Vectors 35
      2.6.1 PDF of a Sum, Difference, Product and Ratio
          of Two Random Variables 37
      2.6.2 Probability Density of the Magnitude and the
          Phase of a Complex Random Vector with
          Jointly Gaussian Components 39
          2.6.2.1 Zero Mean Uncorrelated Gaussian Components
                    of Equal Variance 41

     2.6.2.2 Case of Uncorrelated Components with Equal
         Variances and Non-Zero Mean. 41
   2.6.3 PDF of the Maximum (Minimum) of two Random Variables 42
   2.6.4 PDF of the Maximum (Minimum) of n Independent
     Random Variables. 44
 2.7 Additional Properties of Cumulants. 44
   2.7.1 Moment and Cumulant Brackets. 46
   2.7.2 Properties of Cumulant Brackets. 48
   2.7.3 More on the Statistical Meaning of Cumulants. 49
 2.8 Cumulant Equations 49
   2.8.1 Non-Linear Transformation of a Random Variable:
     Cumulant Method. 52
 Appendix: Cumulant Brackets and Their Calculations. 54

3. Random Processes. 59
 3.1 General Remarks. 59
 3.2 Probability Density Function (PDF). 60
 3.3 The Characteristic Functions and Cumulative
   Distribution Function. 63
 3.4 Moment Functions and Correlation Functions. 64
 3.5 Stationary and Non-Stationary Processes 70
 3.6 Covariance Functions and Their Properties. 71
 3.7 Correlation Coefficient. 74
 3.8 Cumulant Functions. 77
 3.9 Ergodicity 77
 3.10 Power Spectral Density (PSD) 80
 3.11 Mutual PSD 82
   3.11.1 PSD of a Sum of Two Stationary and Stationary
     Related Random Processes. 83
   3.11.2 PSD of a Product of Two Stationary Uncorrelated Processes 84
 3.12 Covariance Function of a Periodic Random Process 85
   3.12.1 Harmonic Signal with a Constant Magnitude 85
   3.12.2 A Mixture of Harmonic Signals 86
   3.12.3 Harmonic Signal with Random Magnitude and Phase 87
 3.13 Frequently Used Covariance Functions 88
 3.14 Normal (Gaussian) Random Processes. 88
 3.15 White Gaussian Noise (WGN) 95

4. Advanced Topics in Random Processes. 99
 4.1 Continuity, Differentiability and Integrability of a Random Process. 99
   4.1.1 Convergence and Continuity. 99
   4.1.2 Differentiability 100
   4.1.3 Integrability 102
 4.2 Elements of System Theory 103
   4.2.1 General Remarks 103
   4.2.2 Continuous SISO Systems 105
   4.2.3 Discrete Linear Systems 107
   4.2.4 MIMO Systems 109

4.2.5   Description of Non-Linear Systems. 110
4.3   Zero Memory Non-Linear Transformation of Random Processes 112
    4.3.1   Transformation of Moments and Cumulants. 112
        4.3.1.1   Direct Method 115
        4.3.1.2   The Rice Method. 116
    4.3.2   Cumulant Method 117
4.4   Cumulant Analysis of Non-Linear Transformation of Random Processes . . 118
    4.4.1   Cumulants of the Marginal PDF. 118
    4.4.2   Cumulant Method of Analysis of Non-Gaussian
        Random Processes 119
4.5   Linear Transformation of Random Processes 121
    4.5.1   General Expression for Moment and Cumulant Functions at the
        Output of a Linear System. 121
        4.5.1.1   Transformation of Moment and Cumulant Functions 122
        4.5.1.2   Linear Time-Invariant System Driven by a
                Stationary Process 125
    4.5.2   Analysis of Linear MIMO Systems. 131
    4.5.3   Cumulant Method of Analysis of
        Linear Transformations 132
    4.5.4   Normalization of the Output Process by a Linear System 137
4.6   Outages of Random Processes 140
    4.6.1   General Considerations 140
    4.6.2   Average Level Crossing Rate and the Average Duration
        of the Upward Excursions 141
    4.6.3   Level Crossing Rate of a Gaussian Random Process 145
    4.6.4   Level Crossing Rate of the Nakagami Process 149
    4.6.5   Concluding Remarks. 152
4.7   Narrow Band Random Processes. 152
    4.7.1   Definition of the Envelope and Phase of
        Narrow Band Processes 154
    4.7.2   The Envelope and the Phase Characteristics. 156
        4.7.2.1   Blanc-Lapierre Transformation. 156
        4.7.2.2   Kluyver Equation. 160
        4.7.2.3   Relations Between Moments of $p_{A_n}(a_n)$ and $p_i(I)$ 161
        4.7.2.4   The Gram–Charlier Series for $p_{\xi_R}(x)$ and $p_i(I)$ 163
    4.7.3   Gaussian Narrow Band Process 166
        4.7.3.1   First Order Statistics. 166
        4.7.3.2   Correlation Function of the In-phase and Quadrature
                Components 168
        4.7.3.3   Second Order Statistics of the Envelope 169
        4.7.3.4   Level Crossing Rate 172
    4.7.4   Examples of Non-Gaussian Narrow Band Random Processes. 173
        4.7.4.1 K Distribution 173
        4.7.4.2   Gamma Distribution. 175
        4.7.4.3   Log-Normal Distribution. 175
        4.7.4.4   A Narrow Band Process with Nakagami
                Distributed Envelope 177

4.8   Spherically Invariant Processes   181
      4.8.1   Definitions   181
      4.8.2   Properties   182
              4.8.2.1   Joint PDF of a SIRV   182
              4.8.2.2   Narrow Band SIRVs   183
      4.8.3   Examples   184

5. Markov Processes and Their Description **189**
   5.1   Definitions   189
         5.1.1   Markov Chains   190
         5.1.2   Markov Sequences   203
         5.1.3   A Discrete Markov Process   207
         5.1.4   Continuous Markov Processes   212
         5.1.5   Differential Form of the Kolmogorov–Chapman Equation   214
   5.2   Some Important Markov Random Processes   217
         5.2.1   One-Dimensional Random Walk   217
                 5.2.1.1   Unrestricted Random Walk   219
         5.2.2   Markov Processes with Jumps   221
                 5.2.2.1   The Poisson Process   221
                 5.2.2.2   A Birth Process   223
                 5.2.2.3   A Death Process   224
                 5.2.2.4   A Death and Birth Process   224
   5.3   The Fokker–Planck Equation   227
         5.3.1   Preliminary Remarks   227
         5.3.2   Derivation of the Fokker–Planck Equation   227
         5.3.3   Boundary Conditions   231
         5.3.4   Discrete Model of a Continuous Homogeneous
                 Markov Process   234
         5.3.5   On the Forward and Backward Kolmogorov Equations   235
         5.3.6   Methods of Solution of the Fokker–Planck Equation   236
                 5.3.6.1   Method of Separation of Variables   236
                 5.3.6.2   The Laplace Transform Method   243
                 5.3.6.3   Transformation to the Schrödinger Equations   244
   5.4   Stochastic Differential Equations   245
         5.4.1   Stochastic Integrals   246
   5.5   Temporal Symmetry of the Diffusion Markov Process   257
   5.6   High Order Spectra of Markov Diffusion Processes   258
   5.7   Vector Markov Processes   263
         5.7.1   Definitions   263
                 5.7.1.1   A Gaussian Process with a Rational Spectrum   270
   5.8   On Properties of Correlation Functions of One-Dimensional
         Markov Processes   271

6. Markov Processes with Random Structures **275**
   6.1   Introduction   275
   6.2   Markov Processes with Random Structure and Their
         Statistical Description   279

6.2.1   Processes with Random Structure and Their Classification.   279
6.2.2   Statistical Description of Markov Processes
       with Random Structure   280
6.2.3   Generalized Fokker–Planck Equation for Random Processes with
       Random Structure and Distributed Transitions   281
6.2.4   Moment and Cumulant Equations of a Markov Process
       with Random Structure   288
6.3   Approximate Solution of the Generalized Fokker–Planck Equations   295
    6.3.1   Gram–Charlier Series Expansion.   296
        6.3.1.1   Eigenfunction Expansion.   296
        6.3.1.2   Small Intensity Approximation   297
        6.3.1.3   Form of the Solution for Large Intensity.   302
    6.3.2   Solution by the Perturbation Method for the Case of Low
        Intensities of Switching   304
        6.3.2.1   General Small Parameter Expansion of Eigenvalues
            and Eigenfunctions.   304
        6.3.2.2   Perturbation of $\Psi_0(x)$ 305
    6.3.3   High Intensity Solution   310
        6.3.3.1   Zero Average Current Condition   310
        6.3.3.2   Asymptotic Solution $P_\infty(x)$.   311
        6.3.3.3   Case of a Finite Intensity v   314
6.4   Concluding Remarks   317

7. Synthesis of Stochastic Differential Equations.   321
7.1   Introduction   321
7.2   Modeling of a Scalar Random Process Using a First Order SDE   322
    7.2.1   General Synthesis Procedure for the First Order SDE   322
    7.2.2   Synthesis of an SDE with PDF Defined on a Part of the
        Real Axis.   326
    7.2.3   Synthesis of λ Processes   329
    7.2.4   Non-Diffusion Markov Models of Non-Gaussian Exponentially
        Correlated Processes   334
        7.2.4.1   Exponentially Correlated Markov Chain—DAR(1) and
            Its Continuous Equivalent   335
        7.2.4.2   A Mixed Process with Exponential Correlation   341
7.3   Modeling of a One-Dimensional Random Process on the
    Basis of a Vector SDE.   347
    7.3.1   Preliminary Comments   347
    7.3.2   Synthesis Procedure of a (λ, ω) Process.   347
    7.3.3   Synthesis of a Narrow Band Process Using a Second
        Order SDE.   351
        7.3.3.1   Synthesis of a Narrow Band Random Process Using a
            Duffing Type SDE   352
        7.3.3.2   An SDE of the Van Der Pol Type   356
7.4   Synthesis of a One-Dimensional Process with a Gaussian Marginal
    PDF and Non-Exponential Correlation.   361

7.5   Synthesis of Compound Processes............................. 364
      7.5.1   Compound Λ Process 365
      7.5.2   Synthesis of a Compound Process with a Symmetrical PDF...... 367
7.6   Synthesis of Impulse Processes.............................. 369
      7.6.1   Constant Magnitude Excitation......................... 370
      7.6.2   Exponentially Distributed Excitation...................... 371
7.7   Synthesis of an SDE with Random Structure 371

8. Applications **377**
8.1   Continuous Communication Channels 377
      8.1.1   A Mathematical Model of a Mobile Satellite
              Communication Channel 377
      8.1.2   Modeling of a Single-Path Propagation 380
              8.1.2.1   A Process with a Given PDF of the Envelope
                        and Given Correlation Interval................... 380
              8.1.2.2   A Process with a Given Spectrum and
                        Sub-Rayleigh PDF 383
8.2   An Error Flow Simulator for Digital Communication Channels 388
      8.2.1   Error Flow in Digital Communication Systems.............. 389
      8.2.2   A Model of Error Flow in a Digital Channel with Fading 389
      8.2.3   SDE Model of a Buoyant Antenna–Satellite Link............ 391
              8.2.3.1   Physical Model 391
              8.2.3.2   Phenomenological Model 392
              8.2.3.3   Numerical Simulation......................... 395
8.3   A Simulator of Radar Sea Clutter with a Non-Rayleigh Envelope....... 397
      8.3.1   Modeling and Simulation of the K-Distributed Clutter......... 397
      8.3.2   Modeling and Simulation of the Weibull Clutter.............. 404
8.4   Markov Chain Models in Communications...................... 408
      8.4.1   Two-State Markov Chain—Gilbert Model 408
      8.4.2   Wang–Moayeri Model 409
      8.4.3   Independence of the Channel State Model on the
              Actual Fading Distribution............................ 418
      8.4.4   A Rayleigh Channel with Diversity...................... 418
      8.4.5   Fading Channel Models............................... 419
      8.4.6   Higher Order Models 421
8.5   Markov Chain for Different Conditions of the Channel 422

Index **433**

As an extra resource we have set up a companion website for our book containing supplementary material devoted to the numerical simulation of stochastic differential equations and description, modeling and simulation of impulse random processes. Additional reference information is also available on the website. Please go to the following URL and have a look: ftp://ftp.wiley.co.uk/pub/books/primak/

1

Introduction

1.1 PREFACE

A statistical approach to consideration of most problems related to information transmission became dominant during the past three decades. This can be easily explained if one takes into account that practically any real signal, propagation media, interference and even information itself all have an intrinsically random nature.[1] This is why D. Middleton, one of the founders of modern communication theory, coined a term "statistical theory of communications" [2]. The recent spectacular achievements in information technology are based mainly on progress in three fundamental areas: communication and information theory, signal processing and computer and related technologies. All these allow the transmission of information at a rate close to that limited by the Shannon theorem [2].

In principle, the limitation on the speed of the transmission of information is defined by the noise and interference which are inherently present in all communication systems. An accurate description of such impairments is very important for a proper organization and noise immunity of communication systems. The choice of a relevant interference model is a crucial moment in design of communication systems and is an important step in their testing and performance evaluation. The requirements, formulated to improve the performance of systems are often conflicting and hard to formulate in a way convenient for optimization. On one side, such models must accurately reflect the main features of the interference under investigation. On the other hand, a maximally simple description is needed to be applicable for the massive numerical simulation required to test modern communication system designs.

Historically, two simple processes, so-called White Gaussian Noise (WGN) and the Poisson Point Process (PPP) have been widely used to obtain first rough estimates of a system performance: the WGN model of the noise is a good approximation of an additive wide band noise, caused by a variety of natural phenomena, while the PPP is a good model for event modeling in a discrete communication channel. Unfortunately, in the majority of realistic situations these basic models of the noise and errors are not adequate. In realistic communication channels, a non-stationary non-Gaussian interference and noise are often present. In addition, these processes are often band-limited, thus showing significant time correlation, which cannot be represented by WGN and PPP. The following phenomena can be considered as examples when the simplest models fail to provide an accurate description:

[1]Having said that, we would like to acknowledge an exponentially increasing body of literature which describes the same phenomena using a rather different approach, based on the chaotic description of signals [1].

Stochastic Methods and Their Applications to Communications.
S. Primak, V. Kontorovich, V. Lyandres
© 2004 John Wiley & Sons, Ltd   ISBN: 0-470-84741-7

fading in communication channels; interchannel/interuser interference; impulsive noise, including man-made noise. Of course this list can be greatly expanded [3]. Description of such phenomena using joint probability densities of order higher than two is difficult since a large amount of *a priori* information on the properties of the noise and interference is required. Such information is usually not available or difficult to obtain [4]. Furthermore, specification of higher order joint probability densities is usually not productive since it results in complicated expressions which cannot be effectively used at the current level of computer simulation. Substitution of non-Gaussian models by their equivalent Gaussian models may result in significant and unpredictable errors. All of this, coupled with increased complexity of the systems and reduced design time, requires simple and adequate models of the blocks, stages, complete systems and networks of the communication systems. Such a variety requires an approach which is flexible enough to cover the majority of these possibilities, and this is a major requirement for models of communication systems.

This book describes techniques which, in the opinion of the authors, are well suited for the task of effective modeling of non-Gaussian phenomena in communication systems and related disciplines. It is based on the Markov approach to modeling of random processes encountered in applications. In particular, non-Gaussian processes are modeled as the solution of an appropriate Stochastic Differential Equation (SDE), i.e. a differential equation with a random excitation.

This approach, in general phenomenological, is built on the idea of the system state, suggested by van Trees [5]. The essence of this method is to describe the process under investigation as a solution of some SDE synthesized based on some *a priori* information (such as marginal distribution and correlation functions). Such synthesis is an attempt to uncover hidden dynamics of the interference formation. Despite the fact that such an approach is very approximate,[2] the SDE approach to modeling of random processes has significant advantages:

- universality: a single structure allows the modeling of a great variety of different processes by simple variation of the system parameters and excitation;

- effectiveness: a single structure allows the modeling and numerical simulation of a spectrum of the characteristics of the random process, such as marginal probability density, correlation function, etc;

- the suggested models suit well computer simulation.

Unfortunately, the SDE approach to modeling is still not widely known as a tool in modeling of communication channels and related issues, despite its effective application in chemistry, physics and other areas [3]. Some work in this area first appeared more than thirty years ago in the Bonch–Bruevich Institute of Telecommunications (St Petersburg, Russia). The first attempt to summarize the results in the area of communication channel modeling based on the SDE approach resulted in a book [6], published in Russian and almost unavailable for international readers. Since then a number of new results have been obtained by a number of researchers, including the authors. A number of old approaches were also reconsidered. The majority of these results are scattered over a number of journal and conference proceedings, partially in Russian and are not always readily available. All of this

---

[2]"Pure" continuous time Markov processes cannot be physically implemented.

gave an impetus for a new book which offers a summary of the SDE approach to modeling and evaluation of communication and related systems, and provides solid background in the applied theory of random processes and systems with random excitation. Alongside classical issues, the book includes the cumulant method of Malakhov [7], often used for analysis of non-Gaussian processes, systems with random and some aspects of numerical simulation of SDE. The authors believe that this composition of the book will be helpful for both graduate students specializing in analysis of communication systems and researchers and practising engineers. Some chapters and sections can be omitted for the first reading; some of the topics formulate problems which have not been solved.[3]

The book is organized as follows. Chapters 2–4 present an introduction to the theory of random processes and form the basis of a first term course for graduate students at the Department of Electrical and Computer Engineering, The University of Western Ontario (UWO). Chapter 5 deals with the theory of Markov processes, stochastic differential equations and the Fokker–Planck equation. Synthesis of models in the form of SDE is described in Chapter 7, while Chapter 8 provides examples of the applications of SDE and other Markov models to practical communications problems. Four appendices detailing the numerical simulation of SDEs, impulse random processes, tables of distributions and orthogonal polynomials are published on the web.

Some of the material presented in this book was originally developed in the State University of Telecommunications (St Petersburg, Russia) and is, to a great degree, augmented by the results of the research conducted in three groups at Ben-Gurion University of the Negev (Beer-Sheva, Israel), University of Western Ontario (London, Canada) and Cinvestav-IPN (Mexico City, Mexico). A large number of people have provided motivation, support and advice. The authors would like to thank Doctors R. Gut, A. Berdnikov, A. Brusentsov, M. Dotsenko, G. Kotlyar, J. LoVetri, J. Roy and D. Makrakis, who contributed greatly to the development of the techniques described in this book. We also would like to recognize the contribution of our graduate students, in particular Mr Mark Shahaf, Mrs My Pham, Mr Chris Snow, Mr Jeff Weaver and Ms Vanja Subotic who have provided contributions through a number of joint publications and valuable comments on the course notes which form the first three chapters of this book. This research has been supported in part by CRC Canada, NSERC Canada, the Ministry of Education of Israel, and the Ministry of Education of Mexico. The authors are also thankful to their families who were very considerate and patient. Without their love and support this book would have never been completed.

1.2 DIGITAL COMMUNICATION SYSTEMS

The following block diagram is used in this book as a generic model of a communication system.[4] A detailed description of the blocks can be found in many textbooks, for example [8,9]. The main focus of this book is to model processes in a communication channel. Referring to Fig. 1.1, there are two ways in which the communication channel can be defined: the channel between points A and A' is known as a continuous (analog) channel; the

---

[3]At least, the authors are not aware of such solutions.

[4]This also includes such electronic systems as radar; the issue of immunity of electronic equipment to interference also can be treated as a communication problem.

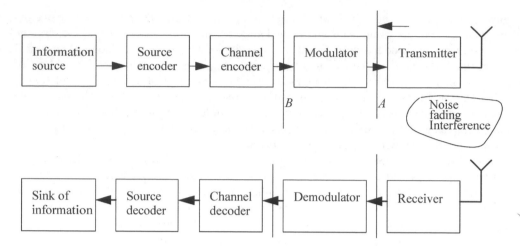

Figure 1.1 Generic communication system.

part of the system between the points B and B' is called a discrete (digital) channel. In general, such separation is very subjective, especially in the light of recent advances in modulation and coding techniques, taking into account properties of the propagation channel already on the modulation stage. However, such separation is convenient, especially for a pedagogical purpose and computer simulation of separate blocks of the communications system [9,10].

In addition to propagation media, the continuous communication channel includes all linear blocks (such as filters and amplifiers) which cannot be identified with the propagation media. Such assignment is somewhat arbitrary, since by varying the parameters of the modulation, number of antennas and their directivity, it is possible to vary properties of the channel.[5] A theoretical description of such a channel could be achieved by modeling the channel as a linear time and space varying filter [10–12]. This allows use of the concept of system functions, which is well suited for both narrow band and wide band channels. Recent advances in communication technology sparked greater interest in wide band channels [9–10,12], since they significantly increase the capacity of communication systems.

In continuous channels one can find a great variety of noise and interference, which limit the information transmission rate subject to certain quality requirements (such as probability of error). Roughly speaking, the interference and noise can be classified as additive (such as thermal noise, antenna noise, man-made and natural impulsive noise, interchannel interference, etc.) and multiplicative (such as fading, intermodulations, etc.). Additive noise (interference) $n(t)$ results in the received signal $r(t)$ being an algebraic sum of the information signal $s(t)$ and noise $n(t)$,

$$r(t) = s(t) + n(t) \tag{1.1}$$

while the multiplicative noise (such as fast and shadowing) changes the energy of the information signal. In the simplest case of frequency non-selective fading, the multiplicative

---

[5]Sometimes modems are also included in the continuous channel. In this case the channel becomes non-linear, depending on the modulation–demodulation technique.

noise is described simply as

$$r(t) = \mu(t)s(t) \tag{1.2}$$

where $\mu(t)$ is the fading, $s(t)$ is the transmitted signal and $r(t)$ is the received signal. Multiplicative noise significantly transforms the spectrum of the received signal and thus severely impairs communication systems. The realistic situation is often much more complicated since different frequencies are transmitted with different gains [12], thus one has to distinguish between so-called flat (non-selective) and frequency selective fading. "Flatness" of fading is closely related to the product of the delay spread of the signal and its bandwidth [13–14].

Large-scale fading, or shadowing, is mainly defined by the absorption of radiowaves, defined by changes in the temperature, composition and other factors in the propagation media. Such changes are very slow compared to the period of the carrier frequency and the signal can be assumed constant for many cycles of the carrier. However, these changes are significant if a relatively long communication session is considered. On the contrary, fast fading is explained by the changing phase conditions on the propagation path, with corresponding destruction or reconstruction of coherency between possible multiple paths of propagation. As a result, the effective magnitude and the phase of the signals varies significantly, which is particularly important in cellular, personal and satellite communications [8,12]. Furthermore, these variations are comparable on the time scale with the duration of the bit, explaining the term "fast fading".

A combination of fast and slow fading results in non-stationary conditions of the channel. Nevertheless, if the performance of a communication system is considered on the level of a bit, code block or sometimes a packet or frame, it could be assumed that the fading is at least locally stationary. However, if long term performance of the system is the concern, for example when evaluating the performance of the communication protocols by means of numerical simulation, the non-stationary nature must be taken into account.

A digital communication channel (Section $B - B'$ in Fig. 1.1) is characterized by a discrete set of the transmitted and received symbols and can usually be described by distribution of the errors, i.e. incorrectly received symbols. There are a great number of statistics which are useful in such description, for example average probability of error, average number of errors in a block of a given length, etc. It can be seen that the continuous channels can be statistically mapped into a discrete channel if the modulation and its characteristics are known. If simple modulation techniques are used, the performance of the modulation technique can be expressed analytically and corresponding statistics of the discrete channel can be derived. In this case, modeling of the continuous channel is unnecessary. However, modern communication algorithms often involve modulation and coding techniques whose performance cannot be described analytically. In this case, it is very important to be able to reproduce proper statistics of the underlying continuous channel. This book discusses techniques for modeling both continuous and discrete channels.

REFERENCES

1. E. Costamagna, L. Favalli, and P. Gamba, Multipath Channel Modeling with Chaotic Attractors, *Proceedings of the IEEE*, vol. 90, no. 5, May 2002, pp. 842–859.

2. D. Middleton, *An Introduction to Statistical Communication Theory*, New York: McGraw-Hill, 1960.

3. C.W. Gardiner, *Handbook of Stochastic Methods for Physics, Chemistry and the Natural Sciences*, Berlin: Springer, 1994, pp. 442.

4. D. Middleton, Non-Gaussian Noise Models in Signal Processing for Telecommunications: New Methods and Results For Class A and Class B Noise Models, *IEEE Trans. Information Theory*, vol. 45, no. 4, May 1999, pp. 1129–1149.

5. H. van Trees, *Detection, Estimation, and modulation theory*, New York: Wiley, 1968–1971.

6. D. Klovsky, V. Kontorovich, and S. Shirokov, *Models of Continuous Communications Channels Based on Stochastic Differential Equations*, Moscow: Radio i sviaz, 1984 (In Russian).

7. A.N. Malakhov, *Kumuliantnyi Analiz Sluchainykh Negaussovykh Protsessov I Ikh Preobrazovanii*, Moscow: Sov. Radio, 1978 (In Russian).

8. M. Jeruchim, *Simulation of Communication Systems: Modeling, Methodology, and Techniques*, New York: Kluwer Academic/Plenum Publishers, 2000.

9. J. Proakis, *Digital Communications*, Boston: McGraw-Hill, 2000.

10. S. Benedetto and E. Biglieri, *Principles of Digital Transmission: with Wireless Applications*, New York: Kluwer Academic/Plenum Press, 1999.

11. R. Steele and L. Hanzo, *Mobile Radio Communications: Second And Third-generation Cellular And Watm Systems*, New York: Wiley, 1999.

12. M. Simon and S. Alouini, *Digital Communication Over Fading Channels: A Unified Approach To Performance Analysis*, New York: Wiley, 2000.

13. T. Rappoport, *Wireless communications: principles and practice*. Upper Saddle River, NJ: London: Prentice Hall PTR, 2001.

14. P. Bello, Characterization of Randomly Time-Variant Linear Channels, *IEEE Trans. Communications*, vol. 11, no. 4, December, 1963, pp. 360–393.

2

Random Variables and Their Description

This chapter briefly summarizes definitions and important results related to the description of random variables and random vectors. More detailed discussions can be found in [1,2]. A number of useful examples, important for applications, is also considered in this chapter.

2.1 RANDOM VARIABLES AND THEIR DESCRIPTION

2.1.1 Definitions and Method of Description

2.1.1.1 Classification

A Random Variable (RV) ξ[1] can be considered as an outcome of an experiment, physical or imaginary, such that this quantity has a different value from one run of the experiment to another, even if the experiment is repeated under the same conditions. The difference in the outcomes may come either as a result of unaccounted conditions (variables), or, as in the quantum theory, from the internal properties of the system under consideration. In order to describe a random variable one needs to specify a range of possible values which this variable can attain. In addition, some numerical measure of probability of these outcomes must be assigned.

Based on the type of values the random variable under consideration can assume, it is possible to distinguish between three classes of random variable: (a) a discrete random variable; (b) a continuous random variable; and (c) a mixed random variable. For a discrete random variable there are a finite or infinite but countable number of values this random variable can attain. These possible values can be enumerated, i.e. a countable set, $\{\xi_n\}$, $n = 0, 1, \ldots$, covers all the possibilities.

Continuous random variables assume values from a single or multiple non-intersecting intervals, $[a_i, b_i]$, $a_i < b_i < a_{i+1} < \ldots$, $i = 1, 2, \ldots$ on the real axis. Both ends of the interval can be at infinity. Finally, a mixed random variable may assume values from both discrete and continuous sets.

There is a number of characteristics which allow a complete description of a random variable ξ. In particular, the Cumulative Distribution Function (CDF), $P_\xi(x)$, Probability

---

[1]In future, the Greek letters are used to denote a random variable. Latin letters a, b, c will be reserved for constants and x, y, z for variables.

Stochastic Methods and Their Applications to Communications.
S. Primak, V. Kontorovich, V. Lyandres
© 2004 John Wiley & Sons, Ltd   ISBN: 0-470-84741-7

Density Function (PDF) $p_\xi(x)$ and the characteristic function $\Theta_\xi(ju)$ are the most important characteristics of random variables, allowing their complete description.

2.1.1.2 Cumulative Distribution Function

The cumulative distribution function $P_\xi(x)$ is defined as the probability of the event that the random variable ξ does not exceed a certain threshold, x, i.e.

$$P_\xi(x) = \text{Prob}\{\xi < x\} \tag{2.1}$$

Any CDF $P_\xi(x)$ is a non-negative, non-decreasing, continuous on the left function which satisfies the following boundary conditions

$$P_\xi(-\infty) = 0, \quad P_\xi(\infty) = 1 \tag{2.2}$$

The converse is also valid: every function which satisfies the above listed conditions is a CDF of some random variable [1]. If CDF $P_\xi(x)$ is known, then the probability that the random variable ξ falls inside a finite interval $[a, b)$ or intersection of non-overlapping intervals $[a_i, b_i)$ is given by

$$\text{Prob}\{a \le \xi < b\} = P_\xi(b) - P_\xi(a)$$
$$\text{Prob}\{\xi \in \cup_i[a_i, b_i)\} = \sum_i [P_\xi(b_i) - P_\xi(a_i)] \tag{2.3}$$

The advantage of the CDF method is the fact that the discrete, continuous and mixed variables are described using the same technique. However, such description is an integral representation and thus it is not easy to interpret.

For a complete description of a discrete random variable ξ, which assumes a set of values, $x_k, k = 0, 1, \ldots$, one needs to know the distribution of probabilities P_k, i.e.

$$P_k = \text{Prob}\{\xi = x_k\} \tag{2.4}$$

It is clear that

$$P_k \ge 0, \quad \sum_k P_k = 1 \tag{2.5}$$

Example: *binomial distribution*. This distribution arises when a given experiment with two possible outcomes, "SUCCESS" and "FAILURE", is repeated N times in a row. The probability of a "SUCCESS" outcome is $0 \le p \le 1$, and the probability of a "FAILURE" is $q = 1 - p$. Probability $P_N(k)$ of k successes in a set of N trials is then given by[2]

$$P_N(k) = C_N^k p^k q^{N-k} = \frac{N!}{k!(N-k)!} p^k q^{N-k}, \quad k = 0, 1, \ldots, N \tag{2.6}$$

---

[2]Here the following notation is used for a combination of k elements from a set of N elements: $C_N^k = \binom{N}{k} = \frac{N!}{k!(N-k)!}$ [5].

It can be seen that the probability $P_N(k)$ is the coefficient in the expansion of $(q + px)^N$ into a power series with respect to x. This explains the term "binomial distribution". Situations described by the binomial law are often encountered in communication theory and many other applications. Examples include the number of errors ("failures") in a group of N bits, the number of corrupted packets in a frame, number of rejected calls during a certain period of time, etc.

Example: *the Poisson random variable*. In the case of the Poisson random variable ξ, the random variable assumes integer numbers as values, with probabilities defined by

$$P_k = \text{Prob}\{\xi = k\} = \frac{\lambda^k}{k!}\exp(-\lambda) \tag{2.7}$$

This distribution can be obtained as a limiting case of the binomial distribution (2.6) when the number of experiments approaches infinity while the product $\lambda = Np$ remains constant [1,3]. This distribution also plays an important role in communication theory, reliability theory and networking.

2.1.1.3 Probability Density Function

A continuous random variable can be described by the probability density function $p_\xi(x)$, defined such that

$$p_\xi(x) = \frac{d}{dx}P_\xi(x), \quad P_\xi(x) = \int_{-\infty}^{x} p_\xi(s)\,ds \tag{2.8}$$

Any PDF satisfies the following conditions, which follow from the definition (2.8):

1. it is a non-negative function

$$p_\xi(x) \geq 0 \tag{2.9}$$

2. it is normalized to 1, i.e.

$$\int_{-\infty}^{\infty} p_\xi(x)dx = 1 \tag{2.10}$$

3. probability $\text{Prob}\{x_1 \leq \xi < x_2\}$ is given by

$$\text{Prob}\{x_1 \leq \xi < x_2\} = \int_{x_1}^{x_2} p_\xi(x)dx = P_\xi(x_2) - P_\xi(x_1) \tag{2.11}$$

Formally, the PDF for a discrete random variable can be defined as a sum of delta functions weighted by the probability of each discrete even, i.e.

$$p_\xi(x) = \sum_k P_k \delta(x - x_k) \tag{2.12}$$

A great variety of probablity densities is used in applications. Discussion of particular distributions is postponed until later. However, it is important to mention the so-called Gaussian (normal) distribution with PDF

$$p(x) = N(x; n, D) = \frac{1}{\sqrt{2\pi D}} \exp\left[-\frac{(x-m)^2}{2D}\right] \tag{2.13}$$

and the PDF of the so-called uniform distribution

$$p(x) = \begin{cases} \frac{1}{b-a} & \text{for } a \leq x \leq b \\ 0 & \text{otherwise} \end{cases} \tag{2.14}$$

A much wider class of PDFs belong to the so-called Pearson family [4], which is described through a differential equation for its PDF

$$\frac{\frac{d}{dx}p(x)}{p(x)} = \frac{d}{dx}\ln p(x) = \frac{x-a}{b_0 + b_1 x + b_2 x^2} \tag{2.15}$$

It follows from eqs. (2.13) and (2.14) that the CDFs of Gaussian and uniform distributions are given by

$$P_N(x) = \int_{-\infty}^{x} N(x; n, D)dx = \Phi\left(\frac{x-m}{\sqrt{D}}\right) \tag{2.16}$$

$$P_u(x) = \begin{cases} 0 & x \leq a \\ \frac{x-a}{b-a} & a \leq x \leq b \\ 1 & x \geq b \end{cases} \tag{2.17}$$

respectively. Here

$$\Phi(x) = \frac{1}{\sqrt{2\pi}} \int_{-\infty}^{x} \exp\left(-\frac{t^2}{2}\right)dt = 1 - \frac{1}{2}\text{erf}\left(\frac{x}{2}\right) \tag{2.18}$$

is the probability integral which in turn can be expressed in terms of the error function [5].

2.1.1.4 The Characteristic Function and the Log-Characteristic Function

Instead of considering the CDF $P_\xi(x)$ or the PDF $p_\xi(x)$ one can consider an equivalent description by means of the Characteristic Function (CF). The characteristic function $\Theta_\xi(ju)$ is defined as the Fourier transform of the corresponding PDF[3]

$$\Theta_\xi(ju) = \int_{-\infty}^{\infty} p_\xi(x)e^{jux}dx, \quad p_\xi(ju) = \frac{1}{2\pi}\int_{-\infty}^{\infty} \Theta_\xi(ju)e^{-jux}du \tag{2.19}$$

Thus, one can restore the PDF $p_\xi(x)$ from the corresponding characteristic function $\Theta_\xi(ju)$.

[3]Note the plus sign in the direct transform and the minus sign in the inverse transform.

There are a number of properties of the CF which follow from the definition (2.19). In particular

$$|\Theta_\xi(ju)| \le \int_{-\infty}^{\infty} |p_\xi(x)||e^{jux}|dx = \int_{-\infty}^{\infty} p_\xi(x)dx = \Theta_\xi(0) = 1 \qquad (2.20)$$

and

$$\Theta_\xi(-ju) = \int_{-\infty}^{\infty} p_\xi(x)e^{-jux}dx = \Theta_\xi^*(ju) \qquad (2.21)$$

Additional properties of the characteristic function follow from the properties of the Fourier transform pair and can be found in [1,6]. Tables of the Fourier transforms of PDFs are also available [7]. In addition to the characteristic function, one can define the so-called log-characteristic function (or cumulant generating function) as

$$\Psi_\xi(ju) = \ln \Theta_\xi(ju) \qquad (2.22)$$

This function is later used to define the cumulants of a random variable.

Example: Gaussian distribution. The characteristic function of a Gaussian distribution can be obtained by using a standard integral (3.323, page 333 in [8])

$$\int_{-\infty}^{\infty} \exp(-p^2x^2 \pm qx)dx = \exp\left(\frac{q^2}{4p^2}\right)\frac{\sqrt{\pi}}{p} \qquad (2.23)$$

with $q = ju$, $p^2 = D/2$. This results in

$$\Theta_N(ju) = \exp\left(jum - \frac{Du^2}{2}\right) \qquad (2.24)$$

The Log-Characteristic Function (LCF) $\Psi_n(ju)$ of the Gaussian distribution is then

$$\Psi_N(ju) = jum - \frac{Du^2}{2} \qquad (2.25)$$

2.1.1.5 Statistical Averages

The CDF, PDF and CF each provide a complete description of any random variable. However, in many practical problems it is possible to limit consideration to a less complete but simpler description of random variables. This can be achieved, for example, through use of statistical averages. Let $\eta = g(\xi)$ be a deterministic function of a random variable ξ, described by the PDF $p_\xi(x)$. The ensemble average of the function $g(\xi)$ is defined as

$$E\{g(\xi)\}_\xi = \langle g(\xi)\rangle_\xi = \int_{-\infty}^{\infty} g(x)p_\xi(x)dx \qquad (2.26)$$

where the subscript indicates over which variable the averaging is performed. In future, the subscript will be dropped if it does not create uncertainty. By specifying a particular form of the function $g(\xi)$ it is possible to obtain various numerical characteristics of the random variable. Because averaging is a linear operation it is possible to interchange the order of averaging and summation, i.e.

$$E\left\{\sum_{k=1}^{K} c_k\, g_k(\xi)\right\}_\xi = \sum_{k=1}^{K} c_k\, E\{g_k(\xi)\}_\xi \tag{2.27}$$

2.1.1.6   Moments

A moment $m_{n\xi} = m_n$ of order n of a random variable ξ is obtained if the function $g(\xi)$ in eq. (2.26) is chosen as $g(\xi) = \xi^n$, thus giving

$$m_{n\xi} = \int_{-\infty}^{\infty} x^n\, p_\xi(x)\, \mathrm{d}x \tag{2.28}$$

The first order moment[4] $m_{1\xi} = m_\xi = m$ is called the average value of the random variable ξ. It follows from its definition that

- the dimension of the average coincides with that of the random variable;

- the average of a deterministic variable coincides with the value of the variable itself;

- the average of a random variable whose PDF is a symmetric function around $x = a$ is equal to $m = a$.

2.1.1.7   Central Moments

A central moment μ_n can be obtained from eq. (2.26) by setting $g(\xi) = (\xi - m_{1\xi})^n$, i.e.

$$\mu_n = \langle (\xi - m_{1\xi})^n \rangle_\xi = \int_{-\infty}^{\infty} (x - m_{1\xi})^n p_\xi(x)\mathrm{d}x \tag{2.29}$$

The first central moment is always zero: $\mu_1 = 0$. The second central moment $\mu_{2\xi} = D_\xi = \sigma_\xi^2$ is called the variance, and represents the degree of variation of the random variable around its mean. The quantity σ_ξ is known as the standard deviation. The following properties of the variance can be easily obtained from definition (2.29)

- the variance D has dimension of the square of the random variable ξ;

- $D_\xi \geq 0$; $D_\xi = 0$ if and only if ξ is deterministic;

---

[4]In the following the subindex indicating a random variable is dropped if it does not cause confusion.

- the variance D_η of $\eta = c\xi$ is equal to $c^2 D_\xi$, where c is a deterministic constant;
- the variance of $\eta = \xi + c$ is equal to $D_\eta = D_\xi$;
- the following inequality is valid (Tschebychev inequality) [3].

$$\text{Prob}\{|\xi - m_{1\xi}| \geq \varepsilon\} \leq \frac{D_\xi}{\varepsilon^2} \tag{2.30}$$

Here ε is an arbitrary positive number. The last property states that the probability of large deviations from the average value are very rare.

It is possible to obtain relations between the moments and the central moments of the same random variable using the binomial expansion. Indeed

$$\mu_n = \int_{-\infty}^{\infty} (x-m)^n p(x)\mathrm{d}x = \int_{-\infty}^{\infty} \sum_{k=0}^{n}[C_n^k(-1)^k x^{n-k} m^k]p(x)\mathrm{d}x = \sum_{k=0}^{n} C_n^k(-1)^k m_{n-k} m^k$$

$$\tag{2.31}$$

Conversely,

$$m_n = \sum_{k=0}^{n} C_n^k \mu_{n-k} m^k \tag{2.32}$$

It is interesting to mention that the moment (central moment) of order n depends on all n central moments (moments) of lower or equal order. In particular

$$m_2 = D + m_1^2 \tag{2.33}$$

Example: *Gaussian PDF.* In the case of the Gaussian distribution the mean value and the variance are

$$m_1 = \int_{-\infty}^{\infty} x \frac{1}{\sqrt{2\pi D}} \exp\left[-\frac{(x-m)^2}{2D}\right] \mathrm{d}x = m \tag{2.34}$$

$$\mu_2 = \int_{-\infty}^{\infty} (x-m)^2 \frac{1}{\sqrt{2\pi D}} \exp\left[-\frac{(x-m)^2}{2D}\right] \mathrm{d}x = D = \sigma^2 \tag{2.35}$$

respectively. Thus, the parameters m and D of the Gaussian distribution are actually the mean and variance of the distribution.

2.1.1.8   *Other Quantities*

There is a great number of other numerical characteristics which are useful in describing a random variable. A few of them are listed below. Much more detailed discussions and applications can be found in [1,2].

If the PDF $p(x)$ has a (local) maximum at $x = x_m$, the value x_m is called a mode of the distribution. A distribution with a single mode is called a unimodal distribution, while a distribution with more than one mode is called multimodal. The median $x_{1/2}$ of the distribution is a value such that

$$P(x_{1/2}) = \int_{-\infty}^{x_{1/2}} p(x)dx = \int_{x_{1/2}}^{\infty} p(x)dx = \frac{1}{2} \tag{2.36}$$

This can be generalized to define $(n-1)$ values $\{x_k\}$, $k = 0, \ldots, n-2$, called the quantiles, which divide the real axis into n equally probable regions

$$\int_{-\infty}^{\infty} p(x)dx = \int_{x_0}^{x_1} p(x)dx = \cdots = \int_{x_{n-2}}^{\infty} p(x)dx = \frac{1}{n} \tag{2.37}$$

Absolute moments can be defined by setting $g(\xi) = |\xi|^n$ and $g(\xi) = |\xi - m|^n$ in eq. (2.26). In addition, the requirement that the power n is an integer can also be dropped to obtain moments of a fractional order.

Finally, the entropy (differential entropy) of the distribution can be defined as

$$H(\xi) = -\sum_{k=1}^{K} p_k \log p_k, \quad H(\xi) = -\int_{-\infty}^{\infty} p_\xi(x) \log p_\xi(x)dx \tag{2.38}$$

The entropy is an important concept which is discussed in detail in textbooks on information theory. It also has significant application as a measure of difference between two PDFs.

2.1.1.9   Moment and Cumulant Generating Functions

The characteristic function $\Theta_\xi(ju)$ was formally defined in Section 2.1.1.4 as the Fourier transform of the corresponding PDF $p_\xi(x)$. It also can be defined in the framework of the statistical averaging of eq. (2.26). Indeed, let $g(\xi)$ be an exponential function with a parameter s: $g(\xi) = \exp(s\,\xi)$. In this case the average value $M(s)$ of $g(\xi)$ is defined as

$$M_\xi(s) = \langle \exp(s\,\xi) \rangle = \int_{-\infty}^{\infty} \exp(s\,x)p_\xi(x)dx \tag{2.39}$$

and is called the moment generating function. Expanding the exponential term $\exp(s\,x)$ into a power series, one obtains the relation between the moment generating function and the moments of the distribution of random variable ξ:

$$M_\xi(s) = \left\langle \sum_{k=0}^{\infty} \frac{\xi^k}{k!} s^k \right\rangle = \sum_{k=0}^{\infty} \frac{m_k}{k!} s^k \tag{2.40}$$

if all moments exist and are finite. In turn, the coefficients of the Taylor expansion (2.40), i.e. the moments m_k, can be found as

$$m_k = \frac{d^k}{d\,s^k} M_\xi(s)\Big|_{s=0} \qquad (2.41)$$

The transform variable s can be a complex one, i.e. $s = \alpha + ju$. In a particular case of $s = ju$, the moment generating function coincides with the characteristic function, i.e. $\Theta_\xi(ju) = M_\xi(s)|_{s=ju}$ and expansion (2.40) can be rewritten as

$$\Theta_\xi(ju) = \left\langle \sum_{k=0}^{\infty} \frac{\xi^k}{k!} s^k \right\rangle = \sum_{k=0}^{\infty} j^k \frac{m_k}{k!} u^k \qquad (2.42)$$

with the moments defined according to

$$m_k = (-j)^k \frac{d^k}{d\,u^k} \Theta_\xi(ju)\big|_{u=0} \qquad (2.43)$$

Thus, under certain conditions, the characteristic function $\Theta_\xi(ju)$ can be restored from its moments. It is interesting to investigate what additional conditions must be imposed on the moments such that the restored characteristic function uniquely defines the distribution. This problem is known as the moments problem. It is possible to show that if all moments m_k are finite and the series (2.42) absolutely converges for some $u > 0$, then the series (2.42) defines a unique distribution [2]. It should be noted that this is not true for an arbitrary distribution. For example, the log-normal distribution, considered below, is not uniquely defined by its moments [2]. Usually, the non-uniqueness arises when the moments m_k increase rapidly with the index k, thus not allowing absolute convergence of the series (2.42). Such problems are often found for distributions with heavy tails. In many cases the higher order moments do not even exist [2].

A slight modification allows the characteristic function to be expressed in terms of central moments by rewriting the series (2.42) as

$$\theta_\xi(ju) = e^{jum_1}\left[1 + \sum_{k=1}^{\infty} \frac{\mu_k}{k!}(ju)^k\right] \qquad (2.44)$$

Here we use the property that the moments of random variable $\eta = \xi - m_1$ are the central moments of the random variable ξ, and the property that the characteristic function of $p_\xi(x - m_1)$ is $\Theta_\xi(ju)\exp(jum_1)$.

2.1.1.10   Cumulants

Instead of considering the Taylor expansion for the characteristic function, one can construct a similar expansion of the log-characteristic function $\Psi(ju)$, i.e. it is possible to formally write

$$\Psi(ju) = \sum_{k=1}^{\infty} j^k \frac{\kappa_k}{k!} u^k \qquad (2.45)$$

Coefficients of this expansion are called cumulants of the distribution $p_\xi(x)$ of the random variable ξ, with κ_k being the cumulants of the k-th order. Since both coefficients m_k and κ_k describe the same function, there is a close relation between the moments and the cumulants. Indeed, provided that both expansions (2.42) and (2.45) are possible, one can write

$$\Theta_\xi(ju) = \sum_{k=0}^{\infty} j^k \frac{m_k}{k!} u^k = \exp\left[\sum_{r=1}^{\infty} j^r \frac{\kappa_r}{r!} u^r\right] = \prod_{r=1}^{\infty} \exp\left(j^r \frac{\kappa_r}{r!} u^r\right) = \prod_{r=1}^{\infty} \sum_{l=0}^{\infty}\left[\left(\frac{\kappa_r j^r u^r}{r!}\right) \frac{1}{s!}\right]$$

(2.46)

Collecting terms with the same power of u on both sides of this expansion, one obtains the expression of the moment of order k in terms of cumulants of order up to k

$$m_k = \sum_{m=1}^{r} \sum \left(\frac{\kappa_{p_1}}{p_1!}\right)^{\alpha_1} \left(\frac{\kappa_{p_2}}{p_2!}\right)^{\alpha_2} \cdots \left(\frac{\kappa_{p_m}}{p_m!}\right)^{\alpha_m} \frac{r!}{\alpha_1!\alpha_2!\ldots\alpha_m!}$$

(2.47)

where the inner summation is taken over all non-negative values of indices α_i, such that

$$\alpha_1 p_1 + \alpha_2 p_2 + \cdots + \alpha_m p_m = r$$

(2.48)

In a similar way, the expression of the cumulant κ_k can be written in terms of the moments of order up to k [2]

$$\kappa_k = r! \sum_{m=1}^{r} \sum \left(\frac{m_{p_1}}{p_1!}\right)^{\alpha_1} \left(\frac{m_{p_2}}{p_2!}\right)^{\alpha_2} \cdots \left(\frac{m_{p_m}}{p_m!}\right)^{\alpha_m} \frac{(-1)^{\rho-1}(\rho-1)!}{\alpha_1!\alpha_2!\ldots\alpha_m!}$$

(2.49)

The inner summation is extended over all the indices α and ρ such that

$$\alpha_1 + \alpha_2 + \cdots + \alpha_m = \rho$$

(2.50)

There are a number of tables which contain explicit expressions for cumulants and moments up to order 12 [2] while [9] provides a convenient means of deriving relations between the cumulants and the moments.

2.2 ORTHOGONAL EXPANSIONS OF PROBABILITY DENSITIES: EDGEWORTH AND LAGUERRE SERIES

In many practical cases one has to deal with a probability density $p_1(x)$ which looks similar to a Gaussian one defined by eq. (2.13). Two characteristic features of such distributions can be summarized as follows:

1. Unimodality, i.e. the PDF has a single maximum, and

2. The PDF has tails extending to infinity on both sides of the maximum, which decay fast when the magnitude of the argument approaches infinity.

In this case it is often possible to approximate such PDFs using series of Hermitian or Laguerre polynomials.

2.2.1 The Edgeworth Series

In this case a PDF $p(x)$ under consideration is approximated by the following series

$$p(x) = p_0(x) \sum_{n=0}^{\infty} \frac{1}{n!} \frac{b_n}{\sigma^n} H_n\left(\frac{x-m}{\sigma}\right) \tag{2.51}$$

Here

$$p_0(x) = \frac{1}{\sqrt{2\pi\sigma^2}} \exp\left[-\frac{(x-m)^2}{2\sigma^2}\right] \tag{2.52}$$

is the Gaussian PDF with the mean value m and variance σ^2, and $H_n(z)$ stands for the Hermitian polynomials [5]

$$H_n(z) = (-1)^n \exp\left[\frac{z^2}{2}\right] \frac{d^n}{dz^n} \exp\left[-\frac{z^2}{2}\right], \qquad n = 0, 1, \ldots \tag{2.53}$$

Since the Hermitian polynomials are orthogonal with weight $p_0(x)$, i.e.

$$\int_{-\infty}^{\infty} H_n(z)H_m(z)p_0(z)dz = n!\delta_{mn} = \begin{cases} n! & m = n \\ 0 & m \neq n \end{cases} \tag{2.54}$$

the coefficients b_n in the expression (2.51), called quasi-moments, can be calculated as

$$b_n = \sigma^n \int_{-\infty}^{\infty} p(x)H_n\left(\frac{x-m}{\sigma}\right)dx = \sigma^n \left\langle H_n\left(\frac{\xi-m}{\sigma}\right)\right\rangle \tag{2.55}$$

This expansion is based on the well known theorem of functional analysis [10] stating that if $p(x)$ is an arbitrary function such that

$$\int_{-\infty}^{\infty} |p(x)|^2 dx < \infty \tag{2.56}$$

then

$$\lim_{N \to \infty} \int_{-\infty}^{\infty} \left| p(x) - p_0(x) \sum_{n=0}^{N} \frac{1}{n!} \frac{b_n}{\sigma^n} H_n\left(\frac{x-m}{\sigma}\right)\right|^2 dx = 0 \tag{2.57}$$

In practice the function $p(x)$ is known only with a certain degree of accuracy. Thus, the sum (2.51) can be truncated after a finite number, N, of terms. The number N depends on the choice of m and σ^2. In most cases the best choice of m and σ^2 for a given N is to choose them to coincide with the first two cumulants derived directly from $p(x)$. Then it is easy to show that

$$b_0 = 1, \quad b_1 = b_2 = 0 \tag{2.58}$$

Indeed, using eq. (2.53) we find that

$$H_0(z) = 1, \quad H_1(z) = z, \quad H_2(z) = z^2 - 1, \quad H_3(z) = z^3 - 3z, \quad H_4(z) = z^4 - 6z^2 + 3 \tag{2.59}$$

Thus, according to eq. (2.55)

$$b_0 = \sigma^0 \left\langle H_0\left(\frac{x-m}{\sigma}\right) \right\rangle = \langle 1 \rangle = 1$$

$$b_1 = \sigma^1 \left\langle H_1\left(\frac{x-m}{\sigma}\right) \right\rangle = \sigma\left\langle \frac{x-m}{\sigma} \right\rangle = 0 \tag{2.60}$$

$$b_2 = \sigma^2 \left\langle H_2\left(\frac{x-m}{\sigma}\right) \right\rangle = \sigma^2 \left\langle \frac{(x-m)^2}{\sigma^2} - 1 \right\rangle = \sigma^2 \left(\frac{\sigma^2}{\sigma^2} - 1\right) = 0$$

The Edgeworth series is obtained from eq. (2.51) by taking a finite number N of terms and choosing m and σ to coincide with those obtained from $p_1(x)$, i.e.

$$p(x) \approx p_0(x)\left[1 + \sum_{n=3}^{N} \frac{b_n}{n!\sigma^n} H_n\left(\frac{x-m}{\sigma}\right)\right] \tag{2.61}$$

The first term in the expansion (2.61) corresponds to a Gaussian distribution. Thus, for a Gaussian distribution all quasi-moments b_n are equal to zero. Coefficients b_3 and b_4 in the series (2.61) describe the departure of $p(x)$ from a Gaussian PDF and are known as the skewness and the curtosis [2]

$$\gamma_1 = \frac{b_3}{\sigma^3} = \frac{\mu_3}{\sigma^3} = \frac{\kappa_3}{\kappa_2^{3/2}}$$

$$\gamma_2 = \frac{b_4}{\sigma^4} = \frac{\mu_4}{\sigma^4} - 3 = \frac{\kappa_4}{\kappa_2^2} \tag{2.62}$$

Here $\kappa_2, \kappa_3, \kappa_4$ are the cumulants of the distribution $p(x)$ as defined by eq. (2.49), and μ_3, μ_4 are the central moments of third and fourth order. The skewness coefficient describes the deviation of the PDF $p(x)$ from a symmetric shape. For any symmetrical distribution this coefficient is equal to zero. Fig 2.1 shows two cases of PDF. The first one (a) has a shallow tail on the right of the mean value, thus in the expression for $(x-m)^3$ the positive deviations overpower the negative deviations and give $\mu_3 > 0$, $\gamma_1 > 0$. A case $\gamma_2 < 0$ leads to a situation where the left tail of the distribution is heavier than the right one.

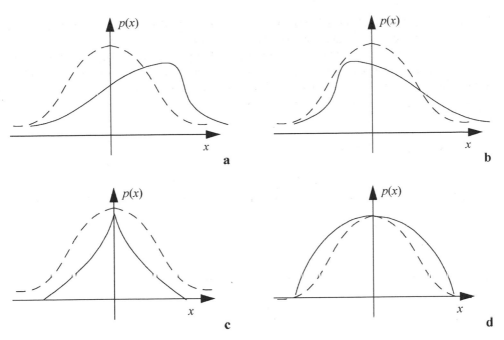

Figure 2.1   Different shapes of PDF similar to a Gaussian PDF (a) $\gamma_1 > 0$, (b) $\gamma_1 < 0$, (c) $\gamma_2 > 0$ and (d) $\gamma_2 < 0$. Dashed line corresponds to a Gaussian distribution.

The curtosis γ_2 describes how flat the PDF is around its peak value. For a Gaussian PDF this coefficient is equal to zero. Positive values of γ_2 show that the PDF curve has a narrower (sharper) top than a Gaussian, while $\gamma_2 < 0$ results in a flatter top. Figures 2.1(c) and 2.1(d) show following curves: $\gamma_2 = 0$ (Gaussian), $\gamma_2 < 0$ and $\gamma_2 > 0$. It is common practice to approximate the distributions with small deviation from Gaussian using only γ_1 and γ_2. In this case

$$p(x) \approx p_0(x)\left[1 + \frac{\gamma_1}{3!} H_3\left(\frac{x-m}{\sigma}\right) + \frac{\gamma_2}{4!} H_4\left(\frac{x-m}{\sigma}\right)\right] \tag{2.63}$$

To obtain such an approximation one needs to estimate the first four moments (or cumulants) of the random process $\xi(t)$.

It is important to mention that for such an approximation (2.63), the non-negativity property of the PDF is often violated. It can be shown that the approximating curve assumes negative values for large values of the argument $|x|$ [9]. This is a consequence of the fact that eq. (2.63) is only an approximation and not an exact PDF.

The characteristic function for the PDF given by the series (2.51) is equal to

$$\theta_1(u) = \exp\left[jmu + \frac{1}{2}(\sigma u)^2\right] \sum_{n=0}^{\infty} \frac{1}{n!} \frac{b_n}{\sigma^n} (-ju)^n \tag{2.64}$$

Expanding the exponent into the Taylor series and comparing with the series (2.61), consisting of the moments of the PDF $p(x)$, one can conclude that the moments are linearly expressed in terms of quasi-moments (and vice versa). This is the reason why the term "quasi-moment" is used. Quasi-moments can be used to describe a random process in exactly the same way as regular moments, correlation functions and cumulants.

It is important to mention that despite the fact that eq. (2.63) provides an approximation which has negative values for large x, the importance of the quasi-moment technique lies in the ability to provide an easy algorithm that can be used to obtain a similar approximation at the output of non-linear and linear systems.

2.2.2 The Laguerre Series

If the PDF $p(x)$ has zero values for $x < 0$, then the Edgeworth series converges slowly to $p(x)$. In this case, a more convenient approximation, called the Laguerre series, can be used

$$p(x) = \sum_{n=0}^{\infty} C_u \exp[-x] x^\alpha L_n^{(\alpha)}(x) \tag{2.65}$$

Here $L_n^{(\alpha)}(z)$ are the generalized Laguerre polynomials [5]

$$L_n^{(\alpha)}(z) = \exp(z) \frac{z^{-\alpha} d^n}{n! d z} [\exp(-z) z^{n+\alpha}], \quad \alpha > -1 \tag{2.66}$$

The first four of these polynomials are

$$\begin{aligned}
L_0^{(\alpha)} &= 1 \\
L_1^{(\alpha)} &= 1 + \alpha - z \\
2 L_2^{(\alpha)} &= (\alpha+1)(\alpha+2) - 2z(\alpha+2) + z^2 \\
6 L_3^{(\alpha)} &= (\alpha+1)(\alpha+2)(\alpha+3) - 3z(\alpha+2)(\alpha+3) + 3z^2(\alpha+3) - z^3
\end{aligned} \tag{2.67}$$

The Laguerre polynomials are orthogonal in the interval $[0, \infty)$ with the weight $p_0(x) = z^\alpha \exp(-z)$, i.e.

$$\int_0^\infty z^\alpha \exp(-z) L_n^\alpha(z) L_m^\alpha(z) dz = \frac{1}{n!} \Gamma(n + \alpha + 1) \delta_{mn} \tag{2.68}$$

Here $\Gamma(z)$ is the Gamma function [5]. Taking the orthogonality (2.68) into account, the coefficients of the expansion (2.65) can be found as

$$C_n = \frac{n!}{\Gamma(n + \alpha + 1)} \int_0^\infty L_n^{(\alpha)} p(x) dx \tag{2.69}$$

If, instead of the original process $\xi(t)$, one considers a normalized process $\eta = \xi/\beta$ (with the coefficient β yet to be defined), then its PDF $p_\eta(y)$ can be expressed in terms of the PDF $p_\xi(x)$ as

$$p_\eta(y) = \frac{1}{\beta} p_\xi\left(\frac{x}{\beta}\right) \tag{2.70}$$

In a similar manner to eq. (2.65) one can write an expansion of the new PDF in terms of the Laguerre polynomials

$$p(y) = \sum_{n=0}^{\infty} b_n \exp[-y] y^{\alpha} L_n^{(\alpha)}(y) \tag{2.71}$$

Here, the coefficients of the expansion can be calculated as

$$b_n = \frac{n!}{\Gamma(n + \alpha + 1)} \int_0^{\infty} L_n^{(\alpha)}(y) p_{\xi}(y) dy \tag{2.72}$$

After substituting eq. (2.67) in eq. (2.72) and taking into account normalization conditions for PDF and the definition of moments, one can obtain

$$b_1 = \frac{1 + \alpha - \dfrac{m}{\beta}}{\Gamma(\alpha + 2)} \tag{2.73}$$

$$b_2 = \frac{1}{\Gamma(\alpha + 3)} \left[(\alpha + 1)(\alpha + 2) - \frac{2m}{\beta}(\alpha + 2) + \frac{m_2}{\beta^2} \right]$$

Since at this moment α and β are arbitrary constants, it is possible to choose them in such a manner that both b_1 and b_2 vanish (i.e. the expansion will have less terms) $b_1 = b_2 = 0$. For this purpose the coefficients α and β have to be chosen as

$$\alpha = \frac{m^2}{m_2 - m^2} - 1 = \frac{m^2}{\sigma^2} - 1 \tag{2.74}$$

$$\beta = \frac{m_2 - m^2}{m} = \frac{\sigma^2}{m}$$

With this choice of α and β the first four coefficients in eq. (2.71) become

$$b_0 = \frac{1}{\Gamma(\alpha + 1)}, \quad b_1 = b_2 = 0, \quad b_3 = \frac{1}{\Gamma(\alpha + 1)} \left[\frac{m_2}{\beta^2}(\alpha + 3) - \frac{m_3}{\beta^3} \right] \tag{2.75}$$

Higher order coefficients have more complicated expressions. The Laguerre series is often used when the first term provides a good approximation, i.e. when the function $p(x)$ is close to

$$p_0(x) = \frac{1}{\beta \Gamma(\alpha + 1)} \left(\frac{x}{\beta} \right)^{\alpha} \exp\left[-\frac{x}{\beta} \right] \tag{2.76}$$

where α and β are defined through the mean and the variance of $p(x)$ as in eq. (2.74). Thus, the first term in the Laguerre series is the Gamma distribution.

Similar orthogonal expansions can be obtained for other weighting functions as long as they generate a complete system of basis functions. Not all probability densities have this property. A detailed discussion of this subject can be found in [2].

2.2.3 Gram–Charlier Series

There is another important method of representation of a PDF $p_\eta(x)$ of some random variable η through a given PDF $p_\xi(x)$ of some standardized random variable ξ. Without loss of generality one can assume that both random variables have zero mean and unit variance. This can always be achieved by considering a normalized and centred random variable

$$\eta_0 = \frac{\eta - m_{1\eta}}{\sigma_\eta} \tag{2.77}$$

The PDFs of η and η_0 are related by

$$p_\eta(x) = \frac{1}{\sigma_\eta} p_{\eta_0}\left(\frac{x - m_{1\eta}}{\sigma_\eta}\right) \tag{2.78}$$

Thus, if the series for p_{η_0} is known then the series for p_η can be easily obtained by changing the variable. In this section the following approximation is considered

$$p_\eta(x) = \sum_{k=0}^{\infty} \alpha_k \frac{\mathrm{d}^k}{\mathrm{d}x^k} p_\xi(x) \tag{2.79}$$

Such expansion is especially convenient when the derivative of the PDF $p_\xi(x)$ can be expressed in the following form

$$\frac{\mathrm{d}^k}{\mathrm{d}x^k} p_\xi(x) = Q_k(x) p_\xi(x) \tag{2.80}$$

where $Q_k(x)$ are well studied functions, for example polynomials[5] of order k. Taking the (inverse) Fourier transform of both sides, one obtains the series relating the characteristic functions $\theta_\eta(ju)$ and $\theta_\xi(ju)$:

$$\theta_\eta(ju) = \left(1 + \sum_{k=1}^{\infty} (ju)^k (-1)^k \alpha_k\right) \theta_\xi(ju) \tag{2.81}$$

Using representation of the characteristic functions through the moments of the corresponding distributions, one obtains the following relation between the unknown coefficients α_k of the expansion (2.79) and the moments of both distributions

$$\sum_{n=0}^{\infty} \frac{(ju)^n}{n!} m_{\eta n} = 1 - \frac{u^2}{2} + \sum_{n=3}^{\infty} \frac{(ju)^n}{n!} m_{\eta n}$$

$$= \left(1 + \sum_{k=1}^{\infty} (ju)^k (-1)^k \alpha_k\right)\left(1 - \frac{u^2}{2} + \sum_{l=3}^{\infty} \frac{(ju)^l}{l!} m_{\xi l}\right) \tag{2.82}$$

[5]The existence of such densities is a consequence of the Rodriguez formula [5].

Finally, equating coefficients of equal powers of u, it is possible to obtain the following expressions for the coefficients α_k:

$$\alpha_1 = \alpha_2 = 0$$

$$\alpha_3 = -\frac{1}{3!}(m_{\eta 3} - m_{\xi 3})$$

$$\alpha_4 = -\frac{1}{4!}(m_{\eta 4} - m_{\xi 4}) \tag{2.83}$$

$$\alpha_5 = -\frac{1}{5!}[m_{\eta 5} - m_{\xi 5} - 10(m_{\eta 3} - m_{\xi 3})]$$

$$\alpha_6 = -\frac{1}{6!}[m_{\eta 6} - m_{\xi 6} - 15(m_{\eta 4} - m_{\xi 4}) - 20\, m_{\xi 3}(m_{\eta 3} - m_{\xi 3})]$$

It can be seen that the Gram–Charlier series coincides with the Edgeworth series if $p_\xi = \dfrac{\exp(-x^2/2)}{\sqrt{2\pi}}$.

Some properties of classical orthogonal polynomials are listed in Appendix B on the web; more detailed discussion can be found in [5,11].

2.3 TRANSFORMATION OF RANDOM VARIABLES

Let a random variable $\xi(t)$ be described by the PDF $p_\xi(x)$ which is assumed to be known. Applying a non-linear memoryless transformation $\eta = g(\xi)$ to this random variable, one obtains a new random variable η, whose PDF $p_\eta(y)$ and moments $m_{n\eta}$ are to be found.

Moments and central moments can be easily found using the definition of a statistical average

$$m_{n\eta} = \langle \eta^n \rangle_\eta = \langle g^n(\xi) \rangle_\xi \int_{-\infty}^{\infty} g^n(x) p_\xi(x) \mathrm{d}x \tag{2.84}$$

$$\mu_{n\eta} = \langle (\eta - m_{1\eta})^n \rangle_\eta = \langle (g(\xi) - m_{1\eta})^n \rangle_\xi = \int_{-\infty}^{\infty} [g(x) - m_{1\eta}]^n p_\xi(x) \mathrm{d}x \tag{2.85}$$

Thus, the calculation of moments of the random variable η can be converted to the calculation of corresponding averages based on the known PDF $p_\xi(x)$. No direct calculation of the PDF $p_\eta(y)$ is required.

The next step is to calculate the PDF $p_\eta(y)$ itself. First, it is assumed that $y = g(x)$ is a monotonic function, and thus $y = g(x)$ is a one-to-one mapping. Furthermore, it should be noted that since $\eta = g(\xi)$ is a deterministic transformation, then if a value of the random variable ξ falls inside a small interval $[x, x + \mathrm{d}x]$, then the random variable η falls insides the region $[y, y + \mathrm{d}y]$, where the boundaries of the interval are defined by the non-linear transformation and its derivative:

$$y = g(x), \quad \mathrm{d}y = \frac{\mathrm{d}}{\mathrm{d}x} g(x) \mathrm{d}x \tag{2.86}$$

This means that probabilities of these two events are equal and thus

$$p_\eta(y)\mathrm{d}y = p_\xi(x)\mathrm{d}x = p_\xi[h(y)]\left|\frac{\mathrm{d}}{\mathrm{d}y}h(y)\right| \tag{2.87}$$

Here $x = h(y)$ is a function inverse to $y = g(x)$. For a monotonic $y = g(x)$ the inverse function is unique, i.e. has only one branch. If the function $f(x)$ is increasing then a positive increment of the argument $\mathrm{d}x > 0$ results in a positive increment of the function $\mathrm{d}y > 0$. However, if $y = g(x)$ is a decreasing function then the increment of the function is negative: $\mathrm{d}y < 0$. However, the area under the curve between y and $y + \mathrm{d}y$, reflecting the probability, is a positive quantity, thus the absolute value must be used in eq. (2.87).

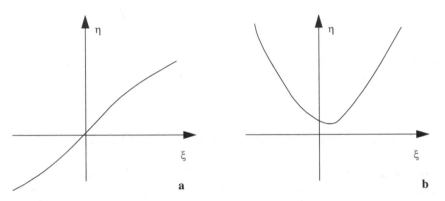

Figure 2.2 Transformation of a random variable: (a) one-to-one transformation; (b) two-to-one transformation.

If the inverse function has more than one branch the situation is more involved. In this case interval $[y, y + \mathrm{d}y]$ is an image of as many non-intersecting intervals as there are branches of the inverse function. For example, if there are two branches $x_1 = h_1(y)$ and $x_2 = g_2(y)$ as in Fig. 2.2(b), then $\eta \in [y, y + \mathrm{d}y]$ is satisfied when $\xi \in [x_1, x_1 + \mathrm{d}x_1]$ or $\xi \in [x_2, x_2 + \mathrm{d}x_2]$. The equality of probabilities thus becomes

$$p_\eta(y)|\mathrm{d}y| = p_\xi(x_1)|\mathrm{d}x_1| + p_\xi(x_2)|\mathrm{d}x_2| = p_\xi(h_1(y))\left|\frac{\mathrm{d}}{\mathrm{d}y}h_1(y)\right| + p_\xi(h_2(y))\left|\frac{\mathrm{d}}{\mathrm{d}y}h_2(y)\right| \tag{2.88}$$

The generalization to a case of M branches $x = h_m(y), m = 1, \ldots, M$ of the inverse function is straightforward

$$p_\eta(y) = \sum_{m=1}^{M} p_\xi(h_m(y))\left|\frac{\mathrm{d}}{\mathrm{d}y}h_m(y)\right| \tag{2.89}$$

It is also worth mentioning that while the probability of the transformed random variable $p_\eta(y)$ can be uniquely defined if the transformation and the input PDF are known, it is not possible to uniquely restore the input distribution by observing the output random variable if the transformation is not one-to-one.

The following examples demonstrate applications of the rule of transformation to a number of problems often encountered in applications.

2.3.1 Transformation of a Given PDF into an Arbitrary PDF

Let ξ be a continuous random variable, uniformly distributed on an interval $[a, b]$. It is often desired to find a transformation $y = g(x)$ which transforms ξ into another random variable whose PDF $p_\eta(y)$ is given. In this case, according to eq. (2.89)

$$p_\eta(y)\mathrm{d}y = \frac{\mathrm{d}x}{b - a} \tag{2.90}$$

or

$$x = c + (b - a) \int_{-\infty}^{y} p_\eta(y)\mathrm{d}y = c + (b - a)P_\eta(y) \tag{2.91}$$

where c is a constant of integration. Since $P_\eta(-\infty)$ and $P_\eta(\infty) = 1$, this constant must be chosen as $c = a$. In this case $a \le \$ \times \$ \le b$ and, consequently

$$P_\eta(y) = \frac{x - a}{b - a} \text{ and } g(x) = P_\eta^{-1}\left(\frac{x - a}{b - a}\right) \tag{2.92}$$

Here P_η^{-1} is a function, inverse to the CDF of the random variable ξ.

On the other hand, if one chooses

$$y = g(x) = a + (b - a)P_\xi(x) \tag{2.93}$$

then eq. (2.89) results in the transformation of an arbitrary distribution $p_\xi(x)$ to a uniform distribution. Thus, in principle, any PDF can be transformed into another PDF. This fact is often used in numerical simulation.

2.3.2 PDF of a Harmonic Signal with Random Phase

The next important example is an ensemble of the harmonic functions with constant amplitude A, angular frequency ω_0 and random phase φ

$$s(t, \varphi) = g(\varphi) = A_0 \sin(\omega_0 t + \varphi) \tag{2.94}$$

Assuming that the PDF $p_\varphi(\varphi)$ is known, the distribution $p_s(x)$ of the signal $s(t)$ can be found for any given fixed moment of time t. However, the calculation is complicated by the fact that the inverse transformation $\varphi = h(x)$ has infinitely many branches

$$\varphi_n = a \sin(x/A_0) - \omega_0 t, \quad \frac{\mathrm{d}\varphi_n}{\mathrm{d}x} = \pm\sqrt{A_0^2 - x^2} \tag{2.95}$$

and thus the expression for the PDF $p_s(x)$ is given by

$$p_s(x) = \begin{cases} \dfrac{1}{\sqrt{A_0^2 - x^2}} \displaystyle\sum_{n=-\infty}^{\infty} p_\varphi(\varphi_n) & |x| \leq A_0 \\ 0 & |x| > A_0 \end{cases} \qquad (2.96)$$

since the resulting signal is confined to the interval $[-A_0, A_0]$.

In many communication applications it is reasonable to assume that the phase φ is uniformly distributed in the interval $[-\pi, \pi]$.

$$p_\varphi(\varphi) = \frac{1}{2\pi}, |\varphi| \leq \pi \qquad (2.97)$$

In this case the inverse transformation is limited to only two branches, each having the distribution (2.97). Thus, the distribution $p_s(x)$ of the signal $s(t)$ is

$$p_s(x) = \frac{1}{\pi A_0 \sqrt{1 - \left[\dfrac{x}{A_0}\right]^2}}, \quad |x| \leq A_0 \qquad (2.98)$$

2.4 RANDOM VECTORS AND THEIR DESCRIPTION

2.4.1 CDF, PDF and the Characteristic Function

A random vector $\xi = [\xi_1, \xi_2, \ldots, \xi_n]^{\mathrm{T}}$ is a set of random variables ξ_i, considered jointly. The number of components n is called the dimension of the random vector. Each component can be continuous, discrete or mixed. In order to describe the random vector one has to describe the range of values each component assumes and probabilities of each value. However, it is not sufficient to provide separate characteristics for each component—all components should be described jointly. Similarly to the case of a scalar random variable, considered above, it is possible to generalize a notion of the cumulative distribution function, probability density function, characteristic function and log-characteristic function for the vector case.

The cumulative distribution function $P_\xi(x)$ and the joint PDF, suitable for description of any type of random vector, are defined as

$$P_\xi(x) = P_\xi(x_1, x_2, \ldots, x_n) = \mathrm{Prob}\{\xi_1 < x_1, \xi_2 < x_2, \ldots, \xi_n < x_n\} \qquad (2.99)$$

$$p_\xi(x) = p_\xi(x_1, x_2, \ldots, x_n) = \frac{\partial^n}{\partial x_1 \partial x_2 \cdots \partial x_n} P_\xi(x_1, x_2, \ldots, x_n) \qquad (2.100)$$

The CDF defined by eq. (2.100) poses the following properties

1. The CDF $P_\xi(x)$ is equal to zero if at least one of its arguments is negative infinity and it approaches one if all the arguments are equal to positive infinity:

$$P_\xi(x_1, \ldots, -\infty, \ldots, x_n) = 0 \qquad (2.101)$$
$$P_\xi(\infty, \ldots, \infty, \ldots, \infty) = 1 \qquad (2.102)$$

2. The CDF is a non-decreasing function in each of its arguments.

3. If the k-th component of the random vector ξ is excluded from consideration, the CDF $P_{\xi_{n-1}}$ of the remaining vector can be obtained as follows

$$\xi_{n-1} = [\xi_1, \ldots, \xi_{k-1}, \xi_{k+1}, \ldots, \xi_n]^T$$
$$P_\xi(x_1, \ldots, \infty, \ldots, x_n) = P_{\xi_{n-1}}(x_1, \ldots, x_{k-1}, x_{k+1}, \ldots, x_n) \tag{2.103}$$

4. The CDF can be expressed as an n-fold repeated integral of the corresponding CDF

$$P_\xi(x) = \int_{-\infty}^{x_1} \cdots \int_{-\infty}^{x_n} p_\xi(x) \mathrm{d}x_1 \, \mathrm{d}x_2 \ldots \mathrm{d}x_n \tag{2.104}$$

The PDF $p_\xi(x)$ of a random vector ξ must satisfy the following conditions, which follow from the properties of the CDF and relation (2.100):

1. The PDF is a non negative function

$$p_\xi(x) \geq 0 \tag{2.105}$$

2. The PDF is normalized to unity, i.e.

$$\int_{-\infty}^{\infty} \cdots \int_{-\infty}^{\infty} p_\xi(x) \mathrm{d}x_1 \, \mathrm{d}x_2 \ldots \mathrm{d}x_n = 1 \tag{2.106}$$

3. The PDF is a symmetric function with respect to its arguments.

4. If integrated over one of the arguments over the real axis, the PDF of the random vector of order n produces a PDF of a random vector of dimension $(n-1)$ obtained by excluding the component over which integration has been performed:

$$p_{\xi_{n-1}}(x_1, \ldots, x_{k-1}, x_{k+1}, \ldots, x_n) = \int_{-\infty}^{\infty} p_\xi(x) \mathrm{d}x_k \tag{2.107}$$

If $\varsigma_n = \xi_n + j\eta_n$ is a vector of complex random variables, this vector can be described as a vector with $2n$ real components, allowing complex vectors to be handled within the framework developed above.

The characteristic function $\theta_\xi(ju)$ of the random vector can be defined as the n-dimensional Fourier transform of the corresponding PDF $p_\xi(x)$

$$\theta_\xi(ju) = \theta_\xi(ju_1, ju_2, \ldots, ju_n)$$
$$= \int_{-\infty}^{\infty} \cdots \int_{-\infty}^{\infty} p_\xi(x_1, \ldots, x_n) \exp[-j(u_1 x_1 + \cdots + u_n x_n)] \mathrm{d}x_1 \cdots \mathrm{d}x_n \tag{2.108}$$

The characteristic function is a symmetric, continuous function of its arguments, $|\theta_\xi(ju)| \leq \theta_\xi(j0) = 1$.

$$\Theta_{m\xi}(ju_1, ju_2, \ldots, ju_m) = \Theta_{n\xi}(ju_1, ju_2, \ldots, ju_m, 0, \ldots, 0), \quad m < n \tag{2.109}$$

The last equation reflects the fact that the characteristic function of a sub-vector of size m can be obtained from the characteristic function of the full vector of size $n > m$ by setting the arguments to zero at coordinates not included in the sub-vector.

If components $\{\xi_1, \xi_2, \ldots, \xi_n\}$ are independent then

$$p_\xi(x) = \prod_{k=1}^{n} p_{\xi_k}(x_k), \quad \text{and} \quad \Theta_\xi(j u) = \prod_{k=1}^{n} \Theta_{\xi_k}(j u_k) \qquad (2.110)$$

The converse statement is also true: if the characteristic function is expressed as a product of characteristic functions of the components then the components of the random vector are independent.

2.4.2 Conditional PDF

Joint consideration of two or more random variables allows the introduction of a new concept not considered in the case of a scalar random variable—the concept of the conditional PDF (CDF or CF). Let ξ be a random variable (or vector) which is observed simultaneously with a certain condition B. Then the probability that ξ does not exceed x, given that the condition B is satisfied, is called the conditional CDF $P_\xi(x \,|\, B)$ of the random variable given the condition B:

$$P_{\xi|B}(x \,|\, B) = \text{Prob}\{\xi < x \,|\, B\} = \frac{\text{Prob}\{\xi < x, B\}}{\text{Prob}(B)} \qquad (2.111)$$

The conditional PDF $p_{\xi|B}(x|B)$ is then defined as the derivative of the conditional CDF with respect to x. Its Fourier transform represents the conditional characteristic function $\Theta_{\xi|B}(j u \,|\, B)$

$$p_{\xi|B}(x \,|\, B) = \frac{\partial}{\partial x} P_\xi(x \,|\, B), \quad \Theta_{\xi|B}(j u \,|\, B) = \int_{-\infty}^{\infty} p_{\xi|B}(x \,|\, B) \exp(-j u x) \mathrm{d}x \qquad (2.112)$$

Conditional CDF, PDF and CF satisfy all the properties obeyed by regular unconditional CDF, PDF and CF, as can be shown from the definitions (2.111) and (2.112).

Upon generalization to the case of random vectors the definition (2.112) becomes

$$p_{\xi|\eta}(x_1, \ldots, x_k | x_{k+1}, \ldots, x_n) = \frac{p_{\xi,\eta}(x_1, \ldots, x_k, x_{k+1}, \ldots, x_n)}{p_\eta(x_{k+1}, \ldots, x_n)} \qquad (2.113)$$

or

$$p_{\xi,\eta}(x_1, \ldots, x_k, x_{k+1}, \ldots, x_n) = p_{\xi|\eta}(x_1, \ldots, x_k, x_{k+1}, \ldots, x_n) p_\eta(x_{k+1}, \ldots, x_n) \qquad (2.114)$$

known as the formula of multiplication of probabilities. Applying it in a chain manner, one obtains

$$
\begin{aligned}
p(x_1, \ldots, x_k, x_{k+1}, \ldots, x_n) &= p(x_1 | x_2, \ldots, x_n) p(x_2, \ldots, x_n) \\
&= p(x_1 | x_2, \ldots, x_n) p(x_2 | x_3, \ldots, x_n) p(x_3, \ldots, x_n) \\
&= p(x_1 | x_2, \ldots, x_n) p(x_2 | x_3, \ldots, x_n) \cdots p(x_{n-1} | x_n) p(x_n) \qquad (2.115)
\end{aligned}
$$

Two rules can be provided in order to eliminate variables from the expression for a conditional PDF.

1. To eliminate elements on the "left of the bar", i.e. to eliminate unconditional variables, one just has to integrate the conditional density over these variables

$$\int_{-\infty}^{\infty} p(x_1, \ldots, x_{j-1}, x_j, x_{j+1}, \ldots, x_k | x_{k+1}, \ldots, x_n) \mathrm{d}\, x_j$$
$$= p(x_1, \ldots, x_{j-1}, x_{j+1}, \ldots, x_k | x_{k+1}, \ldots, x_n) \tag{2.116}$$

2. To eliminate the variable on the "right of the bar", i.e. to eliminate condition variables, one has to multiply the conditional density by the conditional PDF of the conditions to be eliminated conditioned on all other conditions, and integrate over the variables to be eliminated

$$\int_{-\infty}^{\infty} p(x_1, \ldots, x_k | x_{k+1}, \ldots, x_{j-1}, x_j, x_{j+1}, \ldots, x_n) p(x_j | x_{k+1}, \ldots, x_{j-1}, x_{j+1}, \ldots, x_n) \mathrm{d}\, x_j$$
$$= p(x_1, \ldots, x_k | x_{k+1}, \ldots, x_{j-1}, x_{j+1}, \ldots, x_n) \tag{2.117}$$

In particular, the following formula (playing a prominent role in the theory of Markov processes) can be obtained

$$p(x_1 | x_3) = \int_{-\infty}^{\infty} p(x_1 | x_2, x_3) p(x_2 | x_3) \mathrm{d}\, x_2 \tag{2.118}$$

All the considered definitions and rules remain valid for the case of a discrete random variable, with integrals being reduced to sums.

Random variables, $\xi_1, \xi_2, \ldots, \xi_n$ are called mutually independent if events $\{\xi_1 < x_1\}$, $\{\xi_2 < x_2\}, \ldots, \{\xi_n < x_n\}$ are independent for any values of $x_i, i = 1, \ldots, n$. In this case their joint PDF (CDF, CF) is a product of PDF (CDF, CF) of individual components, i.e.

$$p_\xi(x_1, \ldots, x_n) = \prod_{i=1}^{n} p_{\xi_i}(x_i) \tag{2.119}$$

and

$$\Theta_\xi(x_1, \ldots, x_n) = \prod_{i=1}^{n} \Theta_{\xi_i}(x_i) \tag{2.120}$$

respectively. It is also interesting to note that it is possible to find a random vector whose components are pair-wise independent, i.e. $p_{\xi_i \xi_j}(x_i, x_j) = p_{\xi_i}(x_i) p_{\xi_j}(x_j)$, but when considered jointly they are not independent $p_\xi(x_1, x_2, \ldots, x_n) \neq p_{\xi_1}(x_1) p_{\xi_2}(x_2) \ldots p_{\xi_n}(x_n)$.

Two complex variables $\zeta_1 = \xi_1 + j\eta_1, \ldots, \zeta_n = \xi_n + j\eta_n$ are mutually independent if

$$p(\xi_1, \xi_2, \ldots, \xi_n, \eta_1, \eta_2, \ldots, \eta_n) = \prod_{i=1}^{n} p(\xi_i, \eta_i) \tag{2.121}$$

Note that independence of the real and imaginary parts of each variable is not required.

2.4.3 Numerical Characteristics of a Random Vector

The numerical characteristics considered in Sections 2.1.1.5–2.1.1.8 can be extended to the case of random vectors. In this section it is assumed that a random vector $\boldsymbol{\xi} = [\xi_1, \xi_2, \ldots, \xi_n]^{\mathrm{T}}$ is described by the joint PDF $p_{\boldsymbol{\xi}}(\boldsymbol{x})$. In this case the n-dimensional moment $m_{\alpha_1, \alpha_2, \ldots, \alpha_n}$ of order $\alpha = \alpha_1 + \alpha_2 + \cdots + \alpha_n$ is defined as

$$m_{\alpha_1, \alpha_2, \ldots, \alpha_n} = \int_{-\infty}^{\infty} \cdots \int_{-\infty}^{\infty} x_1^{\alpha_1} x_2^{\alpha_2} \ldots x_n^{\alpha_n} p_{\boldsymbol{\xi}}(\boldsymbol{x}) \mathrm{d}x_1 \, \mathrm{d}x_2 \ldots \mathrm{d}x_n \qquad (2.122)$$

Similarly, the central moments can be defined as

$$\mu_{\alpha_1, \alpha_2, \ldots, \alpha_n} = \int_{-\infty}^{\infty} \cdots \int_{-\infty}^{\infty} (x_1 - m_{1\xi_1})^{\alpha_1} (x_2 - m_{1\xi_2})^{\alpha_2} \cdots (x_n - m_{1\xi_n})^{\alpha_n} p_{\boldsymbol{\xi}}(\boldsymbol{x}) \mathrm{d}x_1 \, \mathrm{d}x_2 \ldots \mathrm{d}x_n$$

$$(2.123)$$

where $m_{1\xi_k} = m_{0,0,\ldots,1,\ldots,0} = E\{\xi_k\}$ is the average of the k-th component ξ_k of the random vector $\boldsymbol{\xi}$.

Moments can be obtained as coefficients of the power series expansion of the characteristic function since

$$\Theta_{\boldsymbol{\xi}}(j u_1, j u_2, \ldots, j u_n) = E\left\{ \exp\left[j \sum_{k=1}^{n} u_k x_k \right] \right\}$$

$$= \sum_{\alpha_1 \ldots \alpha_n = 0}^{\infty} \frac{m_{\alpha_1, \alpha_2, \ldots, \alpha_n}}{\alpha_1! \alpha_2! \cdots \alpha_n!} (j u_1)^{\alpha_1} (j u_2)^{\alpha_2} \cdots (j u_n)^{\alpha_n} \qquad (2.124)$$

where, of course,

$$m_{\alpha_1, \alpha_2, \ldots, \alpha_n} = (-j)^{\alpha_1 + \alpha_2 + \cdots + \alpha_n} \left[\frac{\partial^{\alpha_1 + \alpha_2 + \cdots + \alpha_n}}{\partial u_1^{\alpha_1} \partial u_1^{\alpha_1} \cdots \partial u_1^{\alpha_1}} \Theta_n(u_1, u_2, \ldots, u_n) \right]\Bigg|_{u_1 = u_2 = \cdots = u_n = 0}$$

$$(2.125)$$

In a similar manner, n-dimensional cumulants can be defined as coefficients of the power series of the log-characteristic function

$$\kappa_{\alpha_1, \alpha_2, \ldots, \alpha_n} = (-j)^{\alpha_1 + \alpha_2 + \cdots + \alpha_n} \left[\frac{\partial^{\alpha_1 + \alpha_2 + \cdots + \alpha_n}}{\partial u_1^{\alpha_1} \partial u_1^{\alpha_1} \cdots \partial u_1^{\alpha_1}} \ln \Theta_n(u_1, u_2, \ldots, u_n) \right]\Bigg|_{u_1 = u_2 = \cdots = u_n = 0}$$

$$(2.126)$$

A relation between moments and cumulants of a random vector can be established in a way similar to relations obtained for moments and cumulants of a random variable. Finally, conditional densities can be used to define conditional moments and cumulants.

As in the case of a random variable, lower order moments and cumulants play a prominent role in the description of a random vector. In particular, averages of components

$$m_{1\,\xi_k} = E\{\xi_k\} = m_{0,\dots,1,\dots,0}, \alpha_k = 1, \qquad \alpha_l = 0, \quad k \neq l \tag{2.127}$$

their variance $\sigma_{\xi_k}^2$ and $m_{2\,\xi_k}$

$$\sigma_{\xi_k}^2 = D_{\xi_k} = E\{(\xi_k - m_{1\,\xi_k})^2\} = \mu_{0,\dots,2,\dots,0}$$
$$m_{2\,\xi_k} = E\{\xi_k^2\} = m_{0,\dots,2,\dots,0} \tag{2.128}$$

describe the main features of the individual components ξ_k. Statistical dependence between any two components is described by mixed moments, especially by the correlation $R_{\xi_k\,\xi_l}$ and covariance $C_{\xi_k\,\xi_l}$, defined as non-central and central mixed moments of the second order

$$m_{1,1}^{\xi_k,\xi_l} = R_{\xi_k\,\xi_l} = E\{\xi_k\,\xi_l\}$$
$$\mu_{1,1}^{\xi_k,\xi_l} = C_{\xi_k\,\xi_l} = E\{(\xi_k - m_{1\,\xi_k})(\xi_l - m_{1\,\xi_l})\} \tag{2.129}$$

with all sub-indices in the definition (2.122) and (2.123) equal to zero except for the k th and l-th ones.

There is a simple relation between correlation and covariance (resulting from the relation between the central moments and moments)

$$R_{\xi_k\,\xi_l} = C_{\xi_k\,\xi_l} + m_{1\,\xi_k}\,m_{1\,\xi_l} \tag{2.130}$$

Since the expectation of a non-negative random variable is non-negative, it is possible to write that, for any real λ,

$$E\{[\lambda(\xi_k - m_{1\,\xi_k}) + (\xi_l - m_{1\,\xi_l})]^2\} = \lambda^2\sigma_{\xi_k}^2 + 2\lambda C_{\xi_k\,\xi_l} + \sigma_{\xi_l}^2 \geq 0 \tag{2.131}$$

This can be satisfied for an arbitrary λ if and only if the discriminant of the quadratic form in eq. (2.131) is negative, i.e.

$$C_{\xi_k\,\xi_l}^2 - \sigma_{\xi_k}^2\sigma_{\xi_l}^2 \leq 0 \tag{2.132}$$

This allows one to introduce a so-called correlation coefficient r between two components of the random vector

$$r = \frac{C_{\xi_k\,\xi_l}}{\sigma_{\xi_k}\,\sigma_{\xi_l}}, \quad |r| \leq 1 \tag{2.133}$$

Two random variables are called uncorrelated if $r = 0$. This does not imply independence of these variables. It follows from eq. (2.131) that the maximum of the correlation coefficient is achieved when two random variables are linearly dependent, i.e. $\xi_k = a\,\xi_l + b$.

The notion of correlation can be extrapolated to a case of complex random variables ζ_1 and ζ_2 as

$$C_{\zeta_1 \zeta_2} = E\{(\zeta_1 - m_{1\zeta})(\zeta_2 - m_{2\zeta})^*\} \tag{2.134}$$

where the asterisk indicates complex conjugation.

2.5 GAUSSIAN RANDOM VECTORS

Since Gaussian random variables play a prominent role in statistical signal processing and communication theory, their most important properties are summarized in this section. A continuous random variable ξ is called a Gaussian random variable if its PDF is given by

$$p_\xi(x) = \frac{1}{\sqrt{2\pi\sigma^2}} \exp\left[-\frac{(x-m)^2}{2\sigma^2}\right] \tag{2.135}$$

The corresponding characteristic function is then

$$\Theta_\xi(ju) = \exp\left[-jum - \frac{\sigma^2 u^2}{2}\right] \tag{2.136}$$

It is possible to show by direct integration that for a Gaussian random variable and an arbitrary deterministic function $y = g(x)$ [9,12]

$$\frac{\partial^n}{\partial D^n} E\{g(\xi)\} = 2^{-n} E\left\{\frac{d^{2n}}{d\xi^{2n}} g(\xi)\right\} \tag{2.137}$$

$$E\{\xi g(\xi)\} = m E\{g(\xi)\} + D E\left\{\frac{d}{d\xi} g(\xi)\right\} \tag{2.138}$$

Here $D = \sigma^2$. Setting $g(x) = x^n$ in eq. (2.137) allows easy computation of the moments of a Gaussian random variable.

Two random variables ξ_1 and ξ_2 are called jointly Gaussian if their joint PDF has the following form

$$p_{2N}(x_1, x_2) = \frac{1}{2\pi\sqrt{D_1 D_2 - \mu_{11}^2}} \exp\left[-\frac{D_2(x_1 - m_1)^2 - 2\mu_{11}(x_1 - m_1)(x_2 - m_2) + D_1(x_2 - m_2)^2}{2(D_1 D_2 - \mu_{11}^2)}\right]$$

$$= \frac{1}{2\pi\sigma_1\sigma_2\sqrt{1 - r^2}} \exp\left[-\frac{\sigma_2^2(x_1 - m_1)^2 - 2r\sigma_1\sigma_2(x_1 - m_1)(x_2 - m_2) + \sigma_1^2(x_2 - m_2)^2}{2\sigma_1^2\sigma_2^2(1 - r^2)}\right]$$

$$\tag{2.139}$$

The corresponding characteristic function is then

$$\Theta_{2N}(ju_1, ju_2) = \exp\left[j(m_1 u_1 + m_2 u_2) - \frac{1}{2}(D_1 u_1^2 + 2\mu_{11}u_1 u_2 + D_2 u_2^2)\right]$$

$$= \exp\left[j(m_1 u_1 + m_2 u_2) - \frac{1}{2}(\sigma_1^2 u_1^2 + 2r\sigma_1 \sigma_2 u_1 u_2 + \sigma_2^2 u_2^2)\right] \quad (2.140)$$

This distribution has five parameters: the averages of each component m_1, m_2, the variances of each component σ_1^2, σ_2^2, and the correlation coefficient r between the components

$$m_1 = E\{\xi_1\}, \quad m_2 = E\{\xi_2\}, \quad \sigma_1^2 = E\{\xi_1^2\}, \quad \sigma_2^2 = E\{\xi_2^2\} \quad (2.141)$$

$$\mu_{11} = E\{(\xi_1 - m_1)(\xi_2 - m_2)\}, \quad r = \frac{\mu_{11}}{\sigma_1 \sigma_2} \quad (2.142)$$

Taking the derivative of both sides of eq. (2.140) with respect to r, one obtains a useful relationship

$$\frac{\partial^n}{\partial r^n}\Theta(ju_1, ju_2) = (-1)^n(\sigma_1 \sigma_2 u_1 u_2)^n\Theta(ju_1, ju_2) \quad (2.143)$$

Further properties of mutually Gaussian random variables are illustrated by examples of two Gaussian random variables. More detailed discussion can be found in [1,2,13].

1. If ξ_1 and ξ_2 are uncorrelated, i.e. $r = 0$, then according to eqs. (2.139) and (2.140)

$$p_{2N}(x_1, x_2) = p_{\xi_1}(x_1)p_{\xi_2}(x_2)$$
$$\Theta_{2N}(ju_1, ju_2) = \Theta_{\xi_1}(ju_1)\Theta_{\xi_2}(ju_2) \quad (2.144)$$

Thus, two uncorrelated Gaussian random variables are also independent.

2. Two correlated jointly Gaussian random variables can be transformed into a pair of independent zero mean Gaussian random variables by a linear transformation

$$\eta_1 = (\xi_1 - m_1)\cos\alpha + (\xi_2 - m_2)\sin\alpha$$
$$\eta_2 = -(\xi_1 - m_1)\sin\alpha + (\xi_2 - m_2)\cos\alpha \quad (2.145)$$

Here, the angle α can be defined from the condition $E\{\eta_1, \eta_2\} = 0$ to be

$$\tan 2\alpha = \frac{2r\sigma_1 \sigma_2}{\sigma_1^2 - \sigma_2^2} \quad \text{if} \quad \sigma_1 \neq \sigma_2$$
$$\alpha = \pi/4 \quad \text{if} \quad \sigma_1 = \sigma_2 \quad (2.146)$$

3. Any linear transformation of jointly Gaussian random variables results in jointly Gaussian random variables.

4. If two random variables are jointly Gaussian, each variable by itself is also Gaussian. The inverse statement is not necessarily valid.

5. If two random variables ξ_1 and ξ_2 are jointly Gaussian, then both conditional densities $p_{\xi_1|\xi_2}(x_1|x_2)$ and $p_{\xi_2|\xi_1}(x_2|x_1)$ are Gaussian. For example

$$p_{\xi_2|\xi_1}(x_2|x_1) = \frac{p_{\xi_1,\xi_2}(x_1,x_2)}{p_{\xi_1}(x_1)} = \frac{1}{\sigma_2\sqrt{2\pi(1-r^2)}}\exp\left\{-\frac{\left[x_2 - m_2 - r\dfrac{\sigma_2}{\sigma_1}(x_1 - m_1)\right]^2}{2\sigma_2^2(1-r^2)}\right\}$$

(2.147)

It can be seen by inspection of eq. (2.147) that this PDF is a Gaussian PDF with the mean

$$m = m_2 + r(\sigma_2/\sigma_1)(x_1 - m_1)$$

(2.148)

and the variance

$$D = \sigma_2^2(1 - r^2)$$

(2.149)

Interestingly, the mean of the conditional PDF depends on the condition, while the variance does not, except for the correlation coefficient.

6. The joint PDF (2.139) can be expanded into the Miller series [13]

$$p_2(x_1,x_2) = \frac{1}{2\pi\sigma_1\sigma_2}\exp\left[-\frac{(x_1 - m_1)^2}{2\sigma_1^2} - \frac{(x_2 - m_2)^2}{2\sigma_2^2}\right]\sum_{n=0}^{\infty}H_n\left(\frac{x_1 - m_1}{\sigma_1}\right)H_n\left(\frac{x_2 - m_2}{\sigma_2}\right)\frac{r^n}{n!}$$

(2.150)

where $H_n(x)$ are Hermitian polynomials [5]. The advantage of this expansion is the fact that variables x_1, x_2 and r are separated in each term.

Most of the considered properties can be generalized to a case of a jointly Gaussian vector ξ of dimension n. If m_k and $\sigma_k^2 = D_k$ stand for the mean value and the variance of the k-th component and

$$R_{kl} = E\{(\xi_k - m_k)(\xi_l - m_l)\} = r_{kl}\sqrt{D_k D_l} = r_{kl}\sigma_k\sigma_l, \quad R_{kk} = D_k, \quad R_{kl} = R_{lk} \quad (2.151)$$

The marginal joint PDF and the corresponding characteristic function are then given by

$$p_N(x) = \frac{1}{\sqrt{(2\pi)^n|R|}}\exp\left[-\frac{(x-m)^{\mathrm{T}}R^{-1}(x-m)}{2}\right]$$

(2.152)

$$\Theta_N(ju) = \exp\left[jm^{\mathrm{T}}u - \frac{u^{\mathrm{T}}Ru}{2}\right]$$

(2.153)

Here

$$m = \begin{bmatrix} m_1 \\ m_2 \\ \cdots \\ m_n \end{bmatrix}, \quad R = \begin{bmatrix} R_{11} & R_{12} & \cdots & R_{1n} \\ R_{21} & R_{22} & \cdots & R_{2n} \\ \cdots & \cdots & \cdots & \cdots \\ R_{n1} & R_{n2} & \cdots & R_{nn} \end{bmatrix} = \mathrm{diag}\{\sigma_i\}\begin{bmatrix} r_{11} & r_{12} & \cdots & r_{1n} \\ r_{21} & r_{22} & \cdots & r_{2n} \\ \cdots & \cdots & \cdots & \cdots \\ r_{n1} & r_{n2} & \cdots & r_{nn} \end{bmatrix}\mathrm{diag}\{\sigma_i\}$$

(2.154)

are the vector of averages and the covariation matrix, respectively. Vectors x and u are vector-columns of dimension n. R^{-1} represents the inverse matrix and T is used to denote transposition.

Since $r_{kl} = r_{lk}$, the covariation matrix R is a symmetric matrix and contains $n(n+1)/2$ independent elements. Altogether, $n(n+1)/2 + n$ parameters are needed to uniquely define Gaussian distribution.

The conditional probability density can be easily obtained as

$$p(x_1, x_2, \ldots, x_k | x_{k+1}, \ldots, x_n) = p(X_1 | X_2)$$

$$= \frac{1}{\sqrt{(2\pi)^k |R_{X_1|X_2}|}} \exp\left[-\frac{(x - m_{X_1|X_2})^T R_{X_1|X_2}^{-1} (x - m_{X_1|X_2})}{2} \right] \qquad (2.155)$$

where

$$R = \begin{bmatrix} R_{11} & R_{12} \\ R_{21} & R_{22} \end{bmatrix} R_{X_1|X_2} = R_{11} - R_{12} R_{22}^{-1} R_{21}, \quad \text{and } m_{X_1|X_2} = m_1 + R_{12} R_{22}^{-1} (X_2 - m_2)$$

$$(2.156)$$

and the mean vector and the correlation matrix are split according to conditional and non-conditional variables.

2.6 TRANSFORMATION OF RANDOM VECTORS

In this section the question of transformation of random vectors is treated. To simplify derivations, the case of a vector with two components is considered first and the results are generalized to the case of an arbitrary number of components.

Consider two random variables ξ_1 and ξ_2, described by the joint PDF $p_\xi(x_1, x_2)$ which are being transformed by two non-linear functions

$$\eta_1 = g_1(\xi_1, \xi_2)$$
$$\eta_2 = g_2(\xi_1, \xi_2) \qquad (2.157)$$

or, in the vector form $\eta = g(\xi)$, where $\eta = [\eta_1, \eta_2]^T$ and $g = [g_1, g_2]^T$. It is required to find the joint PDF $p_\eta(y_1, y_2)$ of the new vector η.

It is easier to approach this problem by finding the joint CDF $P_\eta(y_1, y_2)$. Let S be an area of the plane $x_1 - x_2$ such that for any point $(x_1, x_2) \in S$ the following conditions are satisfied

$$g_1(x_1, x_2) < y_1, \text{ and } g_2(x_1, x_2) < y_2 \qquad (2.158)$$

Events $\{\eta_1 < y_1, \eta_2 < y_2\}$ and $\{\{\xi_1, \xi_2\} \in S\}$ are then equivalent, since one implies another. Thus, their probabilities are equal as well, i.e.

$$P_\eta(y_1, y_2) = \int_S p_\xi(x_1, x_2) dx_1 dx_2 \qquad (2.159)$$

In general, the area S can be a union of disjointed sets, thus the calculation of the integral may require integration over a few disjointed sets. If $d S_y$ represents an area in the $x_1 - x_2$ plane such that

$$y_1 \leq g_1(x_1, x_2) < y_1 + d y_1, \text{ and } y_2 \leq g_2(x_1, x_2) < y_2 + d y_2 \qquad (2.160)$$

then

$$P_\eta(y_1 + d y_1, y_2 + d y_2) - P_\eta(y_1, y_2) = p_\eta(y_1, y_2) d y_1 d y_2 = \int_{d S_y} p_\xi(x_1, x_2) d x_1 d x_2 \quad (2.161)$$

If the inverse vector function $x = h(y)$ has k branches, $x = h^{(i)}(y)$, $i = 1, \ldots, k$. Let $x_1 - x_2$ and $y_1 - y_2$ be two sets of Cartesian coordinates, then the area of the $x_1 - x_2$ plane corresponding to a small rectangular $d S_y = d y_1 \times d y_2$ is given by

$$d S_x = \sum_{i=1}^{k} |J_k| d S_y = \sum_{i=1}^{k} \begin{vmatrix} \dfrac{\partial}{\partial y_1} h_1^{(i)}(y_1, y_2) & \dfrac{\partial}{\partial y_2} h_1^{(i)}(y_1, y_2) \\ \dfrac{\partial}{\partial y_1} h_2^{(i)}(y_1, y_2) & \dfrac{\partial}{\partial y_2} h_2^{(i)}(y_1, y_2) \end{vmatrix} d S_y \qquad (2.162)$$

where $J = \dfrac{D(x_1, x_2)}{D(y_1, y_2)}$ is a Jacobian of the transformation of each branch of the function $x = h(y)$. Thus, it follows from eq. (2.161) that the transformation of the PDF is

$$p_\eta(y_1, y_2) = \sum_{i=1}^{k} p_\xi(h_1^{(i)}(y_1, y_2), h_2^{(i)}(y_1, y_2)) \begin{vmatrix} \dfrac{\partial}{\partial y_1} h_1^{(i)} & \dfrac{\partial}{\partial y_2} h_1^{(i)} \\ \dfrac{\partial}{\partial y_1} h_2^{(i)} & \dfrac{\partial}{\partial y_2} h_2^{(i)} \end{vmatrix} \qquad (2.163)$$

In general, if a random $\xi = [\xi_1, \xi_2, \ldots, \xi_n]^T$ vector of length n is transformed to another vector $\eta = [\eta_1, \eta_2, \ldots, \eta_n]^T$ by a non-linear transformation $y = g(x)$ having k branches of the inverse transformation $x = h^{(i)}(y)$, then the probability density function of the transformed random variable η is given by

$$p_\eta(y) = \sum_{i=1}^{k} p_\eta(h^{(i)}(y)) \left| \frac{D(x^{(i)})}{D(y)} \right| \qquad (2.164)$$

where

$$\frac{D(x^{(i)})}{D(y)} = \begin{vmatrix} \dfrac{\partial}{\partial y_1} h_1^{(i)}(y) & \cdots & \dfrac{\partial}{\partial y_n} h_1^{(i)}(y) \\ \cdots & \cdots & \cdots \\ \dfrac{\partial}{\partial y_1} h_n^{(i)}(y) & \cdots & \dfrac{\partial}{\partial y_n} h_n^{(i)}(y) \end{vmatrix} \qquad (2.165)$$

is the Jacobian of the i-th branch of the inverse transformation $x = h^{(i)}(y)$.

As in the case of a scalar random variable, it is possible to calculate moments of the transformed random variable without directly calculating the PDF $p_\eta(y)$. Indeed, the definition of moments of an arbitrary order implies that

$$m^{\eta_1,\eta_2,...,\eta_n}_{\alpha_1,\alpha_2,...,\alpha_n} = E\{\eta_1^{\alpha_1}\eta_2^{\alpha_2}\cdots\eta_n^{\alpha_n}\} = \int_{-\infty}^{\infty}\cdots\int_{-\infty}^{\infty} h_1^{\alpha_1}(y)h_1^{\alpha_1}(y)\cdots h_1^{\alpha_1}(y)\mathrm{d}y \quad (2.166)$$

If the transformation maps a vector of dimension n into a vector of dimension $m < n$, the vector $\boldsymbol{\eta}$ can be augmented by $n - m$ dummy variables, often by a sub-vector $[\xi_{l_1}, \ldots, \xi_{l_{n-m}}]$ of the vector $\boldsymbol{\xi}$, to allow use of eq. (2.166). The resulting PDF can then be integrated over dummy variables to obtain the PDF of the original vector. Evaluation of moments and cumulants of non-linear transformation using moment and cumulant brackets are considered in Section 2.8. The application of the technique described in this section is illustrated by a number of examples, which have their own importance in applications.

2.6.1 PDF of a Sum, Difference, Product and Ratio of Two Random Variables

Let the joint PDFs $p_\xi(x_1, x_2)$ of two random variables ξ_1 and ξ_2 be known. The PDF of the sum (difference) of these two variables $\eta = \xi_1 \pm \xi_2$ is to be determined. Introducing a dummy variable $\eta_2 = \xi_2$, one can formulate this problem in a standard form. The inverse transformation is then

$$\begin{aligned}\xi_1 &= \eta \mp \eta_2 \\ \xi_2 &= \eta_2\end{aligned} \qquad J = \begin{vmatrix} 1 & \mp 1 \\ 0 & 1 \end{vmatrix} = 1 \quad (2.167)$$

Thus, according to eq. (2.164), the joint PDF $p_{\eta\eta_2}(y_1, y_2)$ is equal to

$$p_{\eta\eta_2}(y_1, y_2) = p_\xi(y \mp y_2, y_2) \quad (2.168)$$

Integration over the dummy variable η_2 results in the PDF of the sum (difference) of two random variables

$$p_\eta(y) = \int_{-\infty}^{\infty} p_\xi(y \mp y_2, y_2)\mathrm{d}y_2, \text{ and } \eta = \xi_1 \pm \xi_2 \quad (2.169)$$

Equation (2.169) can be further simplified if ξ_1 and ξ_2 are independent. In this case $p_\xi(x_1, x_2) = p_{\xi_1}(x_1)p_{\xi_2}(x_2)$ and

$$p_\eta(y) = \int_{-\infty}^{\infty} p_{\xi_1}(y \mp y_2)p_{\xi_2}(y_2)\mathrm{d}y_2 \quad (2.170)$$

In general, if there are n independent random variables $\xi_1, \ldots, \xi_n$, the characteristic function of the weighted sum,

$$\eta = a_1\xi_1 + a_2\xi_2 + \cdots + a_n\xi_n = \sum_{k=1}^{n} a_k\xi_k \quad (2.171)$$

of these variables can be easily found to be

$$\Theta_\eta(ju) = \langle \exp(-ju\,\eta) \rangle = \left\langle \exp\left(-ju\sum_{k=1}^{n} a_k\,\xi_k\right) \right\rangle = \prod_{k=1}^{n}\langle \exp(-ju\,a_k\,\xi_k)\rangle = \prod_{k=1}^{n}\Theta_{\xi_k}(ju\,a_k)$$

$$(2.172)$$

Similarly, the expression for the log-characteristic function becomes

$$\Psi_\eta(ju) = \ln \Theta_\eta(ju) = \sum_{k=1}^{n} \ln \Theta_{\xi_k}(ju\,a_k) = \sum_{k=1}^{n} \Psi_{\xi_k}(ju\,a_k) \qquad (2.173)$$

The resulting distribution is called the composition of the distributions of each variable. It can be seen from eq. (2.172) that

$$\Psi_\eta(ju) = \sum_{l=1}^{\infty} \frac{\kappa_l^\eta}{l!}(ju)^l = \sum_{k=1}^{n}\sum_{l=1}^{\infty} \frac{\kappa_l^{\xi_k}}{l!}(ju\,a_k)^l = \sum_{l=1}^{\infty} \frac{\sum_{k=1}^{n} a_k^l \kappa_l^{\xi_k}}{l!}(ju)^l, \quad \kappa_l^\eta = \sum_{k=1}^{n} a_k^l \kappa_l^{\xi_k} \quad (2.174)$$

Thus, the cumulant of an arbitrary order of the sum can be expressed as a weighted sum of the cumulants of the same order. In particular, eq. (2.174), written for the average m_η and the variance $D_\eta = \sigma_\eta^2$, produces the following rule

$$m_\eta = \sum_{k=1}^{n} a_k\, m_{\xi_k}, \quad D_\eta = \sum_{k=1}^{n} a_k^2\, D_{\xi_k} \qquad (2.175)$$

If all variables have the same distribution and all the weights are equal, $a_i = a$, then eq. (2.172) is reduced to

$$\Theta_\eta(ju) = \Theta^n(ju\,a), \Psi_\eta(ju) = n\,\Psi(ju\,a), \kappa_k^\eta = n\,a^k\,\kappa_k^\xi \qquad (2.176)$$

Since for an arbitrary Gaussian random variable all the cumulants of order higher than two are equal to zero, eq. (2.174) shows that all the cumulants of the resulting random variable η are also equal to zero, i.e. the random variable η is also a Gaussian random variable.

Finally, eq. (2.174) allows the investigation of the limiting distribution of the average of identically distributed random variables

$$\eta = \frac{\xi_1 + \xi_2 + \cdots + \xi_N}{N} \qquad (2.177)$$

Setting $n = N$ and $a = 1/N$ in eq. (2.176), one obtains

$$\kappa_k^\eta = \frac{N\,\kappa_k^\xi}{N^k} = \frac{\kappa_k^\xi}{N^{k-1}} \qquad (2.178)$$

Thus, it can be seen, that the transformation (2.177) preserves the mean value, since $\kappa_1^\eta = \kappa_1^\xi$. However, all other cumulants are reduced by a factor N^{k-1}. For sufficiently large N

all cumulants of order higher than two can be neglected compared to the two first cumulants, i.e.

$$\Psi_\eta(ju) \approx ju\kappa_1^\xi - \frac{\kappa_2^\xi}{2N}u^2 = jum - \frac{(\sigma/\sqrt{N})^2u^2}{2} + O(N^{-2}) \tag{2.179}$$

Thus, the distribution of the average of a large number of identically distributed random variables differs little from a Gaussian distribution, with the mean value equal to the mean value of each component and the variance reduced by a factor of $\sqrt{N}$.

If random variables η and ζ are the product and ratio of two random variables ξ_1 and ξ_2 whose joint distribution is known, a similar approach allows one to obtain the PDF $p_\eta(y)$ of the result. Indeed, setting a dummy variable $\eta_2 = \xi_2, \zeta_2 = \xi_2$, one obtains

$$\eta = \xi_1\xi_2, \quad \xi_1 = \eta/\eta_2$$
$$\eta_2 = \xi_2, \quad \xi_2 = \eta_2$$

$$J = \left|\frac{D(x_1,x_2)}{D(y_1,y_2)}\right| = \begin{vmatrix} \dfrac{1}{y_2} & -\dfrac{y_1}{y_2^2} \\ 0 & 1 \end{vmatrix} = \frac{1}{|y_2|}, \quad p_\eta(y) = \int_{-\infty}^{\infty} p_\xi\left(\frac{y}{y_2}, y_2\right)\frac{dy_2}{|y_2|} \tag{2.180}$$

$$\zeta = \frac{\xi_1}{\xi_2}, \quad \begin{matrix} \xi_1 = \zeta\zeta_2 \\ \xi_2 - \zeta_2 \end{matrix}$$
$$\zeta_2 = \xi_2,$$

$$J = \left|\frac{D(x_1,x_2)}{D(y_1,y_2)}\right| = \begin{vmatrix} y_2 & y \\ 0 & 1 \end{vmatrix} = |y_2|, \quad p_\zeta(y) = \int_{-\infty}^{\infty} p_\xi(yy_2, y_2)|y_2|dy_2 \tag{2.181}$$

Equations (2.180) and (2.181) are simplified if ξ_1 and ξ_2 are independent. In this case

$$p_\eta(y) = \int_{-\infty}^{\infty} p_{\xi_1}\left(\frac{y}{y_2}\right)p_{\xi_2}(y_2)\frac{dy_2}{|y_2|}, \text{ and } p_\zeta(y) = \int_{-\infty}^{\infty} p_{\xi_1}(yy_2)p_{\xi_2}(y_2)|y_2|dy_2 \tag{2.182}$$

2.6.2 Probability Density of the Magnitude and the Phase of a Complex Random Vector with Jointly Gaussian Components

Let $\zeta = \xi_R + j\xi_I$ be a complex random variable with jointly Gaussian real and imaginary parts. This distribution is described by five parameters: the mean values of each component m_R and m_I, their variances σ_R^2 and σ_I^2 and the correlation coefficient r. The magnitude A and the phase θ of ζ can be obtained using the following non-linear transformation

$$A = \sqrt{\xi_R^2 + \xi_I^2}, \quad \theta = \text{atan}\frac{\xi_I}{\xi_R}, \quad A \geq 0, \quad |\theta| \leq \pi \tag{2.183}$$

Such a problem arises in many cases when a modulated harmonic signal with random phase is studied.

The first step is to transform ξ_R and ξ_I into new variables by a linear transformation (2.145)

$$\eta_1 = \xi_R \cos\alpha + \xi_I \sin\alpha$$
$$\eta_2 = -\xi_R \sin\alpha + \xi_I \cos\alpha \tag{2.184}$$

Here, the angle α is defined as

$$\tan 2\alpha = \frac{2r\sigma_1\sigma_2}{\sigma_1^2 - \sigma_2^2} \qquad \text{if } \sigma_1 \neq \sigma_2$$

$$\alpha = \pi/4 \qquad \text{if } \sigma_1 = \sigma_2$$

(2.185)

In this case the Gaussian components η_1 and η_2 are independent, with the joint distribution

$$p_{\eta_1,\eta_2}(y_1,y_2) = \frac{1}{2\pi\sqrt{D_{\eta_1} D_{\eta_2}}} \exp\left[-\frac{(y_1 - m_{1\eta_1})^2}{2D_{\eta_1}} - \frac{(y_2 - m_{1\eta_2})^2}{2D_{\eta_2}} \right]$$

(2.186)

where the new means and variances are defined as

$$m_{1\eta_1} = m_R \cos\alpha + m_I \sin\alpha$$

$$m_{1\eta_2} = m_I \cos\alpha - m_R \sin\alpha$$

(2.187)

$$D_{\eta_1} = D_R \cos^2\alpha + D_I \sin^2\alpha + r\sqrt{D_R D_I} \sin 2\alpha$$

$$D_{\eta_2} = D_I \cos^2\alpha + D_R \sin^2\alpha - r\sqrt{D_R D_I} \sin 2\alpha$$

(2.188)

The next step is to introduce polar coordinates

$$A = \sqrt{\eta_1^2 + \eta_2^2} = \sqrt{\xi_R^2 + \xi_I^2}, \quad \psi = \theta - \alpha = \operatorname{atan} \frac{\eta_2}{\eta_1}$$

(2.189)

with the inverse transformation given by

$$\eta_1 = A\cos\psi, \qquad J = \begin{vmatrix} \cos\psi & A\sin\psi \\ \sin\psi & -A\cos\psi \end{vmatrix} = -A$$

$$\eta_2 = A\sin\psi,$$

(2.190)

Using eq. (2.164) together with eqs. (2.186), (2.189) and (2.190), one readily obtains the joint PDF $p_{A,\psi}(A,\psi)$ of the magnitude and the phase

$$p_{A,\psi}(A,\psi) = \frac{A}{2\pi\sqrt{D_{\eta_1} D_{\eta_2}}} \exp\left[-\frac{(A\cos\psi - m_{1\eta_1})^2}{2D_{\eta_1}} - \frac{(A\sin\psi - m_{1\eta_2})^2}{2D_{\eta_2}} \right]$$

(2.191)

The PDF $p_A(A)$ of the magnitude and the PDF $p_\psi(\psi) = p_\theta(\theta - \alpha)$ can be obtained by the integration of the joint density $p_{A,\psi}(A,\psi)$ over the phase ψ or magnitude A variables, respectively. Depending on relations between the parameters of the distribution $p_{A,\psi}(A,\psi)$ a number of different distributions can arise from eq. (2.191). Detailed investigation of the particular cases can be found in [12]. Here, only two particularly important cases are considered.

2.6.2.1 Zero Mean Uncorrelated Gaussian Components of Equal Variance

In this case $m_R = m_I = 0$, $r = 0$ and $D_R = D_I = D = \sigma^2$. Using eqs. (2.184) and (2.185) one obtains $m_{1\,\eta_1} = m_{1\,\eta_2} = 0$, $D_{\eta_1} = D_{\eta_2} = D$, $\alpha = \pi/4$. This reduces eq. (2.191) to

$$p_{A,\psi}(A,\psi) = \frac{A}{2\pi\sigma^2}\exp\left[-\frac{A^2}{2\sigma^2}\right] = \frac{A}{\sigma}\exp\left[-\frac{A^2}{2\sigma^2}\right]\frac{1}{2\pi} \tag{2.192}$$

Integration over the phase produces the Rayleigh distribution of the magnitude and the uniform distribution of phase

$$p_A(A) = \frac{A}{\sigma^2}\exp\left[-\frac{A^2}{2\sigma^2}\right], \quad A \geq 0, \qquad p_\theta(\theta-\alpha) = \frac{1}{2\pi}, \quad |\theta| \leq \pi \tag{2.193}$$

In addition, it can be seen from eqs. (2.192) and (2.193) that $p_{A,\psi}(A,\psi) = p_A(A)p_\theta(\theta)$ and thus the phase and the magnitude are independent.

2.6.2.2 Case of Uncorrelated Components with Equal Variances and Non-Zero Mean

In this case $D_{\eta_1} = D_{\eta_2} = D$, $\alpha = \pi/4$, $m = \sqrt{m_R^2 + m_I^2} = \sqrt{m_{1\,\eta_1}^2 + m_{1\,\eta_2}^2} > 0$ and the joint PDF becomes, (2.191),

$$p_{A,\psi}(A,\psi) = \frac{A}{2\pi\sigma^2}\exp\left[-\frac{A^2 + m^2 - 2Am\left(\frac{m_{1\,\eta_1}}{m}\cos\psi + \frac{m_{1\,\eta_2}}{m}\sin\psi\right)}{2\sigma^2}\right]$$

$$= \frac{A}{2\pi\sigma^2}\exp\left[-\frac{A^2 + m^2 - 2Am\cos(\theta-\theta_0)}{2\sigma^2}\right] \tag{2.194}$$

Here

$$\tan\theta_0 = \frac{m_{1\,\eta_2}}{m_{1\,\eta_1}} \tag{2.195}$$

Further integration over the phase variable θ produces the Rice distribution for the magnitude

$$p_A(A) = \int_{-\pi}^{\pi} p_{A,\Psi}(A,\theta)\mathrm{d}\theta = \int_{-\pi}^{\pi}\frac{A}{2\pi\sigma^2}\exp\left[-\frac{A^2 + m^2 - 2Am\cos(\theta-\theta_0)}{2\sigma^2}\right]\mathrm{d}\theta$$

$$\frac{A}{2\pi\sigma^2}\exp\left[-\frac{A^2 + m^2}{2\sigma^2}\right]\int_{-\pi}^{\pi}\exp\left[\frac{Am\cos(\theta-\theta_0)}{\sigma^2}\right]\mathrm{d}\theta = \frac{A}{2\pi\sigma^2}\exp\left[-\frac{A^2 + m^2}{2\sigma^2}\right]I_0\left(\frac{Am}{\sigma^2}\right) \tag{2.196}$$

Similarly, integration of $p_{A,\psi}(A,\theta)$ over A produces a PDF of the phase with the following form

$$p_{\varphi}(\theta) = \frac{1}{2\pi}\exp\left(-\frac{m^2}{2\sigma^2}\right) + \frac{m\cos(\theta-\theta_0)}{2\sqrt{2\pi}\sigma^2}\exp\left[-\frac{m^2\sin^2(\theta-\theta_0)}{2\sigma^2}\right]\left[1+\mathrm{erf}\left(\frac{m\cos(\theta-\theta_0)}{\sqrt{2}\sigma}\right)\right]$$

(2.197)

Here $\mathrm{erf}(x)$ is the error function [5]:

$$\mathrm{erf}(x) = \frac{2}{\sqrt{\pi}}\int_0^x \exp(-t^2)\,dt.$$

(2.198)

It can be seen that if $m = 0$, eq. (2.196) produces the Rayleigh distribution since $I_0(0) = 1$, and eq. (2.197) describes the uniform distribution of the phase. Characteristically, the case of a large m/σ produces approximately Gaussian distributions for both the magnitude and the phase, as can be observed by inspection of Fig. 2.3.

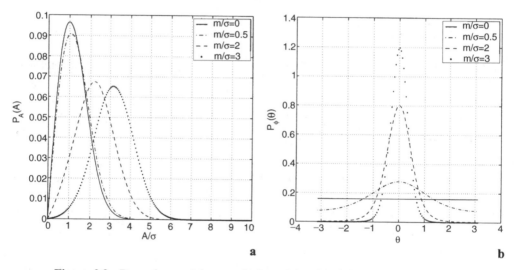

Figure 2.3   Dependence of the magnitude and the phase distribution on the ratio.

2.6.3   PDF of the Maximum (Minimum) of two Random Variables

Once again, let $p_{\xi_1,\xi_2}(x_1,x_2)$ be the joint PDF of two random variables ξ_1 and ξ_2. In some applications, such as combining techniques, it is required to find the distribution of the following random variables

$$\eta = \max\{\xi_1,\xi_2\}$$
$$\zeta = \min\{\xi_1,\xi_2\}$$

(2.199)

In this case, the method of non-linear transformation suggested above does not produce the desired result easily. However, this problem can be solved by finding the CDF of the result first. Indeed, based on the definition of CDF, one can write

$$P_\eta(y) = \text{Prob}\{\eta = \max\{\xi_1, \xi_2\} < y\} = \text{Prob}\{\xi_1 < y, \xi_2 < y\} = \int_{-\infty}^{y} \int_{-\infty}^{y} p_{\xi_1, \xi_2}(x_1, x_2) dx_1 dx_2$$

(2.200)

Taking the derivative with respect to y of both sides of eq. (2.200), the following expression for the PDF of η can be obtained

$$p_\eta(y) = \frac{d}{dy} P_\eta(y) = \frac{d}{dy} \int_{-\infty}^{y} \int_{-\infty}^{y} p_{\xi_1, \xi_2}(x_1, x_2) dx_1 dx_2$$

(2.201)

If ξ_1 and ξ_2 are independent, this expression can be further simplified to produce

$$p_\eta(y) = p_{\xi_1}(y) P_{\xi_2}(y) + p_{\xi_2}(y) P_{\xi_1}(y)$$

(2.202)

where $P_{\xi_i}(y)$ is the CDF of ξ_i, $i = 1, 2$.

In the same manner one can obtain the PDF of the random variable $\zeta = \min\{\xi_1, \xi_2\}$. In this case

$$P_\zeta(y) = \text{Prob}\{\zeta = \min\{\xi_1, \xi_2\} < y\} = \text{Prob}\{\xi_1 < y, x_1 \le x_2\} + \text{Prob}\{\xi_2 < y, x_2 \le x_1\}$$

$$= \int_{-\infty}^{y} dx_1 \int_{x_1}^{\infty} p_{\xi_1, \xi_2}(x_1, x_2) dx_2 + \int_{-\infty}^{y} dx \int_{x_2}^{\infty} p_{\xi_1, \xi_2}(x_1, x_2) dx_1$$

(2.203)

Taking the derivative of CDF, (2.203), the desired PDF is obtained in the following form

$$p_\zeta(y) = \frac{d}{dy} \int_{-\infty}^{y} dx_1 \int_{x_1}^{\infty} p_{\xi_1, \xi_2}(x_1, x_2) dx_2 + \int_{-\infty}^{y} dx_2 \int_{x_2}^{\infty} p_{\xi_1, \xi}(x_1, x_2) dx_1$$

(2.204)

Finally, for independent ξ_1 and ξ_2, eq. (2.204) reduces to

$$p_\zeta(y) = p_{\xi_1}(y)[1 - P_{\xi_2}(y)] + [1 - P_{\xi_1}(y)] p_{\xi_2}(y)$$

(2.205)

Example: selection combining. Two antennas, separated by a relatively large distance, are used to receive a signal from the same source. In each antenna the signal-to-noise ratio (SNR) γ is described by the so-called Gamma distribution

$$p_i(\gamma_i) = \frac{m_i^{m_i} \gamma^{m_i-1}}{\gamma_i^m \Gamma(m_i)} \exp\left(-\frac{m_i \gamma_i}{\gamma_i}\right), \qquad i = 1, 2$$

(2.206)

where m_i and $\bar{\gamma}_i$ are the parameters of the distribution (severity of fading and the average SNR, respectively). The antenna with the highest SNR is chosen to connect to the receiver. It is possible to assume that the levels of signals in both antennas are independent, thus eq. (2.202) can be used to obtain the distribution of the SNR at the input of the receiver. It produces

$$p(\gamma) = \frac{m_1^{m_1} \gamma_{m^{1-1}}}{\bar{\gamma}_1^{m_1} \Gamma(m_1)} \exp\left(-\frac{m_1 \gamma_1}{\bar{\gamma}_1}\right) P\left(m_2, \frac{m_2 \gamma_2}{\bar{\gamma}_2}\right) + \frac{m_2^{m_2} \gamma_{m^{2-1}}}{\bar{\gamma}_2^{m_2} \Gamma(m_2)} \exp\left(-\frac{m_2 \gamma_2}{\bar{\gamma}_2}\right) P\left(m_1, \frac{m_1 \gamma_1}{\bar{\gamma}_1}\right)$$

(2.207)

where

$$P(m, x) = \frac{1}{\Gamma(m)} \int_0^x t^{m-1} \exp(-t) \mathrm{d}t$$

(2.208)

is one of the forms of incomplete Gamma function [5]. For the case of independent Rayleigh fading with equal average SNR in both antennas, eq. (2.207) can be simplified to produce

$$p(\gamma) = \frac{2}{\bar{\gamma}} \exp\left(-\frac{\gamma}{\bar{\gamma}}\right) \left[1 - \exp\left(-\frac{\gamma}{\bar{\gamma}}\right)\right]$$

(2.209)

2.6.4 PDF of the Maximum (Minimum) of n Independent Random Variables

The equations obtained in the previous section can be generalized to the case of n independent random variables $\xi_1, \xi_2, \ldots, \xi_n$, each described by their PDF $p_{\xi_1}(x_1), \ldots, p_{\xi_n}(x_n)$. New variables are defined as

$$\begin{aligned} \eta &= \max\{\xi_1, \xi_2, \ldots, \xi_n\} \\ \zeta &= \max\{\xi_1, \xi_2, \ldots, \xi_n\} \end{aligned}$$

(2.210)

Following the same steps as in Section 2.6.3 one obtains

$$p_\eta(y) = \frac{\mathrm{d}}{\mathrm{d}y} P_\eta(y) = \frac{\mathrm{d}}{\mathrm{d}y} \prod_{k=1}^{n} P_{\xi_k}(y) = \prod_{k=1}^{n} P_{\xi_k}(y) \sum_{k=1}^{n} \frac{p_{\xi_k}(y)}{P_{\xi_k}(y)}$$

(2.211)

Similarly the PDF $p_\zeta(y)$ can be expressed as

$$p_\zeta(y) = \frac{\mathrm{d}}{\mathrm{d}y} P_\zeta(y) = \frac{\mathrm{d}}{\mathrm{d}y} \prod_{k=1}^{n} [1 - P_{\xi_k}(y)] = \prod_{k=1}^{n} [1 - P_{\xi_k}(y)] \sum_{k=1}^{n} \frac{p_{\xi_k}(y)}{1 - P_{\xi_k}(y)}$$

(2.212)

2.7 ADDITIONAL PROPERTIES OF CUMULANTS

It was shown earlier (see eq. (2.42)) that the characteristic function $\theta_\xi(ju)$ of a process with finite moments m_k can be expressed in terms of an infinite series of ju

$$\theta_\xi(ju) = 1 + \sum_{k=1}^{\infty} \frac{m_k}{k!} (ju)^k$$

(2.213)

Here moments m_k are defined by eq. (2.28). Instead of a characteristic function, one can consider the log-characteristic function $\ln \theta_\xi(ju)$ and its expansion in the Taylor series

$$\ln \theta_\xi(ju) = \sum_{k=1}^{\infty} \frac{\kappa_k}{k!}(ju)^k \tag{2.214}$$

Coefficients of this expansion are called the cumulants. Equations (2.213) and (2.214) allow relations between moments and cumulants to be obtained. Furthermore, it is possible to relate central moments and cumulants by noting that

$$\theta_\xi(ju) = e^{jum_1}\left[1 + \sum_{k=1}^{\infty} \frac{\mu_k}{k!}(ju)^k\right] \tag{2.215}$$

Moments and cumulants of random vectors can be introduced in a similar manner by considering corresponding characteristic functions.

$$\Theta_\xi(u) = \sum_{n_1=0}^{\infty}\sum_{n_2=0}^{\infty}\cdots\sum_{n_N=0}^{\infty} \frac{m_{n_1,n_2,\ldots,n_N}}{n_1!n_2!\cdots n_N!}(ju_1)^{n_1}(ju_2)^{n_2}\ldots(ju_N)^{n_N} \tag{2.216}$$

and

$$\ln \Theta_\xi(u) = \sum_{n_1=0}^{\infty}\sum_{n_2=0}^{\infty}\cdots\sum_{n_N=0}^{\infty} \frac{\kappa_{n_1,n_2,\ldots,n_N}}{n_1!n_2!\cdots n_N!}(ju_1)^{n_1}(ju_2)^{n_2}\ldots(ju_N)^{n_N} \tag{2.217}$$

Using the expansion of $\ln(1+z)$ into the series of powers of z, it is possible to find relations between cumulants of any order, and moments, or central moments, of any order. It is worth noting that the elements of the vector ξ can repeat themselves.

It is reasonable to ask why such a variety of descriptions is needed. One of the possible interpretations is the following: if a Gaussian random process is analyzed, then all the descriptions are equally convenient since the first two moments (or cumulants) completely describe a Gaussian random variable; however, description of a non-Gaussian random variable can be often more conveniently accomplished by using cumulant description. The advantages of cumulant description become obvious when orthogonal expansions of unknown densities, such as Edgeworth and Gram–Charlier series, are used. On the other hand, it can be shown, that, in contrast to moments, the relative value of cumulants $\kappa_n/\kappa_2^{n/2}$ is decreasing with the order, thus providing more reasonable approximations for an unknown density.

Yet another important property of the cumulants is the fact that they are better suited to describe statistical dependence between elements of random vectors. This statement can be illustrated by considering the following example. Let $\xi = [\xi_1, \xi_2]^T$ be a vector with two components, described by the joint probability density $p_\xi(x,y)$, such that $m_{1x} = m_{1y} = 0$. Three possible relations between the components are considered here: (a) the components ξ_1 and ξ_2 are completely independent; (b) the components ξ_1 and ξ_2 are linearly dependent, i.e. $\xi_2 = a\xi_1 + b$; (c) the components ξ_1 and ξ_2 are related through a non-linear transformation

$\xi_2 = a\xi_1^2$. It is also assumed that $\xi_1(t)$ is a symmetric random process, thus implying that all its odd moments are equal to zero

$$m_{(2n+1)\xi} = \langle \xi_1^{2n+1} \rangle = 0 \tag{2.218}$$

In the case (a) of independent random variables, the characteristic function of the second is a product of the characteristic function of each component, and thus

$$\ln \Theta_{2\xi}(ju_1, ju_2) = \ln \Theta_{1\xi_1}(ju_1) + \ln \Theta_{1\xi_2}(ju_2) = \sum_{n=0}^{\infty} \frac{\kappa_{n\xi_1}}{n!}(ju_1)^n + \sum_{n=0}^{\infty} \frac{\kappa_{n\xi_2}}{n!}(ju_2) \tag{2.219}$$

Equation (2.219) indicates that all the cross cumulants are equal to zero, since there are no terms containing $(ju_1)^{n_1}(ju_2)^{n_2}$, with both n_1 and n_2 greater than zero. In the case (b) the first mixed cumulant κ_{11} attains its maximum possible value, since

$$\kappa_{11} = m_{11} = \langle \xi_1 \xi_2 \rangle = a\sigma_1^2 \tag{2.220}$$

thus indicating maximum possible correlation. However, the same cumulant is equal to zero in the case (c):

$$\kappa_{11} = \langle \xi_1 \cdot a\xi_1^2 \rangle = a\langle \xi_1^3 \rangle = 0 \tag{2.221}$$

i.e. the first mixed cumulant does not completely describe dependence between two variables. Indeed, such dependence can be traced through the higher order cumulants, for example κ_{21}:

$$\kappa_{21} = \langle \xi_1^2 \cdot a\xi_1^2 \rangle = a\langle \xi_1^4 \rangle > 0 \tag{2.222}$$

The three cases above emphasize the following meaning of cumulants: the first mixed cumulant (or correlation function) measures the degree of linear dependence between two variables. Higher order cumulants are needed to describe non-linear dependence of two random variables.[6]

2.7.1 Moment and Cumulant Brackets

So far the angle brackets have been used as an alternative notation to the averaging operator over all random variables under consideration, i.e.

$$\langle g(x) \rangle = E\{g(\xi)\} = \int_{-\infty}^{\infty} g(x)p_\xi(x)dx, \quad \langle g(x,y) \rangle = E\{g(\xi,\eta)\}$$

$$= \int_{-\infty}^{\infty}\int_{-\infty}^{\infty} g(x,y)p_{\xi\eta}(x,y)dxdy, \text{ etc.} \tag{2.223}$$

---

[6]See [14,18] for more detailed discussions.

The following property of the moment brackets can be obtained from eq. (2.223)

$$\left\langle \sum_{n=1}^{N} \alpha_n f_n(\boldsymbol{x}) \right\rangle = \sum_{n=1}^{N} \alpha_n \langle f_n(\boldsymbol{x}) \rangle \tag{2.224}$$

where α_n are deterministic constants. Up to now such notation has been used mainly to shorten equations. However, it can be extended to allow a convenient tool for manipulation of cumulant and moment equations, considered below. For this purpose the following convention is used for mixed cumulants [9]

$$\kappa_{n_1,n_2,\ldots,n_m}^{\xi_1,\xi_2,\ldots,\xi_m} = \langle \xi_1, \ldots, \xi_1, \xi_2, \ldots, \xi_2, \ldots, \xi_m, \ldots, \xi_m \rangle = \langle \xi_1^{[n_1]}, \xi_2^{[n_2]}, \ldots, \xi_m^{[n_m]} \rangle \tag{2.225}$$

Here, the random variable ξ_1 appears inside the brackets n_1 times, ξ_2 appears n_2 times, etc. In future, brackets which contain at least one comma are called cumulant brackets. Otherwise the term moment brackets is used. For example

$$\kappa_2 = \kappa_2^{[\xi_1]} = \langle \xi_1, \xi_1 \rangle, \qquad \kappa_{12} = \kappa_{1,2}^{\xi_1,\xi_2} = \langle \xi_1, \xi_2, \xi_2 \rangle, \text{ etc.} \tag{2.226}$$

The importance of this notation arises from the fact that it is possible to formalize the relationship between the moment and cumulant brackets in an easy way. In addition, it will be shown later in Section 2.8 that it is relatively easy to obtain systems of equations for cumulant functions of random processes, based on the dynamic equations describing the system under consideration. Moreover, the notation for random variables ξ_1, ξ_2, etc. and their transformations $f(\xi_1)$, $g(\xi_2)$, etc. can be simply substituted by their ordinary number equivalents, as shown below. For example, the following relations between cumulant and moment brackets can be easily formalized through the use of Sratonovich symmetrization brackets [9].

$$\kappa_{1,1}^{\xi_1,\xi_2} = \langle \xi_1, \xi_2 \rangle = \langle \xi_1 \cdot \xi_2 \rangle - \langle \xi_1 \rangle \langle \xi_2 \rangle \equiv \langle 1, 2 \rangle = \langle 1 \cdot 2 \rangle - \langle 1 \rangle \langle 2 \rangle = m_{1,1}^{\xi_1,\xi_2} - m_{1\,\xi_1}\, m_{1\,\xi_2}$$

$$\kappa_{1,1,1}^{\xi_1,\xi_2,\xi_3} = \langle 1, 2, 3 \rangle - 3\{\langle 1 \rangle \langle 2 \cdot 3 \rangle\}_s + 2\langle 1 \rangle \langle 2 \rangle \langle 3 \rangle$$

$$= m_{1,1,1}^{\xi_1,\xi_2,\xi_3} - m_{1\,\xi_1}\, m_{1,1}^{\xi_2,\xi_3} - m_{1,\xi_2} m_{1,1}^{\xi_1,\xi_3} - m_{1\,\xi_3}\, m_{1,1}^{\xi_1,\xi_2} + 2\, m_{1\,\xi_1}\, m_{1\,\xi_2}\, m_{1\,\xi_3}$$

$$\tag{2.227}$$

Symmetrization brackets together with an integer in front of the brackets represent the sum of all possible permutations of the arguments inside the brackets. For example

$$3\{\langle 1 \rangle \langle 2, 3 \rangle\}_s = \langle 1 \rangle \langle 2, 3 \rangle + \langle 2 \rangle \langle 1, 3 \rangle + \langle 3 \rangle \langle 1, 2 \rangle$$
$$3\{\langle 1 \cdot 2 \rangle \langle 3 \cdot 4 \rangle\}_s = \langle 1 \cdot 2 \rangle \langle 3 \cdot 4 \rangle + \langle 1 \cdot 3 \rangle \langle 2 \cdot 4 \rangle + \langle 1 \cdot 4 \rangle \langle 2 \cdot 3 \rangle \tag{2.228}$$

In a similar manner, moment brackets can be expressed in terms of corresponding cumulants. For example

$$m_n = \kappa_N + N \kappa_1 \kappa_{N-1} + \frac{N!}{2!(N-2)!} (\kappa_2 \kappa_{N-2} + \kappa_1^2 \kappa_{N-2})$$

$$+ \frac{N!}{3!(N-3)!} (\kappa_3 \kappa_{N-3} + \kappa_1 \kappa_2 \kappa_{N-3} + \kappa_1^3 \kappa_{N-3}) + \cdots + \kappa_1^N \tag{2.229}$$

If one sets $\kappa_1 = 0$ in eq. (2.229), the relation between central moments and cumulants can be derived. Similar relations can be obtained for the mixed cumulants of two variables. For example, one can show that

$$m_{11} = \kappa_{11} + \kappa_{10}\,\kappa_{01}$$

$$m_{21} = \kappa_{21} + 3\,\kappa_{20}\,\kappa_{01} + 2\,\kappa_{11}\,\kappa_{10} + 3\,\kappa_{10}^2\kappa_{01}$$

$$m_{31} = \kappa_{31} + 3\,\kappa_{20}\,\kappa_{11} + 3\,\kappa_{21}\,\kappa_{10} + 3\,\kappa_{20}\,\kappa_{10}\,\kappa_{01} + 3\,\kappa_{11}\,\kappa_{10}^2 + \kappa_{10}^3\,\kappa_{01} \qquad (2.230)$$

$$m_{22} = \kappa_{22} + \kappa_{20}\,\kappa_{02} + 3\,\kappa_{10}\,\kappa_{12} + 2\,\kappa_{21}\,\kappa_{01} + \kappa_{10}^2\,\kappa_{02} + \kappa_{20}\,\kappa_{01}^2 + 2\,\kappa_{11}^2$$
$$+ 4\,\kappa_{10}\,\kappa_{11}\,\kappa_{01} + \kappa_{10}^2\,\kappa_{01}^2$$

Equations (2.227), (2.229) and (2.230) are useful in the sense that they allow the transformation of a problem of analysis of moments of non-linear memoryless transformation random variables into a problem of evaluating its cumulants. The latter is preferable since the cumulant series converges faster.

It is important to note that the cumulant brackets, in contrast to moment brackets, do not represent averaging of the quantity inside the brackets. Indeed, it can be seen that cumulants are obtained by non-linear transformation of the moments, thus they cannot be expressed as a linear combination of averaged quantities. Since, in any cumulant brackets, there are at least two arguments separated by a comma, there is no confusion between these two. However, of course, both types of brackets are closely related, since it is possible to express cumulants through moments and vice versa. It can also be shown that cumulant brackets are linear functions of their arguments, i.e. functions separated by two sequential commas.

2.7.2 Properties of Cumulant Brackets

In order to be useful, some formal rules of manipulation with cumulant brackets should be obtained. As earlier, Greek letters are used as a notation for random variables, while Latin letters are used for deterministic variables and constants. The following six properties can be easily obtained from the definition of cumulants.

1. $\langle \xi, \eta, \ldots, \omega \rangle$ is a symmetric function of its arguments;
2. $\langle a\,\xi, b\,\eta, \ldots, g\,\omega \rangle = ab \cdots g\langle \xi, \eta, \ldots, \omega \rangle$;
3. $\langle \xi, \eta, \ldots, \theta_1 + \theta_2, \ldots, \omega \rangle = \langle \xi, \eta, \ldots, \theta_1, \ldots, \omega \rangle + \langle \xi, \eta, \ldots, \theta_2, \ldots, \omega \rangle$;
4. $\langle \xi, \eta, \ldots, \theta, \ldots, \omega \rangle = 0$, if θ is independent of $\{\xi, \eta, \ldots\}$; $\qquad$ (2.231)
5. $\langle \xi, \eta, \ldots, a, \ldots, \omega \rangle = 0$;
6. $\langle \xi + a, \eta + b, \ldots, \theta + c, \ldots, \omega + g \rangle = \langle \xi, \eta, \ldots, \theta, \ldots, \omega \rangle$

The first three properties of cumulant brackets coincide with those of moment brackets, however, the last three properties are unique for cumulant brackets. The fourth property shows that if two groups of random processes are independent, any mutual cumulants are equal to zero (it is not true for corresponding moments). This is yet another indication that information about statistical dependence is encoded more evidently in cumulants rather than in moments. The sixth property indicates that the cumulants are invariant with respect to any deterministic shift in the random variables.

2.7.3 More on the Statistical Meaning of Cumulants

It has been shown earlier (see eqs. (2.220)–(2.222) and related discussions) that the first mixed cumulant $\kappa_{1,1}^{\xi_1,\xi_2}$ is a measure of linear correlation between two random processes ξ_1 and ξ_2. Such dependence is also often called first order dependence [14–18].

The next step is to consider cumulant $\kappa_{1,1,1}^{\xi_1,\xi_2,\xi_3} = \langle \xi_1, \xi_2, \xi_3 \rangle$. This cumulant can be expressed through the moment of the same order and mixed cumulants and moments of the lower order as

$$\langle \xi_1, \xi_2, \xi_3 \rangle = \langle \xi_1 \cdot \xi_2 \cdot \xi_3 \rangle - m_{1\xi_1} \langle \xi_2, \xi_3 \rangle - m_{1\xi_2} \langle \xi_1, \xi_3 \rangle - m_{1\xi_3} \langle \xi_1, \xi_2 \rangle - m_{1\xi_1} m_{1\xi_2} m_{1\xi_3} \quad (2.232)$$

Two interesting particular cases of eq. (2.232) can be further considered: all processes are pairwise uncorrelated, i.e. $\langle \xi_1, \xi_2 \rangle = \langle \xi_1, \xi_3 \rangle = \langle \xi_2, \xi_3 \rangle = 0$, i.e. there is no linear dependence between all these elements. If, in addition, these processes are independent, then $\kappa_{1,1,1}^{\xi_1,\xi_2,\xi_3} = 0$. Thus, cumulant $\kappa_{1,1,1}^{\xi_1,\xi_2,\xi_3}$ is not equal to zero if there is dependence between the components of a more complicated nature than just linear dependence. This (non-linear) dependence can be obtained, according to eq. (2.232), by subtracting linear dependence and deterministic average components. This dependence is present or is absent, no matter if there is linear dependence or not. Thus, in some sense, cumulants are independent coordinates which describe a set of dependent random variables. In this sense, a set of cumulants is preferable to an equivalent set of moments.

In general, so-called cumulant coefficients

$$\gamma_{p_1,p_2,\ldots,p_N}^{\xi_1,\xi_2,\ldots,\xi_N} = \frac{\kappa_{p_1,p_2,\ldots,p_N}^{\xi_1,\xi_2,\ldots,\xi_N}}{\sigma_{\xi_1}^{p_1} \sigma_{\xi_2}^{p_2} \ldots \sigma_{\xi_N}^{p_N}} \quad (2.233)$$

of order $p = p_1 + p_2 + \cdots + p_N$, describes a non-linear statistical dependence of $(N - 1)$ between a set of N variables.

It is interesting that there is only a limited freedom in choosing cumulants of a random variable or vector. Indeed, it is shown in [9] that cumulant coefficients γ_n of a random variable

$$\gamma_n = \frac{\kappa_n}{\kappa_2^{n/2}} = \frac{\kappa_n}{\sigma^n} \quad (2.234)$$

must satisfy certain (non-linear) inequalities. For example, skewness γ_3 and curtosis γ_4 must satisfy the condition [9]

$$\gamma_4 - \gamma_3^2 + 2 \geq 0 \quad (2.235)$$

While there is no restriction placed on γ_3, i.e. $\gamma_3 \in (-\infty, \infty)$, γ_4 must exceed -2. Restrictions on higher order cumulants are still an area of active research.

2.8 CUMULANT EQUATIONS

It was shown earlier that the characteristic function $\Theta(ju)$ can be defined if an infinite set of cumulants $\kappa_k, k = 1, 2, \ldots$, is given. In other words, the characteristic function, and thus the

PDF, is a function of an infinite cumulant vector $\kappa = \{\kappa_1, \kappa_2, \ldots\}$:

$$p(x) = p(x; \kappa_1, \kappa_2, \ldots) = p(x, \kappa) \tag{2.236}$$

It was also mentioned that the cumulants can be chosen independently from a certain region of values and this choice can be made independently for each cumulant as long as they satisfy certain inequalities [9]. This fact allows differentiation of the PDF $p(x; \kappa_1, \kappa_2, \ldots)$ with respect to cumulants, i.e. expressions of the type $\frac{\partial}{\partial \kappa_s} p(x; \kappa)$ are justified since it only requires an infinitely small variation of cumulant κ_s around its nominal value. Taking a derivative of both sides of the representation of the PDF through the characteristic function with respect to κ_s one obtains

$$\frac{\partial}{\partial \kappa_s} p(x; \kappa) = \frac{1}{2\pi} \int_{-\infty}^{\infty} e^{-jux} \frac{\partial}{\partial \kappa_s} \Theta(ju; \kappa) du \tag{2.237}$$

Furthermore, using representation (2.19) it is possible to calculate the partial derivative inside the integral to be

$$\frac{\partial}{\partial \kappa_s} \Theta(ju; \kappa) = \frac{(ju)^s}{s!} \Theta(ju, \kappa) \tag{2.238}$$

and, thus eq. (2.237) becomes

$$\frac{\partial^s}{\partial \kappa_s} p(x; \kappa) = \frac{1}{2\pi} \int_{-\infty}^{\infty} (jv)^s e^{-jvx} \Theta(jv; \kappa) dv \tag{2.239}$$

On the other hand, taking derivatives of both sides of eq. (2.19) s times with respect to x one obtains

$$\frac{\partial^s}{\partial x^s} p(x; \kappa) = \frac{1}{2\pi} \int_{-\infty}^{\infty} (-jv)^s e^{-jvx} \Theta(jv; \kappa) dv \tag{2.240}$$

Comparing this with expression (2.239) one can obtain a partial differential equation of the form

$$\frac{\partial}{\partial \kappa_s} p(x; \kappa) = \frac{(-1)^s}{s!} \frac{\partial^s}{\partial x^s} p(x; \kappa) \tag{2.241}$$

This relation between derivatives of the PDF with respect to the argument and the parameters is very useful in the derivation of cumulant equations [9].

The next step is to derive a set of equations which describe evolution of averages of a deterministic function $f(x)$. In this case the average, defined as

$$\langle f(x) \rangle = \int_{-\infty}^{\infty} f(x) p(x; \kappa) dx \tag{2.242}$$

can also be considered as a function of the cumulants of the underlying PDF

$$\langle f(x) \rangle \equiv \langle f(x) \rangle(\kappa) \tag{2.243}$$

Taking a derivative of both sides of eq. (2.242) with respect to κ_s it is easy to obtain

$$\frac{\partial}{\partial \kappa_s} \langle f(x) \rangle = \int_{-\infty}^{\infty} f(x) \frac{\partial}{\partial \kappa_s} p(x; \kappa) dx = \frac{(-1)^s}{s!} \int_{-\infty}^{\infty} f(x) \frac{\partial^s}{\partial x^s} p(x; \kappa) dx \qquad (2.244)$$

Here the relationship (2.242) has been used to substitute the derivative with respect to cumulants by the derivative with respect to the argument x. Further simplification of eq. (2.244) can be achieved by integrating the right hand side by parts s times and taking into account that for most distributions

$$\lim_{x \to \pm\infty} p(x; \kappa) = 0, \qquad \lim_{x \to \pm\infty} \frac{\partial^s}{\partial x^s} p(x; \kappa) = 0 \qquad (2.245)$$

As a result, a simple relation is obtained between the derivative of the average with respect to the s-th cumulant and the average of the s-th derivative of the function $f(x)$

$$\frac{\partial}{\partial \kappa_s} \langle f(x) \rangle = \frac{1}{s!} \int_{-\infty}^{\infty} p(x; \kappa) \frac{d^s}{d x^s} f(x) dx = \frac{1}{s!} \left\langle \frac{d^s}{d x^s} f(x) \right\rangle \qquad (2.246)$$

Taking the derivative of both sides of eq. (2.246) with respect to κ_s once more, one obtains

$$\frac{\partial^2}{\partial \kappa_s^2} \langle f(x) \rangle = \frac{\partial}{\partial \kappa_s} \frac{\partial}{\partial \kappa_s} \langle f(x) \rangle = \frac{1}{s!} \frac{\partial}{\partial \kappa_s} \left\langle \frac{d^s}{d x^s} f(x) \right\rangle = \frac{1}{(s!)^2} \left\langle \frac{d^{2s}}{d x^{2s}} f(x) \right\rangle \qquad (2.247)$$

Repeating this procedure n times, an expression for the n-th derivative can be obtained,

$$\frac{\partial^n}{\partial \kappa_s^n} \langle f(x) \rangle = \frac{1}{(s!)^n} \left\langle \frac{d^{ns}}{d x^{ns}} f(x) \right\rangle \qquad (2.248)$$

By the same token, mixed partial derivatives of the average of a function can be obtained [9] to be

$$\frac{\partial^{(n+m)}}{\partial \kappa_k^n \partial \kappa_p^m} \langle f(x) \rangle = \frac{1}{(k!)^n (p!)^m} \left\langle \frac{d^{(nk+mp)}}{d x^{nk+mp}} f(x) \right\rangle \qquad (2.249)$$

Equation (2.249) are called the cumulant equations [9]. Their usefulness becomes obvious from the examples below.

Example: Relation between the fourth moment m_4 and cumulants. It follows from the definition of averages that $m_4 = \langle x^4 \rangle$. Setting $f(x) = x^4$ in eq. (2.249) one can easily obtain

$$\frac{\partial^2}{\partial \kappa_1 \partial \kappa_3} \langle x^4 \rangle = \frac{1}{(1!)(3!)} \left\langle \frac{d^4}{d x^4} x^4 \right\rangle = \frac{4!}{3!} = 4$$

$$\frac{1}{2!} \frac{\partial^3}{\partial \kappa_1^2 \partial \kappa_2} \langle\langle x^4 \rangle\rangle = \frac{1}{2} \frac{1}{(1!)^2 (2!)} \left\langle \frac{d^4}{d x^4} x^4 \right\rangle = \frac{4!}{2 \cdot 2!} = 6 \qquad (2.250)$$

$$\frac{1}{4!} \frac{\partial^3}{\partial \kappa_1^4} \langle\langle x^4 \rangle\rangle = \frac{1}{4!} \frac{1}{(1!)^4} \left\langle \frac{d^4}{d x^4} x^4 \right\rangle = \frac{4!}{4!} = 1$$

These quantities can easily be identified with the coefficients of the expansion of the fourth moment in terms of corresponding cumulants. In a similar way, expansion of an arbitrary moment can be obtained. Of course, the applications of eq. (2.249) are much wider than just calculation of moments [9].

2.8.1 Non-linear Transformation of a Random Variable: Cumulant Method

Equations (2.249) are particularly important when it is desired to the obtain the relationship between the moments and cumulants of a transformed random variable $\eta = f(\xi)$ given that the cumulants of the input process are known. In order to obtain this dependence, eq. (2.249) must be first modified. For example, if the task is to find the variance κ_2^η of the output process;

$$\kappa_2^\eta = \langle \eta, \eta \rangle = \langle f^2(\xi) \rangle - \langle f(\xi) \rangle^2 \tag{2.251}$$

it can be seen that this equation already contains an expression for the average $\langle f(\xi) \rangle$ of the output variable which itself may depend on all cumulants of ξ. In general, the cumulant of an arbitrary order $\kappa_n^N = \Psi(\xi, \kappa_1^\xi)$ also depends on all cumulants. In order to take this fact into account, the expression for the n-th cumulant can be written as

$$\langle \Psi(\xi, \langle f(\xi) \rangle) \rangle = \int_{-\infty}^{\infty} \Psi(x, \langle f(x) \rangle) p(x; \kappa) \mathrm{d}x \tag{2.252}$$

Taking a derivative with respect to κ_s, one obtains an equation similar to eq. (2.249).

$$\frac{\partial}{\partial \kappa_s} \langle \Psi(\xi, \langle f(\xi) \rangle) \rangle = \int_{-\infty}^{\infty} \frac{\partial}{\partial \kappa_s} \Psi(x, \langle f(x) \rangle) p(x; \kappa) \mathrm{d}x + \int_{-\infty}^{\infty} \Psi(x, \langle f(x) \rangle) \frac{\partial}{\partial \kappa_s} p(x; \kappa) \mathrm{d}x \tag{2.253}$$

Taking advantage of the fact that $\Psi(x, \langle f(x) \rangle)$ depends on the cumulants only through $\langle f(x) \rangle$, and using the relationship (2.241), the equation for the derivative of the output cumulant can be written as

$$\frac{\partial}{\partial \kappa_s} \Psi(\xi, \langle f(\xi) \rangle) = \left\langle \frac{\partial}{\partial \langle f \rangle} \Psi(\xi, \langle f(\xi) \rangle) \right\rangle \frac{\partial}{\partial \kappa_s} \langle f(\xi) \rangle + \frac{1}{s!} \left\langle \frac{\mathrm{d}^s}{\mathrm{d}\xi^s} \Psi(\xi, \langle f(\xi) \rangle) \right\rangle \tag{2.254}$$

In turn, this can be further advanced by using eq. (2.248) to obtain

$$\frac{\partial}{\partial \kappa_s} \Psi(\xi, \langle f(\xi) \rangle) = \frac{1}{s!} \left\langle \frac{\mathrm{d}^s}{\mathrm{d}\xi^s} \left\{ [\Psi(\xi, \langle \langle \xi \rangle \rangle)] + \left\langle \frac{\partial}{\partial \langle f \rangle} \Psi(\xi, \langle f(\xi) \rangle) \right\rangle f(\xi) \right\} \right\rangle \tag{2.255}$$

The latter equation is a generalization of eq. (2.246) and can be used to find cumulants κ_n^η of the output process in terms of the cumulants k_s^ξ of the input process. Using the relation between the cumulant and the moments it is possible to obtain an expression for the differentials of the cumulants as

$$d\kappa_1 = d m_1$$
$$d\kappa_2 = d m_2 - 2 d m_1 \tag{2.256}$$

$$\dots\dots$$

This, coupled with eq. (2.246), produces a set of differential equations for cumulants of the output process

$$\frac{\partial \kappa_1^\eta}{\partial \kappa_s^\xi} = \frac{1}{s!}\left\langle \frac{d^s}{d\xi^s} f(\xi) \right\rangle$$

$$\frac{\partial \kappa_2^\eta}{\partial \kappa_s^\xi} = \frac{1}{s!}\left\langle \frac{d^s}{d\xi^s}[f^2(\xi) - 2\langle f(\xi)\rangle f(\xi)] \right\rangle \tag{2.257}$$

$$\dots\dots$$

Taking into account that $f(\xi) = \eta$, eq. (2.257) can be rewritten as

$$\frac{\partial \kappa_1^\eta}{\partial \kappa_s^\xi} = \frac{1}{s!}\left\langle \frac{d^s\eta}{d\xi^s} \right\rangle$$

$$\frac{\partial \kappa_2^\eta}{\partial \kappa_s^\xi} = \frac{\left\langle \frac{d^s}{d\xi^s}[\eta^2 - 2\langle\eta\rangle\eta] \right\rangle}{s!} \tag{2.258}$$

$$\dots\dots$$

Some important relationships, often used in the following text, can be obtained using eqs. (2.258) and the rules of the simplification of cumulant brackets, described in the appendix. In particular, dependence between lower order cumulants is provided by

$$\frac{\partial m_{1\eta}}{\partial m_{1\xi}} = \left\langle \frac{d\eta}{d\xi} \right\rangle, \quad \frac{\partial D_\eta}{\partial m_{1\xi}} = 2\left\langle \eta, \frac{\partial\eta}{\partial\xi} \right\rangle, \quad \frac{d m_{1\xi}}{d D_\xi^2} = \frac{1}{2}\left\langle \frac{d^2\eta}{d\xi^2} \right\rangle \tag{2.259}$$

$$\frac{\partial D_\eta}{\partial D_\xi} = \left\langle \frac{d\eta}{d\xi}, \frac{d\eta}{d\xi} \right\rangle + \left\langle \eta, \frac{d^2\eta}{d\xi^2} \right\rangle + \left\langle \frac{d\eta}{d\xi} \right\rangle \tag{2.260}$$

Finally, the derived equations can be further simplified if transformation of a Gaussian random variable is considered. In this case the input variable ξ is completely described by its first two cumulants $m_1 = \kappa_1$ and $D_\xi = \kappa_2$. Thus, the output cumulants and any average $\langle\varphi(\xi)\rangle$ will depend only on the two quantities. As a result, eqs. (2.259) and (2.260) can be rewritten as

$$\frac{\partial^n}{\partial m_1^n}\langle f(\xi)\rangle = \langle\varphi^{(n)}(\xi)\rangle$$

$$\frac{\partial^n}{\partial D^n}\langle f(\xi)\rangle = \frac{1}{2^n}\langle\varphi^{(2n)}(\xi)\rangle \tag{2.261}$$

If $D = 0$ then $\xi \equiv m_1$ and the following initial condition can be obtained for eq. (2.261)

$$\frac{\partial^n}{\partial D^n} \langle \varphi(\xi) \rangle \bigg|_{D=0} = \frac{1}{2^n} \langle \varphi^{(2n)}(m_1) \rangle \tag{2.262}$$

Finally, using moments m_n^η as $\langle f(\xi) \rangle$ one obtains

$$\frac{\partial^s m_n^\eta}{\partial D^s} = \frac{1}{2^s} \left\langle \frac{d^{2s}}{d\xi^{2s}} f^{(n)}(\xi) \right\rangle$$

$$\frac{\partial^s m_n^\eta}{\partial D^s} \bigg|_{D=0} = \frac{1}{2^s} \frac{d^{2s}}{dm_1^{2s}} f^{(n)}(m_1) \tag{2.263}$$

Solving differential equations (2.263) with the initial conditions (2.262) it is possible to obtain the expression for the moments of the transformed variable in terms of the mean and variance of the input Gaussian process. It can be seen that such moments are functions of D with coefficients depending only on the average m_1. Thus, an alternative way of calculating the output moments can be suggested. In this case the moments m_n^η can be represented as a series of powers of D with undetermined coefficients

$$m_n^\eta = A_0 + A_1 D + A_2 D^2 + \cdots + A_s D^s + \cdots \tag{2.264}$$

Taking derivatives of eq. (2.264) one obtains

$$\frac{\partial^n m_n^\eta}{\partial D^n} \bigg|_{D=0} = n! A_n \tag{2.265}$$

Comparing eqs. (2.263) and (2.265) the following expansion is obtained

$$m_n^\eta = \sum_{s=0}^{\infty} \frac{1}{s!} \frac{d^{2s}}{dm_1^{2s}} [f^{(n)}(m_1)] \left(\frac{D}{2}\right)^s \tag{2.266}$$

The latter equation can be considered as an alternative for non-linear analysis of non-linear transformations of Gaussian (and only Gaussian) random variables.

APPENDIX: CUMULANT BRACKETS AND THEIR CALCULATIONS [9]

A.1   Evaluation of Averages

A.1.1   Averages of Delta Functions and Their Derivatives

$$\langle \delta(\xi - x) \rangle_\xi = \int_{-\infty}^{\infty} p_\xi(s) \delta(s - x) ds = p_\xi(x) \tag{A.1}$$

$$\langle \delta(\xi - x) \delta(\eta - y) \rangle_{\xi,\eta} = \int_{-\infty}^{\infty} \int_{-\infty}^{\infty} p_{\xi,\eta}(s_1, s_2) \delta(s_1 - x) \delta(s_2 - y) ds_1 ds_2 = p_{\xi,\eta}(x, y) \tag{A.2}$$

$$\left\langle \frac{d^s}{d\xi^s} \delta(\xi - x) \right\rangle_\xi = \int_{-\infty}^{\infty} \frac{d^s}{dy^s} \delta(y - x) p_\xi(y) dy = (-1)^s \frac{d^s}{dx^s} p_\xi(x) \tag{A.3}$$

If ξ and η are jointly Gaussian and both have zero mean $\langle\xi\rangle = \langle\eta\rangle = 0$ then the following equations are valid

$$\left\langle \frac{d^{2s+1}}{d\,\xi^{2s+1}}\delta(\xi-x)\right\rangle_\xi = 0$$

$$\left\langle \frac{d^{2s}}{d\,\xi^{2s}}\delta(\xi-x)\right\rangle_\xi = 2^s \frac{d^s}{d\,D^s}\frac{1}{\sqrt{2\pi D}} = \frac{(-1)^s(2s-1)!!}{D^s}\frac{1}{\sqrt{2\pi D}} \tag{A.4}$$

$$\langle\delta(\xi)\delta(\eta)\rangle_{\xi,\eta} = \frac{1}{2\pi\sqrt{[D_\xi D_\eta - B^2_{\xi\eta}]}} \tag{A.5}$$

A.1.2 Averages of Arbitrary Functions

For an arbitrary function $f(\xi)$ of one variable and $F(\xi,\eta)$ of two variables the following equalities are valid

$$\langle f(\xi)\rangle_\xi = \langle f(\langle\xi\rangle) + \sum_{s=2}^\infty \frac{(x-\langle\xi\rangle)^s}{s!}f^{(s)}(\langle\xi\rangle)\rangle_\xi = f(\langle\xi\rangle) + \sum_{s=2}^\infty \frac{\mu_s}{s!}f^{(s)}(\langle\xi\rangle)$$

$$f(\langle\xi\rangle) + \frac{1}{2!}\frac{d^sf}{d\,\xi^2}\kappa_2 + \frac{1}{3!}\frac{d^3f}{d\,\xi^3}\kappa_3 + \frac{1}{4!}\frac{d^4f}{d\,\xi^4}(\kappa_4 + 3\kappa_2^2) + \frac{1}{5!}\frac{d^5f}{d\,\xi^5}(\kappa_5 + 10\kappa_2\kappa_3) + \cdots \tag{A.6}$$

In particular

$$p_\xi(x) = \sum_{m=0}^\infty (-1)^m \frac{\mu_s}{m!}\delta^{(m)}(x-m) \tag{A.7}$$

$$\langle F(\xi,\eta)\rangle = F(\langle\xi\rangle,\langle\eta\rangle) + \frac{1}{2!}\left(\frac{d^2}{d\,\xi^2}F(\xi,\eta)\kappa_{20} + 2\frac{d^2}{d\xi d\eta}F(\xi,\eta)\kappa_{11} + \frac{d^2}{d\eta^2}F(\xi,\eta)\kappa_{02}\right)$$

$$+ \frac{1}{3!}\left(\frac{d^3}{d\,\xi^3}F(\xi,\eta)\kappa_{30} + 3\frac{d^3}{d\xi^2 d\eta}F(\xi,\eta)\kappa_{21} + 3\frac{d^3}{d\xi d\eta^2}F(\xi,\eta)\kappa_{12}\right.$$

$$\left. + \frac{d^3}{d\eta^3}F(\xi,\eta)\kappa_{03}\right) + \frac{1}{4!}\left(\frac{d^4}{d\,\xi^4}F(\xi,\eta)[\kappa_{40} + 3\kappa_{20}^2]\right.$$

$$+ 4\frac{d^4}{d\xi^3 d\eta}F(\xi,\eta)[\kappa_{31} + 3\kappa_{11}\kappa_{20}]$$

$$+ 6\frac{d^4}{d\xi^2 d\eta^2}F(\xi,\eta)[\kappa_{22} + 2\kappa_{11}^2 + \kappa_{20}\kappa_{02}]$$

$$\times 4\frac{d^4}{d\xi d\eta^3}F(\xi,\eta)[\kappa_{13} + 3\kappa_{11}\kappa_{02}] + \frac{d^4}{d\eta^4}F(\xi,\eta)[\kappa_{04} + 3\kappa_{02}^2]\right) \tag{A.8}$$

In the above equation, derivatives are calculated at $\xi = \langle\xi\rangle$ and $\eta = \langle\eta\rangle$. Both equations are obtained by averaging corresponding Taylor series.

A.2   Rules of Manipulations with Cumulant Brackets

A.2.1   *Arbitrary Distribution at the Same Moment in Time*

$$\langle \xi,\xi^2 \rangle = \kappa_3 + 2\kappa_1\kappa_2 \tag{A.9}$$

$$\langle \xi,\xi,\xi^2 \rangle = \kappa_4 + 2\kappa_1\kappa_3 + 2\kappa_2^2 \tag{A.10}$$

$$\langle \xi,\xi,\xi,\xi^2 \rangle = \kappa_5 + 2\kappa_1\kappa_4 + 6\kappa_2\kappa_3 \tag{A.11}$$

$$\langle \xi,\xi,\xi,\xi,\xi^2 \rangle = \kappa_6 + 2\kappa_1\kappa_5 + 8\kappa_2\kappa_4 + 6\kappa_2^2 \tag{A.12}$$

$$\langle \xi^{[s]},\xi^2 \rangle = \kappa_{s+2} + \sum_{\lambda=0}^{s} C_s^{\lambda} \kappa_{\lambda+1}\kappa_{s-\lambda+1} \tag{A.13}$$

$$\langle \xi^2,\eta \rangle = \langle \xi,\xi,\eta \rangle + 2\langle \xi \rangle\langle \xi,\eta \rangle = \kappa_{21} + 2\kappa_{10}\kappa_{11} \tag{A.14}$$

$$\langle \xi^2,\eta,\eta \rangle = \langle \xi,\xi,\eta,\eta \rangle + 2\langle \xi \rangle\langle \xi,\eta,\eta \rangle + 2\langle \xi,\eta \rangle^2 = \kappa_{22} + 2\kappa_{10}\kappa_{12} + 2\kappa_{11}^2 \tag{A.15}$$

$$\langle \xi^2,\eta,\eta,\eta \rangle = \langle \xi,\xi,\eta,\eta,\eta \rangle + 2\langle \xi \rangle\langle \xi,\eta,\eta,\eta, \rangle + 6\langle \xi,\eta \rangle\langle \xi,\eta,\eta \rangle$$
$$= \kappa_{23} + 2\kappa_{10}\kappa_{13} + 6\kappa_{11}\kappa_{12} \tag{A.16}$$

$$\langle \xi^2,\eta^{[s]} \rangle = \langle \xi,\xi,\eta^{[s]} \rangle + \sum_{\lambda=0}^{s} C_s^{\lambda}\langle \xi,\eta^{[\lambda]} \rangle\langle \xi,\eta^{[s-\lambda]} \rangle = \kappa_{2,s} + \sum_{\lambda=0}^{s} C_s^{\lambda}\kappa_1,\kappa_{1,s-\lambda} \tag{A.17}$$

$$\langle \xi,\xi^3 \rangle = \kappa_4 + 3\kappa_1\kappa_3 + 3\kappa_2^2 + 3\kappa_1^2\kappa_2 \tag{A.18}$$

$$\langle \xi,\xi,\xi^3 \rangle = \kappa_5 + 3\kappa_1\kappa_4 + 9\kappa_2\kappa_3 + 3\kappa_1^2\kappa_3 + 6\kappa_1\kappa_2^2 \tag{A.19}$$

$$\langle \xi,\xi,\xi,\xi^3 \rangle = \kappa_6 + 3\kappa_1\kappa_5 + 12\kappa_2\kappa_4 + 3\kappa_1^2\kappa_4 + 3\kappa_1^2\kappa_4 + 9\kappa_3^2 + 18\kappa_1\kappa_2\kappa_3 + 6\kappa_3^2 \tag{A.20}$$

$$\langle \xi^3,\eta \rangle = \kappa_{3,1} + 3\kappa_{1,0}\kappa_{2,1} + 3\kappa_{2,0}\kappa_{1,1} + 3\kappa_{1,0}^2\kappa_{1,1} \tag{A.21}$$

$$\langle \xi^2,\eta,\eta \rangle = \kappa_{3,2} + 3\kappa_{1,0}\kappa_{2,2} + 3\kappa_{2,0}\kappa_{1,2} + 6\kappa_{1,1}\kappa_{2,1} + 3\kappa_{1,0}^2\kappa_{1,2} + 6\kappa_{1,0}\kappa_{1,1}^2 \tag{A.22}$$

$$\langle \xi\eta^{[k]} \rangle = \kappa_{1,k} + \sum_{s=1}^{k} C_k^s \kappa_s\kappa_{k-s} \tag{A.23}$$

$$\langle \xi\eta,f(\varphi) \rangle = \sum_{s=1}^{\infty} \frac{1}{s!}\left\langle \frac{d^s}{d\varphi^s}f(\varphi) \right\rangle\left[\kappa_{1,1,s} + \sum_{l=0}^{s} C_s^l\kappa_{1,0,l}\kappa_{0,1,s-l} \right] \tag{A.24}$$

$$\langle \xi\eta\varphi,f(\rho) \rangle = \sum_{n=0}^{\infty} \frac{1}{n!}\left\langle \frac{d^n}{d\rho^n}f(\rho) \right\rangle\left[\kappa_{1,1,1,n} + \sum_{k=0}^{s} C_n^k\{3\kappa_{1,0,0,k}\kappa_{0,1,1,s-l}\}_s \right.$$
$$\left. + \sum_{k=0}^{n}\sum_{l=0}^{n-k} C_n^k C_{n-k}^l\kappa_{1,0,0,k}\kappa_{0,1,0,l}\kappa_{0,0,1,n-k-l} \right] \tag{A.25}$$

If $\kappa_1^{\xi} = \langle \xi \rangle = 0$, the following series can be obtained

$$\langle \xi,f(\eta),g(\eta) \rangle = \sum_{n=0}^{\infty} \frac{1}{n!}\left\{ \sum_{k=0}^{n} C_n^k\left\langle \frac{d^{(n-k)}}{d\eta^{n-k}}f(\eta),\frac{d^k}{d\eta^k}g(\eta) \right\rangle \right.$$
$$\left. + \sum_{k=1}^{n-1} C_n^k\left\langle \frac{d^{(n-k)}}{d\eta^{n-k}}f(\eta) \right\rangle\left\langle \frac{d^k}{d\eta^k}g(\eta) \right\rangle \right\}\kappa_{1,n} \tag{A.26}$$

$$\langle \xi, \eta, f(\phi, \rho) \rangle = \sum_{\substack{k=0 \\ k+l>0}}^{\infty} \sum_{l=0}^{\infty} \frac{1}{k!l!} \left\langle \frac{\partial^{k+1}}{\partial \phi^k \, \partial \rho^l} f(\phi, \rho) \right\rangle$$

$$\times \left[\kappa_{1,1,k,l}^{\xi,\eta,\phi,\rho} + \sum_{\substack{i=0 \\ i+j\neq 0, k+1}}^{k} \sum_{j=0}^{l} C_k^i \, C_l^j \, \kappa_{1,i,j}^{\xi,\phi,\rho} \, \kappa_{1,k \ i,l \ j}^{\eta,\phi,\rho} \right] \qquad \text{(A.27)}$$

Here the following shortened notations are used

$$\kappa_n = \kappa_n^{\xi}, \ \kappa_{n,m} = \kappa_{n,m}^{\xi,\eta}, \ \kappa_{m,n,k} = \kappa_{m,n,k}^{\xi,\eta,\phi}, \ \kappa_{n,m,k,l} = \kappa_{n,m,k,l}^{\xi,\eta,\phi,\rho}$$

A.2.2 Two Stationary Related Processes

The following notations are used $\xi = \xi(t), \xi_\tau = \xi(t+\tau), \eta = \eta(t), \eta_\tau = \eta(t+\tau)$.

$$\langle \eta \, \xi_\tau F(\xi_\tau) \rangle = \langle \xi \, F(\xi) \rangle \langle \eta \rangle + \left\langle F(\xi) + \xi \frac{\mathrm{d}}{\mathrm{d}\xi} F(\xi) \right\rangle \langle \eta, \xi_\tau \rangle$$

$$+ \frac{1}{2} \left\langle 2 \frac{\mathrm{d}}{\mathrm{d}\xi} F(\xi) + \xi \frac{\mathrm{d}^2}{\mathrm{d}\xi^2} F(\xi) \right\rangle \langle \eta, \xi_\tau, \xi_\tau \rangle$$

$$+ \frac{1}{6} \left\langle 3 \frac{\mathrm{d}^2}{\mathrm{d}\xi^2} F(\xi) + \xi \frac{\mathrm{d}^3}{\mathrm{d}\xi^3} F(\xi) \right\rangle \langle \eta, \xi_\tau, \xi_\tau, \xi_\tau \rangle + \cdots \qquad \text{(A.28)}$$

$$\langle \xi \, \eta_\tau \, F(\xi_\tau) \rangle = \langle \eta \, F(\xi) \rangle \langle \xi \rangle + \left\langle \frac{\mathrm{d}}{\mathrm{d}\xi} F(\xi) \right\rangle \langle \xi, \xi_\tau \rangle + \langle F(\xi) \rangle \langle \xi, \eta_\tau \rangle$$

$$+ \frac{1}{2} \left\langle \eta \frac{\mathrm{d}^2}{\mathrm{d}\xi^2} F(\xi) \right\rangle \langle \xi, \xi_\tau, \xi_\tau \rangle + \left\langle \frac{\mathrm{d}}{\mathrm{d}\xi} F(\xi) \right\rangle \langle \xi, \xi_\tau, \eta_\tau \rangle$$

$$+ \frac{1}{6} \left\langle \eta \frac{\mathrm{d}^3}{\mathrm{d}\xi^3} F(\xi) \right\rangle \langle \eta, \xi_\tau, \xi_\tau, \xi_\tau \rangle + \frac{1}{2} \left\langle \frac{\mathrm{d}^2}{\mathrm{d}\xi^2} F(\xi) \right\rangle \langle \xi, \xi_\tau, \xi_\tau, \eta_\tau \rangle + \cdots \qquad \text{(A.29)}$$

$$\langle \xi \, \xi_\tau^2 F(\xi_\tau) \rangle = \langle \xi^2 \, F(\xi_\tau) \rangle \langle \xi \rangle + \left\langle 2 \xi F(\xi_\tau) + \xi^2 \frac{\mathrm{d}}{\mathrm{d}\xi} F(\xi) \right\rangle \langle \xi, \xi_\tau \rangle$$

$$+ \frac{1}{2} \left\langle 2 F(\xi) + 4 \xi \frac{\mathrm{d}}{\mathrm{d}\xi} F(\xi) + \xi^2 \frac{\mathrm{d}^2}{\mathrm{d}\xi^2} F(\xi) \right\rangle \langle \xi, \xi_\tau, \xi_\tau \rangle$$

$$+ \frac{1}{6} \left\langle 6 \frac{\mathrm{d}}{\mathrm{d}\xi} F(\xi) + 6 \xi \frac{\mathrm{d}^2}{\mathrm{d}\xi^2} F(\xi) + \xi^2 \frac{\mathrm{d}^3}{\mathrm{d}\xi^3} F(\xi) \right\rangle \langle \xi, \xi_\tau, \xi_\tau, \xi_\tau \rangle \qquad \text{(A.30)}$$

$$\langle \xi_\tau, \xi \, F(\xi_\tau) \rangle = \langle \xi, F(\xi) \rangle \langle \xi \rangle + \left[\langle F(x) \rangle + \left\langle \xi, \frac{\mathrm{d}}{\mathrm{d}\xi} F(\xi) \right\rangle \right] \langle \xi, \xi_\tau \rangle$$

$$+ \frac{1}{2} \left[2 \left\langle \frac{\mathrm{d}}{\mathrm{d}\xi} F(\xi) \right\rangle + \left\langle \xi, \frac{\mathrm{d}^2}{\mathrm{d}\xi^2} F(\xi) \right\rangle \right] \langle \xi, \xi_\tau, \xi_\tau \rangle$$

$$+ \frac{1}{6} \left[3 \left\langle \frac{\mathrm{d}^2}{\mathrm{d}\xi^2} F(\xi) \right\rangle + \left\langle \xi, \frac{\mathrm{d}^3}{\mathrm{d}\xi^3} F(\xi) \right\rangle \right] \langle \xi, \xi_\tau, \xi_\tau, \xi_\tau \rangle \qquad \text{(A.31)}$$

REFERENCES

1. W. Feller, *An Introduction to Probability Theory and Its Applications*, 3rd Edition, New York: Wiley, 1968.
2. M. Kendall and A. Stuart, *The Advanced Theory of Statistics*, 4th Edition, New York: Macmillan, 1977–1983.
3. A. Papoulis, *Probability, Random Variables, and Stochastic Processes*, Boston: McGraw-Hill, 2001.
4. J. Ord, *Families of Frequency Distributions*, London: Griffin, 1972.
5. A. Abramowitz and I. Stegun, *Handbook of Mathematical Functions with Formulas, Graphs, and Mathematical Tables*, New York: Dover, 1972.
6. Iu. Linnik, *Decomposition of Probability Distributions*, New York: Dover, 1965.
7. F. Oberhettinger, *Tables of Fourier Transforms and Fourier Transforms of Distributions*, New York: Springer-Verlag, 1990.
8. I.S. Gradsteyn and I.M. Ryzhik, *Table of Integrals, Series, and Products*, New York: Academic Press, 1980.
9. A.N. Malakhov, *Cumulant Analysis of Non-Gaussian Random Processes and Their Transformations*, Moscow: Sovetskoe Radio, 1978 (in Russian).
10. W. Rudin, *Functional Analysis*, New York: McGraw-Hill, 1991.
11. G. Szego, *Orthogonal Polynomials*, Providence: American Mathematical Society, 1975.
12. D. Middleton, *Topics in Communication Theory*, New York: McGraw-Hill, 1965.
13. K. Miller, *Multidimensional Gaussian Distributions*, New York: Wiley, 1964.
14. O.V. Sarmanov, Maximum Correlation Coefficient (Symmetrical Case), *DAN*, Vol. 120, 1958, pp. 715–718.
15. C. Nikias and A Petropulu, *Higher-Order Spectra Analysis,* New Jersey: Prentice Hall, 1993.
16. D. L. Wallace, Asymptotic Approximations to Distributions. *Ann. Math. Stat.* 29, 635–654, 1958.
17. H. Bateman and A. Erdelyi, *Higher Transcendental Functions*, New York: McGraw-Hill, 1953–1987.
18. O.V. Sarmanov, Maximum Correlation Coefficient (Non-Symmetrical Case), *DAN*, Vol. 121, 1958, pp. 52–55.

3

Random Processes

3.1 GENERAL REMARKS

A random process $\xi(u)$ describes a physical quantity which depends on a parameter u. A specific trace obtained as the result of a simple observation ($u_{min} < u < u_{max}$) is called a realization of this random process, a trajectory or a sample path. In future these terms are used interchangeably.

In communication theory, the parameter u usually represents time t or spacial coordinates x, y, z, or both, depending on the application. For now, discussions will be concentrated on the case of temporal random processes, i.e. $u \equiv t$. As an example of a random process, one can consider the variation of voltage (current) across electric circuit elements in the presence of noise. In this case, the notation $\xi(t)$ is used to emphasize that the random process is a function of time. For every fixed moment of time, $t = t_i$, the value of the random process $\xi_i = \xi(t_i)$ is a random variable, since this value may differ from experiment to experiment, even if they are conducted under the same conditions. The random value ξ_i is called a sample of the random process $\xi(t)$ taken at the moment of time $t = t_i$.

Depending on the shape of the trajectories and the method of description, random processes can be divided into three classes: impulse, fluctuations and a special form. An impulse process can be visualized as a sequence of single pulses, in general of different shapes, which follow each other in time with random intervals between consecutive pulses. Usually, impulse processes are described by piece-wise continuous functions of time. Man-made noise, so-called Electromagnetic Interference (EMI), and a number of natural phenomena, such as lightning, can be described as impulse random processes.

A fluctuation type of random process is usually a result of a combination of a large number of contributing sources, often impulses, which overlap in time. The trajectory of such processes resembles a continuous function of time. Thermal and cosmic noise are the prominent representatives of fluctuations.

The special type of random processes usually involve a combination of impulse or fluctuation random processes mixed with a deterministic signal or a signal of a known deterministic shape and random parameters. As an example one can consider a signal of the following form

$$\xi(t) = A\cos(\omega t + \phi) + n(t) \qquad (3.1)$$

Stochastic Methods and Their Applications to Communications.
S. Primak, V. Kontorovich, V. Lyandres
© 2004 John Wiley & Sons, Ltd   ISBN: 0-470-84741-7

Here, $n(t)$ can be a fluctuation or an impulse process and A, ω, ϕ can be deterministic or random variables.

The next few sections are dedicated to general methods of description of random processes.

3.2 PROBABILITY DENSITY FUNCTION (PDF)

Let one imagine that there is a large number N of completely identical systems (Fig. 3.1), constituting an ensemble system. All the systems work in identical conditions. The output of each system is a sample path of a random process $\xi(t)$ under consideration. If identical registering devices are attached to the systems, we can record each system state at any moment of time t. For a fixed moment of time, $t = t_1$, readings $x_1(t_1), x_2(t_1), \ldots, x_N(t_1)$ from different systems are different from each other since systems are assumed to be described by a random process. In general, values of $x_i(t_i)$ cover all possible regions of variation of $\xi(t)$ if N is sufficiently large. Let n_1 be a number of samples whose values belong to a small interval $[x; x + \Delta x]$. If N is large enough, a relative number of observations inside this interval is proportional to Δx and depends on t_1 as a parameter. In other words,

$$\lim_{N \to \infty} \frac{n_1}{N} = p_{1\xi}(x, t_1)\Delta x \qquad (3.2)$$

The function $p_{1\xi}(x, t_1)$ is called the marginal probability density of the process $\xi(t)$ considered at the moment of time $t = t_1$.

The marginal probability density is a very important, but not a sufficient, description of a random process. It completely describes the statistical properties only at one fixed moment

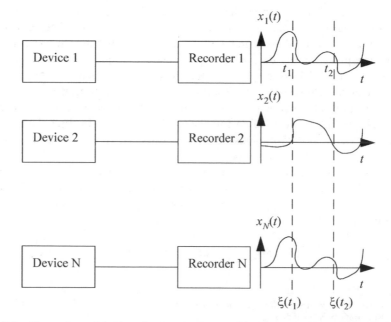

Figure 3.1   Illustration of the imaginary experiment with a large number of identical sources.

of time. However, it does not describe at all how values of the process $\xi(t)$ are related at two different time instants $t = t_1$ and $t = t_2$. It is possible to say that the marginal PDF gives a local view of the random process while ignoring its properties on a whole (even small) time interval.

A more complete description can be achieved if a second order (two time instant) PDF $p_{2\xi}(x_1, x_2; t_1, t_2)$ is considered. It describes relations between any two values of the random process $\xi(t)$ considered at any two separate moments of time t_1 and t_2.

Construction of the second order PDF is similar to that of the marginal PDF. One can observe the state of each of N devices, considered above, at two separate moments of time t_1 and t_2 (see Fig. 3.1) and let $2N$ random variables

$$x_1(t_1), x_2(t_1), \ldots, x_N(t_1); \qquad x_1(t_2), x_2(t_2), \ldots, x_N(t_2) \tag{3.3}$$

be the result of these observations. The next step is to count observations which fell into a small interval $[x_1, x_1 + \Delta x_1]$ at the moment of time $t = t_1$ and into a small interval $[x_2, x_2 + \Delta x_2]$ at the moment of time t_2. Once again it is assumed that if N is large enough, a relative number n_{12} of such events is proportional to the length of each interval and depends on both t_1 and t_2 as parameters. Mathematically speaking

$$\lim_{N \to \infty} \frac{n_{12}}{N} = p_{2\xi}(x_1, x_2; t_1, t_2) \Delta x_1 \Delta x_2 \tag{3.4}$$

The function $p_{2\xi}(x_1, x_2; t_1, t_2)$ is called the joint two-dimensional PDF of the process $\xi(t)$.

In general, the joint second order PDF $p_{2\xi}(x_1, x_2; t_1, t_2)$ also does not provide a complete description of the random process $\xi(t)$. It describes only the relation between two time instants of the process. A more detailed description can be obtained by considering a series of probability density functions of an arbitrary order M $p(x_1, x_2, \ldots, x_M; t_1, t_2, \ldots, t_M)$ constructed as follows: for M fixed moments of time $t_1, t_2, \ldots, t_M$, let $x_i(t_j)$ be a sample of the random variable $\xi(t)$ at the output of the i-th system at the time instant $t = t_j$. For small intervals, $[x_i, x_i + \Delta x_i]$, let n be the number of outcomes such that at t_i the observed quantity fits in the interval $[x_i, x_i + \Delta x_i]$ for every $i = 1, \ldots, M$. In this case

$$\lim_{N \to \infty} \frac{n}{N} = p_{M\xi}(x_1, x_2, \ldots, x_M; t_1, t_2, \ldots, t_M) \Delta x_1 \Delta x_2 \cdots \Delta x_M \tag{3.5}$$

A PDF of order M allows conclusions to be drawn about properties of a random process $\xi(t)$ at M separate moments of time. A higher order of PDF provides additional information about the process with respect to the lower order PDF.

An M-dimensional PDF must satisfy the following conditions

1. Positivity:

$$p_{M\xi}(x_1, x_2, \ldots, x_M; t_1, t_2, \ldots, t_M) \geq 0 \quad \text{for any } x_i \text{ and } t_j \tag{3.6}$$

This follows from the definition (3.5) since $n \geq 0$

2. Normalization:

$$
\int\limits_{-\infty}^{\infty} \cdots \int\limits_{-\infty}^{\infty} p_{M\xi}(x_1, x_2, \ldots, x_M; t_1, t_2, \ldots, t_M) \, dx_1 dx_2 \cdots dx_M = 1 \tag{3.7}
$$

3. Symmetry: $p_{M\xi}$ does not change under any permutation of its arguments $x_1, x_2, \ldots, x_M$.

4. Matching condition: for any $m < M$

$$
p_{m\xi}(x_1, x_2, \ldots, x_m; t_1, t_2, \ldots, t_m) = \int\limits_{R^{M-m}} \int p_{M\xi}(x_1, x_2, \ldots, x_M; t_1, t_2, \ldots, t_M) dx_{m+1} \cdots dx_M
$$

$$\tag{3.8}$$

This equation shows that if the M-dimensional PDF is known, the PDF of a smaller order m can be found by integrating the former over "additional" variables. In other words, if the PDF of order M is known, then a PDF of order $m < M$ can be found by integrating as in eq. (3.8). Since t is continuous, there is no finite order M which allows for a complete description of the process (except a few particular cases such as Gaussian or Markov processes). However, it can be shown that an infinite but countable sequence of PDFs describes a random process completely [1,2].

If a value of the process under consideration is known and fixed at a moment of time $t = t_2$, the conditional PDF $p_{1\xi}(x_1, t_1 | x_2, t_2)$ can be defined as follows

$$
p_{1\eta}(x_1, t_1) = \frac{p_{2\xi}(x_1, x_2; t_1, t_2)}{p_{1\xi}(x_2, t_2)} = p_{1\xi}(x_1, t_1 | x_2, t_2) \tag{3.9}
$$

Here, of course,

$$
p_{1\xi}(x_2, t_2) = \int\limits_{-\infty}^{\infty} p_{2\xi}(x_1, x_2; t_1, t_2) dx_1 \tag{3.10}
$$

The probability density $p_{1\xi}(x_1, t_1 | x_2, t_2)$ is known as the conditional PDF for a given x_2. It can be shown that $p_{1\xi}(x_1, t_1 | x_2, t_2)$ is a proper PDF. Indeed, since $p_{2\xi} \geq 0$ and $p_{1\xi} \geq 0$ then their ratio is also a positive function, thus satisfying one of the conditions. The normalization condition is also satisfied since

$$
\int\limits_{-\infty}^{\infty} p_{1\xi}(x_1, t_1 | x_2, t_2) dx_1 = \frac{1}{p_{1\xi}(x_2, t_2)} \int\limits_{-\infty}^{\infty} p_{2\xi}(x_1, x_2; t_1, t_2) dx_2 = \frac{1}{p_{1\xi}(x_2, t_2)} p_1(x_2, t_2) = 1
$$

$$\tag{3.11}$$

It is important to realize that the conditional PDF contains more (or at least as much) information about the process $\xi(t)$ at the time instant $t = t_1$ than the marginal PDF $p_{1\xi}(x_1, t_1)$.

Indeed, the value of the marginal PDF can be restored from the conditional PDF as follows. If $t_2 \to -\infty$, then the knowledge of $x_2 = \xi(t_2)$ does not produce any new information about $\xi(t_1)$ since they are independent, i.e.

$$\lim_{t_2 \to -\infty} p_{1\xi}(x_1, t_1 | x_2, t_2) = p_{1\xi}(\xi_1, t_1) \tag{3.12}$$

However, it is not possible to restore the conditional PDF from the marginal one. The amount of additional information carried by the conditional PDF is quite different in different cases. If $\xi(t_1)$ and $\xi(t_2)$ are independent samples of a random processes $\xi(t)$, then[1]

$$p_{2\xi}(x_1, x_2; t_1, t_2) = p_{1\xi}(x_1, t_1) p_{1\xi}(x_2, t_2) \tag{3.13}$$

The conditional PDF is then defined as

$$p_{1\xi}(x_1, t_1 | x_2, t_2) = \frac{p_{2\xi}(x_1, x_2; t_1, t_2)}{p_{1\xi}(x_2, t_2)} = p_{1\xi}(x_1, t_2) \tag{3.14}$$

thus stating that the knowledge of x_2 does not contribute any additional knowledge about the value of the process at the previous time. In the other extreme case, two samples $\xi(t_1)$ and $\xi(t_2)$ can be related by a deterministic function dependence, i.e. $\xi(t_1) = g[\xi(t_2)]$. In this case

$$p_{2\xi}(x_1, x_2; t_1, t_2) = \delta(x_1 - g(x_2)) p_{1\xi}(x_2, t_2) \tag{3.15}$$

and the conditional PDF is reduced to a δ function. Most of the realistic situations fall in between these two extremes. Eqs. (3.9)–(3.15) can be generalized to the case of a few variables.

3.3   THE CHARACTERISTIC FUNCTIONS AND CUMULATIVE DISTRIBUTION FUNCTION

Instead of probability densities of an arbitrary order, one can consider corresponding characteristic functions of the same order. The characteristic function $\theta_{n\xi}(u_1, u_2, \ldots, u_n; t_1, t_2, \ldots, t_n)$ of order n is defined as the n-dimensional Fourier transform of the PDF of order n, i.e.

$$\theta_{n\xi}(u_1, u_2, \ldots, u_n; t_1, t_2, \ldots, t_n)$$

$$= \int_{-\infty}^{\infty} \cdots \int_{-\infty}^{\infty} p_{n\xi}(x_1, x_2, \ldots, x_n; t_1, t_2, \cdots, t_n)$$

$$\times \exp[j(u_1 x_1 + u_2 x_2 + \cdots + u_n x_n)] dx_1 dx_2 \cdots dx_n \tag{3.16}$$

---

[1] $p_{1\xi}(x_1, t_1)$ and $p_{1\xi}(x_2, t_2)$ are different functions in general. However, since they represent the same process, the notation is kept the same.

It follows from eq. (3.16) that the characteristic function is an average value of the complex exponent

$$\theta_{n\xi}(u_1, u_2, \ldots, u_n; t_1, t_2, \ldots, t_n) = \langle \exp[j(u_1 x_1 + u_2 x_2 + \cdots + u_n x_n)]\rangle \tag{3.17}$$

The corner brackets here are used as a short-cut for averaging over an assembly. It follows from eqs. (3.16) and (3.17) that

$$\theta_{n\xi}(0, 0, \ldots, 0; t_1, t_2, \ldots, t_n)$$

$$= \int_{-\infty}^{\infty} \cdots \int_{-\infty}^{\infty} p_{n\xi}(x_1, x_2, \ldots, x_n; t_1, t_2, \ldots, t_n) \exp[j0] dx_1 dx_2 \ldots dx_n \equiv 1 \tag{3.18}$$

and

$$\theta_{n\xi}(u_1, u_2, \ldots, u_n; t_1, t_2, \ldots, t_n) = \theta_{(n+m)\xi}(u_1, u_2, \ldots, 0, \ldots, 0; t_1, t_2, \ldots, t_n) \tag{3.19}$$

Cumulative distribution functions (CDF) $P_n(x_1, x_2, \ldots, x_n; t_1, t_2, \ldots, t_n)$ of any order can be defined as

$$\frac{\partial^n P_n(x_1, x_2, \ldots, x_n; t_1, t_2, \ldots, t_n)}{\partial x_1 \partial x_2 \ldots \partial x_n} = p_n(x_1, x_2, \ldots, x_n; t_1, t_2, \ldots, t_n) \tag{3.20}$$

thus leading to

$$P_n(x_1, x_2, \ldots, x_n; t_1, t_2, \ldots, t_n)$$

$$= \int_{-\infty}^{x_1} \int_{-\infty}^{x_2} \cdots \int_{-\infty}^{x_n} p_n(x_1, x_2, \ldots, x_n; t_1, t_2, \ldots, \ldots, t_n) dx_1 dx_2 \ldots dx_n \tag{3.21}$$

Since $p_n(x_1, x_2, \ldots, x_n, t_1, t_2, \ldots, t_n) \geq 0$ then $P_n(x_1, x_2, \ldots, x_n, t_1, t_2, \ldots, t_n)$ is a non-decreasing function in all arguments, which satisfies the following normalization condition

$$P_n(\infty, \infty, \ldots, \infty; t_1, t_2, \ldots, t_n) = 1 \tag{3.22}$$

Considering that there is one-to-one correspondence between $p_n(x)$, $P_n(x)$ and $\theta_n(x)$, either of these functions provides the same description of the random process.

3.4 MOMENT FUNCTIONS AND CORRELATION FUNCTIONS

As was pointed out in Section 3.2, a complete description of a random process is accomplished by a series of PDFs (CDFs, characteristic functions) of increasing order. However, in practice, it is important to find a simpler (though in most cases incomplete) description of random processes for the following reasons:

1. In many cases, one needs to consider a transformation of a random process by a linear or a non-linear system with memory. Even if a complete description of the input of such systems is achieved (i.e. PDF of an arbitrary order is known), it is impossible to provide a general method of deriving PDFs at the output of the system under consideration. The problem can be solved partially by calculating a sufficient number of moments (or cumulants) of the output process instead.

2. Often, a process cannot be described statistically based on some physical (theoretical) model due to its complexity or lack of precise description. In this case, some statistical properties of the process can be obtained from experimental measurements. For example, low order moments of the process can be measured in many cases; however, a complete description in terms of PDF of an arbitrary order is impossible. It is usually possible to determine the marginal PDF and correlation function, but not the higher order PDFs.

3. Some processes are defined by a small number of parameters. For example, knowledge of the mean value and the covariance function completely describe a Gaussian process. Another interesting example is so-called "Spherically Invariant Processes," which are also defined by their marginal PDF and the covariance function. In addition, a widely used parametric technique allows the fitting of a model distribution based on a few lower order moments: the Pearson's class of PDF is widely used in applications and requires only knowledge of the first four moments [3].

4. Answers to many practical questions can be obtained by considering certain particular characteristics of random processes (such as mean, variance, correlation, etc.). For example, analysis of properties of the output of a receiver with an amplitude detector driven by a linearly amplitude modulated signal $A(1 + m\cos\Omega t)\sin(\omega t + \phi)$, usually ignores statistical properties of a random phase ϕ. In this case, only statistical properties of A have impact on the output [4].

5. It is often desired to obtain rough estimates of the output process, rather than its complete and accurate description.

Moment functions, $M_{i_1}(t), M_{i_1 i_2}(t_1, t_2), \ldots$, of a random process $\xi(t)$ are defined as follows. Let $I = [i_1, i_2, \ldots, i_n]$ be a set of integer numbers greater than or equal to zero. Then for different moments of time, $t_1, t_2, \ldots, t_n$, one can define an n-dimensional moment function of order $\alpha = \sum_{i=1}^{n} i_k$ as

$$M_{i_1 i_2 \ldots i_n}(t_1, t_2, \ldots, t_n) = \langle \xi^{i_1}(t_1)\xi^{i_2}(t_2) \ldots \xi^{i_n}(t_n) \rangle$$

$$= \int_{-\infty}^{\infty} \ldots \int_{-\infty}^{\infty} x_1^{i_1} x_2^{i_2} \ldots x_n^{i_n} p_{n\xi}(x_1, x_2, \ldots, x_n; t_1, t_2, \ldots, t_n) dx_1 \ldots dx_n \quad (3.23)$$

For example, the one-dimensional moment of order i_1 is nothing but the statistical average of the i_1-th power of the random process $\xi(t)$ itself

$$M_{i_1}(t) = \langle \xi^{i_1}(t) \rangle = \int_{-\infty}^{\infty} x^{i_1} p_{1\xi}(x) dx \quad (3.24)$$

while the two-dimensional moment $M_{i_1 i_2}(t_1, t_2)$ of order $i_1 + i_2$ is defined as

$$M_{i_1 i_2}(t_1, t_2) = \int_{-\infty}^{\infty} \int_{-\infty}^{\infty} x_1^{i_1} x_2^{i_2} p_{2\xi}(x_1, x_2) dx_1 dx_2 \tag{3.25}$$

and so on. It is important to note that in order to consider moment functions of order n, one needs to know a PDF of order n.

Instead of moment functions $M_{i_1 \cdots i_n}(t_1, \ldots, t_n)$ defined by eq. (3.23), one can consider so-called central moment functions of the same order and dimension, which are defined as follows

$$\mu_{i_1, i_2, \ldots, i_n}(t_1, t_2, \ldots, t_n) = \langle [\xi(t_1) - M_1(t_1)]^{i_1} \ldots [\xi(t_n) - M_n(t_n)]^{i_n} \rangle$$

$$= \int_{-\infty}^{\infty} \cdots \int_{-\infty}^{\infty} (x_1 - M_1(t_1))^{i_1} (x_2 - M_2(t_2))^{i_2} \ldots (x_n - M_n(t_n))^{i_n}$$

$$\times p_\xi(x_1, x_2, \ldots, x_n) dx_1 \ldots dx_n \tag{3.26}$$

There is an obvious relation between the moments and centred moments of the same random process. Indeed, expanding the expression $(x_1 - M_1(t_1))^{i_1} (x_2 - M_2(t_2))^{i_2} \ldots (x_n - M_n(t_n))^{i_n}$ into a power series one can obtain

$$(x_1 - M_1(t_1))^{i_1} (x_2 - M_2(t_2))^{i_2} \ldots (x_n - M_n(t_n))^{i_n} = \sum_{k_1} \sum_{k_2} \cdots \sum_{k_n} C_{k_1 k_2 \ldots k_n} x_1^{k_1} x_2^{k_2} \ldots x_n^{k_n} \tag{3.27}$$

and, thus

$$\mu_{i_1, i_2, \ldots, i_n}(t_1, t_2, \ldots, t_n)$$

$$= \int_{-\infty}^{\infty} \cdots \int_{-\infty}^{\infty} \sum_{k_1} \sum_{k_2} \cdots \sum_{k_n} C_{k_1 k_2 \ldots k_n} x_1^{k_1} x_2^{k_2} \ldots x_n^{k_n} p_{n\xi}(x_1, x_2, \ldots, x_n) dx_1 \ldots dx_n$$

$$= \sum_{k_1} \sum_{k_2} \cdots \sum_{k_n} C_{k_1 k_2 \ldots k_n} \int_{-\infty}^{\infty} \cdots \int_{-\infty}^{\infty} x_1^{k_1} x_2^{k_2} \ldots x_n^{k_n} p_{n\xi}(x_1, x_2, \ldots, x_n) dx_1 \ldots dx_n$$

$$= \sum_{k_1} \sum_{k_2} \cdots \sum_{k_n} C_{k_1 k_2 \ldots k_n} M_{i_1, i_2, \ldots, i_n}(t_1, t_2, \ldots, t_n) \tag{3.28}$$

Here, $C_{k_1 k_2 \ldots k_n} = C_{k_1 k_2 \ldots k_n}(t_1, t_2, \ldots, t_n)$ is a function of time only and it can be obtained by collecting terms with the same power of x_i. For example, the central moment of order k, considered at a single moment of time t, is related to the moments of order up to k as

$$\mu_k(t) = \int_{-\infty}^{\infty} (x - m_1(t))^k p_{1\xi}(x, t) dx = \sum_{i=1}^{k} \frac{k!}{i!(k-i)!} (-1)^i m_1^i(t) M_{k-i}(t) \tag{3.29}$$

Similarly

$$\mu_{1,1}(t_1, t_2) = \int_{-\infty}^{\infty} \int_{-\infty}^{\infty} [x_1 - m_1(t_1)][x_2 - m_1(t_2)] p_{2\xi}(x_1, x_2; t_1, t_2) dx_1 dx_2$$

$$= M_{1,1}(t_1, t_2) - m_1(t_1) m_1(t_2) \tag{3.30}$$

and so on. It is interesting to note that in order to find an expression for the central moment of order α, moments of the order up to α (but not higher) must be considered. The reverse statement is also easy to verify. This can be accomplished by considering the following identity

$$x_1^{i_1} x_2^{i_2} \dots x_n^{i_n} = ([x_1 - m_1(t_1)] + m_1(t_1))^{i_1} ([x_2 - m_1(t_2)]$$

$$+ m_1(t_2))^{i_2} \dots ([x_n - m_1(t_n)] + m_1(t_n))^{i_n} \tag{3.31}$$

Definition (3.23) can be extended to produce generalized averages. In particular, instead of powers of x_i in eq. (3.23), one can use a general function $g(x_1, x_2, \dots, x_n)$ of n variables:

$$M_g(t_1, t_2, \dots, t_n)$$

$$= \langle g(x_1, x_2, \dots, x_n) \rangle = \int_{-\infty}^{\infty} \dots \int_{-\infty}^{\infty} g(x_1, x_2, \dots, x_n) p_{n\zeta}(x_1, x_2, \dots, x_n; t_1, t_2, \dots, t_n)$$

$$\times dx_1 \cdots dx_n \tag{3.32}$$

Obviously, this definition includes the definition of moments, central moments and fractional moments (obtained when indices i_1 are not necessary integer). It also covers the characteristic function, absolute and factorial moments and many other quantities used in statistics and signal processing [4,5].

The moment functions can be obtained by differentiating the corresponding characteristic function, rather than by integration of the PDF. Indeed, using the Taylor series of the exponent, one can obtain

$$\theta_{1\xi}(u, t) = \int_{-\infty}^{\infty} p_{1\xi}(x, t) \exp[jux] dx = \int_{-\infty}^{\infty} p_{1\xi}(x, t) \left(\sum_{n=0}^{\infty} \frac{(jux)^n}{n!} \right) dx$$

$$= \left(\sum_{n=0}^{\infty} \frac{(ju)^n}{n!} \right) \int_{-\infty}^{\infty} x^n p_{1\xi}(x, t) dx = \sum_{n=0}^{\infty} \frac{j^n M_n(t)}{n!} u^n \tag{3.33}$$

Since the above eq. (3.33) is the Taylor expansion of $\theta_{1\xi}(u, t)$, its coefficients can be obtained through the values of its derivatives at $u = 0$ as

$$M_n(t) = (-j)^n \frac{d^n \theta_{1\xi}(u, t)}{du^n} \bigg|_{u=0} \tag{3.34}$$

It is possible to show that similar expressions are valid for k-dimensional moment functions [5].

It is possible to select a narrow group of moments which provide significant information about the random process under consideration, so called covariance functions. Cumulant functions $K_1(t_1)$, $K_2(t_1, t_2)$, $K_3(t_1, t_2, t_3)$ are defined through expansion of $\ln[\theta_{n\xi}(u_1, u_2, \ldots, u_n, t_1, \ldots, t_n)]$ into the MacLaurin series. All of them are symmetric functions of their arguments. In the one-dimensional case, cumulant functions are known as cumulants, or semi-invariants [5]. It is also possible to express covariance functions through related moment functions of similar order. In order to show this, one can use the following expansion of the $\ln(1 + z)$ into MacLaurin series [6]

$$\ln(1 + z) = z - \frac{1}{2}z^2 + \frac{1}{3}z_3 - \frac{1}{4}z^4 + \cdots + \frac{(-1)^n}{n}z^n \tag{3.35}$$

Setting

$$z = \theta_1(u) - 1 \tag{3.36}$$

and using the expansion of $\theta_1(u)$ in terms of the moment functions $M_u(t)$, one obtains

$$\ln \theta_1(u) = (\theta_1 - 1) - \frac{1}{2}(\theta_1 - 1)^2$$

$$+ \cdots \sum_{\nu=1}^{\infty} \frac{M_\nu}{\nu!}(ju)^\nu - \frac{1}{2}\left(\sum_{\nu=1}^{\infty} \frac{M_\nu}{\nu!}(ju)^\nu\right)^2 + \frac{1}{3}\left(\sum_{\nu=1}^{\infty} \frac{M_\nu}{\nu!}(ju)^\nu\right)^3 + \cdots \tag{3.37}$$

since

$$\theta_1(0) = \int_{-\infty}^{\infty} p_1(x) \exp[ju0]dx = \int_{-\infty}^{\infty} p_1(x)dx = 1 \tag{3.38}$$

The right-hand side of expression (3.37) represents a polynomial with respect to (ju). Rearranging terms in eq. (3.37), this equality can be rewritten as

$$\ln \theta_1(u) = \sum_{\nu=1}^{\infty} \frac{\kappa_\nu}{\nu!}(ju)^\nu \tag{3.39}$$

or

$$\theta_1(u) = \exp\left(\sum_{\nu=1}^{\infty} \frac{\kappa_\nu}{\nu!}(ju)^\nu\right) \tag{3.40}$$

Expansion (3.39) defines cumulants κ_ν of order ν, or semi-invariants of order ν.

It can be seen from eq. (3.37) that κ_ν is a polynomial of n variables $M_1, M_2, \ldots, M_n$ and vice versa. An implicit expression relating cumulant and moment functions can be obtained by equating terms of the same order of (ju) in eqs. (3.39) and (3.37). As a result

$$\begin{aligned}
\kappa_1 &= M_1 \\
\kappa_2 &= M_2 - M_1^2 = \mu_2 = \sigma^2 \\
\kappa_3 &= M_3 - 3M_1M_2 + 2M_1^2 = \mu_3 \\
\kappa_4 &= M_4 - 3M_2^2 - 4M_1M_3 + 12M_1^2M_2 - 6M_1^4 = \mu_4 - 3\mu_2^2
\end{aligned} \tag{3.41}$$

and so on. Expressions for higher order cumulants can be found in [5,7].

It is important to mention that the first cumulant coincides with the average of the random process:

$$\kappa_1(t) = M_1(t) = \langle \xi(t) \rangle \tag{3.42}$$

while the second cumulant is its variance

$$\kappa_2(t) = M_2(t) - M_1^2(t) = \langle \xi(t) - M_1(t)^2 \rangle = \sigma^2(t) \tag{3.43}$$

Higher order cumulants are also very important in describing the shape of the marginal PDF $p_{1\xi}(x)$. Let us just mention that the normalized third and fourth cumulants are known as normalized skewness and excess [7,8]

$$\gamma_1 = \frac{\kappa_3}{\kappa_2^{3/2}}, \quad \gamma_2 = \frac{\kappa_4}{\kappa_2^2}, \quad \gamma = \frac{\kappa_4}{\kappa_2^2} \tag{3.44}$$

It is important to notice that higher order cumulants do not coincide with corresponding central moments. The difference starts with κ_4, as can be seen from eq. (3.41).

Higher order covariance (cumulant) functions $K_1(t_1), K_2(t_1, t_2), K_3(t_1, t_2, t_3)$ are defined by expanding $\ln \theta_n(u_1, \ldots, u_n)$ into the MacLaurin series. The first three covariance functions are listed below:

$$\begin{aligned}
K_1(t) &= \kappa_1 = M_1(t) = \langle \xi(t) \rangle \\
K_2(t_1, t_2) &= \langle [\xi(t_1) - M_1(t_1)][\xi(t_2) - M_1(t_2)] \rangle \\
&= M_2(t_1, t_2) - M_1(t_1)M_2(t_2) = \kappa_{1,1}(t_1, t_2) \\
K_3(t_1, t_2, t_3) &= M_3(t_1, t_2, t_3) - M_1(t_1)K_2(t_2, t_3) - M_1(t_2)K_2(t_1, t_3) \\
&\quad - M_1(t_3)K_2(t_1, t_2) + 2M_1(t_1)M_1(t_2)M_1(t_3) = \kappa_{1,1,1}(t_1, t_2, t_3)
\end{aligned} \tag{3.45}$$

It is easy to see that $t_1 = t_2 = t_3$, and eq. (3.45) becomes eq. (3.41).

Knowledge of covariance functions allows one to restore characteristic functions and thus PDFs. This means that an adequate description of the random process can be achieved by a series of PDFs, a series of characteristic functions, moments or cumulants, as indicated in Fig. 3.2. *It is important to note that the decision of which characteristic of a random process to use heavily depends on the type of random process. For a Gaussian process, which is*

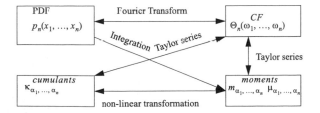

Figure 3.2 Relations between PDF, CF, moments and cumulants.

completely described by its mean and covariance function, all descriptions can be applied with the same complexity.

In future, the first two cumulant functions $M_1(t)$ and $K_2(t_1, t_2)$ will play a central role in the analysis of random processes and their transformation by linear and non-linear systems. A branch of the theory of random processes based on $M_1(t)$ and $K_2(t_1, t_2)$ is called "the correlation theory" of random processes.

Detailed investigation of cumulants can be found in [8]. Some discussions, closely following the approach taken in [8] can also be found in Section 3.8.

3.5   STATIONARY AND NON-STATIONARY PROCESSES

One of the important classes of random processes is so-called stationary random processes. A random process $\xi(t)$ is called stationary in a strict sense if the PDF $p_{n\xi}(x_1, x_2, \ldots, x_n; t_1, \ldots, t_n)$ of any arbitrary order n does not change by a time translation of the time instants $t_1, t_2, \ldots, t_n$ along the time axes. In other words, for any t_0 and n

$$p_{n\xi}(x_1, \ldots, x_n; t_1, \ldots, t_n) = p_{n\xi}(x_1, \ldots, x_n; t_1 - t_0, t_2 - t_0, \ldots, t_n - t_0) \tag{3.46}$$

This means that the stationary process is homogeneous in time. Of course, eq. (3.46) implies that the same property is true for characteristic functions, moments and cumulants.

Stationary processes are similar to steady-state processes in deterministic systems. All other processes are called non-stationary. As an example of a non-stationary process, one can consider a process of switching load of the power system (random number of users) before it reaches a steady state.

It can be seen from eq. (3.46) that the marginal PDF $p_{1\xi}(x, t)$ of a stationary process does not depend on time at all, i.e.

$$p_{1\xi}(x, t) = p_{1\xi}(x, t - t) = p_{1\xi}(x, 0) = p_{1\xi}(x) \tag{3.47}$$

At the same time the PDF of the second order depends only on the time difference, $\tau = t_2 - t_1$, between two instances of time

$$p_{2\xi}(x_1, x_2; t_1, t_2) = p_{2\xi}(x_1, x_2; t_1 - t_2 - t_1) = p_{2\xi}(x_1, x_2; 0, \tau) = p_{2\xi}(x_1, x_2; \tau) \tag{3.48}$$

In general, the PDF of order n depends only on $(n - 1)$ time intervals $\tau_k = t_{k+1} - t_k$ between $t_1, t_2, \ldots, t_n$.

Since the marginal PDF of a stationary process does not depend on time, neither do the characteristic function, moments, and cumulants of the first order. In particular, changing of a time scale also does not change these quantities. In a sense, a description using only the marginal PDF is similar to a description of a harmonic signal $s(t) = A\cos(\omega t + \varphi)$, using only its magnitude A. It provides knowledge about the range of the process (from $-A$ to A) but does not tell how fast the process changes.

Since the probability density of the second order $p_{2\xi}(x_1, x_2; t_1, t_2)$ depends only on the time difference $\tau = t_2 - t_1$ (time lag τ) the covariance[2] function $C(t_1, t_2) = K_2(t_1, t_2)$ also depends only on the time lag τ

$$C(\tau) = \langle [\xi(t) - m][\xi(t_2) - m] \rangle = \langle \xi(t_1)\xi(t_1 + \tau) \rangle - m^2$$

$$= \int_{-\infty}^{\infty} \int_{-\infty}^{\infty} (x_1 - m)(x_2 - m)p_{2\xi}(x_1, x_2; \tau)dx_1 dx_2 \qquad (3.49)$$

Variance of the random process σ^2 then coincides with the value of the covariance function at zero lag:

$$\sigma^2 = C(0) = \int_{-\infty}^{\infty} (x - m)^2 p_{1\xi}(x)dx \qquad (3.50)$$

In many practical cases, one is limited to considering probability densities of order two. Moreover, only the mean value and the variance of the process are often sufficient to obtain an answer or a rough estimate. Due to this fact, a more relaxed definition of stationarity is usually used.

A random process $\xi(t)$ is called a wide sense stationary random process if its mean does not depend on time $(m = M_1(t) = const.)$ and its covariance function $C_{\xi\xi}(t_1, t_2)$ depends only on the time lag $\tau = t_2 - t_1$, i.e. $C_{\xi\xi}(\tau) = C_{\xi\xi}(t_2 - t_1)$. A strict sense stationary process is also stationary in the wide sense. The reverse is not always true.

It is important to mention that there is an important class of random processes for which these two definitions are equivalent: any wide sense Gaussian process is a strict sense stationary process as well. This follows from the fact that knowledge of the mean and covariance function completely describes any Gaussian process [1,4]. Since Gaussian processes are often encountered in applications, it is important to learn how to calculate or measure the mean value and the covariance function. On the other hand, no further information is required to describe such a process. This fact emphasizes the importance of the correlation theory for applications.

3.6 COVARIANCE FUNCTIONS AND THEIR PROPERTIES

A covariance function between samples of the same process at two moments of time is called the autocovariance function. The general definition of autocovariance is given by eq. (3.45)

$$C_{\xi\xi}(t_1, t_2) = \langle [\xi(t_1) - m_{1\xi}(t_1)][\xi(t_2) - m_{1\xi}(t_2)] \rangle \qquad (3.51)$$

which is reduced to

$$C_{\xi\xi}(\tau) = \langle \xi(t)\xi(t + \tau) \rangle - m^2 \qquad (3.52)$$

in the case of a stationary process $\xi(t)$.

---

[2]Here and in the future, the following special notations are used: $C_{\xi\xi}(t_1, t_2) = \langle [\xi(t_1) - m_{1\xi}(t_1)][\xi(t_2) - m_{1\xi}(t_2)] \rangle$ for the covariance function and $R_{\xi\xi}(t_1, t_2) = C_{\xi\xi}(t_1, t_2) + m_{1\xi}(t_1)m_{1\xi}(t_2) = \langle \xi(t_1)\xi(t_2) \rangle$ for the correlation function. Subindex ξ indicating the random process will often be omitted if it does not lead to confusion.

The definition can be extended to consider covariance between two or more random processes. For simplicity, a case of two stationary random processes $\xi(t)$ and $\eta(t)$ with averages m_ξ and m_η respectively is considered. It is possible to define a mutual (joint) covariance function between these two processes as

$$C_{\xi\eta}(t_1, t_2) \triangleq \langle [\xi(t_1) - m_\xi][\eta(t_2) - m_\eta] \rangle$$
$$C_{\eta\xi}(t_1, t_2) \triangleq \langle [\eta(t_1) - m_\eta][\xi(t_2) - m_\xi] \rangle \tag{3.53}$$

If both $C_{\xi\eta}$ and $C_{\eta\xi}$ depend only on the time lag $\tau = t_2 - t_1$, i.e.

$$C_{\xi\eta}(t_1, t_2) = C_{\xi\eta}(\tau)$$
$$C_{\eta\xi}(t_1, t_2) = C_{\eta\xi}(\tau) \tag{3.54}$$

then $\xi(t)$ and $\eta(t)$ are called stationary related processes. For a stationary related process, one can obtain that

$$C_{\xi\eta}(\tau) = \langle [\xi(t) - m_\xi][\eta(t + \tau) - m_\eta] \rangle = \langle [\xi(t - \tau) - m_\xi][\eta(t) - m_\eta] \rangle = C_{\eta\xi}(-\tau) \quad (3.55)$$

Equations (3.52)–(3.53) can be generalized to the case of complex random processes as follows

$$C_{\xi\xi}(\tau) = \langle [\xi(t) - m_\xi][\xi^*(t_1 + \tau) - m_\xi^*] \rangle \tag{3.56}$$
$$C_{\xi\eta}(\tau) = \langle [\xi(t) - m_\xi][\eta^*(t_1 + \tau) - m_\eta^*] \rangle \tag{3.57}$$

where $*$ stands for the complex conjugation.

The importance of the mutual covariance function in describing the relation between two random processes can be explained by considering two extreme cases of dependence between two processes: (a) $\xi(t)$ and $\eta(t)$ are independent, and (b) $\xi(t)$ and $\eta(t)$ are linearly related, i.e. $\xi(t) = \pm\alpha\eta(t) + \beta, \alpha > 0$. In case (a), the joint probability density is just a product of marginal PDFs

$$p_{2\xi\eta}(x_1, x_{2\tau}) = p_{1\xi}(x_1)p_{1\eta\tau}(x_{2\tau}) \tag{3.58}$$

where $p_{1\xi}(x_1)$ is the PDF of $\xi(t)$ and $p_{1\eta\tau}(x_{2\tau})$ is the PDF of $\eta(t + \tau)$. In this case

$$C_{\xi\eta} = \int_{-\infty}^{\infty} \int_{-\infty}^{\infty} (x_1 - m_\xi)(x_{2\tau} - m_\eta)p_{1\xi}(x_1)p_{1\eta\tau}(x_{2\tau})dx_1 dx_{2\tau} = 0 \tag{3.59}$$

Thus, the mutual covariance function between two independent processes is equal to zero (i.e. there is no correlation between $\xi(t)$ and $\eta(t)$). On the contrary, under the linear dependence as specified in case (b), one can write

$$C_{\xi\eta}(t_1, t_2) = \langle [\xi(t_1) - m_\xi][\eta(t_2) - m_\eta] \rangle = \pm\alpha C_{\xi\xi}(\tau) \tag{3.60}$$

since

$$\langle \eta(t) \rangle = \langle \pm \alpha \xi(t) + \beta \rangle = \pm \alpha \langle \xi(t) \rangle + \langle \beta \rangle = \pm \alpha m_\xi + \beta \qquad (3.61)$$

Setting $t_2 = t_1 (\tau = 0)$, one obtains

$$C_{\xi\eta}(t_1, t_1) = \pm \alpha C_{\xi\xi}(0) = \pm \alpha \sigma_\xi^2 = \pm \sigma_\xi \sigma_\eta \qquad (3.62)$$

It will be shown later that eq. (3.62) is the maximum correlation which can be achieved. In other words, the covariance function shows how much knowledge about the process $\eta(t)$ can be obtained by observing the process $\xi(t)$.

The following are some properties of autocovariance functions which will be often used

1. The covariance function $C(\tau)$ of a stationary process is an even function, i.e.

$$C(\tau) = C(-\tau) \qquad (3.63)$$

Indeed, since the process is stationary, one can shift the time origin by $-\tau$ to obtain

$$C(\tau) = \langle \xi(t) \xi(t+\tau) \rangle - m_\xi^2 = \langle \xi(t-\tau) \xi(t) \rangle - m_\xi^2 = C(-\tau) \qquad (3.64)$$

2. The absolute value of the covariance function does not exceed its value at $\tau = 0$, i.e. the variance of the process

$$|C(\tau)| \leq C(0) = \sigma^2 \qquad (3.65)$$

It is clear that the average of a non-negative function is a non-negative quantity, i.e.

$$E\{[\xi(t) - m_\xi] \pm [\xi(t+\tau) - m_\xi]\}^2 \geq 0 \qquad (3.66)$$

The last inequality implies that

$$\langle (\xi(t) - m_\xi)^2 \pm 2[\xi(t) - m_\xi][\xi(t+\tau) - m_\xi] + (\xi(t+\tau) - m_\xi)^2 \rangle = 2[\sigma_\xi^2 \pm C(\tau)] \geq 0 \qquad (3.67)$$

and thus

$$|C(\tau)| \leq \sigma^2 \qquad (3.68)$$

For many practical random processes, the following condition is also satisfied

$$\lim_{\tau \to \infty} C(\tau) = 0 \qquad (3.69)$$

From a physical point of view, this result can be explained by the fact that most of the physically possible systems have an impulse response which approaches zero (stable systems). As a consequence, the effect of the current value of the system state on the future is limited by the time when the impulse response is significant and disappears when the time lag τ approaches infinity.

To summarize, the covariance function $C(t)$ of a stationary process is an even function, having a maximum at $\tau = 0$. This maximum is equal to the variance σ^2, and the covariance function vanishes as $\tau \to \infty$. Some commonly used covariance functions are listed in Table 3.1.

Not every function which satisfies the conditions 1–3 can be a proper covariance function, an autocovariance function must satisfy an additional condition.

3. The covariance function is a positively defined function; i.e. for any a_i and a_j and any N moments of time

$$\sum_{i,k=1}^{N} a_i a_j^* C(t_i, t_j) \geq 0 \tag{3.70}$$

This is an obvious consequence of the fact that an average of a non-negative random variable is a non-negative quantity

$$\left\langle \left| \sum_{i=1}^{N} a_i \xi(t_i) \right|^2 \right\rangle = \left\langle \sum_{i=1}^{N} a_i a_j^* \xi(t_i) \xi^*(t_j) \right\rangle = \sum_{i,k=1}^{N} a_i a_j^* C(t_i, t_j) \geq 0 \tag{3.71}$$

An equivalent form of this definition is the fact that the cosine transform of the covariance function is also a non-negative function:

$$\int_0^\infty C(\tau) \cos \omega \tau \, d\tau \geq 0 \tag{3.72}$$

It is worth mentioning that mutual covariance functions do not satisfy the conditions above. Some of the properties of autocovariance can be generalized to a case of non-stationary processes. In this case, eqs. (3.63) and (3.65) become

$$C(t_1, t_2) = C(t_2, t_1)$$
$$|C(t_1, t_2)| \leq \sigma(t_1) \sigma(t_2) \tag{3.73}$$

3.7   CORRELATION COEFFICIENT

Equations (3.53) and (3.62) show that the correlation between functions describes not only a level of dependence between two random processes, but it also takes into account their variance. Indeed if one of the functions $\xi(t)$ or $\eta(t)$ deviates a little from its mean, the absolute value of the covariance function would be small, no matter how closely related $\xi(t)$

Table 3.1   Common covariance functions and their Fourier transforms [20]

Normalized covariance function $C(\tau)/\sigma^2 = R(\tau)$	Normalized PSD $S(\omega) = \int_{-\infty}^{\infty} R(\tau)e^{-j\omega\tau}d\tau$	Description or form-filter	Correlation interval (τ_{corr}) $\tau_{corr} = \int_0^\infty |R(\tau)|d\tau$	$-R''(0)$	$\dfrac{\Delta f_{eff}}{\Delta f}$
$\dfrac{N_0}{2}\delta(\tau)$	$\dfrac{N_0}{2}$	White noise	0	∞	1
$\exp[-\alpha\tau]$	$\dfrac{2\alpha}{\alpha^2+\omega^2}$	First order low-pass filter	$\dfrac{1}{\alpha}$	∞	1.571
$(1+\alpha|\tau|)\exp(-\alpha\tau)$	$\dfrac{4\alpha^3}{(\alpha^2+\omega^2)^2}$	Second order low-pass filter	$\dfrac{2}{\alpha}$	α^2	1.221
$\exp[-\alpha^2\tau^2]$	$\sqrt{\dfrac{\pi}{\alpha^2}}\exp\left(-\dfrac{\omega^2}{4\alpha^2}\right)$	Gaussian filter	$\dfrac{\sqrt{\pi}}{2\alpha}$	$2\alpha^2$	1.065
$\exp[-\alpha|\tau|]\cos(\omega_0\tau)$	$\alpha\left[\dfrac{1}{\alpha^2+(\omega-\omega_0)^2}+\dfrac{1}{\alpha^2+(\omega+\omega_0)^2}\right]$	Resonant contour	$\dfrac{1}{\alpha}$	∞	1.571
$\exp[-\alpha|\tau|]\big[\cos(\omega_0\tau)+\dfrac{\alpha}{\omega_0}\sin\omega_0|\tau|\big]$	$4\alpha(\omega_0^2+\alpha^2)\left[\dfrac{1}{\alpha^2+(\omega-\omega_0)^2}+\dfrac{1}{\alpha^2+(\omega+\omega_0)^2}\right]$	Resonant contour	$\dfrac{1}{\alpha}$	$\omega_0^2+\alpha^2$	1.571
$\dfrac{\sin\Delta\omega\tau}{\Delta\omega\tau}$	$\begin{cases}\dfrac{\pi}{\Delta\omega} & \text{if } |\omega|\le\Delta\omega\\ 0 & \text{if } |\omega|>\Delta\omega\end{cases}$	Ideal low-pass filter	$\dfrac{\pi}{2\Delta\omega}$	$\dfrac{(\Delta\omega)^2}{3}$	1
$J_0(2\pi f_D|\tau|)$	$\dfrac{2}{\sqrt{(2\pi f_D)^2-\omega^2}}$	Jake's fading	$\dfrac{1}{2f_D}$	$2\pi^2 f_D^2$	1
$\dfrac{A_0^2}{2}\cos\omega_0\tau$	$\dfrac{\pi}{2}A_0^2[\delta(\omega-\omega_0)+\delta(\omega+\omega_0)]$	Sinusoid with a random phase	∞	$\dfrac{A_0^2\omega_0^2}{2}$	1
$\dfrac{\sin\dfrac{\Delta\omega\tau}{2}}{\dfrac{\Delta\omega\tau}{2}}$	$\begin{cases}\dfrac{\pi}{\Delta\omega} & \text{if } |\omega\pm\omega_0|\le\dfrac{\Delta\omega}{2}\\ 0 & \text{if } |\omega\pm\omega_0|>\dfrac{\Delta\omega}{2}\end{cases}$	Ideal bandpass filter	$\dfrac{\pi}{\Delta\omega}$	$\omega_0^2+\dfrac{(\Delta\omega)^2}{12}$	1
$\exp(-\alpha^2\tau^2)\cos(\omega_0\tau)$	$\sqrt{\dfrac{\pi}{4\alpha^2}}\times\left[e^{-\frac{(\omega-\omega_0)^2}{4\alpha^2}}+e^{-\frac{(\omega+\omega_0)^2}{4\alpha^2}}\right]$	Gaussian bandpass filter	$\dfrac{\sqrt{\pi}}{2\alpha}$	$\omega_0^2+2\alpha^2$	1.065

and $\eta(t)$ are.[3] For a quantitative description of linear[4] correlation, it is reasonable to introduce a normalized function, called the correlation coefficient $R(\tau)$

$$R(\tau) = \frac{C(\tau)}{\sigma^2}, R_{\xi\eta}(t_1, t_2) = \frac{C_{\xi\eta}(t_1, t_2)}{\sigma_\xi(t_1)\sigma_\xi(t_2)} \tag{3.74}$$

Functions $R_{\xi\eta}$ and R are called the mutual correlation coefficient and autocorrelation coefficient respectively. Alternative notation $\rho(\tau)$ for the correlation coefficient will also be used here.

It follows from eq. (3.62) that if two random functions are related by a deterministic function, then the coefficient of mutual correlation between them in any instant of time is equal to ± 1; if these functions are independent, the correlation coefficient becomes zero. Thus, the correlation coefficient describes linear (but not an arbitrary [10,11]) dependence between two processes.

Two stationary processes $\xi(t)$ and $\eta(t + \tau)$ with the correlation coefficient between $\xi(t)$ and $\eta(t + \tau)$ equal to zero for all values of $\tau > 0$, (i.e. $R_{\xi\eta}(\tau) = 0$) are called uncorrelated.

It was shown above that any two independent processes are also uncorrelated. The reverse statement is not generally true. Indeed, $R(\tau) = 0$ does not generally provide enough grounds to state that any higher order moments (non-linear covariance functions) $\langle \xi^{i_1}(t)\eta^{i_2}(t_2)\rangle$ all equal zero in the case of non-Gaussian processes. One of the most important examples which shows that uncorrelation does not imply independence can be constructed as follows. Let the joint PDF of the second order be given by

$$p_{2\xi\eta}(x_1, x_2) = \frac{p_s\left(\sqrt{x_1^2 + x_2^2}\right)}{\sqrt{x_1^2 + x_2^2}} \tag{3.75}$$

where $p_s(x)$ is a PDF of an arbitrary positive random variable. It can be seen that, due to symmetry, the cross-correlation function is equal to zero, i.e. $\xi(t)$ and $\eta(t)$ are uncorrelated. However, $p_{2\xi\eta}(x_1, x_2)$ can be represented as a product of $p_{1\xi}(x_1)$ and $p_{1\eta}(x_2)$ only in the case of Rayleigh PDF $p_2(s)$.

It follows from eqs. (3.63)–(3.72) that the autocorrelation coefficient has the following properties

$$R(\tau) = R(-\tau), |R(\tau)| \leq R(0) = 1, \qquad \lim_{\tau \to \infty} R(\tau) = 0, \qquad \int_0^\infty R(\tau)\cos\omega\tau\,d\tau \geq 0 \quad (3.76)$$

In the following, a notion of the correlation interval τ_{corr} will be often used for both the cases of practically important correlation coefficients: autocorrelation coefficient with a monotonic decay $R(\tau) = \rho(\tau)$ and fast oscillating decaying correlation coefficient, for example in the form $R(\tau) = \rho(\tau)\cos\omega_0\tau$. In both cases, correlation interval τ_{corr} is defined as

$$\tau_{\text{corr}} = \frac{1}{2}\int_{-\infty}^{\infty} |\rho(\tau)|d\tau = \int_0^\infty |\rho(\tau)|d\tau \tag{3.77}$$

---

[3]If one sets $\xi(t) = \alpha\eta(t)$ then $C_{\xi\eta}(\tau) = \alpha C_{\eta\eta}(t) \leq \alpha\sigma_\eta^2$. Arbitrarily small α would make $C_{\xi\eta}(\tau)$ arbitrarily small.
[4]For non-linear correlation see [10,11]

Geometrically, the correlation interval τ_{corr} is equal to the base of a rectangle of height 1 and having the same area as under the curve $|\rho(\tau)|$ (see Table 3.1). The quantity τ_{corr} gives a rough estimate of the maximum time lag between two samples of the random process which exhibit a significant correlation between them. It is also important to mention that the definition (3.77) is not unique and an alternative definition can be introduced, for example based on the average square spread [12], or at -3 dB level.

3.8 CUMULANT FUNCTIONS

Cumulant functions $\kappa_{x,\ldots,z}(t_1^{[n_1]},\ldots,t_N^{[n_N]})$ of a random process, or a group of random processes, can be introduced using the definition of cumulants of the random variable or a group of random variables and by sampling the random processes under consideration at the moments of time $t_1,\ldots,t_N$. First, the definitions are given for a case of a single random process $\xi(t)$ and then extended to the case of a group of processes. The cumulant and moment brackets technique is applied below to the case of random processes. As a result the cumulants become functions of moments of time $t_1,\ldots,t_N$.

Let $\xi(t)$ be a scalar random process considered at N moments of time $t_1,\ldots,t_N$. Then $\{\xi(t_1),\xi(t_2),\ldots,\xi(t_N)\}$ is a random vector which, in turn, can be described using the joint PDF, characteristic function, joint moments, or joint cumulants as discussed in Chapter 2. For example, in the case of marginal PDF $p_\xi(x,t_1)$ depending on time t_1, the characteristic function $\theta_\xi(u,t_1)$ also depends on the time instant t_1. As a result so do the cumulants

$$\theta_\xi(u,t_1) = \langle e^{jv\xi(t)}\rangle = \exp\left(\sum_{\nu=1}^{\infty}\frac{\kappa_\nu(t_1)}{\nu!}(ju)^\nu\right) \tag{3.78}$$

In a similar way, higher order cumulants may depend on more than one moment of time. The following notations are adopted here. Let a random process $\xi(t)$ be sampled at N moments of time $t = t_i$, $1 \le i \le N$, then $\kappa_{\alpha_1,\alpha_2,\ldots,\alpha_N}^{\xi,\xi,\ldots,\xi}(t_1,t_2,\ldots,t_N)$ is the cumulant of order $\sum\alpha_i$, which, in turn, is the coefficient of expansion of the log-characteristic function $\theta_{\xi N}(ju_1,ju_2,\ldots,ju_N)$ into a power series

$$\ln\theta_{\xi N}(ju_1,ju_2,\ldots,ju_N) = \sum_{i_1=0}^{\infty}\sum_{i_2=0}^{\infty}\cdots\sum_{i_N=0}^{\infty}(-j)^{(i_1+i_2+\cdots+i_N)}\frac{\kappa_{\alpha_1,\alpha_2,\ldots,\alpha_N}^{\xi,\xi,\ldots,\xi}(t_1,t_2,\ldots,t_N)}{i_1!i_2!\ldots i_N!}u_1^{i_1}u_2^{i_2}\ldots u_N^{i_N}$$

$$\tag{3.79}$$

3.9 ERGODICITY

Until now the description of a random process has been achieved through corresponding statistical averages over an ensemble of realizations, i.e. over responses of a large number of identical systems. In many cases, for some strict sense stationary processes, it is possible to obtain the same characteristics by averaging a single realization over time. This possibility is justified by the fact that a stationary process is homogeneous in time. Thus, a single realization which is long enough would contain all the information about the process. A process which can be described based only on a single realization is called an ergodic process. All ergodic processes are obviously stationary but the reverse statement is not true.

Leaving detailed mathematical considerations outside the scope of this book, it can be noted that sufficient and necessary conditions for a stationary process to be ergodic is the fact that its covariance function vanishes at $\tau = \infty$, i.e.

$$\lim_{\tau \to \infty} C(\tau) = 0 \tag{3.80}$$

A few particular results can be derived from the ergodicity of a random process $\xi(t)$. First of all, let $Z(t)$ be a random function of an ergodic process $\xi(t)$. It is assumed that $Z(t)$ is also stationary and ergodic. Then, with the probability 1

$$\langle Z(t) \rangle = \lim_{T \to \infty} \frac{1}{T} \int_0^T Z(t) dt \tag{3.81}$$

In order to prove this formula, one can consider a random variable $\overline{Z}_T$, defined as an average of $Z(t)$ over the time interval $[0, T]$[5]

$$\overline{Z}_T = \frac{1}{T} \int_0^T Z(t) dt \tag{3.82}$$

Taking the statistical average of both sides, changing the order of integration and statistical averaging, it can be seen that the average of $\overline{Z}$ coincides with that of the original process:

$$\langle \overline{Z}_T \rangle = \frac{1}{T} \int_0^T \langle Z(t) \rangle dt = \frac{\langle Z(t) \rangle}{T} \int_0^T dt = \langle Z(t) \rangle \tag{3.83}$$

Here the fact that $\langle Z(t) \rangle$ does not depend on time for a stationary process has been used.

The next step is to show that the variance $\sigma^2(T)$ of $\overline{Z}_T$ approaches zero when T approaches infinity. It can be done by subtracting $\langle \overline{Z}_T \rangle$ from both sides of eq. (3.82) and squaring the difference

$$(\overline{Z}_T - \langle \overline{Z}_T \rangle)^2 = \left[\frac{1}{T} \int_0^T \{Z(t) - \langle Z(t) \rangle\} dt \right]^2 = \frac{1}{T^2} \int_0^T \int_0^T \{Z(t_1) - \langle Z(t) \rangle\}\{Z(t_2) - \langle Z(t) \rangle\} dt_1 dt_2 \tag{3.84}$$

Taking the statistical average of both sides of eq. (3.84), one obtains

$$\sigma^2(T) = \langle (\overline{Z}_T - \langle \overline{Z}_T \rangle)^2 \rangle = \frac{1}{T^2} \int_0^T \int_0^T C_Z(t_2 - t_1) dt_1 dt_2 \tag{3.85}$$

---

[5]In the following, an overbar over a function indicates averaging over time, i.e. $\overline{x(t)} = \frac{1}{T} \int_0^T x(t) dt$

Here,

$$C_Z(t_2 - t_1) = \sigma_Z^2 R_Z(t_2 - t_1) = E(\{Z(t_1) - \langle Z(t) \rangle\}\{Z(t_2) - \langle Z(t) \rangle\}) \tag{3.86}$$

is the covariance function of the process $Z(t)$ and $R_z(t_2 - t_1)$ is its correlation coefficient. By assumption $Z(t)$ is stationary process and thus both of these functions depend only on $t_2 - t_1$. Changing variables to

$$\tau = t_2 - t_1 \quad t_0 = \frac{t_1 + t_2}{2}$$

$$t_2 = t_0 + \frac{\tau}{2} \quad t_1 = t_0 - \frac{\tau}{2} \tag{3.87}$$

and taking into account that $C_z(\tau)$ is an even function $C_z(\tau) = C_z(-\tau)$, one obtains

$$\sigma^2(\tau) = \frac{2}{T} \int_0^T \left(1 - \frac{\tau}{T}\right) C_Z(\tau) d\tau = \frac{2\sigma_z^2}{T} \int_0^T \left(1 - \frac{\tau}{T}\right) R_Z(\tau) d\tau \tag{3.88}$$

The last equation shows that in order to calculate $\sigma^2(T)$ one needs to know the correlation coefficient of $Z(t)$. In two particular cases (small and large T), a good approximation of these values can be obtained. Let τ_{corr} be a correlation interval of the process $Z(t)$. Then, for $T \ll \tau_K$ we can assume that $R_Z \approx 1$ and thus

$$\sigma^2(T) \approx \sigma_Z^2 \tag{3.89}$$

If $T \gg \tau_{\text{corr}}$, then eq. (3.88) can be simplified to produce the following estimate

$$\sigma^2(T) \leq \frac{2\sigma_Z^2}{T} \int_0^\infty |R_Z(\tau)| d\tau = \frac{2\sigma_Z^2 \tau_{\text{corr}}}{T} \tag{3.90}$$

Thus, if a stationary process $Z(t)$ has a finite variance, correlation interval and, in addition, its covariance function approaches zero, $\lim_{\tau \to \infty} C_Z(\tau) \to 0$, then $\sigma^2(T)$ vanishes. This means that $\bar{Z}_T$ approaches a deterministic quantity which coincides with the average value of the process $\langle Z(t) \rangle$, according to eq. (3.83)

$$\langle Z(t) \rangle = \lim_{T \to \infty} \frac{1}{T} \int_0^T Z(t) dt \tag{3.91}$$

At the same time, one can estimate how fast expression (3.91) converges. Indeed, it follows from eq. (3.90) that

$$\left| \frac{1}{T} \int_0^T Z(t) dt - \langle Z(t) \rangle \right| \leq \sigma_Z \left(\frac{2\tau_{\text{corr}}}{T}\right)^{1/2} \tag{3.92}$$

This means that averaging over time converges at least as fast as the arithmetic mean

$$\bar{Z}_T = \frac{1}{N} \sum_{i=1}^N Z(i\Delta) \tag{3.93}$$

of identically distributed and independent random variables $Z(i\Delta)$ if $N = T/(2\tau_{corr})$. Thus, to simplify evaluation of the average $\langle Z(t) \rangle$, it is reasonable to use the sum (eq. 3.93) instead of the integral (eq. 3.82), assuming that the discretization step Δ is greater than $2\tau_{corr}$!

In a very similar way, one can obtain different moments by averaging an ergodic process over time.

3.10 POWER SPECTRAL DENSITY (PSD)

The power spectral density (PSD) of a random process $\xi(t)$ is defined as the Fourier transform of its covariance function $C(\tau)$, i.e.

$$S(\omega) \triangleq \int_{-\infty}^{\infty} C(\tau) \exp[-j\omega\tau] d\tau \tag{3.94}$$

The inverse Fourier transform restores the covariance function from its PSD

$$C(\tau) = \frac{1}{2\pi} \int_{-\infty}^{\infty} S(\omega) \exp[j\omega\tau] d\omega \tag{3.95}$$

Equations (3.94)–(3.95) are known as the Wiener–Khinchin theorem.

Setting $\tau = 0$ in eq. (3.95), one can obtain that the variance of the process is equal to the area under the $S(\omega)$ curve, i.e.

$$\sigma^2 = C(0) = \frac{1}{2\pi} \int_{-\infty}^{\infty} S(\omega) \exp[-j\omega 0] d\omega = \frac{1}{2\pi} \int_{-\infty}^{\infty} S(\omega) d\omega \tag{3.96}$$

Since $C(\tau)$ is a symmetric positively definite function, its Fourier transform $S(\omega)$ is always a positive (and thus real) symmetric function. If one considers $\xi(t)$ to be a random voltage (current) then σ^2 can be considered as the power dissipated by a resistor of $R = 1\,\Omega$. A differential power $\frac{1}{2\pi} S(\omega) d\omega$ can be attributed to the spectral components located between ω and $\omega + d\omega$. This is why $S(\omega)$ is called the power spectral density or the energy spectrum.

The covariance function $C(\tau)$ and PSD $S(\omega)$ of a stationary process $\xi(t)$ "inherit" all the properties of a Fourier pair. In particular, if $S(\omega)$ is "wider" than some test function $S_0(\omega)$ then $C(\tau)$ must be "narrower" than the corresponding covariance function $C_0(\tau)$ and vice versa. This is dictated by the so-called uncertainty principle of the Fourier transform [13]. It also imposes conditions between smoothness of the covariance function and the rate of roll off of the PSD [14].

The effective spectral width $\Delta\omega_{eff}$ of the PDF can be defined according to the following equation

$$\Delta\omega_{eff} = \frac{1}{S_0} \int_0^{\infty} S(\omega) d\omega \tag{3.97}$$

where S_0 is the value of the PSD at a selected characteristic frequency ω_0. It is common to choose ω_0 such that $S(\omega)$ achieves its maximum $S_{max} = S(\omega_0)$ at ω_0. In this case

$$\tau_{corr} \cdot \Delta\omega_{corr} \approx \text{const.} \tag{3.98}$$

Alternative definitions are also possible [12]; for example spectral width $\Delta\omega_{0.5}$ can be defined on the level of $0.5S_{\max}$. Of course, $\Delta\omega_{\text{eff}} > \Delta\omega_{0.5}$.

Sometimes instead of PSD, defined by eq. (3.94), normalized PSD $s(\omega)$ is considered

$$s(\omega) = \sigma^{-2}S(\omega) = \frac{1}{\sigma^2}\int_{-\infty}^{\infty} C(\tau)\exp[-j\omega\tau]d\tau \qquad (3.99)$$

It follows from eq. (3.99) that

$$s(\omega) = \int_{-\infty}^{\infty} R(\tau)\exp[-j\omega\tau]d\tau, \quad R(\tau) = \frac{1}{2\pi}\int_{-\infty}^{\infty} s(\omega)\exp[j\omega\tau]d\omega \qquad (3.100)$$

Using the fact that the covariance function and the correlation coefficient are even functions, one obtains the equivalent form of the Wiener–Khinchin theorem:

$$\begin{aligned}
S(\omega) &= \int_{-\infty}^{\infty} C(\tau)\exp[-j\omega\tau]d\tau = \int_{-\infty}^{0} C(\tau)\exp[-j\omega\tau]d\tau + \int_{0}^{\infty} C(\tau)\exp[-j\omega\tau]d\tau \\
&= \int_{0}^{-\infty} C(-\tau)\exp[-j\omega\tau]d(-\tau) + \int_{0}^{\infty} C(\tau)\exp[-j\omega\tau]d\tau = 2\int_{0}^{\infty} C(\tau)\cos\omega\tau d\tau
\end{aligned} \qquad (3.101)$$

and, conversely

$$C(\tau) = 2\int_{0}^{\infty} S(\omega)\cos\omega\tau \frac{d\omega}{2\pi} \qquad (3.102)$$

It is important to note that the PSD $S(\omega)$ in eqs. (3.94), (3.95), (3.101), and (3.102) is defined for both positive and negative frequencies. In addition, $S(\omega) = S(-\omega)$. Instead of a two-sided "mathematical" spectrum, it is possible to consider a one-sided "physical" spectrum $S^{+}(\omega)$ defined as

$$S^{+}(\omega) = \begin{cases} 0 & \omega < 0 \\ S(\omega) + S(-\omega) = 2S(\omega) & \omega \geq 0 \end{cases} \qquad (3.103)$$

The Wiener–Khinchin theorem thus becomes

$$S^{+}(\omega) = 4\int_{0}^{\infty} C(\tau)\cos\omega\tau d\tau, \quad C(\tau) = \frac{1}{2\pi}\int_{0}^{\infty} S^{+}(\omega)\cos\omega\tau d\omega \qquad (3.104)$$

Equations (3.104) are usually more practical in calculations.

In contrast to the spectral analysis of deterministic signals, the PSD of a random process $\xi(t)$ does not provide information for restoring a specific time realization of this random signal, since it does not contain any phase information of each spectral component. It is possible to find a number of functions with the same magnitude spectrum but different phase spectrum, thus having the same covariance function.

While definition (3.94)–(3.95) seems to be artificial, there are practical considerations justifying such a definition. Indeed, a spectral function $F(\omega)$ can be defined for a particular realization of a random process $\xi(t)$ known for a significant interval of time T, i.e.

$$F(\omega) = \int_0^T \xi(t) \exp[-j\omega t] dt \qquad (3.105)$$

If $F^*(\omega)$ is the complex conjugate of $F(\omega)$, then

$$|F(\omega)|^2 = F(\omega)F^*(\omega) = \int_0^T \int_0^T \xi(t)\xi(t') \exp[-j\omega(t - t')] dt\, dt' \qquad (3.106)$$

Taking the statistical average of both sides and keeping in mind that $\xi(t)$ is a stationary process, it can be seen that

$$\langle |F(\omega)|^2 \rangle = \int_0^T \int_0^T \langle \xi(t)\xi(t') \rangle \exp[-j\omega(t - t') dt\, dt' = \int_0^T \int_0^T C(t - t') \exp[-j\omega(t - t')] dt\, dt' \qquad (3.107)$$

After a substitution $\tau = t - t'$ and evaluation of the integral over t', eq. (3.107) can be reduced to

$$\langle |F(\omega)|^2 \rangle = \int_0^T dt' \int_{-T}^T C(\tau) \exp[-j\omega\tau] d\tau = T \int_{-T}^T K(\tau) \exp[-j\omega\tau] d\tau \qquad (3.108)$$

Finally, dividing both sides by T, taking the limit for $T \to \infty$ and recalling eq. (3.94), one finally arrives at the conclusion that

$$S(\omega) = \int_{-\infty}^\infty C(\tau) \exp[-j\omega\tau] d\tau = \lim_{T \to \infty} \frac{\langle |F(\omega)|^2 \rangle}{T} \qquad (3.109)$$

This formula can be considered an equivalent definition of PSD, and it can be used to determine the PSD of a random sequence of pulses.

3.11 MUTUAL PSD

Let $\xi(t)$ and $\eta(t)$ be two stationary processes, with mutual covariance functions $C_{\xi\eta}(\tau)$ and $K_{\eta\xi}(\tau)$ as defined by eq. (3.94). Similar to the definition of PSD, the mutual PSD can be defined as the Fourier transformation of the mutual covariance function

$$\begin{aligned} S_{\xi\eta}(\omega) &= \int_{-\infty}^\infty C_{\xi\eta}(\tau) \exp[-j\omega\tau] d\tau \\ S_{\eta\xi}(\omega) &= \int_{-\infty}^\infty C_{\eta\xi}(\tau) \exp[-j\omega\tau] d\tau \end{aligned} \qquad (3.110)$$

Conversely, mutual covariance can be expressed in terms of the mutual PSD as

$$C_{\xi\eta}(\tau) = \frac{1}{2\pi} \int_{-\infty}^{\infty} S_{\xi\eta}(\omega) \exp[j\omega\tau] d\omega, \qquad C_{\eta\xi}(\tau) = \frac{1}{2\pi} \int_{-\infty}^{\infty} S_{\eta\xi}(\omega) \exp[j\omega\tau] d\omega \qquad (3.111)$$

Since the mutual covariance function is not necessarily even, the corresponding PSD is not necessarily real. However, if both $\xi(t)$ and $\eta(t)$ are real, so are $C_{\eta\xi}(\tau)$ and $C_{\xi\eta}(\tau)$. This implies that

$$S_{\xi\eta}^*(\omega) = S_{\xi\eta}(-\omega), \qquad S_{\eta\xi}^*(\omega) = S_{\eta\xi}(-\omega) \qquad (3.112)$$

The fact that $C_{\eta\xi}(\tau) = C_{\xi\eta}(-\tau)$ also results in

$$S_{\xi\eta}(\omega) = S_{\eta\xi}(-\omega), \qquad S_{\xi\eta}^*(\omega) = S_{\eta\xi}(\omega) \qquad (3.113)$$

The following two examples play an important role in analysis theory and applications: (1) sum of two stationary related processes and (2) product of two independent processes.

3.11.1 PSD of a Sum of Two Stationary and Stationary Related Random Processes

In many practical cases, it is important to obtain an expression for PSD $S_\zeta(\omega)$ of a sum

$$\zeta(t) = \xi(t) + \eta(t) \qquad (3.114)$$

of two stationary and stationary related random processes $\xi(t)$ and $\eta(t)$. It is assumed that auto- and mutual covariance functions $C_\xi(\tau)$, $C_\eta(\tau)$, $C_{\xi\eta}(t)$, and $C_{\eta\xi}(\tau)$ of these processes are known. The covariance function of $\zeta(t)$ is then

$$\begin{aligned} C_\zeta(\tau) &= \langle [\xi(t) + \eta(t)][\xi(t+\tau) + \eta(t+\tau)] \rangle \\ &= \langle \xi(t)\xi(t+\tau) \rangle + \langle \xi(t)\eta(t+\tau) \rangle + \langle \eta(t)\xi(t+\tau) \rangle + \langle \eta(t)\eta(t+\tau) \rangle \\ &= C_\xi(\tau) + C_\eta(\tau) + C_{\eta\xi}(\tau) + C_{\xi\eta}(\tau) \end{aligned} \qquad (3.115)$$

which immediately implies that

$$S_\zeta(\omega) = S_\xi(\omega) + S_\eta(\omega) + S_{\xi\eta}(\omega) + S_{\eta\xi}(\omega) \qquad (3.116)$$

Here, $S_{\xi\eta}(\omega)$ and $S_{\eta\xi}(\omega)$ form a mutual PSD, defined as in eq. (3.110).

In the case of real random functions $\xi(t)$ and $\eta(t)$, eq. (3.116) can be rewritten with use of eq. (3.113) to produce

$$S_\zeta(\omega) = S_\xi(\omega) + S_\eta(\omega) + [S_{\xi\eta}(\omega) + S_{\xi\eta}(-\omega)] \qquad (3.117)$$

Further simplification can be achieved in the case of uncorrelated (but not necessarily independent) processes $\xi(t)$ and $\eta(t)$. In this case, $C_{\eta\xi}(\tau) = C_{\xi\eta}(\tau) \equiv 0$ and $S_{\xi\eta}(\omega) =$

$S_{\eta\xi}(\omega) = 0$, thus implying that the PSD of the sum is just the sum of the PSDs of the components.

It is interesting to note that even if $\xi(t)$ and $\eta(t)$ are non-stationary in the wide sense, their sum could be a stationary process. For example, if $A_c(t)$ and $A_s(t)$ are independent, stationary, zero-mean processes with identical covariance function, then neither one of the processes $\xi(t) = A_c(t)\cos\omega t$, $\eta(t) = A_s(t)\sin\omega t$ is stationary. However, their sum is stationary in a wide sense.

3.11.2 PSD of a Product of Two Stationary Uncorrelated Processes

Let a random process $\zeta(t, \tau_0)$ be a product of two stationary zero mean[6] processes $\xi(t)$ and $\eta(t + \tau_0)$:

$$\zeta(t, \tau_0) = \xi(t)\eta(t + \tau_0) \tag{3.118}$$

where τ_0 is a fixed time lag. The covariance function of this process can be found to be

$$\begin{aligned} C_\zeta(t, \tau_0, \tau) &= \langle \xi(t)\eta(t + \tau_0)\xi(t + \tau)\eta(t + \tau + \tau_0)\rangle \\ &= \langle \xi(t)\xi(t + \tau)\rangle\langle \eta(t + \tau_0)\eta(t + \tau + \tau_0)\rangle \\ &= C_\xi(\tau)C_\eta(\tau) = C_\zeta(\tau) \end{aligned} \tag{3.119}$$

The corresponding PSD is then

$$S_\zeta(\omega) = \int_{-\infty}^{\infty} C_\xi(\tau)C_\eta(\tau)\exp[-j\omega\tau]d\tau \tag{3.120}$$

Using the expression of $C_\eta(\tau)$ in terms of the corresponding PSD

$$C_\eta(\tau) = \frac{1}{2\pi}\int_{-\infty}^{\infty} S_\eta(\omega')\exp[-j\omega'\tau]d\omega' \tag{3.121}$$

one can see that

$$\begin{aligned} S_\zeta(\omega) &= \int_{-\infty}^{\infty} C_\xi(\tau)\frac{1}{2\pi}\int_{-\infty}^{\infty} S_\eta(\omega')\exp[j\omega'\tau]d\omega'\exp[-j\omega\tau]d\tau \\ &= \frac{1}{2\pi}\int_{-\infty}^{\infty} S_\eta(\omega')d\omega'\int_{-\infty}^{\infty} C_\xi(\tau)\exp[-j(\omega - \omega')\tau]d\tau = \frac{1}{2\pi}\int_{-\infty}^{\infty} S_\eta(\omega')S_\xi(\omega - \omega')d\omega' \end{aligned}$$

$$\tag{3.122}$$

Thus, the PSD of a product of two stationary uncorrelated functions is equal to the convolution of their PSDs.

---

[6]In this case the correlation function coincides with the covariance function.

3.12   COVARIANCE FUNCTION OF A PERIODIC RANDOM PROCESS

In this section, a few examples of the covariance function of periodic signals are considered. Such signals play an important role in a wide range of applications such as radar detection, communication systems with amplitude modulation, etc. The common factor of such processes is the fact that the covariance function of the signal does not vanish at infinity, a property which can be exploited in detection of signals embedded in additive noise.

3.12.1   Harmonic Signal with a Constant Magnitude

As a first example, a covariance function of a random harmonic signal

$$s(t) = A_0 \cos(\omega_0 t + \varphi) \tag{3.123}$$

is considered. The magnitude A_0 and the angular frequency ω_0 are assumed to be fixed deterministic quantities, while φ is a random, uniformly distributed on the interval $[-\pi; \pi]$ phase. In this case,

$$p_\varphi(\varphi) = \begin{cases} \dfrac{1}{2\pi} & -\pi \leq \varphi \leq \pi \\ 0 & \text{otherwise} \end{cases} \tag{3.124}$$

Since the statistical average value of this signal is equal to zero

$$m_s = \langle s(t) \rangle = \int_{-\pi}^{\pi} s(t)\,d\varphi = \frac{A_0}{2\pi} \int_{-\pi}^{\pi} \cos(\omega_0 t + \varphi)d\varphi = 0 \tag{3.125}$$

its covariance function is just

$$C_s(t, \tau) = \langle s(t)s(t+\tau) \rangle = A_0^2 \langle \cos(\omega_0 t + \varphi) \cos(\omega_0(t+\tau) + \varphi) \rangle$$

$$= \frac{A_0^2}{2} [\langle \cos(\omega_0 t + \omega_0 \tau + \varphi - \omega_0 t - \varphi) \rangle + \langle \cos(\omega_0 t + \varphi + \omega_0 t + \omega_0 \tau + \varphi) \rangle]$$

$$= \frac{A_0^2}{2} \cos \omega_0 \tau + \frac{A_0^2}{4\pi} \int_{-\pi}^{\pi} \cos(2\omega_0 t + \omega_0 \tau + 2\varphi)d\varphi = \frac{A_0^2}{2} \cos \omega_0 \tau = C_s(\tau) \tag{3.126}$$

As can be seen, the covariance function is a periodic function with the same period as the signal itself. However, it does not approach zero when $\tau \to \infty$, and, thus, is not an ergodic process[7]. This property could be used to identify a weak but long signal $s(t)$ embedded in a strong colored noise, whose covariance function approaches zero for relatively large time lags.

Indeed, let signal $s(t)$ be mixed with a stationary noise $n(t)$

$$\eta(t) = s(t) + n(t) \tag{3.127}$$

[7]It is not possible to extract the statistics of the phase from a single realization of the random process.

If the signal and noise are independent (which is often the case), then the covariance function of the sum is just a sum of the covariance functions of the components, according to eq. (3.115)

$$C_\eta(\tau) = C_s(\tau) + C_n(\tau) \tag{3.128}$$

If $C_\eta(\tau)$ approaches zero when $\tau \to \infty$, then for large values of τ the eq. (3.128) becomes

$$C_\eta(\tau) \approx C_s(\tau) = \frac{A_0^2}{2} \cos \omega_0 \tau \tag{3.129}$$

Thus, it is often possible to obtain information about the presence or absence of an information signal by analysing its covariance function. If, for a fairly large τ, the measured covariance function exhibits periodic properties, then the mixture $\eta(t)$ contains the information signal. Otherwise, only noise is present.

The PSD of such a signal is, according to eq. (3.94)

$$\begin{aligned}
S(\omega) &= \frac{1}{2} A_0^2 \int_{-\infty}^{\infty} \cos \omega_0 \tau \exp[-j\omega\tau] d\tau \\
&= \frac{A_0^2}{4} \int_{-\infty}^{\infty} (\exp[-j\omega_0\tau] + \exp[j\omega_0\tau]) \exp[-j\omega\tau] d\tau \\
&= \frac{\pi}{2} A_0^2 [\delta(\omega - \omega_0) + \delta(\omega + \omega_0)]
\end{aligned} \tag{3.130}$$

since

$$\int_{-\infty}^{\infty} \exp[j\omega\tau] d\tau = 2\pi\delta(\omega) \tag{3.131}$$

Equivalently, the one-sided density is then

$$S^+(\omega) = 2S(\omega) = \pi A_0^2 \delta(\omega - \omega_0) = \frac{A_0^2}{2} \delta(f - f_0) \tag{3.132}$$

This result is quite logical, since eq. (3.132) reflects the fact that all the energy is concentrated at a single frequency $f = f_0$. The magnitude of this spectral line is $A_0^2/2$ and is the square of its RMS.

3.12.2　A Mixture of Harmonic Signals

A random signal $s(t)$ which is a mixture of harmonic waveforms of different frequencies

$$s(t) = \sum_{k=1}^{n} A_k \sin(\omega_k t + \varphi_k) \tag{3.133}$$

is often used as a model of multitone signalling systems, such as OFDM systems, with frequency diversity [15], etc. Here A_k and ω_k are deterministic quantities, and only phases φ_k

are allowed to fluctuate. As a further limitation, it is assumed that φ_k and φ_l are independent and uniformly distributed over the interval $[-\pi; \pi]$.

Repeating the previous calculations and taking into account that due to the assumption on the phases

$$p(\varphi_1, \varphi_2, \ldots, \varphi_n) = p(\varphi_1)p(\varphi_2)\ldots p(\varphi_n) = \left(\frac{1}{2\pi}\right)^n \qquad (3.134)$$

one can obtain the expression for the covariance function and one-sided PSD of such a process as

$$C_s(\tau) = \int_{-\infty}^{\infty} \cdots \int_{-\infty}^{\infty} s(t)s(t+\tau)p(\varphi_1, \varphi_2, \ldots, \varphi_n)d\varphi_1 d\varphi_2 \ldots d\varphi_n = \frac{1}{2}\sum_{k=1}^{n} A_k^2 \cos \omega_k \tau$$

$$\qquad (3.135)$$

and

$$S^+(\omega) = 2\pi \sum_{k=1}^{n} \frac{A_k^2}{2}\delta(\omega - \omega_k) \qquad (3.136)$$

respectively. The last two expressions can be easily obtained from the expression for a single harmonic signal, noting that all the components are independent.

3.12.3 Harmonic Signal with Random Magnitude and Phase

The next task is to find the covariance function of a harmonic signal with a random magnitude.

$$s(t) = A_0\xi(t)\cos(\omega_0 t + \varphi) \qquad (3.137)$$

Here $\xi(t)$ is a stationary random process with the covariance function $C_\xi(\tau)$, φ is an independent random variable uniformly distributed over $[-\pi; \pi]$. Such a process corresponds to a signal modulated by a random information signal as in ASK or AM [16]. In this case

$$C_s(\tau) = \langle s(t)s(t+\tau)\rangle = \langle A_0\xi(t)\cos(\omega_0 t + \varphi)A_0\xi(t+\tau)\cos(\omega_0 t + \omega_0\tau + \varphi)\rangle$$

$$= A_0^2 \langle s(t)s(t+\tau)\rangle_\xi \langle \cos(\omega_0 t + \varphi)\cos(\omega_0 t + \omega_0\tau + \varphi)\rangle_\phi = \frac{A_0^2}{2} C_\xi(\tau)\cos\omega_0\tau \quad (3.138)$$

Here a subscript is used to indicate the variable over which the statistical averaging is being performed. Thus, the covariance function of a modulated signal carries information about the covariance function of the modulated signal and the carrier frequency. Covariance functions of various information signals are considered in [16].

It is important to mention that if the initial phase ϕ in eq. (3.137) is deterministic, then the signal $s(t)$ will be a periodic non-stationary process, often called a cyclo-stationary process. Such processes are treated in detail in [17].

3.13   FREQUENTLY USED COVARIANCE FUNCTIONS

It is possible to list a number of covariance functions and PSDs that are often used in applications. Many of them can be obtained by considering the filtering of White Gaussian Noise (WGN)[8] with a spectral density $N_0/2$. If the transfer function of the filter is $H(j\omega)$ then the PSD of the output signal is

$$S(\omega) = \frac{N_0}{2}|H(j\omega)|^2 \tag{3.139}$$

and thus

$$C(\tau) = \frac{1}{2\pi}\int_{-\infty}^{\infty} S(\omega)\exp[j\omega\tau]d\tau \tag{3.140}$$

Thus, for any particular system the expression for the covariance function is unique. However, in engineering calculations a set of typical correlation functions and PSDs is used to simplify calculations. Some of these functions are listed in Table 3.1. All of the listed examples are obtained by filtering WGN; the corresponding shaping filters (form-filters) are also shown. In addition, Table 3.1 also lists the effective bandwidth $\Delta\omega_{\text{eff}}$ as defined by eq. (3.97), ratio $\Delta\omega_{\text{eff}}/\Delta\omega_{0.5}$ and the second derivative of the covariance function $C''(\tau)\,|_{\tau=0}$, calculated at zero lag and playing an important role in level crossing problems (see Chapter 4).

3.14   NORMAL (GAUSSIAN) RANDOM PROCESSES

A random process $\xi(t)$ is called a normal (Gaussian) random process if, for an arbitrary n moments of time $t_1, t_2, \ldots, t_n$, the n-dimensional joint PDF of $\xi_\mu = \xi(t_\mu)$, $\mu = 1, 2, \ldots, n$ is a normal (Gaussian) PDF

$$p_{n\xi}(x_1, x_2, \ldots, x_n) = \frac{1}{\sqrt{(2\pi)^n|C|}}\exp\left[-\frac{1}{2}(x-m)^{\mathrm{T}}C^{-1}(x-m)\right] \tag{3.141}$$

Here

$$m = [m_1, m_2, \ldots, m_n]^{\mathrm{T}}, \qquad m_\mu = \langle\xi(t_\mu)\rangle \tag{3.142}$$

is the average value of the process at time instant $t = t_\mu$,

$$\sigma_\mu^2 = \langle(\xi(t_\mu) - m_\mu)^2\rangle \tag{3.143}$$

---

[8]See section 3.15 for more details.

is the variance of the process $\xi(t)$ at the same moment of time. Finally, C is the covariance matrix defined as

$$
C = \begin{bmatrix} C_{11} & C_{12} & \cdots & C_{1n} \\ C_{21} & C_{22} & \cdots & \cdots \\ \cdots & \cdots & \cdots & \cdots \\ C_{n1} & \cdots & \cdots & C_{nn} \end{bmatrix}, \qquad C_{\mu\nu} = C_{\xi\xi}(t_\mu, t_\nu) = \langle (\xi(t_\mu) - m_\mu)(\xi(t_\nu) - m_\nu) \rangle
$$

(3.144)

and $|C|$ stands for the determinant of the matrix C.

The characteristic function, which corresponds to the PDF (eq. 3.141) is given by the following expression [19]

$$
\theta_{n\xi}(u_1, u_2, \ldots, u_n; t_1, t_2, \ldots, t_n) = \exp\left[j \sum_{\mu=1}^{n} m_\mu u_\mu - \frac{1}{2} \sum_{\mu,\nu=1}^{n} C_{\xi\xi}(t_\mu, t_\nu) u_\mu u_\nu \right]
$$

(3.145)

It can be seen from eqs. (3.141) and (3.145) that the expression for the PDF of a characteristic function is completely defined by values of the average at any time instant and the covariance function $C(t_1, t_2)$ between any two instances of time. Thus, if for a physical reason it is known that the process is Gaussian, then in order to have its complete description one just needs to measure its mean and covariance function. Thus the correlation theory provides a complete description of such processes.

For a Gaussian process, all higher order cumulants are equal to zero, as can be seen from the following expansion

$$
\ln \theta_{n\xi} = j \sum_{\mu=1}^{n} m_\mu u_\mu - \frac{1}{2} \sum_{\mu,\nu=1}^{n} C_{\xi\xi}(t_\mu, t_\nu) u_\mu u_\nu + \ldots + 0 u_\mu^k u_\nu^m
$$

(3.146)

Thus, two Gaussian processes can be different only by mean values and the shape of the covariance function (or PSD).

If values of a Gaussian random process $\xi(t)$ are uncorrelated at the time instants $t_1, t_2, \ldots, t_n$ then

$$
p_{n\xi}(x_1, x_2, \ldots, x_n; t_1, t_2, \ldots, t_n) = \sum_{\mu=1}^{n} p_{1\xi}(x_\mu, t_\mu)
$$

$$
= \frac{1}{(2\pi)^{n/2} \sigma_1, \sigma_2 \ldots \sigma_n} \exp\left[-\frac{1}{2} \sum_{\mu=1}^{n} \frac{(x_\mu - m_\mu)^2}{\sigma_\mu^2} \right]
$$

(3.147)

since all $R_{ij} = 0$ for $i \neq j$. This means that if two Gaussian processes are uncorrelated they are also independent.

Setting $n = 1$ and $n = 2$ in eqs. (3.141) and (3.145), one obtains

$$p_{1\xi}(x_1) = \frac{1}{\sqrt{2\pi}\sigma_1} \exp\left[-\frac{(x_1 - m_1)^2}{2\sigma_1^2}\right], \qquad \theta_{1\xi}(x_1) = \exp\left[jm_1u_1 - \frac{1}{2}\sigma_1^2 u_1^2\right] \tag{3.148}$$

$$p_{2\xi}(x_1, x_2) = \frac{1}{2\pi\sigma_1\sigma_2\sqrt{1 - R^2}}$$

$$\exp\left[-\frac{1}{2(1 - R^2)}\left\{\frac{(x_1 - m_1)^2}{\sigma_1^2} - 2R\frac{(x_1 - m_1)(x_2 - m_2)}{\sigma_1\sigma_2} + \frac{(x_2 - m_2)^2}{\sigma_2^2}\right\}\right]$$

$$\theta_{2\xi}(x_1, x_2) = \exp\left[j(m_1u_1 + m_2u_2) - \frac{1}{2}[\sigma_1^2 u_1^2 + 2R\sigma_1\sigma_2 u_1 u_2 + \sigma_2^2 u_2^2]\right] \tag{3.149}$$

where

$$R = R_{12}(t_1, t_2) = \frac{C_{\xi\xi}(t_1, t_2)}{\sigma_1\sigma_2} = \frac{C_{\xi\xi}(t_2, t_1)}{\sigma_1\sigma_2} = R_{21}(t_1, t_2) \tag{3.150}$$

is the correlation coefficient.

Equations (3.145)–(3.148) allow an important conclusion to be reached. By definition, the characteristic function of an ensemble of n random variables $\xi_1, \xi_2, \ldots, \xi_n$ is

$$\theta_n(u_1, u_2, \ldots, u_n; t_1, \ldots, t_n) = \langle \exp(ju_1\xi_1 + ju_2\xi_2 + \cdots + ju_n\xi_n) \rangle \tag{3.151}$$

Thus, for the characteristic function $\theta_1(u)$ of the sum

$$\xi = \xi_1 + \xi_2 + \cdots + \xi_n \tag{3.152}$$

one can write

$$\theta(u, t_1, \ldots, t_n) = \langle \exp(ju\xi) \rangle = \langle \exp[ju(\xi_1 + \xi_2 + \cdots + \xi_n)] \rangle = \theta_n(u, u, \ldots, u; t_1, t_2, \ldots, t_n) \tag{3.153}$$

If all the random variables ξ_i are Gaussian random variables described by the characteristic function eq. (3.145), then

$$\theta(u; t_1, \ldots, t_n) = \exp\left[jMu - \frac{1}{2}\sigma^2 u^2\right] \tag{3.154}$$

where

$$M = \sum_{\mu=1}^{n} M_\mu, \qquad \sigma^2 = \sum_{\mu,\nu=1}^{n} C(t_\mu, t_\nu) \tag{3.155}$$

This equation has the same form as eq. (3.148), but now m and σ^2 depend on n instants of time. Since eq. (3.148) describes a Gaussian process, then the random process ξ is also Gaussian. In other words, a sum of a finite number of Gaussian random processes is a Gaussian random process. This statement can be generalized to the case of an arbitrary linear transformation, including differentiation, integration and filtering.

The next important property of a Gaussian process is the fact that wide sense stationarity implies strict sense stationarity. It can be easily seen that the conditional PDF of a Gaussian random process is also Gaussian and is given by

$$
p_2(x_2 \mid x_1; \tau) = \frac{p_2(x_2, x_1; \tau)}{p(x_1)}
$$

$$
= \frac{1}{\sigma_2\sqrt{2\pi[1 - R^2(\tau)]}} \exp\left[-\frac{1}{2[1 - R^2(\tau)]} \left\{ \frac{x_1 - m_1}{\sigma_1} R(\tau) - \frac{x_2 - m_2}{\sigma^2} \right\}^2 \right]
$$

(3.156)

If the process under consideration is a wide sense stationary process, then the following conditions are also satisfied

$$
m_\mu = m = \text{const}, \qquad R(t_\mu, t_\nu) = R(|t_\mu - t_\nu|), \qquad \mu, \nu = 1, \ldots, n \tag{3.157}
$$

That is, the average of the process is a constant and the covariance function depends only on the time lag between two time instants. In this case, the PDF and the characteristic function (3.141) and (3.145) would also depend only on $|t_\mu - t_\nu|$, meaning that the process is also strict sense stationary. Thus, for a Gaussian random process, these two types of stationarity are equivalent.

For a Gaussian stationary process, eqs. (3.141) and (3.145) can be written as

$$
p_n(x_1, x_2, \ldots, x_n) = \frac{1}{\sqrt{(2\pi\sigma^2)^n D}} \exp\left[-\frac{1}{2\sigma^2 D} \sum_{\mu,\nu=1}^{n} D_{\mu\nu}(x_\mu - m)(x_\nu - m) \right] \tag{3.158}
$$

$$
\theta_n(u_1, u_2, \ldots, u_n) = \exp\left[jm \sum_{\mu=1}^{n} x_\mu - \frac{1}{2}\sigma^2 \sum_{\mu,\nu=1}^{n} R(\tau_{\mu\nu})u_\mu u_\nu \right] \tag{3.159}
$$

where

$$
\sigma^2 = \langle (\xi(t) - m)^2 \rangle \tag{3.160}
$$

is the variance of the random process and

$$
\tau_{\mu\nu} = |t_\mu - t_\nu| \tag{3.161}
$$

Implicit expressions for the PDF and characteristic functions of a Gaussian stationary process can be easily obtained from eqs. (3.148)–(3.149) to produce

$$p_1(x) = \frac{1}{\sqrt{2\pi}\sigma} \exp\left[-\frac{(x-m)^2}{2\sigma^2}\right], \qquad \theta_1(u) = \exp\left[jmu - \frac{1}{2}\sigma^2 u^2\right] \qquad (3.162)$$

$$p_2(x_1, x_2) = \frac{1}{2\pi\sigma^2\sqrt{1 - R^2(\tau)}}$$

$$\times \exp\left[-\frac{(x_1 - m)^2 - 2R(\tau)(x_1 - m)(x_2 - m) + (x_2 - m)^2}{2\sigma^2(1 - R^2(\tau))}\right] \qquad (3.163)$$

$$\theta_2(u_1, u_2) = \exp\left[jm(u_1 + u_2) - \frac{1}{2}\sigma^2[u_1^2 + 2R(\tau)u_1 u_2 + u_2^2]\right]$$

If the mean value of the process is zero, $m = 0$, then eq. (3.162) is simplified even further

$$\theta_1(u) = \exp\left[-\frac{1}{2}\sigma^2 u^2\right] \qquad (3.164)$$

Using eq. (3.34) one can obtain expressions for all the moments of a marginal PDF

$$\exp\left[-\frac{1}{2}\sigma^2 u^2\right] = \sum_{n=0}^{\infty} \frac{\left(-\frac{1}{2}\sigma^2 u^2\right)^n}{n!} = \sum_{n=0}^{\infty} \frac{(-1)^n \sigma^{2n} u^{2n}}{(2n)!} \frac{(2n)!}{n!2^n} \qquad (3.165)$$

and thus

$$M_n = \begin{cases} 1 \cdot 3 \cdot 5 \cdot \cdots \cdot (n-1)\sigma^n & \text{for even } n \\ 0 & \text{for odd } n \end{cases} \qquad (3.166)$$

For a stationary Gaussian process with zero mean $m = 0$, all moments of an odd order are equal to zero. All moments of an even order can be expressed in terms of the covariance function[9] $C(\tau)$ using the expression for its probability density (3.141) or its characteristic function (3.145)

$$m_{2n-1}(t_1, \ldots, t_{2n-1}) = 0$$

$$m_{2n}(t_1, \ldots, t_{2n}) = \sum_{\text{all pairs}} \left[\prod_{i \neq j} \langle \xi(t_i)\xi(t_j)\rangle\right]$$

$$= \sum_{\text{all pairs } i \neq j, k \neq l, p \neq q} \left[\langle \xi(t_i)\xi(t_j)\rangle\langle \xi(t_k)\xi(t_l)\rangle \ldots \langle \xi(t_p)\xi(t_q)\rangle\right] \qquad (3.167)$$

---

[9]This also follows from the fact that the covariance function completely defines the PDF of any order of a zero mean Gaussian process, i.e. the results must not come as a "surprise".

In particular

$$m_{1111}(t_1, t_2, t_3, t_4) = \langle \xi(t_1)\xi(t_2)\xi(t_3)\xi(t_4) \rangle$$
$$C(t_2 - t_1)C(t_4 - t_3) + C(t_3 - t_1)C(t_4 - t_2) + C(t_4 - t_1)C(t_3 - t_2) \qquad (3.168)$$
$$m_{22}(t_1, t_2) = m_{1111}(t_1, t_1, t_2, t_2) = \sigma^4 + 2C^2(t_2 - t_1)$$

PDF eq. (3.163) of the second order of a zero mean Gaussian process $(m = 0)$ can be represented as a series

$$p_2(x_1, x_2) = \sigma^{-2} \sum \Phi^{(n+1)}\left(\frac{x_1}{\sigma}\right)\Phi^{(n+1)}\left(\frac{x_2}{\sigma}\right)\frac{R^n(\tau)}{n!} \qquad (3.169)$$

Here, $\Phi(z)$ is the probability integral closely related to the error function [6]

$$\Phi(z) = \frac{1}{\sqrt{2\pi}}\int_{-\infty}^{z} \exp\left[-\frac{1}{2}x^2\right]dx = \frac{1}{2}\left[1 + \mathrm{erf}\left(\frac{x}{2}\right)\right], \qquad \Phi^{(n)}(z) = \frac{d^n}{dz^n}\Phi(z) \qquad (3.170)$$

Derivatives of the probability integral can also be expressed in terms of Hermitian polynomials as

$$\Phi^{(n)}(z) = \frac{d^n}{dz^n}\Phi(z) = \frac{(-1)^{n-1}}{2^{n-1}\sqrt{\pi}}H_{n-1}\left(\frac{x}{2}\right)\exp\left[-\frac{x^2}{4}\right] \qquad (3.171)$$

Expansion (3.169) is often used when considering the non-linear transformation of a Gaussian stationary process and other calculations involving Gaussian and Rayleigh processes. For example, one can obtain expressions for joint moments of a Gaussian process

$$M_{\mu\nu}(\tau) = \langle \xi^\mu(t)\xi^\nu(t+\tau) \rangle = \int_{-\infty}^{\infty}\int_{-\infty}^{\infty} x_1^\mu x_2^\nu p_2(x_1, x_2)dx_1 dx_2 = \sigma^{\mu+\nu}\sum_{k=0}^{\infty}N_{\mu,k}N_{\nu,k}\frac{R^k(\tau)}{k!} \qquad (3.172)$$

where

$$N_{i,k} = \int_{-\infty}^{\infty} x^i \phi^{(k+1)}(x)dx = \frac{(-1)^n}{2^{n-i-1}\sqrt{\pi}}\int_{-\infty}^{\infty} x^i H_n(x)\exp[-x^2]dx \qquad (3.173)$$

A set of coefficients $N_{i,k}$ forms a matrix, shown in Table 3.2.
It is possible to point out rules for calculating $N_{i,k}$:

1. All elements above the main diagonal $(i < k)$ are equal to 0.

2. Below the main diagonal and on the main diagonal $(i \geq k)$, the entries which are non-zero have indices of the same parity (i.e. they are either both even or both odd). All other entries are equal to zero.

Table 3.2

i	$k = 0$	1	2	3	4	5	6
0	1	0	0	0	0	0	0
1	0	-1	0	0	0	0	0
2	1	0	2	0	0	0	0
3	0	-3	0	-6	0	0	0
4	3	0	12	0	24	0	0
5	0	-15	0	0	0	-120	0
6	15	0	90	-60	360	0	720

3. Element $N_{i,k}$ can be constructed as follows:
   The first k multipliers are $i(i-1)\ldots(i-k+1)$. The next group of multipliers are all odd numbers between $(i-k+1)$ to 1, excluding $(i-k+1)$ itself. For example, for $i=6$, $k=2$.

$$|N_{6,2}| = (6 \cdot 5) \cdot (3 \cdot 1) = 90 \tag{3.174}$$

4. The sign of the element $N_{i,k}$ depends on the parity of k: it is negative for odd k and positive for even k.
   Using this rule, it is easy to obtain that

$$m_{24}(\tau) = \sigma^6[3 + 12R^2(\tau)]$$
$$m_{44}(\tau) = \sigma^8[9 + 72R^2(\tau) + 24R^4(\tau)] \tag{3.175}$$

It is possible to show that the three-dimensional PDF can be represented by a series, similar to eq. (3.169). Such a representation allows one to calculate three-dimensional moments.

It should be stressed again that any linear transformation of a Gaussian process produces a Gaussian process again. Thus, if at the input of a linear filter there is a Gaussian process, a Gaussian process will appear at the output as well. In other words, a Gaussian process is stable with respect to linear transformations. In addition, it is interesting that the PDF of a weighted sum of similarly distributed and weakly correlated random variables approaches the Gaussian PDF, thus emphasizing the importance of the Gaussian distribution.

In the case of non-linear transformation, the Gaussian property is lost. If a Gaussian process $\xi(t)$ is transformed according to a non-linear rule $\eta(t) = f(t, \xi)$, then $\eta(t)$ is not a Gaussian process. In particular, a product of two Gaussian processes is non-Gaussian.

However, a non-Gaussian process with the correlation interval τ_{corr}, acting at the input of a narrow band system with a time constant $\tau_c > \tau_{corr}$ and the impulse response $h(t)$ produces an output $\eta(t)$ which approaches a Gaussian process. As the time constant τ_s increases, the output processes becomes closer to a Gaussian process. This is due to the fact that the input process can be considered as a large sum of samples of the input processes which are almost independent:

$$\eta(t) = \int_{-\infty}^{\infty} h(\tau - t)\xi(\tau)d\tau \approx \Delta\tau \sum_{k=-N}^{N} h(k\Delta\tau - t)\xi(k\Delta\tau) \tag{3.176}$$

If $\tau_s \gg \tau_{corr}$ then one can choose $\tau_{corr} < \Delta\tau < \tau_c$ such that $\xi(k\Delta\tau)$ are almost independent while $h(k\Delta\tau - t)$ does not change much, i.e. the weights are approximately equal. The greater τ_{corr}, the more uniformity in weights one can observe and a larger number N of terms is needed to approximate the integral in eq. (3.176).

3.15 WHITE GAUSSIAN NOISE (WGN)

One of the most important processes used in applications is a stationary Gaussian process $n(t)$, whose covariance function is a delta function scaled by a positive constant $N_0/2$

$$C(\tau) = \frac{N_0}{2}\delta(\tau) \tag{3.177}$$

Since the delta function is equal to zero everywhere except $\tau = 0$, the process defined by the covariance function (3.177) is uncorrelated at any two separate instants of time. Such a process is called absolutely random. The expression for the PSD of this process can be easily obtained from the sifting property of the delta function

$$S(\omega) = \int_{-\infty}^{\infty} C(\tau)\exp[-j\omega\tau]d\tau = C(\tau)|_{\tau=0} = \frac{N_0}{2} = \text{const}, S^+(\omega) = 2 \cdot \frac{N_0}{2} = N_0 \tag{3.178}$$

Thus, the spectral density of an absolutely random process has a spectral density which is a constant at all frequencies. Such a process is also known as a "white" process, similar to the white light which is composed of all possible chromatic lines with approximately equal power (see Fig. 3.3).

For the white Gaussian noise, eq. (3.96) produces a singular result

$$\sigma^2 = \frac{1}{2\pi}\int_{-\infty}^{\infty} S(\omega)d\omega = \infty \tag{3.179}$$

In other words, the variance of a white noise is equal to infinity, $\sigma^2 = \infty$. This can be explained by the fact that a white noise is rather a mathematical abstraction than a real physical process. In real life, PSD usually decays with frequency, thus providing for a finite correlation interval, $\tau_{corr} \neq 0$, and a limited variance. However, white noise is a useful abstraction, which can be used in a case where the correlation interval of the process τ_{corr} is

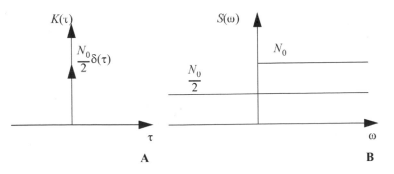

Figure 3.3 Covariance function and PSD of a white process.

much smaller than a characteristic time constant of the system under consideration. In an equivalent statement, one can use white noise if the bandwidth of the system under consideration is much smaller than the bandwidth of the noise.

A simple rule of thumb can be provided for substitution of a real process by a white noise $n(t)$ in analysis of systems under random excitation. Let a system with a characteristic time constant τ_s be considered. This system is excited by a random process with covariance function $C(\tau) = \sigma^2 R(\tau)$ which has a relatively wide spectrum, and thus $\tau_{\text{corr}} \ll \tau_s$. In this case, one can assume that a white noise model is a suitable model. An equivalent spectral density of noise $N_0/2$ can be chosen such that $S(\omega = 0) = S(0)$. In turn, this can be calculated from the covariance function $C(\tau)$ as

$$\frac{N_0}{2} = S(0) = \int_{-\infty}^{\infty} C(\tau)d\tau = 2 \int_0^{\infty} C(\tau)d\tau = 2\sigma^2 \tau_{\text{corr}} \tag{3.180}$$

One of the prominent examples of white noise is so-called thermal noise, caused by random movement of a large number of electrons inside metals. It is known that the spectral density of thermal noise of a resistor R is defined by the so-called Nyquist formula

$$S_a^+(\omega) = 4kTR, \qquad N_0 = 4kTR \tag{3.181}$$

where $k = 1.38 \cdot 10^{-23}$ J/K is the Boltzman constant (for $T = 200°$, $kT = 4 \cdot 10^{-21}$ W/Hz).

It can be seen from the last equation that the PSD of thermal noise is constant and it may appear that thermal noise is an ideal example of white noise. However, one has to take into account that eq. (3.181) is valid only for low frequencies. It is obtained from a quantum formula

$$S_u(f) = 4kTR\frac{hf}{kT}\left[\exp\left(\frac{hf}{kT}\right) - 1\right]^{-1} \tag{3.182}$$

where $h = 6.62 \cdot 10^{-34}$ J $\cdot$ s is the Planck constant. If $hf/kT \ll 1$, i.e.

$$f \ll \frac{kT}{h} \approx \frac{4 \cdot 10^{-21} \text{ W/Hz}}{6.62 \cdot 10^{-34} \text{ J} \cdot \text{s}} \approx 6 \cdot 10^{12} \text{ Hz} \tag{3.183}$$

(i.e. eq. (3.181) works well for frequencies up to 100 GHz and normal temperature). However, in order to calculate variance of the thermal noise, one must use the exact expression (3.182) in order to obtain a finite result

$$\sigma_u^2 = 4kTR\int_0^{\infty} \frac{hf}{kT}\left[\exp\left(\frac{hf}{kT}\right) - 1\right]^{-1}dt = 4kTR\frac{kT}{h}\int_0^{\infty}\frac{xdx}{\exp x - 1} = \frac{2\pi^2}{3h}(kT)^2R \tag{3.184}$$

As another example, one can consider the covariance function of the random process

$$s(t) = n(t)\cos(\omega_0 t + \varphi) \tag{3.185}$$

obtained by modulation of a harmonic signal with a random uniformly distributed phase by white noise, which can be obtained from eq. (3.138) and is given by the equation

$$C_s(\tau) = \frac{1}{2} N_0 \delta(\tau) \frac{A_0^2}{2} \cos \omega\tau = \frac{N_0}{4} A_0^2 \delta(\tau) \tag{3.186}$$

That is, such a process is also white.

REFERENCES

1. J. Doob, *Stochastic Processes*, New York: Wiley, 1990, 1953.
2. A. Kolmogorov, *Foundations of the Theory of Probability*, New York: Chelsea Pub. Co., 1956.
3. J.K. Ord, *Families of Frequency Distributions*, London: Griffin, 1972.
4. D. Middleton, *An Introduction to Statistical Communication Theory*, New York: McGraw-Hill, 1960.
5. A. Stuart and K. Ord, *Kendall's Advanced Theory of Statistics* London: Charles Griffin & Co., 1987.
6. A. Abramowitz and I. Stegun, *Handbook of Mathematical Functions with Formulas, Graphs, and Mathematical Tables*, New York: Dover, 1972.
7. C. Nikias and A. Petropulu, *Higher-Order Spectra Analysis: A Non-Linear Processing Framework*, Upper Sadle River: Prentice Hall, 1993.
8. A.N. Malakhov, *Cumulant Analysis of Non-Gaussian Random Processes and Their Transformations*, Moscow: Sovetskoe Radio, 1978 (in Russian).
9. A. Papoulis, *Probability, Random Variables, and Stochastic Processes*, New York: McGraw-Hill, 3rd edition, 1991.
10. O.V. Sarmanov, Maximum Correlation Coefficient (Symmetrical Case), *DAN*, Vol. 120, 1958, pp. 715–718.
11. O.V. Sarmanov, Maximum Correlation Coefficient (Non-Symmetrical Case), *DAN*, Vol. 121, 1958, pp. 52–55.
12. P. Bello, Characterization of Randomly Time-Variant Linear Channels, *IEEE Trans. Communications*, vol. 11, no. 4, December, 1963, pp. 360–393.
13. W. Siebert, *Circuits, Signals and Systems*, Boston: MIT Press, 1985.
14. S. Mallat, *A Wavelet Tour of Signal Processing*, San Diego: Academic Press, 1999.
15. L. Hanzo, W. Webb and T. Keller, *Single- and Multi-Carrier Quadrature Amplitude Modulation: Principles and Applications for Personal Communications, WLANs and Broadcasting*, New York: John Wiley & Sons, 2000.
16. J. Proakis, *Digital Communications*, Boston: McGraw-Hill, 2001.
17. H. Sakai, Theory of Cyclostationary Processes and Its Applications, in: Tohru Katayama, Sueo Sugimoto, Eds., *Statistical Methods in Control and Signal Processing*, New York: Marcel Dekker, 1997.
18. F. Riesz and B. Sz.-Nagy, *Functional Analysis*, New York: Ungar, 1955.
19. K. Miller, *Multidimensional Gaussian Distributions*, New York: Wiley, 1964.
20. V. Tikhonov, *Statistical Radiotechnique*, Moscow: 1966 (in Russian).

4

Advanced Topics in Random Processes

4.1 CONTINUITY, DIFFERENTIABILITY AND INTEGRABILITY OF A RANDOM PROCESS

Analysis and design of systems is often based on their representation by a system of differential or difference equations with random forcing functions. In order to provide a proper mathematical foundation for such analysis one has to operate with properties such as continuity, differentiability and integrability in a way similar to the deterministic case. Some basic definitions are presented below. More detailed considerations can be found in the literature [1–3].

4.1.1 Convergence and Continuity

Let $\{\xi_n\} = \xi_1, \xi_2, \ldots, \xi_n, \ldots$ be a sequence of random variables. In order to define a notion of the limit ξ of a sequence one has to define a proper measure of the difference $\xi - \xi_n$. Based on the chosen measure, the following types of convergence can be defined:

1. Convergence with probability 1. In this case

$$\text{Prob}\{\xi_n \to \xi\} = 1, \quad n \to \infty \tag{4.1}$$

2. Convergence in probability. In this case for an arbitrary $\varepsilon > 0$, the probability of deviations larger than ε is equal to zero

$$\text{Prob}\{|\xi_n - \xi| > \varepsilon\} = 0, \quad n \to \infty \tag{4.2}$$

3. Mean square convergence. In this case it is required that the mean square deviation is zero

$$\lim_{n \to \infty} E\{|\xi_n - \xi|^2\} = 0 \tag{4.3}$$

These definitions are related: convergence in probability follows from both convergence in the mean square sense and from the convergence with probability 1.

Stochastic Methods and Their Applications to Communications.
S. Primak, V. Kontorovich, V. Lyandres
© 2004 John Wiley & Sons, Ltd   ISBN: 0-470-84741-7

A similar definition can be extended to the case of a random process $\xi(t)$ by considering samples of the sequence of random processes $\{\xi_n(t)\}$ at a given moment of time t. In addition, the notion of continuity could be introduced by considering values of the random process at two closely spaced moments of time t and $t + h$. In particular, it is said that a random process $\xi(t)$ is continuous with probability 1 at the moment of time t, if $\xi(t + h) \to \xi(t)$ with probability 1 as $h \to 0$. It is important to mention though that continuity with probability 1 does not imply that each realization of the random process $\xi(t)$ is continuous at the same point itself[1].

In the following, mainly convergence (and continuity) in the mean square sense is used. Such an approach is applicable to so-called random processes of the second order, i.e. to processes whose second moment is finite $m_2^\xi(t) = E\{\xi^2(t)\} < \infty$ at any given moment of time t. A random process of the second order is continuous in the mean square sense at the moment of time t if

$$\lim_{h \to 0} E\{[\xi(t + h) - \xi(t)]^2\} = 0 \tag{4.4}$$

In order to check the validity of this condition, it is necessary and sufficient to know the covariance function $C_{\xi\xi}(t_1, t_2)$ of the random process $\xi(t)$: a random process $\xi(t)$ is continuous in the mean square sense if and only if its mean is continuous for the moment of time t and the covariance function $C_{\xi\xi}(t_1, t_2)$ is continuous at $t_1 = t_2 = t$. Indeed, if the conditions above are satisfied then

$$\lim_{h \to 0} E\{[\xi(t + h) - \xi(t)]^2\}$$
$$= \lim_{h \to 0} E\{[(\xi(t + h) - m_\xi(t + h)) - (\xi(t) - m_\xi(t))]^2\}$$
$$= \lim_{h \to 0} [C_{\xi\xi}(t + h, t + h) - 2C_{\xi\xi}(t + h, t) + C_{\xi\xi}(t, t) + (m(t + h) - m(t))^2] = 0 \tag{4.5}$$

and, thus, these conditions are necessary. Sufficiency follows from a simple estimate

$$E\{[\xi(t + h) - m(t + h) - \xi(t) + m(t)]^2\} = C_{\xi\xi}(t + h, t + h) - 2C_{\xi\xi}(t + h, t) + C_{\xi\xi}(t, t) \geq 0 \tag{4.6}$$

If the left-hand side converges to zero, each term in this expression must converge to zero, thus proving sufficiency of the conditions.

4.1.2 Differentiability

The random process $\xi(t)$ of the second order is called a differentiable in the mean square sense random process at the moment of time $t = t_0$ if the limit

$$\eta(t_0) = \lim_{h \to 0} \frac{\xi(t_0 + h) - \xi(t_0)}{h} \tag{4.7}$$

---

[1]The definition based on the probability measure allows a certain number of realizations to have discontinuity, as long as the number of such realizations is "small", or more accurately, the number of deviations has measure zero.

exists in the mean square sense. The new random process $\eta(t)$ is called the derivative of the random process $\xi(t)$ in the mean square sense if the derivative exists for all moments of time t considered. The necessary and sufficient conditions for differentiability of a process $\xi(t)$ can be formulated as follows: a second order random process $\xi(t)$ is differentiable at the moment of time t in the mean square sense if and only if its mean $m_\xi(t)$ is differentiable at the moment of time t and the second mixed derivative

$$\frac{\partial^2}{\partial t_1 \partial t_2} C_{\xi\xi}(t_1, t_2)\bigg|_{t_1 = t_2 = t} \tag{4.8}$$

exists and is finite.

In order to prove necessity of these conditions, one can consider the average of the following quantity

$$\left[\frac{\xi(t_0 + h_1) - \xi(t_0)}{h_1} - \frac{\xi(t_0 + h_2) - \xi(t_0)}{h_2}\right]^2$$

$$= \left[\frac{\xi(t_0 + h_1) - \xi(t_0)}{h_1}\right]^2 - 2\frac{\xi(t_0 + h_1) - \zeta(t_0)}{h_1} \cdot \frac{\zeta(t_0 + h_2) - \zeta(t_0)}{h_2} + \left[\frac{\zeta(t_0 + h_2) - \zeta(t_0)}{h_2}\right]^2 \tag{4.9}$$

Averaging of the middle term on the right-hand side results in

$$E\left\{\frac{\xi(t_0 + h_1) - \xi(t_0)}{h_1} \cdot \frac{\xi(t_0 + h_2) - \xi(t_0)}{h_2}\right\} = \frac{m_\xi(t_0 + h_1) - m_\xi(t_0)}{h_1} \cdot \frac{m_\xi(t_0 + h_2) - m_\xi(t_0)}{h_2}$$

$$+ \frac{C_{\xi\xi}(t_0 + h_1, t_0 + h_2) - C_{\xi\xi}(t_0 + h_1, t_0) - C_{\xi\xi}(t_0, t_0 + h_2) + C_{\xi\xi}(t_0, t_0)}{h_1 h_2} \tag{4.10}$$

Taking the limit on the both sides of eq. (4.10), one obtains

$$\lim_{h_1, h_2 \to 0} E\left\{\left[\frac{\xi(t_0 + h_1) - \xi(t_0)}{h_1} - \frac{\xi(t_0 + h_2) - \xi(t_0)}{h_2}\right]^2\right\}$$

$$= \left[\frac{\partial}{\partial t} m_\xi(t)\right]^2 + \frac{\partial^2}{\partial t_1 \partial t_2} C_{\xi\xi}(t_1, t_1)\bigg|_{t_1 = t_2 = t} \tag{4.11}$$

thus indicating that the average of the right-hand side of eq. (4.9) is finite. After calculating all the terms and taking the limit for $h_1, h_2 \to 0$, one verifies that

$$\lim_{h_1, h_2 \to 0} E\left\{\left[\frac{\xi(t_0 + h_1) - \xi(t_0)}{h_1} - \frac{\xi(t_0 + h_2) - \xi(t_0)}{h_2}\right]^2\right\} = 0 \tag{4.12}$$

Thus, the derivative of the random process $\xi(t)$ exists in the mean square sense. Necessity also can be easily proven by observing that the right-hand side of eq. (4.9) is a sum of two non-negative terms. If the left-hand side approaches zero, then each of these terms approaches zero separately which is equivalent to the requirements of existence of the derivatives of the mean and covariance functions.

4.1.3 Integrability

Finally, an integral of a random process in the mean square sense can be defined as follows. Let a time interval $[a, b]$ be divided into sub-intervals by points $a = t_0 < t_1 < \cdots < t_n = b$, and $f(t)$ be a deterministic function defined on the same interval $[a, b]$. A partial sum

$$\eta_n = \sum_{i=1}^{n} \xi(t_i) f(t_i)(t_i - t_{i-1}) \tag{4.13}$$

is a sequence of random variables $\{\eta_n\}$. If this sequence has a finite limit J in the mean square sense when $\max(t_j - t_{j-1}) \to 0$

$$J = \lim_{\max(t_j - t_{j-1}) \to 0} \sum_{i=1}^{n} \xi(t_i) f(t_i)(t_i - t_{i-1}) \tag{4.14}$$

this is called the Rieman mean square integral of the random process $\xi(t)$ over the time interval $[a, b]$.

$$J = J[a, b] = \int_a^b f(t)\xi(t)dt = \lim_{\max(t_j - t_{j-1}) \to 0} \sum_{i=1}^{n} \xi(t_i) f(t_i)(t_i - t_{i-1}) \tag{4.15}$$

Improper integrals also can be defined by considering the mean square limit of the integral over a finite but increasing time interval $[a, b]$. For example

$$\int_a^\infty f(t)\xi(t)dt = \lim_{b \to \infty} \int_a^b f(t)\xi(t)dt \tag{4.16}$$

The mean square integral of the Rieman type exists if the following two integrals

$$I_1 = \int_a^b f(t)m_\xi(t)dt$$

$$I_2 = \int_a^b \int_a^b f(t_1)f(t_2)C_{\xi\xi}(t_1, t_2)dt_1 dt_2 \tag{4.17}$$

exist. In this case

$$E\left\{ \int_a^b f(t)\xi(t)dt \right\} = \int_a^b f(t)m_\xi(t)dt \tag{4.18}$$

and

$$E\left\{ \int_a^b \int_a^b f(t_1)f(t_2)\xi(t_1)\xi(t_2)dt_1 dt_2 \right\} = \int_a^b \int_a^b f(t_1)f(t_2)C_{\xi\xi}(t_1, t_2)dt_1 dt_2 + \left[\int_a^b f(t)m_\xi(t)dt \right]^2 \tag{4.19}$$

4.2 ELEMENTS OF SYSTEM THEORY

4.2.1 General Remarks

Analytical investigation of any realistic system requires an adequate mathematical model. There are three main principles which guide choice of the model: (a) it must reflect real properties of the system; (b) it must be reasonably simple; and (3) it must allow well developed mathematical techniques to be applied to the analysis and the design. In order to build such a model, certain *a priori* information must be available. The amount of such information defines two significantly different approaches to modeling: (a) in some cases there is enough information to model a system's behaviour quite accurately; (b) only minimal information is available. In the former case, it is usually assumed that the mathematical model of the system is known (analysis) or its parameters must be identified (estimation), while the latter requires that such a model be synthesized. One of the most prominent examples of such synthesis is phenomenological modeling of the communication channel, considered in this book.

From the mathematical perspective, operation of any system can be thought of in terms of the input–output relationships, i.e. for an arbitrary input $\xi(t)$ the output $\eta(t)$ is defined as some function of the input

$$\eta(t) = \boldsymbol{T}[\xi(t)] \tag{4.20}$$

If such description is possible the mathematical description of the system is achieved by providing a proper functional $\boldsymbol{T}$. The function $\eta(t)$ is referred to as the reaction to the stimulus $\xi(t)$. Depending on the nature of the input stimulus and the reaction, a great variety of systems can be considered. Of course, the stimulus can depend not only on time but on other variables such as frequency, space coordinates, etc. It is important to note that the structure of the operator $\boldsymbol{T}$ does not necessarily reflect physical processes taking place inside the real system. Such a description only allows accurate reproduction of the output of the system once the input is specified.

There are a great variety of possible structures of the system and the corresponding operator $\boldsymbol{T}$ which allow input–output description. The input signal (stimulus) and the reaction can be vectors of dimension n and m respectively, thus describing multiple-input, multiple-output (MIMO) systems, as shown in Fig. 4.1(A). Such description incorporates a simple case of single-input single-output (SISO) systems $m = n = 1$ and autonomous systems $(n - 0, m - 1)$.

The operator $\boldsymbol{T}$ can be deterministic or stochastic. The deterministic operator describes systems which produce the same response to the same stimulus if the same initial conditions

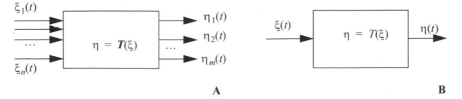

Figure 4.1 (A) MIMO systems, (B) SISO systems.

are enforced. In other words, the randomness in the response $\eta(t)$ may appear only due to the randomness in the input signal $\xi(t)$. On the contrary, the operator T is called stochastic if the randomness in the response $\eta(t)$ appears also due to the internal structure of the system. For example, it can be attributed to the random initial conditions, changing structure of the system, etc. If all physical phenomena are to be taken into account, most realistic systems are described by stochastic operators due to such internal factors as thermal noise, fluctuating external conditions (temperature, humidity), etc. However, in many cases the influence of these random factors is negligible and the system can be sufficiently described by a deterministic operator with randomness attributed only to the input stimuli. In many cases, internal randomness can be also properly represented by an equivalent random input, etc. The majority of the discussion in this book deals with deterministic systems with random, inputs, with the notable exception of systems with random structure, described in Chapter 6.

Further classification of operators can be achieved based on the type of output process $\eta(t)$. In particular, one can distinguish between the analog and discrete and digital systems, continuous systems with jumps, etc. It is also possible to distinguish between zero-memory non-linear (ZMNL) systems, linear systems, non-linear systems with memory, causal and non-causal[2], etc. The system is called a zero-memory non-linear system if the input–output relationship can be represented in the following form

$$\eta(t) = g[\xi(t), t] \tag{4.21}$$

where $[\xi(t), t]$ is a non-linear deterministic vector function. In other words, the output of a ZMNL system at the moment of time t can be predicted based on the value of the input signal $\xi(t)$ at the same moment of time and does not depend on the values of the input and the output signals at any other previous moment of time $t' < t$.

A casual system is a system which transforms current and past values of the input signals to the current value of the system output: the future values of the system input have no effect on the current output value. In other words, the output always lags the excitation. While the causal systems are the only systems which can be physically implemented, non-causal systems are often a useful tool in theoretical analysis. As an example, one can mention an ideal low-pass filter [14].

It is often possible to represent a general non-linear system with memory as a cascaded connection of ZMNL and linear sub-systems as shown in Fig. 4.2. In this case, non-linear effects of the original system are reproduced by ZMNL blocks while the memory effects are attributed to the impact of the linear block.

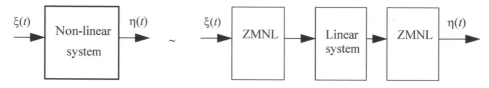

Figure 4.2   Representation of a non-linear system with memory.

---

[2]It will be seen from the following consideration that there is a group of methods well tailored for each separate class of systems, thus justifying such classification.

Strictly speaking, an operator $T = L$ is called a linear operator if the so-called superposition principle is valid for this operator, i.e. the following equation must be satisfied for any set of the input stimuli $\xi_k(t)$

$$\eta(t) = L\left[\sum_{k=1}^{n} c_k \xi_k(t)\right] = \sum_{k=1}^{n} c_k L[\xi_k(t)] \tag{4.22}$$

Here c_k are arbitrary constants, which are independent of time. All other operators are treated as non-linear.

4.2.2 Continuous SISO Systems

The deterministic linear operator L for a SISO system can be given in the form of a differential or difference equation with corresponding initial conditions. For a continuous system with constant parameters, the differential equation can be written in the form

$$\sum_{k=0}^{m} a_k \frac{d^k}{dt^k} \eta(t) = \sum_{k=0}^{n} b_k \frac{d^k}{dt^k} \xi(t), \qquad n < m \tag{4.23}$$

or, in the symbolic form

$$A_m(s)\eta(t) = B_n(s)\xi(t), \tag{4.24}$$

where

$$s = \frac{d}{dt}, A_m(s) = \sum_{k=0}^{m} a_k s^k, B_n(s) = \sum_{k=0}^{n} b_k s^k \tag{4.25}$$

are polynomials in s defined by the coefficients of differential equation (4.23). Equation (4.23) must be augmented by a set of initial conditions in the form

$$\eta_0 = \eta(t_0), \eta'_0 = \frac{d}{dt}\eta(t)\bigg|_{t=t_0}, \ldots, \eta_0^{(m-1)} = \frac{d^{m-1}}{dt^{m-1}}\eta(t)\bigg|_{t=t_0} \tag{4.26}$$

Instead of differential equations one can use the following alternative but equivalent descriptions of the same linear system and its operator L.

1. The impulse response of the system $h(t, \tau)$ is the reaction of a linear system to a delta pulse at time $t = \tau$:

$$h(t, \tau) = L[\delta(t - \tau)] \tag{4.27}$$

The response to an arbitrary stimulus $\xi(t)$ can be then found as

$$\eta(t) = L[\xi(t)] = L\left[\int_{-\infty}^{\infty} \xi(\tau)\delta(t-\tau)d\tau\right] = \int_{-\infty}^{\infty} \xi(\tau)L[\delta(t-\tau)]d\tau = \int_{-\infty}^{\infty} \xi(\tau)h(t,\tau)d\tau \tag{4.28}$$

Here the sifting property of the delta function is used in the third term and the superposition principle is used to exchange the order of integration and the operator L action. For a causal system, no response appears before the excitation, i.e.

$$h(t, \tau) = 0, \quad \text{for } t < \tau \tag{4.29}$$

thus transforming eq. (4.28) to

$$\eta(t) = \int_{-\infty}^{t} \xi(\tau) h(t, \tau) \mathrm{d}\tau \tag{4.30}$$

Linear systems can be further divided into time-variant and time-invariant systems. For a time-invariant system, the impulse response is a function of difference of the arguments, i.e. $h(t, \tau) = h(t - \tau)$ and, thus

$$\eta(t) = \int_{-\infty}^{t} \xi(\tau) h(t - \tau) \mathrm{d}\tau \tag{4.31}$$

In other words, a time-invariant system is invariant to a shift of the origin of time:

$$\eta(t - t_0) = L[\xi(t - t_0)] \tag{4.32}$$

All other systems are called time-variant.

The duration of the impulse response defines the memory of a system since it reflects for how long the perfectly localized input affects the output.

A linear system is called stable if its impulse response is absolutely integrable, i.e.

$$\int_{0}^{\infty} |h(t)| \mathrm{d}t < \infty \tag{4.33}$$

In this case a bounded input would always produce a bounded response. Stability is widely investigated in the control theory with numerous criteria of stability [5,6]. In particular, a system is stable if all roots of the characteristic polynomial

$$\sum_{k=0}^{m} a_k \lambda^k = 0 \tag{4.34}$$

have negative real parts.

2. An equivalent description of a stationary linear system can be achieved by means of the frequency response, or the transfer function, $H(j\omega)$, defined as the Fourier transform of the impulse response

$$H(j\omega) = \int_{-\infty}^{\infty} h(\tau) \exp(-j\omega\tau) \mathrm{d}\tau = |H(j\omega)| \exp[j\Phi(j\omega)] \tag{4.35}$$

or, equivalently,

$$h(t) = \frac{1}{2\pi} \int_{-\infty}^{\infty} H(j\omega) \exp(j\omega\tau) d\tau \tag{4.36}$$

The reaction of a linear system to a harmonic signal of frequency ω is the harmonic signal of the same frequency scaled by the value of the transfer function of the system, i.e.

$$\eta(t) = L[\exp(j\omega t)] = \int_{-\infty}^{t} \xi(\tau) \exp[j\omega(t - \tau)] d\tau = H(j\omega) \exp(j\omega t) \tag{4.37}$$

Instead of the Fourier transform, one can use the Laplace transform to include signals which are not properly described by the Fourier transform. In this case

$$H(s) = \int_{0}^{\infty} h(\tau) \exp(-s\tau) d\tau \tag{4.38}$$

A detailed discussion of the advantages of both methods can be found in [7].

4.2.3 Discrete Linear Systems

Similar characteristics can be obtained for discrete systems. It is also important to mention an important property of linear systems: characteristics of a discrete system can be obtained from similar characteristics of an equivalent continuous system, while the inverse is not always possible [8]. However, the analysis of discrete systems is important in itself, without relation to the underlying continuous systems.

Let $\zeta_{-n} = \zeta(t - n\Delta t)$ be samples of a random process $\zeta(t)$ at the time instants separated by a fixed discretization interval Δt, and let δ be the shift (delay) operator

$$\delta\zeta_{-n} = \delta\zeta_{-(n+1)}, \delta^m\zeta_{-n} = \zeta_{-n-m} \tag{4.39}$$

The difference operator ∇ can be defined as

$$\nabla\zeta_{-n} = \zeta_{-n} - \zeta_{-(n+1)} = (1 - \delta)\zeta_{-n} \tag{4.40}$$

Using this notation, a formal equivalent to eq. (4.23) is then given by

$$(1 + \alpha_1\nabla + \cdots + \alpha_p\nabla^p)\eta = (\beta_0 + \beta_1\nabla + \cdots + \beta_q\nabla^q)\xi \tag{4.41}$$

In applied problems, a few particular forms of eq. (4.41) are used: (1) autoregressive (AR); (2) moving average (MA); (3) autoregressive with moving average (ARMA); and (4) autoregressive integrated moving average (ARIMA).

An AR model of the order M, or $AR(M)$, produces the current sample η_0 of the centred output process[3] as a linear combination of M samples of the output process and the sample of the input process ξ_0, i.e.

$$\eta_0 = -\sum_{m=1}^{M} \alpha_m \eta_{-m} + \xi_0 \tag{4.42}$$

This model has M parameters α_m plus freedom in choosing the distribution of the input signal. Due to analytical difficulties with considering higher order difference equations, it is common to limit considerations only to $AR(1)$ and $AR(2)$ models. Despite such limitations, a large number of useful results can be obtained [9–12].

An MA model of order N, or $MA(N)$, produces the current value of a centred output signal as a weighted sum of the current and N past values of the input signal, i.e.

$$\eta_0 = \xi_0 + \sum_{n=1}^{N} \beta_n \xi_{-n} \tag{4.43}$$

This model is also defined by values of N coefficients β_n and by choice of the distribution of the input process ξ.

Both AR and MA models can be treated as discrete filters. The AR model is described by an Infinite Impulse Response (IIR) filter, while the MA model is described by a Finite Impulse Response (FIR) filter. The MA model is well suited for description of relatively smooth processes, while the AR model is used when a certain impulse behaviour has to be properly represented.

The $ARMA(M,N)$ model is a generalization which has features of both the MA and AR models defined above. In this case, the current sample of the output signal is obtained as a sum of weighted averages of the past N samples of the input process, past M samples of the output process and the current input process. Mathematically, the equation describing $ARMA(M,N)$ is thus

$$\eta_0 = -\sum_{m=1}^{M} \alpha_m \eta_{-m} + \xi_0 + \sum_{n=1}^{N} \beta_n \xi_{-n} \tag{4.44}$$

ARMA models also can be treated as discrete linear filters.

Finally, the $ARIMA(M,N,K)$ model is a further generalization of the $ARMA(M,N)$ with AR part driven by the increments

$$\zeta_n = \nabla^K \eta_n = (1 - \delta)^K \eta_n \tag{4.45}$$

of the output process:

$$\eta_0 = -\sum_{m=1}^{M} \alpha_m \zeta_{-m} + \xi_0 + \sum_{n=1}^{N} \beta_n \xi_{-n} \tag{4.46}$$

---

[3]A non-zero mean value m_η can be added later. The variance of the sequence η_n can also be adjusted in such a way that the sample of the external excitation has the weight 1 in eq. (4.42).

Analysis of a discrete system is most conveniently performed using either the discrete Fourier transform (DFT) [13]

$$\Xi(e^{j\omega}) = \sum_{n=-\infty}^{\infty} \xi_n e^{-j\omega n} \tag{4.47}$$

or the so-called z transform method [13]

$$\Xi(z) = \sum_{n=-\infty}^{\infty} \xi_n z^{-n} \tag{4.48}$$

$$\Xi(e^{j\omega}) = \Xi(z)|_{z=\exp(j\omega)} \tag{4.49}$$

Both methods transform difference equations of the form (4.42)–(4.44) to an algebraic equation in the ω or z domain. Indeed, if $\psi_n = \xi_{n-k}$, its z transform is given by

$$\Psi(z) = \sum_{n=-\infty}^{\infty} \psi_n z^{-n} = \sum_{n=-\infty}^{\infty} \psi_{n-k} z^{-n} = z^{-k} \sum_{n=-\infty}^{\infty} \xi_{n-k} z^{-(n-k)} = z^{-k} \Xi(z), \quad \Phi(e^{j\omega}) = e^{-j\omega k} \Xi(e^{j\omega}) \tag{4.50}$$

which allows the ARMA equation to be rewritten as

$$\Theta(z)\left[1 + \sum_{m=1}^{M} \alpha_m z^{-m}\right] = \Xi(z)\left[1 + \sum_{n=1}^{N} \beta_n z^{-n}\right] \tag{4.51}$$

and thus

$$\Theta(z) = \frac{1 + \sum_{n=1}^{N} \beta_n z^{-n}}{1 + \sum_{m=1}^{M} \alpha_m z^{-m}} \Xi(z) \tag{4.52}$$

Here $\Theta(z)$ is the z transform of the sequence η_n.

4.2.4 MIMO Systems

Analysis of MIMO systems can be conducted within the same framework considered in the previous section. In general, it is impossible to provide an expression for joint probability densities of the output vector processes but it is feasible to obtain expressions for moments and cumulants. Let $\xi_n(t)$, $1 \le n \le N$, be a random process acting at the n-th input of an N inputs, M outputs MIMO system. If the impulse response between the n-th input and m-th output is denoted by $h_{nm}(t)$, the response $\eta_n(t)$ of the system at the m-th output is then given by

$$\eta_m(t) = \sum_{n=1}^{N} \int_0^t h_{nm}(\tau_n) \xi_n(t - \tau_n) d\tau_n \tag{4.53}$$

Thus, the average value $m_{\eta_m}(t)$ of the m-th output and the mutual correlation function $R_{\eta_{m_1},\eta_{m_2}}(t_1, t_2)$ between two outputs can be calculated as follows

$$m_{\eta_m}(t) = \sum_{n=1}^{N} \int_0^t h_{nm}(\tau_n) m_{\xi_n}(t - \tau_n) d\tau_n \tag{4.54}$$

$$
\begin{aligned}
R_{\eta_{m_1}\eta_{m_2}}(t_1, t_2 &= E\{\eta_{m_1}(t)\eta_{m_2}(t)\} \\
&= \sum_{n_2=1}^{N}\sum_{n_1=1}^{N} \int_0^{t_1}\int_0^{t_2} h_{m_1 n_1}(\tau_{n_1}) h_{m_2 n_2}(\tau_{n_2}) R_{\eta_{n_1}\eta_{n_2}}(t_1 - \tau_{n_1}, t_2 - \tau_{n_2}) d\tau_{n_1} d\tau_{n_2}
\end{aligned} \tag{4.55}
$$

The same equations hold for the centred processes and thus for the mutual covariance functions $C_{\eta_{m_1}\eta_{m_2}}(t_1, t_2)$:

$$C_{\eta_{m_1}\eta_{m_2}}(t_1, t_2) = \sum_{n_2=1}^{N}\sum_{n_1=1}^{N} \int_0^{t_1}\int_0^{t_2} h_{m_1 n_1}(\tau_{n_1}) h_{m_2 n_2}(\tau_{n_2}) C_{\eta_{n_1}\eta_{n_2}}(t_1 - \tau_{n_1}, t_2 - \tau_{n_2}) d\tau_{n_1} d\tau_{n_2} \tag{4.56}$$

If the input signals are stationary related, the expression for the mutual covariance function is somewhat simplified to read

$$C_{\eta_{m_1}\eta_{m_2}}(t_1, t_2) = \sum_{n_2=1}^{N}\sum_{n_1=1}^{N} \int_0^{t_1+\tau}\int_0^{t_1} h_{m_1 n_1}(\tau_{n_1}) h_{m_2 n_2}(\tau_{n_2}) C_{\eta_{n_1}\eta_{n_2}}(\tau - \tau_{n_1} + \tau_{n_2}) d\tau_{n_1} d\tau_{n_2} \tag{4.57}$$

Finally, if $t_1 \to \infty$, the expression for the asymptotically stationary mutual covariance function becomes

$$C_{\eta_{m_1}\eta_{m_2}}(\tau) = \sum_{n_2=1}^{N}\sum_{n_1=1}^{N} \int_0^{\infty}\int_0^{\infty} h_{m_1 n_1}(\tau_{n_1}) h_{m_2 n_2}(\tau_{n_2}) C_{\eta_{n_1}\eta_{n_2}}(\tau - \tau_{n_1} + \tau_{n_2}) d\tau_{n_1} d\tau_{n_2}. \tag{4.58}$$

4.2.5 Description of Non-Linear Systems

As was mentioned earlier, two classes of non-linear system can be distinguished: zero memory (non-inertial) and non-zero memory (inertial) systems. Zero memory non-linearities are described by an operator of the type

$$\eta(t) = T[\xi(t)] = g(\xi(t)) \tag{4.59}$$

where $y = g(x)$ is a deterministic (or random) function which does not depend on time. Often a polynomial

$$y(x) \approx \sum_{n=0}^{N} a_n (x - x_0)^n, \tag{4.60}$$

piece-wise linear

$$y(x) \approx a_i x + b_i, x \in [x_i, x_{i+1}] \tag{4.61}$$

or transcendental approximations are used. In the latter case, the non-linearity $g(x)$ is approximated by some simple and well studied transcendental functions such as exponent, tangent, logarithm, etc. Each of these approximations has its advantages and disadvantages. Often it is difficult to balance accuracy of approximation and simplicity of calculations.

Mathematical models of non-linear systems with memory vary greatly. Often, they are described by a system of non-linear differential equations (state–space equations), written in the Cauchy form

$$\frac{d}{dt}\eta(t) = f[\eta(t), t] + g[\eta(t), t, \xi(t)] \tag{4.62}$$

where $f(\bullet)$ and $g(\bullet)$ are deterministic functions, which may depend on time.

In contrast to linear systems, when the general structure of the solution is known, the form of solution for a non-linear equation (4.62) depends on the order, type, excitation and many other factors. In most cases, it is impossible to obtain an exact solution even for the autonomous case $\xi(t) = 0$. This is why the analysis of non-linear systems with memory is fundamentally different from analysis of linear systems and zero memory non-linearity: each non-linear system must be analyzed separately. However, it will be seen that for some classes of non-linear system it is possible to obtain significant information in a very general form.

Instead of using differential equations, one can rewrite the state–space equations of type (4.62) using the so-called Volterra representation [14]

$$\eta(t) = h_0(t) + \sum_{i=1}^{\infty} \int_{R^i} h_i(t, \tau_1, \ldots, \tau_i)\xi(\tau_1)\cdots\xi(\tau_i)d\tau_1\cdots d\tau_i \tag{4.63}$$

Such representation is legitimate if the function $f(x)$ is well approximated by the Taylor series and the function $g(x)$ does not depend on $\eta(t)$. Furthermore, if functions $f(x)$ and $g(x)$ do not depend explicitly on time, the eq. (4.63) can be recast as

$$\eta(t) = h_0(t) + \sum_{i=1}^{\infty} \int_{R^i} h_i(\tau_1, \ldots, \tau_i)\xi(t - \tau_1)\cdots\xi(t - \tau_i)d\tau_1\cdots d\tau_i \tag{4.64}$$

Functions $h_i(\tau_1, \ldots, \tau_i)$ are called the Volterra kernels of the system. The first term of the sum $h_0(t)$ can be interpreted as the internal dynamics of the system since it can be obtained by setting the forcing function $\xi(t)$ to zero. The second term is the reaction of a linear system with the impulse response $h_1(t)$ to the external signal $\xi(t)$. The third term is more complicated and can be viewed as a two-dimensional convolution. It also can be treated as a combination of a linear system and a ZMNL system. The same is true for higher order terms in eq. (4.64) [14]. Thus, in some cases, a non-linear system with memory can be represented by a combination of linear blocks and a ZMNL block as shown in Fig. 4.3. As a result, analysis, and sometimes synthesis, of a non-linear system can be reduced to analysis of linear systems and ZMNL systems.

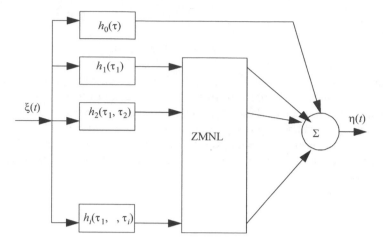

Figure 4.3   Representation of non-linear systems using the Volterra series.

4.3   ZERO MEMORY NON-LINEAR TRANSFORMATION OF RANDOM PROCESSES

In this section, dependence of moments and cumulants of the marginal PDF $p_\eta(x)$ and the joint PDF $p_\eta(x, y)$ of the second order of a random process $\eta = f(\xi)$ on the characteristic of the input process $\xi(t)$ is studied. A zero memory non-linear transformation

$$y = f(x) \tag{4.65}$$

is assumed to be known.

Transformation of a random process $\xi(t)$ by a ZMNL system can be reduced to a problem of transformation of a random vector as considered in section 2.3. Indeed, if the joint PDF $p_\eta(y_1, \ldots, y_n; t_1, \ldots, t_n)$ of the output process $\eta(t) = f[\xi(t)]$ considered at n moments of time is to be found, one can approach the problem in the following manner. Let $\xi = [\xi(t_1), \xi(t_2), \ldots, \xi(t_n)]^T$ be a vector formed by the samples of the input random process $\xi(t)$ and $\eta = [\eta(t_1), \eta(t_2), \ldots, \eta(t_n)]^T$ be the vector formed by the samples of the output process $\eta(t)$. Since the transformation has no memory, each component is transformed independently and according to the same rule, i.e.

$$y = [y_1, y_2, \ldots, y_n]^T = f(x) = [f(x_1), f(x_2), \ldots, f(x_n)]^T \tag{4.66}$$

Thus, if the joint PDF $p_\xi(x)$ is known one can calculate the joint PDF $p_\eta(y)$ using rules of transformation of a random vector.

4.3.1   Transformation of Moments and Cumulants

At a fixed moment of time t values of the random processes $\eta(t)$ and $\xi(t)$ are random variables, and, thus, the relation for the moments and the central moments of the output process are just given by

$$m_n^\eta(t) = \int_{-\infty}^{\infty} y^n p_\eta(y, t) dy = \int_{-\infty}^{\infty} f^n(x) p_\xi(x, t) dx \qquad (4.67)$$

$$\mu_n^\eta(t) = \int_{-\infty}^{\infty} [y - m_1^\eta(t)]^n p_\eta(y, t) dy = \int_{-\infty}^{\infty} [f(x) - m_1^\eta(t)]^n p_\xi(x, t) dx \qquad (4.68)$$

In principle, eqs. (4.67) and (4.68) solve the problem of finding the moments of the marginal PDF of the output process $\eta(t)$ at any given instant of time. Of course, for a stationary process, all these quantities do not depend on time. In general, a complete knowledge of the marginal PDF $p_\xi(x, t)$ is required. Further simplification can be achieved for a polynomial non-linear transformation. In this case, the transformation is a sum of powers of the input, i.e.

$$f(x) = \sum_{i=0}^{N} a_i x^i \qquad (4.69)$$

Using this expression in (4.67) and setting $n = 1$, one obtains the expression for the average $m_1^\eta(t_1)$ of the output process

$$m_1^\eta(t) = \int_{-\infty}^{\infty} \left(\sum_{i=0}^{N} a_i x^i \right) p_\xi(x, t) dx = \sum_{i=0}^{N} a_i m_i^\xi(t) \qquad (4.70)$$

Similarly, the expression for the second moment can be obtained by noting that

$$f^2(x) = \left(\sum_{i=0}^{N} a_i x^i \right)^2 = \sum_{i=0}^{N} a_i^2 x^{2i} + 2 \sum_{j=0, j\neq i}^{N} a_i a_j x^{i+j} \qquad (4.71)$$

and, thus

$$m_2^\eta(t) = \sum_{i=0}^{N} a_i^2 m_{2i}^\xi(t) + 2 \sum_{j=0, j\neq i}^{N} a_i a_j m_{i+j}^\xi(t) \qquad (4.72)$$

In general, the expression $f^k(x)$ would be a polynomial of the order kN and, thus, only the moments $m_n^\xi(t_1)$ of order up to kN are needed to calculate the k-th moment of the output:

$$f^k(x) = \sum_{i=0}^{kN} b_i x^i, m_k^\eta(t_1) = \sum_{i=0}^{kN} b_i m_i^\xi(t_1) \qquad (4.73)$$

This approach works well for calculation of low order moments or computer simulations. However, sometimes a simpler way of calculating moments can be found based on application of eqs. (2.255)–(2.257) which allow differential equations for $\langle f^n(x) \rangle$ in terms of the cumulants of the input process $\xi(t)$ to be obtained. Indeed, plugging $f^n(x)$ into eqs. (2.257), one obtains

$$
\begin{aligned}
\frac{\partial}{\partial \kappa_s^\xi(t)} m_k^\eta(t) &= \frac{1}{s!} \left\langle \frac{d^s}{dx^s} f^k(x) \right\rangle_\xi \\
\frac{\partial^2}{\partial \kappa_s^\xi \partial \kappa_p^\xi} m_k^\eta(t) &= \frac{1}{s!p!} \left\langle \frac{d^{s+p}}{dx^{s+p}} f^k(x) \right\rangle_\xi
\end{aligned}
\tag{4.74}
$$

Example [15]: As an example an exponential transformation is considered here, i.e.

$$
\eta(t) = \exp[a\xi(t)] \tag{4.75}
$$

In this case

$$
\frac{\partial}{\partial \kappa_s^\xi} \langle e^{ax} \rangle = \frac{a^s}{s!} \langle e^{ax} \rangle \tag{4.76}
$$

Integrating this equation, one obtains

$$
\langle e^{ax} \rangle = A_s \exp\left(\frac{a^s \kappa_s^\xi}{s!} \right) \tag{4.77}
$$

where A_s is a constant to be determined. Since s in eq. (4.77) is an arbitrary integer, a constant A should exist such that

$$
\langle e^{ax} \rangle = A \exp\left(\sum_{s=1}^\infty \frac{a^s \kappa_s}{s!} \right) \tag{4.78}
$$

Furthermore, if the random variable ξ is identically zero, then $\kappa_s = 0$ for any s and thus $A = \langle e^0 \rangle = 1$. Thus, finally

$$
\langle e^{ax} \rangle = \exp\left(\sum_{s=1}^\infty \frac{a^s \kappa_s}{s!} \right) \tag{4.79}
$$

This result also allows one to calculate moments m_n^η. Indeed,

$$
m_\eta^k = \langle e^{kax} \rangle = \exp\left(\sum_{s=1}^\infty \frac{a^s k^s}{s!} \kappa_s \right) \tag{4.80}
$$

If the input process $\xi(t)$ is a Gaussian process then the calculation of the moments can be further simplified since the higher order cumulants vanish[4]. In this case it can be shown [15] that

$$
m_2^\eta(t) = \sum_{s=0}^\infty \frac{1}{s!} \left[\frac{d^{2s}}{dx^{2s}} f^2(x) \right]\Bigg|_{x=m_1^\xi} \cdot \left(\frac{D_\xi}{2} \right)^s \tag{4.81}
$$

---

[4]While the cumulants vanish, even moments do not. Thus, the application of the approach given by eqs. (4.67) and (4.68) is not "optimal".

Example [15]:   Let a Gaussian random process be transformed by a square law detector. In this case $\eta = f(\xi) = \xi^2$. Using eq. (4.81), one can easily obtain expressions for the first and second moments

$$m_1^\eta(t) = m_{1\xi}^2(t) + d_\xi(t) = m_2^\xi(t) \tag{4.82}$$

$$m_2^\eta(t) = m_{1\xi}^4(t) + 6m_{1\xi}^2(t)D_\xi(t) + 3D_\xi^2(t) \tag{4.83}$$

Finally, the variance of the output process is just

$$D_\eta(t) = m_2^\eta(t) - m_{1\eta}^2(t) = 4m_{1\xi}^2(t)D_\xi(t) + 2D_\xi^2(t) \tag{4.84}$$

The next task is to find an expression for the correlation function $R_{\xi\xi}(t_1, t_2)$ of the transformed process $\eta(t)$

$$
\begin{aligned}
R_{\eta\eta}(t_1, t_2) &= \int_{-\infty}^{\infty} \int_{-\infty}^{\infty} y_1 y_2 p_{\eta\eta}(y_1, y_2; t_1, t_2) dy_1 dy_2 \\
&= \int_{-\infty}^{\infty} \int_{-\infty}^{\infty} f(x_1) f(x_2) p_{\xi\xi}(x_1, x_2; t_1, t_2) dx_1 \, dx_2
\end{aligned} \tag{4.85}
$$

If both processes $\xi(t)$ and $\eta(t)$ are stationary processes then eq. (4.85) is simplified to

$$R_\eta(\tau) = \int_{-\infty}^{\infty} \int_{-\infty}^{\infty} f(x) f(x_\tau) p_{\xi\xi}(x, x_\tau; \tau) dx \, dx_\tau \tag{4.86}$$

Thus, in order to evaluate the correlation function at the output of a non-linear zero memory device, it is sufficient to know either the joint PDF $p_{\xi\xi}(x_1, x_2; t_1, t_2)$ at the input or its characteristic function $\Theta_{\xi\xi}(ju_1, ju_2; t_1, t_2)$. As a result, there are two main methods developed for the calculation of the correlation function [5]: (a) direct method based on the known PDF and (b) the Rice method, based on the orthogonal expansion of the characteristic function. Here both methods are briefly summarized.

4.3.1.1   Direct Method

In this section, the following orthogonal expansion of the joint PDF $p_{\xi\xi}(x, x_\tau; \tau)$, studied in section 2.2, is widely used

$$p_{\xi\xi}(x, x_\tau; \tau) = p(x)p(x_\tau) \sum_{n=0}^{\infty} \alpha_n(\tau) Q_n(x) Q_n(x_\tau) \tag{4.87}$$

Here $\{Q_n(x)\}_1^\infty$ is a set of orthogonal polynomials generated[5] by the PDF $p_\xi(x)$ and the coefficients $C_n(\tau)$ can be found as

$$\alpha_n(\tau) = \int_{-\infty}^{\infty} \int_{-\infty}^{\infty} Q_n(x) Q_n(x_\tau) p_{\xi\xi}(x, x_\tau; \tau) dx \, dx_\tau \tag{4.88}$$

[5]While it is always possible to build a system of orthogonal polynomials with a given weight $p_\xi(x)$, this system is not always complete. It can be shown that for the Pearson system of PDFs, such a system is indeed complete. For more details see [16].

Using expansion (4.87), the expression for the correlation function can be rewritten as

$$p_{\xi\xi}(x, x_\tau; \tau) = \sum_{n=0}^{\infty} \alpha_n(\tau) \left[\int_{-\infty}^{\infty} f(x)Q_n(x)p(x)dx \cdot \int_{-\infty}^{\infty} f(x_\tau)Q_n(x_\tau)p(x_\tau)dx_\tau \right] = \sum_{n=0}^{\infty} \alpha_n(\tau)a_n^2$$

(4.89)

In this expansion, the coefficients α_n can be calculated through the marginal PDF $p(x)$ as

$$a_n = \int_{-\infty}^{\infty} f(x)Q_n(x)p(x)dx$$

(4.90)

It is important to mention here that both $\alpha_n(\tau)$ and a_n depend on a particular choice of $p(x)$, which, in turn, defines the set of the orthogonal polynomials $Q_n(x)$. The most popular choice is a Gaussian PDF and the Gamma PDF, as discussed in section 2.2. As an alternative, $p(x)$ can be chosen as the marginal PDF of the input process $p(x) = p_\xi(x)$.

4.3.1.2 The Rice Method

Let

$$F(ju) = \int_{-\infty}^{\infty} f(x)e^{-jux}dx$$

(4.91)

be the Fourier transform of the non-linearity $f(x)$ under consideration

$$f(x) = \frac{1}{2\pi} \int_{-\infty}^{\infty} F(ju)e^{jux}du$$

(4.92)

As a result, the expression for the correlation function $R_{\eta\eta}(\tau)$ can be rewritten as

$$R_{\eta\eta}(\tau) = \frac{1}{4\pi^2} \cdot \int_{-\infty}^{\infty} \int_{-\infty}^{\infty} \left[\int_{-\infty}^{\infty} F(ju)e^{jux}du \cdot \int_{-\infty}^{\infty} F(ju_\tau)e^{ju_\tau x_\tau}du_\tau \right] p_{\xi\xi}(x, x_\tau; \tau)dx\,dx_\tau$$

$$= \frac{1}{4\pi^2} \cdot \int_{-\infty}^{\infty} \int_{-\infty}^{\infty} F(ju)F(ju_\tau) \left[\int_{-\infty}^{\infty} \int_{-\infty}^{\infty} p_{\xi\xi}(x, x_\tau; \tau)e^{j(ux+u_\tau x_\tau)}dx\,dx_\tau \right] du\,du_\tau$$

(4.93)

Since the expression in the square brackets is the characteristic function of the second order $\Theta_{\xi\xi}(ju, ju_\tau; \tau)$, eq. (4.93) becomes

$$R_{\eta\eta}(\tau) = \frac{1}{4\pi^2} \cdot \int_{-\infty}^{\infty} \int_{-\infty}^{\infty} F(ju)F(ju_\tau)\Theta_{\xi\xi}(ju, ju_\tau; \tau)du\,du_\tau$$

(4.94)

Calculations using eq. (4.94) can be simplified if $\xi(t)$ is a zero mean Gaussian process. In this case

$$
\begin{aligned}
\Theta_{\xi\xi}(u, u_\tau) &= \exp\left(-\frac{\sigma^2}{2}[u_1^2 + 2\rho_{\xi\xi}(\tau)u_1 u_2 + u_2^2]\right) \\
&= \exp\left(-\frac{\sigma^2}{2}u_1^2\right)\exp\left(-\frac{\sigma^2}{2}u_2^2\right)\sum_{n=0}^{\infty}\frac{(-1)^n}{n!}(\sigma^2\rho(\tau)u_1 u_2)^n
\end{aligned} \tag{4.95}
$$

As a result, eq. (4.94) becomes

$$
C_{\eta\eta}(\tau) = \frac{1}{4\pi^2}\cdot\sum_{n=0}^{\infty}\frac{(-1)^n}{n!}\sigma^{2n}\rho_{\xi\xi}^n(\tau)\left[\int_{-\infty}^{\infty}F(ju)u^{2n}\exp\left(-\frac{\sigma^2 u^2}{2}\right)du\right]^2 = \sum_{n=0}^{\infty}\frac{(-1)^n}{n!}d_n^2\rho_{\xi\xi}^n(\tau) \tag{4.96}
$$

where

$$
d_n = \frac{\sigma^n}{2\pi}\int_{-\infty}^{\infty}F(ju)u^{2n}\exp\left(-\frac{\sigma^2 u^2}{2}\right)du \tag{4.97}
$$

Relations (4.89)–(4.90) and (4.96)–(4.97) in principle allow calculation of the correlation function of the output of a non-linear device. They are especially helpful for numerical simulations. However, for the purpose of analytical analysis, these equations may become intractable. Especially, it is difficult to trace the effect of the input correlation function to the shape of the output correlation function. An alternative approach, based on the cumulant equations is considered in the following section.

4.3.2 Cumulant Method

The cumulant method is based on the application of the following identity, known as the Price formula [4,14,15,17,18]

$$
\frac{\partial^n}{\partial[R_\xi(\tau)]^n}R_\eta(\tau) = \langle f^{(n)}(x)f^{(n)}(x_\tau)\rangle \tag{4.98}
$$

This equation, which is valid for an arbitrary marginal PDF of the input process[6], is most useful for analysis of transformation of Gaussian random processes. Without loss of generality, it is assumed that both the input and the output processes have zero mean, i.e. $m_1^\eta = m_1^\xi = 0$. Also, for convenience, the following notation is used

$$
g_n(\tau) \equiv \langle f^{(n)}(x)f^{(n)}(x_\tau)\rangle \tag{4.99}
$$

[6]Originally it was derived only for the case of Gaussian processes.

In the general case, the output correlation function depends on all cumulants of the input joint density $p_{\xi\xi}(x, x_\tau; \tau)$, i.e.

$$g_n(\tau) = F_n[\kappa_2^\xi(0, \tau), \kappa_3^\xi(0, \tau, \tau) \cdots \kappa_{p+q}^\xi(0^{[p]}, \tau^{[q]}) \cdots] \tag{4.100}$$

Combining the Price formula (4.98) with (4.100), one obtains a differential equation written now in terms of the cumulants

$$\frac{\partial^n}{\partial[R_\xi(\tau)]^n} R_\eta(\tau) = F_n[R_\xi(\tau), \kappa_3^\xi(0, \tau^{[2]}), \ldots] \tag{4.101}$$

Let $\overset{\circ}{g}_n(\tau)$ be the value of the function $F_n(\bullet)$ at zero value of $R_\xi(\tau)$, i.e.

$$\overset{\circ}{g}_n(\tau) = F_n[R_\xi(\tau), \kappa_3^\xi(0, \tau^{[2]}), \ldots]\Big|_{R_\xi(\tau)=0} \tag{4.102}$$

As was shown in [17], the correlation function $R_\eta(\tau)$ can be represented as a power series in terms of $R_\xi(\tau)$. In particular

$$R_\eta(\tau) = \sum_{k=1}^\infty \frac{1}{k!} \frac{d^n(R_\eta(\tau))}{d(R_\xi(\tau))^n}\Big|_{R_\xi(\tau)=0} \cdot R_\xi^n(\tau) = \sum_{k=1}^\infty \frac{1}{k!} \langle f^{(k)}(x) f^{(k)}(x_\tau)\rangle\big|_{R_\xi(\tau)=0} \cdot R_\xi^n(\tau) + \overset{\circ}{g}_0(\tau)$$

$$\tag{4.103}$$

The unknown function $\overset{\circ}{g}_0(\tau)$ can be defined from the condition

$$R_\eta(\tau)\big|_{R_\xi=0} \equiv 0 \tag{4.104}$$

Equation (4.103) represents a general form of dependence of the output correlation function on the input correlation function. This series is significantly simplified if the non-linear transformation is a polynomial function. In this case, an infinite series (4.103) is reduced to a sum of a finite number of terms.

Equation (4.103) also allows an asymptotic value of the correlation function $R_\eta(\tau)$ for large time lags to be obtained. Indeed, for a sufficiently large τ, the value of the correlation function is small, thus only the first term in expansion (4.103) could be kept to allow for an appropriate accuracy. In this case

$$\lim_{\tau\to\infty} R_\eta(\tau) \cong \langle f(x)\rangle^2 R_\xi(\tau) \tag{4.105}$$

4.4 CUMULANT ANALYSIS OF NON-LINEAR TRANSFORMATION OF RANDOM PROCESSES

4.4.1 Cumulants of the Marginal PDF

The cumulants of the marginal PDF of the output process can be found using a technique considered in section 2.8. A few examples are considered below.

Example 1:   Let $\eta = f(\xi)$ be given a deterministic ZMNL transformation. It is required to find the dependence of the variance $D_\eta = \kappa_2^\eta$ of the output process on the input cumulants κ_s^ξ. It follows from eq. (4.98) that

$$\frac{\partial D_\eta}{\partial \kappa_s^\xi} = \frac{1}{s!}\left\langle \frac{d^s}{dx}[f^2(x) - 2\langle f(x)\rangle f(x)] \right\rangle \tag{4.106}$$

Setting the order s sequentially to $1, 2, \ldots$, one obtains the following system of equations

$$\frac{\partial D_\eta}{\partial m_1^\xi} = 2\left\langle f(x), \frac{d}{dx}f(x) \right\rangle$$

$$\frac{\partial D_\eta}{\partial D_\xi} = \left\langle \frac{d}{dx}f(x), \frac{d}{dx}f(x) \right\rangle + \left\langle f(x), \frac{d^2}{dx^2}f(x) \right\rangle + \left\langle \frac{d}{dx}f(x) \right\rangle^2$$

$$\frac{\partial D_\eta}{\partial \kappa_3^\xi} = \left\langle \frac{d}{dx}f(x), \frac{d^2}{dx^2}f(x) \right\rangle + \frac{1}{3}\left\langle f(x), \frac{d^3}{dx^3}f(x) \right\rangle \tag{4.107}$$

$$\frac{\partial D_\eta}{\partial \kappa_4^\xi} = \frac{1}{4}\left\langle \frac{d^2}{dx^2}f(x) \right\rangle + \frac{1}{3}\left\langle f(x), \frac{d^3}{dx^3}f(x) \right\rangle + \frac{1}{12}\left\langle f(x), \frac{d^4}{dx^4}f(x) \right\rangle$$

Example 2:   Let $\eta(t) = A\xi^2(t)$. In this case, equations for derivatives of the variance D_η with respect to low order cumulants can be easily obtained from eq. (4.107) by noticing that $f'(x) = 2Ax$, $f''(x) = 2A$ and all the higher order derivatives are equal to zero. Using the rules of manipulating cumulant brackets, listed in the appendix of chapter 2, one obtains

$$\frac{\partial D_\eta}{\partial m_1^\xi} = 2\langle Ax^2, 2Ax\rangle = (4\kappa_3^\xi + 8\kappa_1^\xi \kappa_2^\xi)A^2$$

$$\frac{\partial D_\eta}{\partial D_\xi} = \langle 4a^2x^2\rangle + \langle Ax^2, 2A\rangle = 4A^2[\kappa_2^\xi + (\kappa_1^\xi)^2]$$

$$\frac{\partial D_\eta}{\partial \kappa_2^\xi} = \langle 2Ax, 2A\rangle + \frac{1}{3}\langle Ax^2, 0\rangle = 4A^2\kappa_1^\xi \kappa_3^\xi = 0 \tag{4.108}$$

$$\frac{\partial D_\eta}{\partial \kappa_4^\xi} = \frac{1}{4}\langle 2A\rangle + \frac{1}{3}\langle Ax^2, 0\rangle + \frac{1}{12}\langle f(x), \frac{d^4}{dx^4}f(x)\rangle = \frac{A^2}{2}\kappa_4^\xi$$

Solving this system of equations starting from the equation which responds to the fourth cumulant, onc obtains

$$D_\eta = A^2[\kappa_4^\xi + 4\kappa_1^\xi \kappa_3^\xi + 2\kappa_2^\xi + 4(\kappa_1^\xi)^2\kappa_2^\xi] \tag{4.109}$$

It is worth noting that the expansion (4.109) does not depend on the particular form of the PDF $p_\xi(x)$ of the input process, which emphasizes the power of the cumulant method.

4.4.2   Cumulant Method of Analysis of Non-Gaussian Random Processes

In the previous section, cumulant equations were used to obtain relations between the cumulants of the marginal PDF at the input and the output of a zero memory non-linear transformer. Application of the Price formula allows calculation of the correlation function. It is possible to generalize this equation [4,14,15,17,18] to the case of higher order correlation (cumulant) functions. For example, it can be shown [15] that

$$\frac{D^n[\kappa_2^\eta(\tau)]}{D\kappa_{p+q}^\xi(0^{[p]}, \tau^{[q]})^n} = \frac{1}{p!q!} \langle f^{(pn)}(x) f^{(qn)}(x_\tau) \rangle, \quad p \neq 0, q \neq 0 \tag{4.110}$$

It can be seen that the Price equation (4.98) is a particular case of eq. (4.110) with $p = q = 1$. However, its application to the case of a non-Gaussian driving process $\xi(t)$ is limited since it does not provide a relation with all the summands of the input process $\xi(t)$. In order to illustrate this statement, the example of a square law transformation considered above can be used. Once again the transformation $\eta(t) = A\xi^2(t)$ is considered. Since the right-hand side of eq. (4.110) differs from zero only for $p = q = 1$, and $n = 2$ or p, $q = 1$ and $n = 1$, then, eq. (4.110) produces the following five differential equations with respect to cumulants

$$\frac{D^2[C_\eta(\tau)]}{D[C_\xi(\tau)]^2} = \frac{A^2}{(1!1!)^2} \left\langle \frac{d^2 x^2}{dx^2} \cdot \frac{dx_\tau^2}{dx_\tau^2} \right\rangle = 4A^2$$

$$\frac{D[C_\eta(\tau)]}{D[\kappa_4^\xi(0^{[2]}, \tau^{[2]})]} = \frac{1}{(2!)^2} \langle 4A^2 \rangle = A^2$$

$$\frac{D[C_\eta(\tau)]}{D[C_\xi(\tau)]} = \frac{1}{1!1!} \langle 2Ax \cdot 2Ax_\tau \rangle = 4A^2[C_x^\xi(\tau) + m_1^2] \tag{4.111}$$

$$\frac{D[C_\eta(\tau)]}{D[\kappa_3^\xi(0, \tau^{[2]})]} = \frac{1}{2!1!} \langle 2Ax \cdot 2A \rangle = 2A^2 m_1$$

$$\frac{D[C_\eta(\tau)]}{D[\kappa_3^\xi(0^{[2]}, \tau)]} = 2A^2 m_1$$

If all cumulants κ_s, $s \geq 1$ are equal to zero, then the random processes $\eta(t)$ and $\xi(t)$ are equal to zero $\eta(t) = \xi(t) = 0$. Having this in mind, the following expression for the covariance function can be obtained from eq. (4.111)

$$C_\eta(\tau) = A^2\{\kappa_4^\xi(0^{[2]}, \tau^{[2]}) + 2m_1[\kappa_3^\xi(0^{[2]}, \tau) + \kappa_3^\xi(0, \tau^{[2]})] + 4m_1^\xi C_\xi(\tau) + 2[C_\xi(\tau)]^2\} \tag{4.112}$$

If the input process is a Gaussian process then all the cumulants of order $s \geq 3$ are equal to zero and eq. (4.112) can be further simplified to

$$R_\eta(\tau) = A^2\{2[R_\xi(\tau)]^2 + 4m_1^\xi R_\xi(\tau)\} \tag{4.113}$$

Of course, the same result can be obtained from the Price formula. Equation (4.113) remains valid if some non-Gaussian process has the skewness and the curtosis equal to zero, i.e. $\kappa_3^\xi = \kappa_4^\xi = 0$ and all other higher order cumulants are non-zero.

   Example: Square law detection of a telegraph signal. Let process $\xi(t)$ be a random telegraph signal with the average bit rate R_b and equal probability of 1s and 0s, encoded by voltage values $\pm a$. This signal has a symmetric PDF and the covariance function of the exponential type

$$C_{\xi\xi}(\tau) = a^2 e^{-2R_b|\tau|} \tag{4.114}$$

The cumulant function of the fourth order is obtained in [17] as

$$\kappa_4^\xi(0^{[2]}, \tau^{[2]}) = -2a^4 e^{-4R_b|\tau|} \tag{4.115}$$

Application of eq. (4.112) results in the following expression for the output covariance function:

$$C_{\eta\eta}(\tau) = A^2\{-2a^4 e^{-4R_b|\tau|} + 2a^4 e^{-4R_b|\tau|}\} \equiv 0 \tag{4.116}$$

This result is easy to understand, since the output process is a constant process $\eta(t) = Aa^2$. Thus, the compensation of cumulant functions is possible only for non-Gaussian processes. For an input Gaussian process, the output process is always random and, thus, $C_{\eta\eta}(\tau) \geq 0$.

Example: exponential detector. In this case the transformation under consideration is

$$\eta(t) = \exp[\beta\xi(t)], \quad f(x) = \exp(\beta x). \tag{4.117}$$

It follows from eq. (4.110) that

$$\frac{D[C_{\eta\eta}(\tau)]}{D[\kappa_{p+q}^\xi(0^{[p]}, \tau^{[q]})]} = \frac{\beta^{p+q}}{p!q!} C_\eta(\tau) \tag{4.118}$$

and its solution is given by

$$C_{\eta\eta}(\tau) = \exp\left\{\sum_{s=1}^{\infty} \frac{\beta^s}{s!} \sum_{i=0}^{s} C_s^i \kappa_s^\xi(0^{[s-i]}, \tau^{[i]})\right\} \tag{4.119}$$

Thus, in contrast to a square law detector, the correlation function of the output process $\eta(t)$ depends on all cumulant functions $\kappa_s^\xi(0^{[s-i]}, \tau^{[i]})$ of the input process $\xi(t)$. This can be explained by the fact that the exponential function has an infinite number of terms in its Taylor series; as a result, all the moments of the input process contribute to the resulting covariance function.

It can be concluded that the method of cumulant equations is a very effective tool for investigation of non-linear transformations of random processes.

4.5   LINEAR TRANSFORMATION OF RANDOM PROCESSES

4.5.1   General Expression for Moment and Cumulant Functions at the Output of a Linear System

The following quite general problem is to be solved in the analysis of transformation of random processes by a linear (time-invariant) system: given the system impulse response

$h(t)$ and the statistical characteristics of the input process $\xi(t)$, such as joint PDF $p_\xi(x_1, x_2, \ldots, x_n)$ of order n, find statistical characteristics of the output process $\eta(t) = L[\xi(t)]$, such as joint PDF $p_\eta(y_1, y_2, \ldots, y_k)$ of order $k < n$.

There is no known general method to solve this problem, unless the transformation of a Gaussian, a Spherically Invariant or a Markov process is considered[7]. Instead, the problem of evaluation of moment and cumulant functions could be studied.

4.5.1.1   Transformation of Moment and Cumulant Functions

Solution of the general problem for a Gaussian driving process $\xi(t)$ is significantly simplified since it is known that the output process $\eta(t)$ is also Gaussian and one has to find only the corresponding average $m_\eta(t)$ and the output covariance function $C_{\eta\eta}(t_1, t_2)$.

Let $\xi(t)$ be the input of an LTI system[8] with the impulse response $h(t)$ and zero initial conditions. Then the output random process is given by the following integral

$$\eta(t) = \int_0^t h(t-s)\xi(s)\mathrm{d}s = \int_0^t h(s)\xi(t-s)\mathrm{d}s \qquad (4.120)$$

This integral can be approximated by a partial sum as

$$
\begin{aligned}
\eta(t) &\approx \sum_{k=1}^K h(t-s_k)\xi(s_k)\Delta s_k \\
&= \sum_{k=1}^K h(s_k)\xi(t-s_k)\Delta s_k, \\
&= t, \Delta s_k = s_k - s_{k-1} \qquad 0 < s_1 < s_2 < \cdots < s_k
\end{aligned}
\qquad (4.121)
$$

Averaging both sides and taking into account that the average of a sum is a sum of the averages of the summands, one obtains

$$m_\eta(t) \approx \sum_{k=1}^K h(t-s_k)m_\xi(s_k)\Delta s_k = \sum_{k=1}^K h(s_k)m_\xi(t-s_k)\Delta s_k \qquad (4.122)$$

Since in this relation only deterministic functions are present, one can easily take a limit of both sides assuming that all $\Delta s_k \to 0$

$$m_\eta(t) = \int_0^t h(t-s)m_\xi(s)\mathrm{d}s = \int_0^t h(s)m_\xi(t-s)\mathrm{d}s \qquad (4.123)$$

This property reflects the fact that operations of statistical averaging and time integration are interchangeable.

---

[7]There are also a number of particular examples which can be solved completely.
[8]It is assumed here that $\xi(t) = 0$ if $t < 0$.

Eq. (4.120) can be written for η moments of time, thus producing the following chain of equations

$$E\{\eta(t_1)\eta(t_2)\cdots\eta(t_n)\}$$

$$= E\left\{\int_0^{t_1} h(t_1-s_1)\xi(s_1)ds_1 \int_0^{t_2} h(t_2-s_2)\xi(s_2)ds_2\cdots\int_0^{t_n} h(t_n-s_n)\xi(s_n)ds_n\right\}$$

$$= \int_0^{t_1}\int_0^{t_2}\cdots\int_0^{t_n} h(t_1-s_1)h(t_2-s_2)\cdots h(t_n-s_n)E\{\xi(s_1)\xi(s_2)\cdots\xi(s_n)\}ds_1ds_2\cdots ds_n$$

$$(4.124)$$

or, equivalently

$$m_{1,1,\ldots,1}^{\eta,\eta,\ldots,\eta}(t_1,t_2,\ldots,t_n)$$

$$= \int_0^{t_1}\int_0^{t_2}\cdots\int_0^{t_n} h(t_1-s_1)h(t_2-s_2)\cdots h(t_n-s_n)m_{1,1,\ldots,1}^{\xi,\xi,\ldots,\xi}(s_1,s_2,\ldots,s_n)ds_1ds_2\cdots ds_n$$

$$= \int_0^{t_1}\int_0^{t_2}\cdots\int_0^{t_n} h(s_1)h(s_2)\cdots h(s_n)m_{1,1,\ldots,1}^{\xi,\xi,\ldots,\xi}(t_1-s_1,t_2-s_2,\ldots,t_n-s_n)ds_1ds_2\cdots ds_n$$

$$(4.125)$$

If all the moments of time are the same, $t_1 = t_2 = \cdots = t_n = t$, an expression for the n-th moment $m_n^\eta(t)$ of the output process $\eta(t)$ can be obtained as

$$m_n^\eta(t) = \int_0^t\int_0^t\cdots\int_0^t h(t-s_1)h(t-s_2)\cdots h(t-s_n)m_{1,1,\ldots,1}^{\xi,\xi,\ldots,\xi}(s_1,s_2,\ldots,s_n)ds_1ds_2\cdots ds_n$$

$$= \int_0^t\int_0^t\cdots\int_0^t h(s_1)h(s_2)\cdots h(s_n)m_{1,1,\ldots,1}^{\xi,\xi,\ldots,\xi}(t-s_1,t-s_2,\ldots,t-s_n)ds_1ds_2\ldots ds_n$$

$$(4.126)$$

Characteristically, this moment depends on the mixed moment function $m_{1,1,\ldots,1}^{\xi,\xi,\ldots,\xi}(s_1,s_2,\ldots,s_n)$ considered at n moments of time rather than on the moment $m_n^\xi(t)$, thus illustrating the effect of memory. The effect of linearity is emphasized by the fact that the output moment still depends only on a single moment function, rather than a combination of moments of different orders. Equation (4.126) allows one to understand the level of complexity of the problem of evaluation of the moments of the output process. Indeed, it indicates that if the moment function of order n is to be found, one has to measure or derive a joint moment function of the order n and perform n-fold integration. In other words, the complexity of this problem is very high unless one limits considerations only to a few lower order moments.

The expression for the correlation function can be obtained from eq. (4.126) by setting $n = 2$, thus producing

$$R_{\eta\eta}(t_1,t_2) = m_{1,1}^{\eta,\eta}(t_1,t_2) = \int_0^{t_1}\int_0^{t_2} h(s_1)h(s_2)R_{\xi\xi}(t_1-s_1,t_2-s_2)ds_1ds_2 \qquad (4.127)$$

In a very similar manner, equations for mutual moments between the input and the output processes can be defined as

$$m_{1,\ldots,1;1,\ldots,1}^{\xi,\ldots,\xi;\eta,\ldots,\eta}(t_1',\ldots,t_m';t_1,\ldots,t_n)$$

$$= E\{\xi(t_1')\cdots\xi(t_m')\eta(t_1)\cdots\eta(t_n)\}$$

$$= \int_0^{t_1}\int_0^{t_2}\cdots\int_0^{t_n} h(s_1)h(s_2)\cdots h(s_n)m_{1,1,\ldots,1}^{\xi,\xi,\ldots,\xi}(t_1'\ldots,t_m';t_1-s_1,t_2-s_2,\ldots,t_n-s_n)ds_1ds_2\cdots ds_n$$

(4.128)

In particular, the mutual correlation function $R_{\xi\eta}(t_1,t_2)$ can be expressed as

$$R_{\xi\eta}(t_1,t_2) = m_{1,1}^{\xi,\eta}(t_1;t_2) = \int_0^{t_2} h(s)R_{\xi\xi}(t_1,t_2-s)ds \qquad (4.129)$$

It is also possible to obtain expressions for centred moments. The first step is to subtract eq. (4.123) from eq. (4.120) to obtain processes $\eta_0(t) = \eta(t) - m_\eta(t)$ and $\xi_0(t) = \xi(t) - m_\xi(t)$ with zero mean

$$\eta_0(t) = \int_0^t h(t-s)\xi(s)ds - \int_0^t h(t-s)m_\xi(s)ds = \int_0^t h(t-s)[\xi(s) - m_\xi(s)]ds$$

$$= \int_0^t h(t-s)\xi_0(s)ds$$

(4.130)

Thus, the relation between centred processes obeys the same equation as the non-centred processes. This mean that all derivations for moments and mutual moments remain valid for centred moments. The only difference is that centred moments must be used on the right-hand side of eqs. (4.124)–(4.129). In particular, the covariance $C_{\eta\eta}(t_1,t_2)$ and the mutual covariance $C_{\xi\eta}(t_1,t_2)$ can be obtained as follows

$$C_{\eta\eta}(t_1,t_2) = \int_0^{t_1}\int_0^{t_2} h(t_1-s_1)h(t_2-s_2)C_{\xi\xi}(s_1,s_2)ds_1ds_2$$

$$= \int_0^{t_1}\int_0^{t_2} h(s_1)h(s_2)C_{\xi\xi}(t_1-s_1,t_2-s_2)ds_1ds_2 \qquad (4.131)$$

$$C_{\xi\eta}(t_1,t_2) = \int_0^{t_1} h(t_2-s)C_{\xi\xi}(t_1,s)ds = \int_0^{t_1} h(s)C_{\xi\xi}(t_1,t_2-s)ds \qquad (4.132)$$

Thus, the variance of the output process is given by

$$D_\eta(t) = \int_0^t\int_0^t h(t-s_1)h(t-s_2)C_{\xi\xi}(s_1,s_2)ds_1ds_2 = \int_0^t\int_0^t h(s_1)h(s_2)C_{\xi\xi}(t-s_1,t-s_2)ds_1ds_2$$

(4.133)

If a linear time-variant system is considered, the above remains valid if the impulse response $h(t)$ is substituted by the system function $h(t,\tau)$.

4.5.1.2 Linear Time-Invariant System Driven by a Stationary Process

The results obtained in the previous section can be further simplified if the input process $\xi(t)$ is a stationary process. In this case, expressions for the mean, covariance function, mutual covariance and the variance functions can be written as

$$m_\eta(t) = m_\xi \int_0^t h(t)dt \qquad (4.134)$$

$$C_{\eta\eta}(t_1, t_2) = \int_0^{t_1} \int_0^{t_2} h(s_1)h(s_2)C_{\xi\xi}(t_2 - t_1 - s_2 + s_1)ds_1 ds_2 \qquad (4.135)$$

$$C_{\xi\eta}(t_1, t_2) = \int_0^{t_2} h(s)C_{\xi\xi}(t_2 - t_1 - s)ds \qquad (4.136)$$

$$D_\eta(t) = \int_0^t \int_0^t h(s_1)h(s_2)C_{\xi\xi}(s_1 - s_2)ds_1 ds_2 \qquad (4.137)$$

These equations show that even if the input process is stationary, the output process could be non-stationary. This is very similar to the deterministic case when a constant input, applied at $t = 0$, produces a transient process and the steady state is achieved only at $t = \infty$. However, in most realistic applications the impulse response $h(t)$ decays fast enough to justify the steady state and the stationarity achieved after a time greater than the time scale of the impulse response. For example, if the impulse response has an exponential form $h(t) = \exp(-t/\tau_{corr})$, $t \geq 0$, it is reasonable to assume that at the moment of time $t \gg \tau_{corr}$ the transient process has been completed and the steady state, or stationary regime, is in place. Indeed, in this case, the finite limit of integration in eq. (4.134) can be substituted by infinite limits, thus producing

$$m_\eta \approx m_\xi \int_0^\infty h(s)ds = m_\xi H(j\omega)|_{\omega=0} \qquad (4.138)$$

That is, the mean value does not depend on time. In order to obtain the expression for the covariance function, one can introduce a time lag $\tau = t_2 - t_1$ and assume that $t_1 \gg \tau_{corr}$. Thus, changing the variable of integration and assuming infinite limits of integration, one obtains

$$C_{\eta\eta}(t_1, t_2) \approx \int_0^\infty \int_0^\infty h(s_1)h(s_2)C_{\xi\xi}(\tau - s_1 + s_2)ds_1 ds_2$$

$$= \int_0^\infty h(s)ds \int_0^{s+\tau} h(\tau + s - t)C_{\xi\xi}(t)dt = C_{\eta\eta}(\tau) \qquad (4.139)$$

Therefore it can be said that after a transitional period, the output of a linear system driven by a wide sense stationary random process also becomes a wide sense stationary process. In other words, the output process is an asymptotically wide sense stationary process.

Equation (4.139) also allows the relation between power spectral densities of the input and the output processes to be obtained. Indeed, taking the Fourier transform of both sides one obtains the required relation

$$
\begin{aligned}
S_\eta(\omega) &= \int_{-\infty}^{\infty} C_{\eta\eta}(\tau) \exp(-j\omega\tau) dt \\
&= \int_{-\infty}^{\infty} \exp(-j\omega\tau) d\tau \\
&\quad \int_0^{\infty} \exp[-j\omega(s_1 - s_2)]h(s_1)h(s_2)C_{\xi\xi}(\tau - s_1 + s_2)\exp[-j\omega(s_2 - s_1)]ds_1 ds_2 \\
&= \int_0^{\infty} \exp[-j\omega s_1]h(s_1)ds_1 \int_0^{\infty} \exp(j\omega s_2)h(s_2)ds_2 \\
&\quad \times \int_{-\infty}^{\infty} C_{\xi\xi}(\tau - s_1 + s_2)\exp[-j\omega(\tau - s_1 + s_2)]ds
\end{aligned}
\tag{4.140}
$$

It also can be rewritten in the equivalent form in terms of the transfer function $H(j\omega)$

$$
S_\eta(\omega) = H(j\omega)H(-j\omega)S_\xi(\omega) = |H(j\omega)|^2 S_\xi(\omega)
\tag{4.141}
$$

Thus, the spectral density of the output process is that of the input multiplied by the squared absolute value of the transfer function. This fundamental result allows one to effectively use the PSD for calculation of the parameters of the output process in linear systems. It is interesting to note that the transformation of the spectral densities is invariant with respect to the phase characteristic of the linear system. In other words, it is impossible, in general, to restore the transfer function of the system by measuring the PSD of the input and the output. Such limitation is related to the fact that the PSD is defined through the averaging of the Fourier transform of the input process and does not completely represent the property of the linear system. However, in many cases the phase characteristic can also be restored if there is some additional information about the properties of the system [13].

Equations (4.136) and (4.137) can be simplified to produce the following expression for the variance and mutual covariance function in the stationary regime:

$$
D_\eta = \int_0^{\infty}\int_0^{\infty} h(s_1)h(s_2)C_{\xi\xi}(s_1 - s_2)ds_1 ds_2
\tag{4.142}
$$

$$
C_{\xi\eta}(\tau) = \int_0^{\infty} h(s)C_{\xi\xi}(\tau - s)ds
\tag{4.143}
$$

The last equation indicates that the input and the output process are stationary related.

Example: filtering of WGN. One of the most often considered problems is filtering of WGN by a filter with the impulse response $h(t)$ and the transfer function $H(j\omega)$. In this case the output process is a zero mean Gaussian process, since $m_\xi = 0$ in eq. (4.138). The PSD of the output process can be obtained from eq. (4.141) by noting that the PSD of WGN is a constant $N_0/2$:

$$
S_\eta(\omega) = |H(j\omega)|^2 S_\xi(\omega) = \frac{N_0}{2}|H(j\omega)|^2
\tag{4.144}
$$

The covariance function of the output process $\eta(t)$ can be obtained either by taking the inverse Fourier transform of $S_\eta(\omega)$ or by applying eq. (4.139) and taking into account the fact that the WGN is delta correlated

$$D_\eta = \int_0^\infty h(s)ds \int_0^{s+\tau} h(\tau+s-t)\frac{N_0}{2}\delta(t)dt = \frac{N_0}{2}\int_0^\infty h(s)h(\tau+s)ds \tag{4.145}$$

Example: sliding averaging of a stationary random process. Let the output of a linear system be given as an integral of the input process over a finite interval of time, so-called sliding averaging of the process

$$\eta(t) = \frac{1}{2\Delta}\int_{t-\Delta}^{t+\Delta}\xi(s)ds \tag{4.146}$$

The random process $\xi(t)$ is assumed to be wide sense stationary with the average m_ξ and the covariance function $C_{\xi\xi}(\tau)$. The mean value of the process $\eta(t)$ is thus

$$m_\eta = \frac{1}{2\Delta}\int_{t-\Delta}^{t+\Delta}E\{\xi(s)\}ds = m_\xi \tag{4.147}$$

The expression for the covariance function can be found to be

$$C_{\eta\eta}(\tau) = E\{(\eta(t)-m_\eta)(\eta(t+\tau)-m_\eta)\} = \frac{1}{4\Delta^2}\int_{t-\Delta}^{t+\Delta}\int_{t+\tau-\Delta}^{t+\tau+\Delta}C_{\xi\xi}(s_1-s_2)ds_1 ds_2 \tag{4.148}$$

Using the expression of the covariance function in terms of the PSD

$$C_{\xi\xi}(s_1-s_2) = \frac{1}{2\pi}\int_{-\infty}^\infty S_\xi(\omega)\exp[-j\omega(s_1-s_2)]d\omega \tag{4.149}$$

Equation (4.148) can be rewritten as

$$C_{\eta\eta}(\tau) = \frac{1}{2\pi}\frac{1}{4\Delta^2}\int_{-\infty}^\infty S_\xi(\omega)d\omega \int_{t-\Delta}^{t+\Delta}\exp[-j\omega s_1]ds_1 \int_{t+\tau-\Delta}^{t+\tau+\Delta}\exp[j\omega s_2]ds_2$$

$$= \frac{1}{2\pi}\int_{-\infty}^\infty S_\xi(\omega)\left(\frac{\sin\Delta\omega}{\Delta\omega}\right)^2\exp(j\omega\tau)d\omega \tag{4.150}$$

This also implies that the PSD $s_\eta(\omega)$ of the output signal is equal to

$$S_\eta(\omega) = \left(\frac{\sin\Delta\omega}{\Delta\omega}\right)^2 S_\xi(\omega) \tag{4.151}$$

Since the function $\sin\Delta\omega/\Delta\omega$ is concentrated around small values of the frequency $\Delta\omega < \pi$, the sliding averaging removes high frequency components of the signal spectrum, thus eliminating fast, noise-like components.

The same results can be obtained by noting that the impulse response of the system described by eq. (4.147) is

$$
h(t) = \begin{cases} \dfrac{1}{2\Delta} & |t| < \Delta \\[2mm] 0 & |t| > \Delta \end{cases}
\tag{4.152}
$$

and the corresponding transfer function is thus

$$
H(j\omega) = \frac{\sin \Delta\omega}{\Delta\omega}
\tag{4.153}
$$

Having this in mind, eq. (4.139) can be now written as

$$
C_{\eta\eta}(\tau) = \frac{1}{2\Delta} \int_{-2\Delta}^{2\Delta} \left[1 - \frac{|s|}{2\Delta} \right] C_{\xi\xi}(\tau - s) ds
\tag{4.154}
$$

Setting $\tau = 0$, the expression for the variance of the output process can be obtained as

$$
D_\eta = C_{\eta\eta}(0) = \frac{1}{2\Delta} \int_{-2\Delta}^{2\Delta} \left[1 - \frac{|s|}{2\Delta} \right] C_{\xi\xi}(s) ds
\tag{4.155}
$$

since $C_{\eta\eta}(\tau)$ is an even function.

A slight modification of eq. (4.147) shows that the variance of the integrated process

$$
\eta(t) = \frac{1}{T} \int_0^T \xi(s) ds
\tag{4.156}
$$

is given by

$$
D_\eta(T) = \frac{2}{T} \int_0^T \left(1 - \frac{\tau}{T} \right) C_{\xi\xi}(\tau) d\tau = \frac{2D_\xi}{T} \int_0^T \left(1 - \frac{\tau}{T} \right) r_{\xi\xi}(\tau) d\tau
\tag{4.157}
$$

This equation is instrumental in the ergodic theory of random processes. In particular, it allows one to find if the variance of the average process approaches zero, i.e. the values of the average converge to the value of the quantity under consideration. Equation (4.157) indicates that if the covariance function decays fast enough, the resulting process is ergodic in the wide sense.

   Example: reaction of an RC circuit to the WGN.   Let WGN drive an *RC* circuit shown in Fig. 4.4. The WGN $\xi(t)$, applied at the moment of time $t_0 = 0$, has constant PSD $N_0/2$ and the covariance function given by

$$
C_{\xi\xi}(\tau) = \frac{N_0}{2} \delta(\tau)
\tag{4.158}
$$

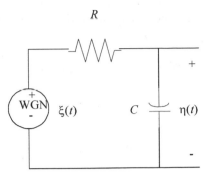

Figure 4.4   Reaction of RC circuit.

The output voltage $\eta(t)$ is measured across the capacitor C and it obeys the following differential equation

$$\frac{\partial}{\partial t}\eta(t) + \alpha\eta(t) = \alpha\xi(t), \quad \alpha = \frac{1}{RC} \tag{4.139}$$

The general solution, given the initial condition $\eta_0 = \eta(t_0)$ is well known and is given by the following equation

$$\eta(t) = \eta_0 \exp(-\alpha t) + \alpha \exp(-\alpha t) \int_0^t \exp(\alpha s)\xi(s)ds \tag{4.160}$$

Since the output process $\eta(t)$ is a Gaussian process for a fixed initial condition, it is only necessary to calculate its mean and covariance functions to completely describe properties of the process $\eta(t)$. In order to account for the random initial condition, $\eta(t_0) = \eta_0$, it is assumed that the distribution of the random variable η_0 is given by a known PDF $p_0(\eta)$. A deterministic initial condition can be considered as a particular case with $p_0(\eta) = \delta(\eta - \eta_0)$. The first step is to solve the problem for a deterministic η_0. In this case, taking the expectation of both sides of the solution (4.160), one obtains an expression for the mean value $m_\eta(t \mid \eta_0)$ of the output process $\eta(t)$ as

$$m_\eta(t \mid \eta_0) = \eta_0 \exp(-\alpha t) \tag{4.161}$$

since the average of the WGN is zero. Using the definition of the covariance function, one calculates

$$\begin{aligned}
C_{\eta\eta}(t_1, t_2) &= E\{[\eta(t_1) - m_\eta(t_1 \mid \eta_0)][\eta(t_2) - m_\eta(t_2 \mid \eta_0)]\} \\
&= \frac{\alpha^2 N_0}{2}\exp[-\alpha(t_1 + t_2)] \int_0^{t_1}\int_0^{t_2} \exp[\alpha(s_1 + s_2)](s_2 - s_1)ds_1 ds_2
\end{aligned} \tag{4.162}$$

Setting $t_2 = t_1 + \tau = t + \tau$ and using the sifting properties of the delta function, expression (4.162) becomes

$$C_{\eta\eta}(t, t + \tau) = \frac{N_0 \alpha}{4}\exp(-\alpha|\tau|)[1 - \exp(-2\alpha t)] \tag{4.163}$$

This immediately produces an expression for the variance of the output process as a function of time

$$D_\eta(t) = \frac{N_0 \alpha}{4}[1 - \exp(-2\alpha t)] \tag{4.164}$$

For the asymptotically stationary regime, i.e. for $t \to \infty$, eqs. (4.161) and (4.163)–(4.164) are reduced to

$$m_\xi(t \mid \eta_0) = 0, \quad D_\eta = \frac{N_0 \alpha}{4}, \quad C_{\eta\eta}(\tau) = \frac{N_0 \alpha}{4}\exp(-\alpha|\tau|) \tag{4.165}$$

Thus, it can be seen that the stationary mean and variance do not depend on the value of the initial condition. However, the transient characteristics do depend on η_0. In order to account for the effect of random initial conditions, one can split solution (4.160) into two parts $\eta(t) = \eta_1(t) + \eta_2(t)$ where

$$\eta_1(t) = \eta_0 \exp(-\alpha t), \text{ and } \eta_2(t) = \alpha \exp(-\alpha t) \int_0^t \exp(\alpha s)\xi(s)ds \tag{4.166}$$

The process $\eta_2(t)$ is a linear transformation of a Gaussian process and thus is a Gaussian process itself. It is easy to see from the analysis above that this process has zero mean and the covariance function defined by eq. (4.165). The process η_1 has the PDF defined by

$$p_{\eta_1}(x) = \exp(\alpha t)p_0[\exp(\alpha t)x] \tag{4.167}$$

which follows the shape of the PDF of the initial condition but has a different scale, which varies in time. The mean and the variance are given by

$$m_1(t) = m_0 \exp(-\alpha t), \quad D_1(t) = D_0 \exp(-2\alpha t) \tag{4.168}$$

respectively. Here m_0 and D_0 are the mean and the variance of the initial condition η_0. The resulting distribution can be found as a composition of the distributions of $\eta_1(t)$ and $\eta_2(t)$.

It is important to mention here that the situation described above is common for all linear systems driven by a Gaussian process: the stationary solution does not depend on the initial state of the system and it is a Gaussian process.

Example: derivative of a Gaussian process. Properties of the derivative $\eta(t) = \xi'(t)$ of a random process $\xi(t)$ have been considered above in section 4.1.2. To complete the considerations, the spectral domain analysis is presented here. Since the transfer function of the ideal differentiator is given by

$$H(j\omega) = j\omega \tag{4.169}$$

the expression for the PSD $S_\eta(\omega)$ of the output process can be written as

$$S_\eta(\omega) = |H(j\omega)|^2 S_\xi(\omega) = \omega^2 S_\xi(\omega) = -(j\omega)^2 S_\xi(\omega) \tag{4.170}$$

It is easy to see that this condition is equivalent to eq. (4.10) derived by different means.

4.5.2 Analysis of Linear MIMO Systems

Analysis of MIMO systems can be conducted within the same framework which was used in the previous section. In general, it is impossible to provide an expression for joint probability densities of the output vector processes, but it is feasible to obtain expressions for moments and cumulants. Let $\xi_n(t)\,1 \leq n \leq N$ be a random process acting at the n-th input of an N inputs, M outputs MIMO system. If the impulse response between the n-th input and m-th output is denoted by $h_{nm}(t)$, the response $\eta_m(t)$ of the system at the m-th output is then given by

$$\eta_m(t) = \sum_{n=1}^{N} \int_0^t h_{nm}(\tau_n)\xi_n(t - \tau_n)\mathrm{d}\tau_n \tag{4.171}$$

Thus, the average value $m_{\eta_m}(t)$ of the m-th output and the mutual correlation function $R_{\eta_{m_1},\eta_{m_2}}(t_1, t_2)$ between two outputs can be calculated as follows

$$m_{\eta_m}(t) = \sum_{n=1}^{N} \int_0^t h_{nm}(\tau_n)m_{\xi_n}(t - \tau_n)\mathrm{d}\tau_n \tag{4.172}$$

and

$$R_{\eta_{m_1}\eta_{m_2}}(t_1, t_2) = E\{\eta_{m_1}(t)\eta_{m_2}(t)\}$$
$$= \sum_{n_2=1}^{N} \sum_{n_1=1}^{N} \int_0^{t_1} \int_0^{t_2} h_{m_1 n_1}(\tau_{n_1})h_{m_2 n_2}(\tau_{n_2})R_{\eta_{n_1}\eta_{n_2}}(t_1 - \tau_{n_1}, t_2 - \tau_{n_2})\mathrm{d}\tau_{n_1}\mathrm{d}\tau_{n_2} \tag{4.173}$$

The same equation will hold for the centred processes and, thus, for the mutual covariance functions $C_{\eta_{m_1}\eta_{m_2}}(t_1, t_2)$:

$$C_{\eta_{m_1}\eta_{m_2}}(t_1, t_2) = \sum_{n_2=1}^{N} \sum_{n_1=1}^{N} \int_0^{t_1} \int_0^{t_2} h_{m_1 n_1}(\tau_{n_1})h_{m_2 n_2}(\tau_{n_2})C_{\eta_{n_1}\eta_{n_2}}(t_1 - \tau_{n_1}, t_2 - \tau_{n_2})\mathrm{d}\tau_{n_1}\mathrm{d}\tau_{n_2} \tag{4.174}$$

If the input signals are stationary related, the expression for the mutual covariance function is somewhat simplified to read

$$C_{\eta_{m_1}\eta_{m_2}}(t_1, t_2) = \sum_{n_2=1}^{N} \sum_{n_1=1}^{N} \int_0^{t_1+\tau} \int_0^{t_1} h_{m_1 n_1}(\tau_{n_1})h_{m_2 n_2}(\tau_{n_2})C_{\eta_{n_1}\eta_{n_2}}(\tau - \tau_{n_1} + \tau_{n_2})\mathrm{d}\tau_{n_1}\mathrm{d}\tau_{n_2} \tag{4.175}$$

Finally, if $t_1 \to \infty$, the expression for the asymptotically stationary mutual covariance function becomes

$$C_{\eta_{m_1}\eta_{m_2}}(\tau) = \sum_{n_2=1}^{N} \sum_{n_1=1}^{N} \int_0^{\infty} \int_0^{\infty} h_{m_1 n_1}(\tau_{n_1})h_{m_2 n_2}(\tau_{n_2})C_{\eta_{n_1}\eta_{n_2}}(\tau - \tau_{n_1} + \tau_{n_2})\mathrm{d}\tau_{n_1}\mathrm{d}\tau_{n_2} \tag{4.176}$$

4.5.3 Cumulant Method of Analysis of Linear Transformations

It has been shown in section 4.2.2 that a transformation of a random process $\xi(t)$ by a linear system with constant parameters can be described by linear differential equations of the form

$$N(p)\eta(t) = M(p)\xi(t), \quad N(p) = \sum_{k=0}^{n} a_k p^k, \quad M(p) = \sum_{k=0}^{m} b_k p^k \tag{4.177}$$

Here $N(p)$ and $M(p)$ are polynomials of order n and m of the differential operators. If $y(t)$ and $x(t)$ are deterministic signals then an auxiliary variable $z(t)$, defined as

$$z(t) = N(p)y(t) = M(p)x(t) = \sum_{k=0}^{n} a_k \frac{d^k}{dt^k} y(t) = \sum_{k=0}^{m} b_k \frac{d^k}{dt^k} x(t) \tag{4.178}$$

is also a deterministic signal. It is known from the theory of ordinary differential equations that the solution of eq. (4.177) is a sum of a general solution $y_0(t)$ of the homogeneous equation $z(t) = 0$ and a particular (forced) solution of the non-homogeneous equation $y_p(t)$:

$$y(t) = y_0(t) + y_p(t) \tag{4.179}$$

If the linear system is stable[9], then the part of the response $y_0(t)$ induced by a non-zero initial condition $y_0 = y(t_0)$ vanishes after a period of time greater than the characteristic time constant τ_s of the system. Thus, for $t - t_0 \gg \tau_s$, the output of the system consists only of the forced solution

$$y(t) = y_p(t) = \int_0^{\infty} z(t - \tau)h(\tau)d\tau \tag{4.180}$$

Here $h(t)$ is the impulse response of the system $z(t) = N(p)y(t)$. The function $h(t)$ is also Green's function of the operator $N(p)$, since it satisfies the condition $\delta(t) = N(p)y(t)$. Thus, the operator (4.180) describes a stationary regime of a process transformed by a linear system.

By the same argument, the stationary solution of the operator (4.177) can be written as

$$y(t) = \int_0^{\infty} h_M(\tau)x(t - \tau)d\tau \tag{4.181}$$

where

$$h_M(t) \equiv M\frac{d}{dt}h(t) = \sum_{k=0}^{m} b_k \frac{d^k}{dt^k} h(t) \tag{4.182}$$

if $m < n$. At this moment, it is important to mention that introduction of a auxiliary function $z(t)$ does not play a significant role in the derivation of eqs. (4.181)–(4.182). However, it will be extremely useful in the calculation of cumulants of the transformed process.

---

[9] See [5] for discussion of different criteria of stability.

The correlation function of the systems (4.177)–(4.178) can be defined as

$$R_h(\tau) = \int_{-\infty}^{\infty} h(t)h(t + \tau)d\tau$$

$$R_{h_M}(\tau) = \int_{-\infty}^{\infty} h_M(t)h_M(t + \tau)dt$$

(4.183)

respectively. These and similar characteristics are often used to describe communication channels [4,17]. Similarly, moment functions $m_k^h(\tau_1, \tau_2, \ldots \tau_s)$ of linear systems can be defined as

$$m_k^h(\tau_1, \tau_2, \ldots \tau_k) \equiv \int_{-\infty}^{\infty} h(t)h(t + \tau_1)h(t + \tau_2) \cdots h(t + \tau_s)dt$$

(4.184)

The next step is to quantitatively define the characteristic time τ_s. While this definition is not unique, here the energy-based definition is considered:

$$\tau_s = \frac{1}{h_0^2} \int_0^{\infty} h^2(t)dt$$

(4.185)

where $h_0 = \max[h(t)]$. Similarly, the correlation interval τ_h can be defined as

$$\tau_h = \frac{1}{R_h^2(0)} \int_0^{\infty} R_h^2(\tau)d\tau$$

(4.186)

Instead of considering a time-domain description of a linear system, one can use its description in the frequency domain. In this case, the system is completely described by its transfer function $H(j\omega)$, which, in turn, is the Fourier transform of the impulse response $h(t)$

$$H(j\omega) = \int_{-\infty}^{\infty} h(t) \exp(-j\omega t)dt$$

(4.187)

The transfer function completely describes the system response in the steady state; however, some additional information (such as initial conditions) is needed to describe the reaction to non-zero initial conditions.

Assuming that the transfer function is known, the next task is to consider calculation of the moment and cumulant functions. At this point, it is assumed that the moment and cumulant functions $m_s^\zeta(t_1, t_2, \ldots t_s)$, $\kappa_s^\zeta(t_1, t_2, \ldots t_s)$ of the auxiliary process $\zeta(t) = N(p)\xi(t)$ are also known. If the system has deterministic non-zero initial conditions, then it will produce a deterministic relaxation process[10]. The random component is thus only a result of the external driving force. Thus, in both stationary and non-stationary cases, the random output signal $\eta(t)$ can be represented as

$$\eta(t) = \int_0^t h(\tau)\zeta(t - \tau)d\tau$$

(4.188)

---

[10] This deterministic process would contribute only to the first moment (cumulant) $m_1^\eta(t) = \kappa_1^\eta(t)$.

The latter allows one to immediately write expressions for the moment and cumulant functions, $m_s^\eta(t)$ and $\kappa_s^\eta(t)$, of the marginal distribution. Indeed, using the definition of the moments it is possible to write

$$m_s^\eta(t_1, t_2, \ldots, t_s)$$

$$= E\{\eta(t_1)\eta(t_2) \cdots \eta(t_s)\}$$

$$= E\left\{\int_0^{t_1} h(\tau_1)\zeta(t - \tau_1)d\tau_1 \int_0^t h(\tau_2)\zeta(t - \tau_2)d\tau_2 \cdots \int_0^t h(\tau_s)\zeta(t - \tau_s)d\tau_s\right\}$$

$$= \int_0^\infty \cdots \int_0^\infty h(\tau_1)h(\tau_2) \cdots h(\tau_s)m_s^\zeta(t_1 - \tau_1, t_2 - \tau_2, \ldots, t_s - \tau_s)d\tau_1 d\tau_2 \cdots d\tau_s \qquad (4.189)$$

The cumulant functions can be found in many ways. First of all one can calculate them from the moments, obtained through the use of eq. (4.189), using relations between cumulants and moments. This works well for the lower order cumulants. In the general case, it is more convenient to use the relation between moment and cumulant brackets, as discussed in [15]. This approach produces the following set of equations

$$\kappa_1^\eta(t_1) = \int_0^{t_1} h(\tau_1)\kappa_1^\zeta(t_1 - \tau_1)d\tau_1 \qquad (4.190)$$

$$\kappa_2^\eta(t_1, t_2) = \int_0^{t_1}\int_0^{t_2} h(\tau_1)h(\tau_2)\kappa_2^\zeta(t_1 - \tau_1, t_2 - \tau_2)d\tau_1 d\tau_2 \qquad (4.191)$$

$$\kappa_s^\eta(t_1, t_2, \ldots, t_s) = \int_0^{t_1} \cdots \int_0^{t_s} h(\tau_1)h(\tau_2)\ldots h(\tau_s)\kappa_s^\zeta(t_1 - \tau_1, t_2 - \tau_2, \ldots, t_s - \tau_s)d\tau_1 d\tau_2 \ldots d\tau_s$$

$$(4.192)$$

It follows from eqs. (4.189) and (4.192) that the relationship between moment and cumulant functions at the input and the output of a linear system are linear and inertial. This means that the values of the cumulants κ_s^η of the marginal PDF depend only on the cumulant of the same order but considered in separate time instants. Indeed, setting all the limits of integration to a fixed moment of time $t_k = t$, one obtains

$$\kappa_s^\eta(t) = \int_0^t \cdots \int_0^t h(\tau_1)h(\tau_2) \cdots h(\tau_s)\kappa_s^\zeta(t - \tau_1, t - \tau_2, \ldots, t - \tau_s)d\tau_1 d\tau_2 \cdots d\tau_s \qquad (4.193)$$

thus indicating non-zero memory of linear systems. This leads to a significant change of the PDF of the process $\eta(t)$ compared to the PDF of the input signal $\zeta(t)$ with only one remarkable exception[11]: linear transformation of a Gaussian random process remains Gaussian! This follows from the fact that all higher order cumulant functions of the input process are zero. In general, if some cumulant function is zero for the process $\zeta(t)$, the cumulant function of the same order of the process $\eta(t)$ is also equal to zero. This constitutes so-called cumulant invariance of linear transformation [15] and emphasizes the importance of the model approximation: if only a first few cumulants are sufficiently non-zero (two for

[11]This is also true for so-called Spherical Invariant Processes considered in Section 4.8.

the Gaussian approximation and four for the model approximation), the same would be valid for the output process, i.e. both these models are sufficient to describe both the input and the output process. However, the cumulant functions themselves can be changed significantly, as will be seen from the example below.

A stationary regime is obtained from eq. (4.189) by setting $t \to \infty$. In this case

$$\kappa_s^\eta(0, \tau_2, \tau_3, \ldots, \tau_s)$$
$$= \int_0^\infty \cdots \int_0^\infty h(u_1) \cdots h(u_s) \kappa_s^\zeta(0, \tau_2 - u_2 + u_1, \ldots, \tau_s - u_s + u_1) du_1 \cdots du_s \quad (4.194)$$

A similar expression can be derived for the moment functions $m_s^\eta(0, \tau_2, \tau_3, \ldots, \tau_s)$. In order to obtain particular results, one has to specify the structure of the input random process. This will be considered below.

It was assumed in eqs. (4.192)–(4.194) that the characteristics of the auxiliary process $\zeta(t)$ are known. However, it is worth recalling that

$$\zeta(t) = M\left(\frac{d}{dt}\right)\xi(t) = \sum_{k=1}^m b_k \frac{d^k}{dt^k}\xi(t) \qquad (4.195)$$

That is, the properties of the processes $\eta(t)$ and $\zeta(t)$ should be expressed in terms of cumulant and moment functions of the input process $\xi(t)$. This can be easily done if one recalls that the order of operations of statistical averaging and differentiation with respect to time is interchangeable. Having this in mind, the following relation can be derived

$$m_s^\zeta(t_1, t_2, \ldots, t_s) = \langle \zeta(t_1) \cdot \zeta(t_2) \cdots \zeta(t_s) \rangle$$
$$= \left\langle M\left(\frac{\partial}{\partial t_1}\right)\zeta(t_1) M\left(\frac{\partial}{\partial t_2}\right)\zeta(t_1) \cdots M\left(\frac{\partial}{\partial t_s}\right)\zeta(t_s) \right\rangle$$
$$= M\left(\frac{\partial}{\partial t_1}\right) \cdots M\left(\frac{\partial}{\partial t_s}\right) \langle \zeta(t_1)\zeta(t_2) \cdots \zeta(t_s) \rangle$$
$$= \prod_{i=1}^s M\left(\frac{\partial}{\partial t_i}\right) m_s^\xi(t_1, \ldots, t_s) \qquad (4.196)$$

Thus, the moments of the auxiliary process $\zeta(t)$ are the partial derivatives of the corresponding moments of the input random process. The same can be derived for the cumulants

$$\kappa_s^\zeta(t_1, t_2, \ldots, t_s) = \prod_{i=1}^s M\left(\frac{\partial}{\partial t_i}\right) \kappa_s^\xi(t_1, \ldots, t_s) \qquad (4.197)$$

Finally, substituting eqs. (4.196) and (4.197) into eqs. (4.189) and (4.192) one obtains the desired expressions for the moments and the cumulants of the output process $\eta(t)$ in terms of the moments and cumulants of the input process $\xi(t)$. It can be seen that if a cumulant of the input process $\xi(t)$ is equal to zero, so will be the corresponding cumulant of the auxiliary process $\zeta(t)$. Thus the cumulant invariance is preserved for the process $\zeta(t)$. If the input process $\xi(t)$ is Gaussian, then the output process is a Gaussian process as well and it is sufficient to find only second order cumulants.

Example: response of a linear circuit to a Poisson pulse. Poisson pulse flow is described by a flow of pulses of a deterministic or random shape. Time instants of the impulse appearance obey the Poisson distribution: the probability that there are exactly N pulses on the interval of duration t is given by

$$P_n = \frac{(\nu t)^N}{N!} \exp(-\nu t) \tag{4.198}$$

Here, ν is the average number of pulses per unit interval. Thus, the Poisson pulse flow can be represented as

$$\xi(t) = \sum_{i=-\infty}^{\infty} a_i F_i(t - t_i) \tag{4.199}$$

where $F_i(t)$ is the so-called elementary pulse (random or deterministic) and a_i is the pulse magnitude (random or deterministic). It is shown in [15] that the cumulants of this process are given by

$$\kappa_s^\xi(0, \tau_2, \ldots, \tau_s) = \nu \langle a^s \rangle \int_{-\infty}^{\infty} F(u) \, F(u + \tau_2) \ldots F(u + \tau_s) \mathrm{d}u \tag{4.200}$$

Now, the problem is reduced to calculation of the cumulants of the output process using the relation (4.193). If the linear system has the impulse response $h(t)$, then its response to an elementary pulse is

$$I(t) = \int_{-\infty}^{\infty} h(\tau) \, F(t - \tau) \mathrm{d}\tau \tag{4.201}$$

It can be clearly seen that, using this notation, eq. (4.193) for the output cumulants can be rewritten as

$$\kappa_s^\eta(0, \tau_1, \ldots, \tau_s) = \nu \langle a^s \rangle \int_{-\infty}^{\infty} I(u)I(u + \tau_1) \cdots I(u + \tau_s) \mathrm{d}u \tag{4.202}$$

Thus, it can be deducted by comparing eqs. (4.200) and (4.202), that these equations are identical in structure and the only difference is that instead of the pulse shape function $F(t)$, the response waveform $I(t)$ of the linear system to the pulse function $F(t)$ is used. This implies that the output process is also a Poisson process defined as

$$\eta(t) = \sum_{i=-\infty}^{\infty} a_i I(t - t_i) \tag{4.203}$$

The cumulant functions κ_s^η of the output process $\eta(t)$ are thus given by eq. (4.202). Setting all the lag variables to zero, $\tau_i = 0$, one obtains

$$\begin{aligned}
\kappa_s^\xi &= \nu \langle a^s \rangle \int_{-\infty}^{\infty} F^s(t) \mathrm{d}t \\
\kappa_s^\eta &= \nu \langle a^s \rangle \int_{-\infty}^{\infty} I^s(t) \mathrm{d}t
\end{aligned} \tag{4.204}$$

A characteristic time constant for the cumulant function $\kappa_s^\eta(\bullet)$ can be introduced in the very same manner as it was introduced for the second order cumulant function. Specifically, they can be defined by the following equations

$$T_s^\xi = \frac{1}{F_0^2} \int_{-\infty}^\infty F^s(t)dt$$
$$T_s^\eta = \frac{1}{I_0^2} \int_{-\infty}^\infty I^s(t)dt$$

(4.205)

where $F_0 = \max\{F(t)\}$ and $I_0 = \max\{I(t)\}$. Using this notation, the corresponding cumulants can be written as

$$\kappa_s^\xi = \nu\langle a^s\rangle F_0^s T_s^\xi, \quad \kappa_s^\eta = \nu\langle a^s\rangle I_0^s T_s\eta$$

(4.206)

It is clear that eq. (4.206) can be used to define normalized cumulant coefficients γ_s, similarly to the definition introduced in section 2.2. It also can be shown that the relation between the input cumulant coefficients γ_s^{in} and the output cumulant coefficient γ_s^{out} under the assumption that $\langle a\rangle = 0$

$$\gamma_s^{out} = \gamma_s^{in} \frac{T_s^\eta [T_2^\eta]^{\frac{s}{2}}}{[T_2^\eta]^{\frac{s}{2}} T_s^\xi}$$

(4.207)

It is shown in [15] that

$$\gamma_{2s}^{out} \approx \gamma_2^{in} \left[\frac{T_2^\xi}{T_2^\eta}\right]^{s-1}$$

(4.208)

and

$$\gamma_{2s-1}^{out} \approx \gamma_{2s-1}^{in} \frac{T_1^\eta}{T_1^\xi} \left[\frac{T_2^\xi}{T_2^\eta}\right]^{s-\frac{1}{2}}$$

(4.209)

It can be seen by inspection of expression (4.207) that the main factor contributing to change of the PDF of the output process is the ratio between duration of pulses $F(t)$ and $I(t)$. Since this ratio is defined by the length of the impulse response of the linear system under consideration, one can conclude that the systems with long memory (or, equivalently, with long response time) would significantly change the PDF of the output process.

4.5.4 Normalization of the Output Process by a Linear System

In this section, it is shown that a linear system tends to normalize the output process, independently of the distribution of the process at the input. For simplicity, the case of a symmetric distribution of pulses is considered. This implies that all odd moments of a_i are equal to zero, i.e.

$$\langle a^{2s+1} \rangle = 0 \tag{4.210}$$

In turn, eq. (4.210) implies that all odd moments of the output process are also equal to zero: $\langle \eta^{2s+1} \rangle = 0$ and the output PDF is also a symmetric curve. In this case, it is possible to rewrite eqs. (4.208)–(4.209) to obtain the following expression for the even cumulant coefficients

$$\gamma_{2s}^{\text{out}} = \gamma_{2s}^{\text{in}} \left(\frac{T_2^{\xi}}{T_2^{\eta}} \right)^{s-1} \tag{4.211}$$

keeping in mind that the odd coefficients are equal to zero: $\gamma_{2s+1}^{\text{out}} = 0$. Since, for any linear system with memory, the impulse response has non-zero duration, then $T_2^{\eta} < T_2^{\xi}$, which immediately implies that the relative weight of the higher order cumulants decreases with number:

$$\gamma_{2s}^{\text{out}} < \gamma_{2s}^{\text{in}}, \frac{\gamma_{2s}^{\text{out}}}{\gamma_{2s+2}^{\text{out}}} = \frac{\gamma_{2s}^{\text{in}}}{\gamma_{2s+2}^{\text{in}}} \cdot \left(\frac{T_2^{\xi}}{T_2^{\eta}} \right)^2 \leq \frac{\gamma_{2s}^{\text{in}}}{\gamma_{2s+2}^{\text{in}}} \tag{4.212}$$

This fact shows that the linear system with memory and $T_2^{\eta} > T_2^{\xi}$ suppresses higher order cumulants[12], i.e. it normalizes the output. If the memory of the linear system is long, i.e. $T_2^{\eta} \gg T_2^{\xi}$, the output process is certainly very close to a Gaussian random process.

The normalization property is clearly related to the scale of the time constant τ_s of the linear system with respect to the correlation interval of the input signal if $\tau_s \gg T_2^{\xi}$. Then, during the correlation interval of the input signal the impulse response of a linear system remains virtually unchanged, and, thus

$$I(t) = \int_0^{\infty} h(\tau) F(t - \tau) \mathrm{d}\tau = h(t) \int_0^{\infty} F(u) \mathrm{d}u = h(t) A \tag{4.213}$$

and

$$T_2^{\eta} = \frac{1}{E^2 h_0^2} \int_{-\infty}^{\infty} A^2 h^2(t) \mathrm{d}t = \tau_s \gg T_2^{\xi} \tag{4.214}$$

Here $E = \int_0^{\infty} F^2(t) \mathrm{d}t$ is the energy of a single pulse. Thus, the bigger τ_s is with respect to T_2^{ξ} the bigger the normalization of the output process is as well. This statement can also be recast in terms of the effective bandwidth of the linear system

$$\Delta f_{\text{eff}} = \frac{\int_0^{\infty} |H(j\omega)|^2 \frac{\mathrm{d}\omega}{2\pi}}{K_0^2} \tag{4.215}$$

---

[12]Since the duration of pulses is defined in an equal area then it is possible that a linear system with memory would shorten the duration of the pulse defined in an equal area sense.

Since in all of the physical systems the following relationship is valid [19]

$$T_s \sim \frac{1}{\Delta f_{\text{eff}}} \tag{4.216}$$

it follows from eq. (4.214) that a deeper normalization is achieved by a system with a narrower bandwidth.

Usually the normalization property of a linear system is explained in terms of the Central Limit Theorem [1]. Indeed, the number of impulses exciting the system during a time interval T_2^{η} is on average equal to νT_2^{η}. During the same period of time, the input process can be represented as a sum of T_2^{ξ} terms on average. If $T_2^{\eta} \gg T_2^{\xi}$, and $\nu T_2^{\eta} \gg 1$, then the conditions of the Central Limit Theorem are satisfied and the output process approaches a Gaussian one.

Example: normalization of a process by a linear RC circuit. The impulse response of a linear *RC* circuit is an exponential function

$$h(t) = e^{-\alpha t}, \quad t \geq 0 \tag{4.217}$$

where

$$\alpha = \Delta f_{\text{eff}} = \frac{1}{RC} \tag{4.218}$$

is the effective bandwidth of the filter. It is assumed that $\langle a_n \rangle = 0$ and the excitation pulses in eq. (4.199) are the delta functions, i.e.

$$F(t) = \delta(t) \tag{4.219}$$

It is known that in this case, the cumulants of the input process are delta functions as well [15]

$$\kappa_1^{\xi}(t) = 0, \kappa_s^{\xi}(t_1, t_2, \dots, t_s) = D_s \delta(t_2 - t_1) \cdots \delta(t_2 - t_1) \tag{4.220}$$

where

$$D_s = \langle a^s \rangle \tag{4.221}$$

Taking into account that all the cumulants are either zero or delta functions, the integration of eq. (4.202) results in

$$\kappa_1^{\eta} = 0, \quad \text{and} \quad \kappa_s^{\eta} = \frac{D_s}{2\alpha s}, \quad s = 2, 3, \dots \tag{4.222}$$

It seems that there is no normalization, since all the cumulants are still non-zero, while the time constant $\tau_s = 1/\alpha$ is much greater than $T_2^{\xi} = 0$. However, the cumulants of the filtered process are finite, while the cumulants of the original process are equal to infinity. It is also characteristic that the normalized magnitude of cumulants decays with order. Thus, it could be claimed that the filtered process $\eta(t)$ is indeed closer to Gaussian than the original one

$$\gamma_s^\eta = \frac{(2\alpha)^{s/2} D_s}{s\alpha D_d^{s/2}}, \quad s \geq 3 \tag{4.223}$$

4.6 OUTAGES OF RANDOM PROCESSES

4.6.1 General Considerations

The problem of a boundary attainment existed for a relatively long period of time. One of the first results in this area was the work of A. Pontryagin [20], who considered calculations of the distribution of the first passage time by a Markov random process, and the pioneering work of S.O. Rice [21,22]. Essentially, the latter laid a foundation for the applied analysis of the characteristics of random processes related to a level crossing problem. A comprehensive review of the recent advances in the theory can be found in [20,23]. In particular, level crossing problems have a great importance in communication problems. The following are just a few examples from a long list: calculation of the average duration of the outage of the fading carrier in channels with fading [24], loss of synchronization in phase-lock loops [25], the average level crossing rate and fade duration in channels with fading [26], evaluation of the average length of bursts of errors in discrete communication channels [27], etc.

Let $\xi_1(t)$ be a realization of a random continuous process $\xi(t)$ defined on a finite time interval $[0, T]$ and $\xi_{min} < H < \xi_{max}$ be a fixed (for now) level, as shown in Fig. 4.5.

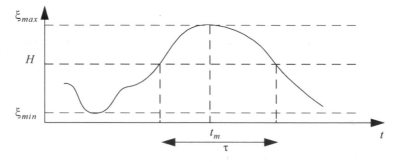

Figure 4.5 Definition of the first passage time and the average level crossing rate.

Let τ be the duration of the segment of the realization $\xi_1(t)$ which exceeds the chosen level H. This interval of time is called a duration of the upward excursion, and the portion of the trajectory above the threshold H is called a positive pulse. Similarly, the duration downward excursion θ is defined as the duration of the portion of the trajectory below the threshold H (negative pulse). In addition, let t_m be the moment of time when the trajectory achieves its maximum value ξ_{max}. Finally, let n be the number of pulses (both negative and positive).

Distributions of the number of extrema and related statistics are studied in extreme value theory [26] and are not considered in this book. Here, the main focus of discussion is kept on the evaluation of some statistical characteristics of the random variables τ, θ and n due to their applied importance.

4.6.2 Average Level Crossing Rate and the Average Duration of the Upward Excursions

Let random process $\xi(t)$ be differentiated in a mean square sense (see Section 4.1.2 for more details). It is also assumed for now that the joint probability density $p_{\xi\dot{\xi}}(x, \dot{x}, t)$ of the process $\xi(t)$ and its mean square sense derivative $\dot{\xi}(t) = d\xi(t)/dt$ at the same moment of time are also known. Details of its evaluation from the joint PDF are considered below. The process $\xi(t)$ will cross the level H in the vicinity $[t_i - \Delta t/2, t_i + \Delta t/2]$ of the moment of time t_i if the following condition is met

$$\left[\xi\left(t_i - \frac{\Delta t}{2} \right) - H \right]\left[\xi\left(t_i + \frac{\Delta t}{2} \right) - H \right] < 0 \qquad (4.224)$$

This case is illustrated in Fig. 4.6. For now, the consideration is limited to the upward excursion, i.e. process $\zeta(t)$ is assumed to cross the threshold H from below. In this case, the

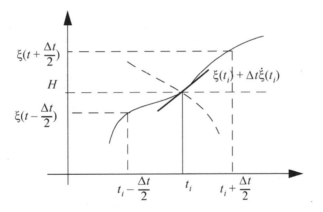

Figure 4.6 Calculation of the LCR.

derivative $\dot{\xi}(t)$ of the process must be positive, i.e. $\dot{\xi}(t_i) > 0$. This is also equivalent to the condition that $\xi(t - \Delta t/2) < H$ and $\xi(t + \Delta t/2) > H$. Let the interval $[0, T]$ be split into m non-intersecting sub-intervals of equal duration Δt. The number m is to be chosen in such a way that there is no more than one upward excursion on each of the intervals. In other words, the length of the interval Δt is so short that the probability of more than one intersection is much smaller than the probability of a single upward excursion [21]. Inside this small interval, a continuous process $\xi(t)$ can be approximated as

$$\xi^{(i)}(t') \approx \xi^{(i)}(t) + \dot{\xi}^{(i)}(t)(t' - t), \qquad t < t' < t + \Delta t \qquad (4.225)$$

Here $\xi^{(i)}(t)$ is the i-th realization of the random process $\xi(t)$. Thus, on the interval $[t, t + \Delta t]$ there are two possibilities: there is, with the probability q_0, a level crossing inside this interval, or, with the probability $p_0 = 1 - q_0$, there is none. Having this in mind, a binary sequence of m pulses $\delta_1^{(i)}, \delta_2^{(i)}, \ldots, \delta_m^{(i)}$, where $\delta_j^{(i)} = 1$ if $\dot{\xi}(\Delta t_j) > 0$ and $\delta_j^i = 0$ otherwise, can be constructed. Thus, the number of upward excursions on the i-th realization can be calculated as

$$n^+(H, T) = \sum_{i=1}^{m} \delta_j^{(i)} \tag{4.226}$$

Since this number varies from realization to realization, it is more appropriate to calculate the average number of level crossings as

$$N^+ = \langle n^+(H, T) \rangle = \sum_{i=1}^{m} \langle \delta_j^{(i)} \rangle \tag{4.227}$$

The average value of an individual pulse can be calculated using the probability of each pulse. This produces

$$\langle \delta_j^{(i)} \rangle = 0 \cdot \text{Prob}\{\delta_j^i = 0\} + 1 \cdot \text{Prob}\{\delta_j^i = 1\} = q_0 \tag{4.228}$$

Thus, the problem of finding an average number of upward excursions is reduced to finding the probability q_0. It can be seen by inspection of Fig. 4.6 and eq. (4.224) that the conditions of the appearance of the positive pulse can be rewritten on the interval $[t_j, t_j + \Delta t_j]$ as

$$H - \Delta\xi_j < \xi(t_j) \leq H, \text{ and } \dot{\xi}(t_j) > 0 \tag{4.229}$$

where

$$\Delta\xi_j = \dot{\xi}(t_j)\Delta t_j \tag{4.230}$$

This immediately implies that

$$q_0 = \text{Prob}\{H - \Delta\xi_j < \xi(t_j) \leq H, \quad \dot{\xi}(t_j) > 0\} \tag{4.231}$$

Assuming that $\Delta t_j, \Delta\xi_j \to 0$ and the derivative of the process varies in a narrow interval $d\dot{x}$ around a fixed value $\dot{x}$, it is possible to approximate the right-hand side of eq. (4.231) by

$$\text{Prob}(H - \Delta\xi_j < \xi_j \leq H, \dot{x} < \dot{\xi}(t_j) \leq \dot{x} + d\dot{x}) \approx p_{\xi\dot{\xi}}(H, \dot{x}; t_j)\Delta\xi_j d\dot{x} \approx p_{\xi\dot{\xi}}(H, \dot{x}; t_j)\Delta t_j \dot{x} dx \tag{4.232}$$

where $p_{\xi\dot{\xi}}(H, \dot{x}; t_j)$ is the joint PDF of the process $\xi(t)$ and its time derivative. Integrating eq. (4.232) over all possible positive slopes (only such trajectories will be able to achieve the boundary), one concludes that

$$q_0 = \Delta t \int_0^{\infty} \dot{x} p_{\xi\dot{\xi}}(H, \dot{x}; t) d\dot{x} \tag{4.233}$$

Finally, the average number of positive pulses N^+ can be obtained by setting the number of intervals $m \to \infty$ and adding contributions from each interval:

$$N^+ = \int_0^T dt \int_0^\infty \dot{x} p_{\xi\dot{\xi}}(H, \dot{x}; t) d\dot{x} \tag{4.234}$$

In the very same manner, the average number N^- of negative pulses is

$$N^- = -\int_0^T dt \int_{-\infty}^0 \dot{x} p_{\xi\dot{\xi}}(H, \dot{x}; t) d\dot{x} \tag{4.235}$$

Ultimately, the average number of level crossings is then

$$N = N^+ + N^- = \int_0^T dt \int_{-\infty}^\infty |\dot{x}| p_{\xi\dot{\xi}}(H, \dot{x}; t) d\dot{x} \tag{4.236}$$

Furthermore, expressing the joint probability density $p_{\xi\dot{\xi}}(H, \dot{x}; t)$ in terms of conditional density $p_{\dot{\xi}|H}(\dot{x}; t \mid H, t)$ as

$$p_{\xi\dot{\xi}}(H, \dot{x}; t) = p_{\dot{\xi}|H}(\dot{x}; t \mid H, t) p_\xi(H, t) \tag{4.237}$$

expression (4.236) can be rewritten as

$$N = \int_0^T p_\xi(H, t) dt \int_{-\infty}^\infty |\dot{x}| p_{\dot{\xi}|H}(\dot{x}; t \mid H, t) d\dot{x} \tag{4.238}$$

These equations were first obtained in [21,22].

If the process under consideration is a stationary process then its joint PDF does not depend on time, i.e.

$$p_{\xi\dot{\xi}}(x, \dot{x}; t) = p_{\xi\dot{\xi}}(x, \dot{x}) = p_{\dot{\xi}|H}(\dot{x} \mid x) p_\xi(x) \tag{4.239}$$

In this case, eq. (4.238) can be further simplified to

$$N = \int_0^T p_\xi(H) dt \int_{-\infty}^\infty |\dot{x}| p_{\dot{\xi}|H}(\dot{x} \mid x) d\dot{x} = T p_\xi(H) \int_{-\infty}^\infty |\dot{x}| p_{\dot{\xi}|H}(\dot{x} \mid x) d\dot{x} \tag{4.240}$$

Dividing both parts of the latter by the duration of the observation interval T, the average number n of crossings per unit time can be obtained as

$$n = \frac{N}{T} = p_\xi(H) \int_{-\infty}^\infty |\dot{x}| p_{\dot{\xi}|H}(\dot{x} \mid x) d\dot{x} \tag{4.241}$$

Similarly, the average number of upward n^+ and downward n^- excursions are given by

$$n^+ = \frac{N^+}{T} = p_\xi(H) \int_0^\infty \dot{x} p_{\dot{\xi}|H}(\dot{x} \mid x) d\dot{x} \tag{4.242}$$

$$n^- = \frac{N^-}{T} = -p_\xi(H) \int_{-\infty}^0 \dot{x} p_{\dot{\xi}|H}(\dot{x} \mid x) d\dot{x} \tag{4.243}$$

It was shown earlier that the values of a random process $\xi(t)$ and its derivative $\dot{\xi}(t)$ are uncorrelated at the same moment of time. This means that the values of the process and its velocity at the same level are only loosely related. In many cases, an even stronger statement is true: the process and its derivative are independent, i.e.

$$p_{\dot{\xi}|H}(\dot{x} \mid x) = p_{\dot{\xi}}(\dot{x}) \tag{4.244}$$

and, thus, eqs. (4.240)–(4.243) can be further simplified to produce

$$n = \frac{N}{T} = p_\xi(H) \int_{-\infty}^\infty |\dot{x}| p_{\dot{\xi}}(\dot{x}) d\dot{x} = C p_\xi(H)$$

$$n^+ = \frac{N^+}{T} = p_\xi(H) \int_0^\infty \dot{x} p_{\dot{\xi}}(\dot{x}) d\dot{x} = C^+ p_\xi(H) \tag{4.245}$$

$$n^- = \frac{N^-}{T} = -p_\xi(H) \int_{-\infty}^0 \dot{x} p_{\dot{\xi}}(\dot{x}) d\dot{x} = C^- p_\xi(H)$$

Thus, for the case of an independent derivative the level crossing rates (LCR) of a level H n, n^+, and n^- are proportional to the values of the PDF of the process considered at the level H. This fact can be used in order to experimentally measure the PDF of the process by counting a number of crossings at different levels of the process.

The next step is to calculate an average duration of upward (downward) excursions. Without loss of generality, only the case of upward excursions is considered. For a given time interval $[0, T]$, the average time spent by the process above a given threshold H is approximately given by

$$T^p = T \int_H^\infty p_\xi(x) dx \tag{4.246}$$

Since, during the same time interval, there are approximately $N^+ = n^+ T$ positive pulses, the average duration T^+ of a positive pulse is

$$T^+ = \frac{T^p}{N^+} = \frac{\int_H^\infty p_\xi(x) dx}{p_\xi(H) \int_0^\infty \dot{x} p_{\dot{\xi}|H}(\dot{x} \mid x) d\dot{x}} \tag{4.247}$$

In the very same manner, the average duration of a negative pulse is

$$T^- = \frac{T^n}{N^-} = \frac{\int_{-\infty}^H p_\xi(x) dx}{-p_\xi(H) \int_{-\infty}^0 \dot{x} p_{\dot{\xi}|H}(\dot{x} \mid x) d\dot{x}} \tag{4.248}$$

It is common to refer to the quantities defined by eqs. (4.245) and (4.248) as an average level crossing rate (LCR) and an average fade duration (AFD). These statistics are often misleadingly called higher order statistics (HOS) of signals. However, it is commonly agreed in statistical signal processing that the higher order statistics refer to the cumulants, moments and their spectra of order higher than two [28].

The calculations above assume that the expression for the joint PDF $p_{\xi\dot\xi}(x,\dot x)$ of the process and its derivative is known. In the case of a stationary process, it can be shown that this function can be calculated from the joint PDF $p_{\xi\xi}(x_1,x_2;\tau)$ of the process at two moments of time. Indeed, choosing

$$t_1 = t - \frac{\Delta t}{2}, \quad t_2 = t + \frac{\Delta t}{2}, \quad \Delta t = t_2 - t_1 \tag{4.249}$$

and using continuity of the process $\xi(t)$, one can write

$$\begin{aligned}
\xi\left(t + \frac{\Delta t}{2}\right) &= \xi(t) + \frac{\Delta t}{2}\dot\xi(t), \quad x_2 = x + \frac{\Delta t}{2}\dot x \\
\xi\left(t - \frac{\Delta t}{2}\right) &= \xi(t) - \frac{\Delta t}{2}\dot\xi(t), \quad x_1 = x - \frac{\Delta t}{2}\dot x
\end{aligned} \tag{4.250}$$

This transformation allows the joint PDF of $\xi(t)$ and $\dot\xi(t)$ to be obtained as

$$p_{\xi\dot\xi}(x,\dot x) = \lim_{\Delta t \to 0} \Delta t p_{\xi\xi}\left(x - \frac{\Delta t}{2}\dot x, x + \frac{\Delta t}{2}\dot x; \Delta t\right) \tag{4.251}$$

Since the joint PDF $p_{\xi\xi}(x_1,x_2)$ is a symmetric function of its arguments, eq. (4.251) implies that

$$p_{\xi\dot\xi}(x,\dot x) = p_{\xi\dot\xi}(x,-\dot x) \tag{4.252}$$

Thus, if a process $\xi(t)$ is stationary and has a derivative, then the distribution of the velocity $\dot\xi(t)$ is symmetric with respect to zero. If it were not so, the bias in velocity distribution would lead to non-stationarity, since the process will tend either to drift "upward" or "downward," thus causing a change in the mean value of the process in time.

4.6.3 Level Crossing Rate of a Gaussian Random Process

The problem of level crossing can be solved completely for a Gaussian case. Since $\xi(t)$ and $\dot\xi(t)$ are related by a linear operation of differentiation, the Gaussian nature of the process $\xi(t)$ would imply that its derivative is also a Gaussian process. Due to the symmetry prescribed by eq. (4.252), the mean value of the process $\dot\xi(t)$ is equal to zero and it is only required to find its correlation function[13] $R_{\dot\xi\dot\xi}(\tau)$. Using eqs. (4.131)–(4.132), it is easy to show that

$$R_{\xi\dot\xi}(\tau) = \frac{\partial}{\partial\tau}R_{\xi\xi}(\tau), \quad R_{\dot\xi\dot\xi}(\tau) = -\frac{\partial^2}{\partial\tau^2}R_{\xi\xi}(\tau) \tag{4.253}$$

---

[13]Since the process $\xi(t)$ has zero mean, the correlation function and the covariance function are equal: $R_{\xi\xi}(\tau) = C_{\xi\xi}(\tau)$.

Since the correlation function $R_{\xi\dot\xi}(\tau)$ has maximum at $\tau = 0$, this means that

$$R_{\xi\dot\xi}(0) = 0 \tag{4.254}$$

For a Gaussian process, this implies that the process and its derivatives are also independent and, thus,

$$p_{\xi\dot\xi}(x, \dot x) = p_\xi(x)p_{\dot\xi}(\dot x) \tag{4.255}$$

The variance of the derivative $\dot\xi(t)$ can be found as the values of its correlation function at zero

$$D_{\dot\xi} = -\frac{\partial^2}{\partial\tau^2}R_{\xi\xi}(\tau)\bigg|_{\tau=0} = -R''_{\xi\xi}(0) = -D_\xi\rho''_{\xi\xi}(0) \tag{4.256}$$

Here $\rho_{\xi\xi}(\tau)$ is the normalized covariance function (coefficient) of the process $\xi(t)$ and D_ξ is the variance of this process. Combining eqs. (4.255) and (4.256), one obtains an expression for the joint PDF of a Gaussian process and its derivative.

$$p_{\xi\dot\xi}(x, \dot x) = p_\xi(x)p_{\dot\xi}(\dot x) = \frac{1}{2\pi D_\xi\sqrt{-\rho''_\xi(0)}}\exp\left\{-\frac{1}{2D_\xi}\left[x^2 - \frac{\dot x^2}{\rho''_\xi(0)}\right]\right\} \tag{4.257}$$

Finally, the level crossing rate of a Gaussian process with the covariance function $R_{\xi\xi}(\tau)$ can be obtained from eq. (4.245)

$$LCR(H) = \frac{\sqrt{-\rho''_\xi(0)}}{2\pi}e^{-\frac{H^2}{2D_\xi}}\int_0^\infty se^{-\frac{s^2}{2}}ds, \tag{4.258}$$

where

$$s = \frac{\dot x}{D_\xi\sqrt{-\rho''_\xi(0)}} \tag{4.259}$$

Since the last integral in eq. (4.258) is equal to unity, one obtains

$$LCR(H) = \frac{\sqrt{-\rho''_\xi(0)}}{2\pi}e^{-\frac{H^2}{2D_\xi}} = \frac{\sqrt{-\rho''_\xi(0)}}{2\pi}e^{-\frac{H^2}{2D_\xi}} = \frac{\sqrt{-\rho''_\xi(0)}}{2\pi}e^{-\frac{(H/\sqrt{D_\xi})^2}{2}} \tag{4.260}$$

The values of $-\rho''_\xi(0)$ for some important correlation functions are summarized in Table 3.1. Some of these values are infinite, thus indicating that the process with such a correlation function has a derivative with infinite variance.

The expression for the level crossing rate (4.260) can now be used to obtain an expression for an average duration of the positive outage. Indeed, using the Laplace function $\Phi(x)$ [29]

$$\int_H^\infty p_\xi(x)dx = 1 - \Phi\left(\frac{H}{\sqrt{D_\xi}}\right) \tag{4.261}$$

expression (4.260) can be transformed to produce

$$
T^{+} = \frac{2\pi \left[1 - \Phi\left(\frac{H}{\sqrt{D_\xi}}\right)\right] \exp\left(\frac{H^2}{2D_\xi}\right)}{\sqrt{-\rho_\xi''(0)}}
\tag{4.262}
$$

It can be seen from eqs. (4.260) and (4.262) that the expressions for the LCR and AFD are more conveniently expressed in terms of the normalized threshold

$$
h = \frac{H}{\sqrt{D_\xi}}
\tag{4.263}
$$

Dependence of the LCR and AFD on the normalized threshold h is shown in Fig. 4.7 for two cases of covariance function

$$
\rho_{\xi\xi}(\tau) = J_0(2\pi f_D \tau), \quad -\rho_\xi''(0) = 2\pi^2 f_D^2
\tag{4.264}
$$

and

$$
\rho_{\xi\xi}(\tau) = \exp(-\tau^2/\tau_{\text{corr}}^2), \quad -\rho_\xi''(0) = \frac{2}{\tau_{\text{corr}}^2}
\tag{4.265}
$$

It can be seen that the LCR and AFD are expressed in terms of the time characteristic of the random process, in particular both quantities are functions of the values of the second

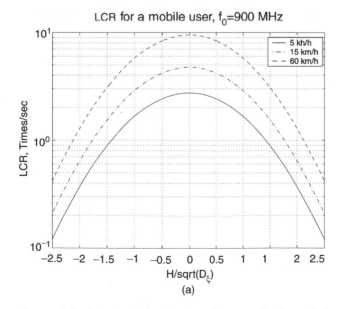

LCR for a mobile user, f_0=900 MHz

(a)

Figure 4.7 Dependence of LCR (a) and AFD (b) on the normalized threshold and the type of correlation function with the same correlation interval.

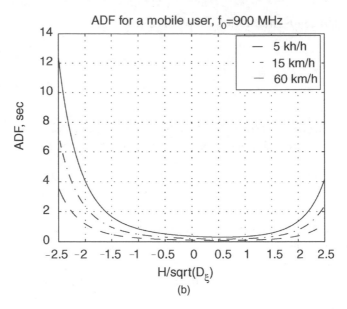

Figure 4.7 (*Continued*)

derivative of the correlation function $-\rho''_\xi(0)$. The same calculation could be conducted in the frequency domain as well. Indeed, using the Parseval theorem and the properties of the Fourier transform one obtains

$$-\rho''_\xi(0) = \frac{\frac{1}{2\pi}\int_0^\infty (j\omega)^2 G(\omega)d\omega}{\sigma^2} = \frac{4\pi}{\sigma^2}\int_0^\infty f^2 G(f)df \qquad (4.266)$$

Example: Bandlimited (pink) noise. In this case of an ideal baseband filter with the cut-off frequency ΔF, the one-sided PSD of filtered WGN is given by

$$\text{PSD}(f) = \begin{cases} N_0 & 0 \le f \le \Delta F \\ 0 & f > \Delta F \end{cases} \qquad (4.267)$$

The value of $-\rho''_\xi(0)$ for this process is thus

$$-\rho''_\xi(0) = \frac{4\pi \int_0^\infty f^2 G(f)df}{\int_0^\infty G(f)df} = \frac{4\pi \int_0^{\Delta F} f^2 N_0 df}{\int_0^{\Delta F} N_0 df} = \frac{4\pi N_0}{N_0 \Delta F}\frac{(\Delta F)^3}{3} = \frac{4\pi(\Delta F)^2}{3} \qquad (4.268)$$

Example: A Gaussian process with a Gaussian PSD. Let a Gaussian process $\xi(t)$ with zero mean be described by a Gaussian PSD:

$$\text{PSD}(f) = N_0 \exp\left[-\frac{\pi}{4}\left(\frac{f}{\Delta f_{\text{eff}}}\right)^2\right] \qquad (4.269)$$

This PSD corresponds to a Gaussian correlation function

$$\rho(\tau) = \exp(-\pi \Delta f_{\text{eff}}^2 \tau^2) \tag{4.270}$$

It is easy to check that both time-domain and frequency-domain methods produce the same answer:

$$-\rho_\xi''(0) = 2\pi(\Delta f_{\text{eff}})^2 \tag{4.271}$$

Example: Clark's PSD. It is commonly accepted that the one-sided PSD of the signal received by a moving receiver is provided by the following expression [30]

$$\text{PSD}(f) = \frac{2}{f_D} \frac{1}{\sqrt{1 - f^2/f_D^2}}, \qquad 0 \le f \le f_D \tag{4.272}$$

where $f_D = \frac{v}{c} f \cos \theta$ is the Doppler frequency of the signal moving with velocity v and the angle of arrival of the signal at the receiver's antenna is θ; c is the speed of light in the air. Calculations using both eqs. (4.253) and (4.266) produce

$$-\rho_\xi''(0) = 2\pi f_D^2 \tag{4.273}$$

4.6.4 Level Crossing Rate of the Nakagami Process

The Nakagami process is often used to describe non-Rayleigh fading [31]. As was pointed out in section 4.6.2, the LCR can be found using the joint PDF of the process at two separate moments of time. In contrast to a Gaussian random process and the Rayleigh process as an envelope of a complex Gaussian process, the Nakagami process in general cannot be described completely by the joint PDF of the second order. In addition, a different joint PDF can be derived for the same marginal PDF and the correlation function. However, the following model is the most popular in practical applications [31]:

$$p_{\xi\xi}(x_1, x_2; \tau)$$
$$= \frac{4(x_1 x_2)^m m^{m+1}}{\Gamma(m)(1 - \rho_0^2)\rho_0^{m-1}\Omega^{m+1}} \exp\left[-\frac{mx_1^2 + mx_2^2}{\Omega(1 - \rho_0^2)}\right] I_{m-1}\left(\frac{2m\rho_0}{\Omega(1 - \rho_0^2)} x_1 x_2\right), \qquad m \ge \frac{1}{2} \tag{4.274}$$

Here $\rho_0(\tau)$ is the correlation function of the underlying Gaussian process [31], $\rho(\tau) \approx \rho_0^2(\tau)$ is the normalized covariance function of the squared process $\xi^2(t)$ (square of the envelope) and $I_n(x)$ is the modified Bessel function of the first type of the order n [29]. Thus, eq. (4.274) can be rewritten as

$$p_{\xi\xi}(x_1, x_2; \tau) = \frac{4(x_1 x_2)^m}{\Gamma(m)(1 - \rho)\rho^{(m-1)/2}\Omega^{m+1}} \exp\left[-\frac{mx_1^2 + mx_2^2}{\Omega(1 - \rho)}\right] I_{m-1}\left[\frac{2m\sqrt{\rho}}{\Omega(1 - \rho)} x_1 x_2\right] \tag{4.275}$$

The next step is calculation of the joint PDF of the process $\xi(t)$ and its derivative $\dot{\xi}(t)$ by means of eq. (4.251). At this stage, the time lag τ can be assumed to be a small quantity, thus allowing for the following approximate relations, derived from the Taylor expansion of the correlation coefficient and the fact that $\rho'(0) = 0$:

$$\rho(\tau) \cong 1 + \frac{\rho''(0)}{2}\tau^2 + O(\tau^3), \qquad 1 - \rho(\tau) \cong -\rho''(0)\tau^2 + O(\tau^3),$$

$$\sqrt{\rho(\tau)} \cong 1 + \frac{\rho''(0)}{4}\tau^2 + O(\tau^3)$$

(4.276)

Furthermore, it can be seen that the argument of the Bessel function approaches infinity since $1 - \rho$ approaches zero. In this case, the following asymptotic values of the Bessel function can be used [29]

$$I_{m-1}(x) \sim \frac{e^x}{\sqrt{2\pi x}}$$

(4.277)

Having this in mind and using eq. (4.251), one easily obtains

$$p_{\xi\dot{\xi}}(x, \dot{x})$$

$$= \lim_{\tau \to 0}\left\{\tau \exp\left[-m\frac{\left(x - \frac{1}{2}\dot{x}\tau\right)^2 + \left(x + \frac{1}{2}\dot{x}\tau\right)^2}{-\frac{\rho''(0)}{2}\tau^2\Omega}\right]I_{m-1}\left[\frac{2m\left(1 + \frac{\rho''(0)}{4}\tau^2\right)\left(x - \frac{1}{2}\dot{x}\tau\right)\left(x + \frac{1}{2}\dot{x}\tau\right)}{-\frac{\rho''(0)}{2}\tau^2\Omega}\right]\right\}$$

$$= \lim_{\tau \to 0}\left\{\frac{1}{\tau}\exp\left[-\frac{2m\left(x^2 + \frac{1}{4}\dot{x}^2\tau^2\right)}{-\frac{\rho''(0)}{2}\Omega}\right]\frac{\exp\left[\frac{2m\left(1 + \frac{\rho''(0)}{4}\tau^2\right)\left(x^2 - \frac{1}{4}\dot{x}^2\tau^2\right)}{-\frac{\rho''(0)}{2}\tau^2\Omega}\right]}{\sqrt{2\pi\frac{2m\sqrt{\rho}\left(x - \frac{1}{2}\dot{x}\tau\right)\left(x + \frac{1}{2}\dot{x}\tau\right)}{-\frac{\rho''(0)}{2}\tau^2\Omega}}}\right\}$$

(4.278)

$$= \frac{\exp\left(-\frac{2m}{-\frac{\rho''(0)}{2}\Omega}\frac{1}{2}\dot{x}^2 - \frac{\rho''(0)}{4}x^2\right)}{\sqrt{\frac{4\pi m x^2}{-\frac{\rho''(0)}{2}\Omega}}}$$

It follows from the latter that the Nakagami process $\zeta(t)$ and its derivative $\dot\xi(t)$ are independent and the derivative itself has a Gaussian PDF:

$$p_{\xi\dot\xi}(x,\dot x)$$

$$= \frac{\exp\left(-\dfrac{2m}{-\dfrac{\rho''(0)}{2}\Omega}\dfrac{1}{2}\dot x^2 - \dfrac{\rho''(0)}{4}x^2\right)}{\sqrt{-\dfrac{4\pi mx^2}{\dfrac{\rho''(0)}{2}\Omega}}} = p_{\xi\dot\xi}(x,\dot x) = \frac{\exp\left(-\dfrac{2m}{-\dfrac{\rho''(0)}{2}\Omega}\dfrac{1}{2}\dot x^2 - \dfrac{\rho''(0)}{4}x^2\right)}{\sqrt{-\dfrac{4\pi mx^2}{\dfrac{\rho''(0)}{2}\Omega}}}$$

$$(4.279)$$

Correctness of this equation is guaranteed since $-\rho''(0) > 0$ implies that the expression under the square root is positive[14]. Finally, the LCR and AFD can be expressed as

$$N_1^+ = \frac{\sqrt{-\rho''(0)}}{\Gamma(m)} C^{2m-1} e^{-C^2} \qquad (4.280)$$

$$T_1^+ = \frac{\Gamma(m, C^2)}{C^{2m-1}\sqrt{-\rho''(0)}} e^{C^2} \qquad (4.281)$$

$$N_1^+ = \sqrt{-\rho''(0)} C e^{-C^2} \qquad (4.282)$$

where

$$C = \frac{\sqrt{m}}{\sqrt{\Omega}} H \qquad (4.283)$$

It can be noted that for $m = 1$ these equations coincide with eq. (4.257) derived for the Rayleigh process.

It has been noted already that the Nakagami m distribution is widely used to describe fading in communication channels. This also poses a problem to the investigation of LCR and AFD in systems with combining. The combining methods can be viewed as techniques allowing formation of a continuous communication channel which allows much smaller variance of the PDF, describing the received signal. This question was investigated as far back as in [31]. It was shown there that if all L branches of the combining scheme are

---

[14]It is interesting to note that eq. (4.279) was probably first published in [32,33] the latter of which unfortunately is not widely available in English. Since then this equation has been rediscovered in a number of publications.

identical, the resulting process is also approximately a Nakagami m process with the following parameters

$$m_L = mL, \qquad \Omega_L = L^2\Omega \tag{4.284}$$

This fact can be used to obtain rough estimates of the performance of the combining scheme under consideration. More detailed and accurate results are described in [35].

4.6.5   Concluding Remarks

All the attention in the previous section has been devoted to calculation of only the first moments of the number of crossings, duration of fades and overshoots, etc. However, such description does not provide complete information about the mentioned quantities, which can be provided only by a marginal PDF. Rice [23] outlined an approach allowing calculation of the marginal PDF $p_F(\tau)$ of the fade duration. However, a general approach to this problem is still unknown. Even calculation of the variance D_τ appears to be a difficult problem. Often this quantity is estimated from experiments or numerical simulations. However, for the case of rare fades[15] $(C \gg 1)$, it can be assumed [20,22] that the outages obey the Poisson law and, thus

$$p_\tau(\tau) = \frac{1}{T^+}\exp\left(-\frac{\tau}{T^+}\right), \qquad \tau > 0 \tag{4.285}$$

In the other limiting case $C \to 0$ of the Gaussian process, it can be shown that the fade duration obeys the Rayleigh law [22]. A number of authors suggest a similar representation for other distributions of the process $\xi(t)$, however, there are no convincing arguments to support such approximation. Good approximations for a sum of Gaussian noise and a harmonic signal are obtained in [20,23]. However, these expressions are quite complicated and difficult to use in analysis and synthesis. A situation when such equations can be directly used is rarely found.

4.7   NARROW BAND RANDOM PROCESSES

Most communication systems work at a very high frequency (the carrier frequency f_0) and utilize bandwidth Δf (measured at -3 dB level) which is usually much smaller than the carrier[16]

$$\frac{\Delta f}{f} \ll 1 \tag{4.286}$$

---

[15]That is, the outage process is an ordinary process.

[16]However, recently great advances have been achieved in the area of so-called ultra-wide band systems, which utilize a relatively wide spectral band. For more detailed discussion see [36].

If a stationary wide band excitation such as WGN acts on the input of such systems, the asymptotically stationary output process has spectral density which is concentrated in a narrow band around the carrier frequency f_0, according to eq. (4.144).

It will be shown below that the covariance function of a narrow band random process $\xi(t)$ can be written in the following form

$$C_{\xi\xi}(\tau) = D_\xi \rho(\tau) \cos[\omega_0 \tau + \gamma(\tau)] \tag{4.287}$$

Here $\omega_0 = 2\pi f_0$ and $\rho(\tau)$ and $\gamma(\tau)$ are functions which vary slowly compared to $\cos \omega_0 \tau$. In order to justify representation (4.288), one can introduce a new variable f_D, which is sometimes referred to as the Doppler shift, by the rule

$$f_D = f - f_0 \tag{4.288}$$

Then, using the one-sided PSD $S^+(f)$, one can write

$$C_{\xi\xi}(\tau) = \int_0^\infty S^+(f) \cos(2\pi f \tau) df = \int_{-f_0}^\infty S^+(f_0 + f_D) \cos[2\pi(f_0 + f_D)\tau] df_D \tag{4.289}$$

Using the following notation

$$D_\xi \rho_c(\tau) = \int_{-f_0}^\infty S^+(f_0 + f_D) \cos[2\pi f_D \tau] df_D$$
$$D_\xi \rho_s(\tau) = \int_{-f_0}^\infty S^+(f_0 + f_D) \sin[2\pi f_D \tau] df_D \tag{4.290}$$

one can rewrite eq. (4.289) in the following form

$$C_{\xi\xi}(\tau) = D_\xi[\rho_c(\tau) \cos(2\pi f_0 \tau) - \rho_s(\tau) \sin(2\pi f_0 \tau)] = D_\xi \rho(\tau) \cos[\omega_0 \tau + \gamma(\tau)] \tag{4.291}$$

where

$$\rho(\tau) = \sqrt{\rho_c^2(\tau) + \rho_s^2(\tau)}, \tan \gamma(\tau) = \frac{\rho_s(\tau)}{\rho_c(\tau)} \tag{4.292}$$

Since $\rho_c(\tau)$ is an even function and $\rho_s(\tau)$ is an odd function, $\rho(\tau)$ is always an even function and $\gamma(\tau)$ is always an odd function of τ.

If the spectral density $S^+(f)$ is a symmetric function around f_0, the lower limit of integration in eq. (4.290) can be substituted by $-\infty$, thus producing $\rho_s(\tau) = 0$ and $\gamma(\tau) = 0$.

$$C_{\xi\xi}(\tau) = D_\xi \rho(\tau) \cos \omega_0 \tau = \cos \omega_0 \tau \int_{-f_0}^\infty S^+(f_0 + \tau) \cos(2\pi f_D \tau) d\tau \tag{4.293}$$

Thus, the covariance function of a random narrow band process with symmetric PSD can be represented by a modulated cosine function (4.293). Since the spectral density $S^+(f_0 + \tau)$ is almost completely located at the low frequency band (compared to f_0), functions $\rho_c(\tau)$, and $\rho_s(\tau)$, and as a result $\rho(\tau)$ and $\gamma(\tau)$, are slow functions of time-shift τ compared to $\cos \omega_0 \tau$.

4.7.1 Definition of the Envelope and Phase of Narrow Band Processes

It could be also said that a realization of a random process looks like a modulated harmonic signal, that is why it is often referred to as a quasiharmonic signal or modulated signal. It is convenient to write a quasiharmonic process $\xi(t)$ as

$$\xi(t) = A(t) \cos[\omega_0 t + \phi(t)], \qquad A(t) \geq 0 \tag{4.294}$$

where $A(t)$ and $\phi(t)$ are random slowly varying functions with respect to $\cos \omega_0 t$. The random process $A(t)$ is called the envelope and the process $\phi(t)$ is called the random phase. The representation (4.294) is not unique, unless additional conditions are specified. This stems from the fact that for a very short observation interval, it is impossible to estimate the central frequency ω_0 of the random signal, thus causing uncertainties in the value of the phase and the magnitude. Moreover, the form of the equation will not be changed if the phase shift is $2\pi n$. In order to eliminate such uncertainties, one has to apply some meaningful restrictions on the phase $\phi(t)$ and the magnitude. First of all, it will be assumed that the phase is confined to an interval $[0, 2\pi]$ (or, equivalently $[-\pi, \pi]$). The restriction on the magnitude is more complex and requires definition of the analytical signal. In this case, the original process $\xi(t)$ is augmented by a conjugate process $\eta(t)$ defined by the so-called Hilbert transform [4]

$$\eta(t) = \frac{1}{\pi} \int_{-\infty}^{\infty} \frac{\xi(s)}{t - s} \, ds$$
$$\xi(t) = -\frac{1}{\pi} \int_{-\infty}^{\infty} \frac{\eta(s)}{t - s} \, ds \tag{4.295}$$

The Hilbert transform of a signal $\xi(t)$ can be considered as the reaction of a filter with the impulse response $h(t) = 1/\pi t$ to the signal $\xi(t)$. In the frequency domain, this transform is described by the following transfer function

$$H(j\omega) = \frac{1}{\pi} \int_{-\infty}^{\infty} \frac{1}{t} \exp(-j\omega t) dt = -j \mathrm{sgn}\omega \tag{4.296}$$

Since $|H(j\omega)| = 1$, this filter shifts the phase of all negative frequencies up by $\pi/2$ and shifts the phase of all positive frequencies down by $\pi/2$. Equivalently, any component of the form $\cos \omega t$ is transformed into $\sin \omega t$. The Hilbert filter is often called the quadrature filter, since it transforms in-phase components of signals into quadrature components of the same signal. The Hilbert transform filter is also an all-pass system, since the magnitude spectrum is unchanged at the output of the system with respect to one at the input of the system. This immediately implies that the PSDs of both $\xi(t)$ and $\eta(t)$ are the same

$$S_\eta(\omega) = |H(j\omega)|^2 S_\xi(\omega) = S_\xi(\omega) \tag{4.297}$$

so are the covariance functions:

$$C_{\xi\xi}(\tau) = C_{\eta\eta}(\tau) \tag{4.298}$$

The mutual PSD and the covariance function can be directly calculated from eq. (4.143)

$$S_{\xi\eta}(\omega) = H(j\omega)S_\xi(\omega) = -jS_\xi(\omega)\text{sgn}\omega \tag{4.299}$$

$$C_{\xi\eta}(\tau) = -C_{\eta\xi}(\tau) = \frac{1}{2\pi}\int_{-\infty}^{\infty} -jS_\xi(\omega)\text{sgn}\,\omega\exp[j\omega\tau]d\omega = \frac{1}{\pi}\int_0^\infty S_\xi(\omega)\sin\omega\tau\,d\omega \tag{4.300}$$

The latter shows that the processes $\eta(\tau)$ and $\xi(\tau)$ are uncorrelated if considered at the same moment of time, since

$$E\{\eta(t)\xi(t)\} = C_{\xi\eta}(0) = 0 \tag{4.301}$$

For a Gaussian process, this also means independence, however, for non-Gaussian processes it is not so.

The in-phase component $\xi(t)$ and the quadrature component $\eta(t)$ obtained through the Hilbert transform can be combined in a single complex signal called a pre-envelope [4]

$$\zeta(t) = \xi(t) + j\eta(t) = A(t)\exp[j\Phi(t)] \tag{4.302}$$

Thus, the envelope and the full phase can be defined as

$$A(t) = \sqrt{\xi^2(t) + \eta^2(t)}, \Phi(t) = \text{atan}\frac{\eta(t)}{\xi(t)} = \omega_0 t + \phi(t) \tag{4.303}$$

It can be seen that $A(t) \geq |\xi(t)| \geq 0$ and $A(t) = |\xi(t)|$ if $\eta(t) = 0$ and, thus, the process $A(t)$ envelopes the process $\xi(t)$. Equation (4.248) also allows definition of a low frequency equivalent $\vartheta(t)$ of the complex pre-envelope as

$$\zeta(t) = A(t)\exp[j(\omega_0 t + \phi(t))] = A(t)\exp[j\phi(t)]\exp(j\omega_0 t) = \vartheta(t)\exp(j\omega_0 t) \tag{4.304}$$

The frequency content of the process $\vartheta(t)$ is concentrated around zero frequency if the original process $\xi(t)$ is narrow band and is concentrated around the frequency f_0. Furthermore, the low frequency equivalent can be obtained through a demodulation process as

$$\vartheta(t) = \zeta(t)\exp(-j\omega_0 t) = [\xi(t) + j\eta(t)][\cos\omega_0 t + j\sin\omega_0 t] = A_c(t) + jA_s(t) \tag{4.305}$$

and, thus

$$\begin{aligned} A_c(t) &= A(t)\cos\phi(t) = \xi(t)\cos\omega_0 t + \eta(t)\sin\omega_0 t \\ A_s(t) &= A(t)\sin\phi(t) = -\xi(t)\sin\omega_0 t + \eta(t)\cos\omega_0 t \end{aligned} \tag{4.306}$$

In order to find characteristics of the envelope and the phase of a narrow band process, it is necessary to find a joint distribution of the in-phase and quadrature components $p_{\xi\eta}(x_1, x_2)$.

Such calculations are relatively simple to perform for a Gaussian process, as shown in the example below.

An alternative definition of the conjugate process and the envelope can be obtained following [37]. In this case

$$\eta(t) \approx \frac{1}{\omega_0} \frac{d}{dt} \xi(t) = \frac{1}{\omega_0} \dot{\xi}(t), A(t) = \sqrt{\xi^2(t) + \frac{\dot{\xi}^2(t)}{\omega_0^2}} \tag{4.307}$$

It can be seen that, for a deterministic harmonic signal $x(t) = A \cos \omega_0 t$, eq. (4.307) produces an exact result $y(t) = A \sin \omega_0 t$. However, for a finite bandwidth of the signal $\xi(t)$ this definition is an approximate one. Both definitions will be used in the considerations below.

4.7.2 The Envelope and the Phase Characteristics

One of the most important questions in the theory of narrow band processes is the relation between the characteristics of the random process itself and the characteristics of its envelope, especially relations between the marginal densities and the covariance functions.

4.7.2.1 Blanc-Lapierre Transformation

Let a narrow band process be represented in the following form

$$\xi(t) = \xi_R(t) = A(t) \cos[\omega_0 t + \phi(t)] = A(t) \cos \Psi(t), \tag{4.308}$$

where the envelope $A(t)$ and the phase $\phi(t)$ are slowly varying random functions with respect to $\cos(\omega_0 t)$ and are defined by eqs. (4.303). Jointly, the envelope $A(t)$ and the total phase $\Psi(t)$ can be described by the joint PDF $p_{A,\Psi}(a, \psi)$. The marginal PDF $p_A(a)$ of the envelope and the marginal PDF $p_\Psi(\psi)$ of the phase can be obtained by integration of the joint PDF over the auxiliary variable. The process $\xi_R(t)$ is described by its PDF $p_{\xi_R}(x)$. The goal of this section is to derive the relationship between $p_A(a)$ and $p_{\xi_R}(x)$ assuming that the process $\xi_R(t)$ is a stationary process. As a by-product of this analysis, it will be shown that the phase distribution must be uniform to obtain a stationary process:

$$p_\Psi(\psi) = \frac{1}{2\pi} \tag{4.309}$$

Following the technique discussed in section 4.7.1, the conjugate $\xi_I(t)$ and the pre-envelope $\zeta(t)$ processes can be defined as

$$\xi_I(t) = A(t) \sin[\omega_0 t + \phi(t)] = A(t) \sin \Psi(t) \tag{4.310}$$
$$\zeta(t) = \xi_R(t) + j\xi_I(t) \tag{4.311}$$

The complex process $\zeta(t)$ can be described by means of the joint PDF of $\xi_R(t)$ and $\xi_I(t)$, as discussed in section 2.1

$$p_\zeta(x_1, x_2) = \frac{\partial^2}{\partial x_1 \partial x_2} \text{Prob}\{\xi_R(t) \le x_1, \xi_I(t) \le x_2\} \tag{4.312}$$

Conversantly, both $p_{\xi_R}(x)$ and $p_{\xi_I}(x)$ can be obtained from $p_\zeta(x_1, x_2)$ by integration, that is

$$p_{\xi_R}(x) = \int_{-\infty}^{\infty} p_\xi(x, x_2) dx_2 \qquad (4.313)$$

$$p_{\xi_I}(x) = \int_{-\infty}^{\infty} p_\xi(x_1, x) dx_1 \qquad (4.314)$$

An alternative description of $\zeta(t)$ can be achieved through its characteristic function

$$\Theta_\zeta(u, v) = \int_{-\infty}^{\infty} \int_{-\infty}^{\infty} p_\xi(x_1, x) \exp[j(ux_1 + vx_2)] dx_1 dx_2 \qquad (4.315)$$

As a particular case of eq. (4.315), one can see that

$$\Theta_\zeta(u, 0) = \int_{-\infty}^{\infty} \int_{-\infty}^{\infty} p_\xi(x_1, x_2) \exp(jux_1) dx_1 dx_2$$

$$= \int_{-\infty}^{\infty} \exp(jux_1) dx_1 \int_{-\infty}^{\infty} p_\xi(x_1, x_2) dx_2 = \int_{-\infty}^{\infty} \exp(jux_1) p_{\xi_R}(x_1) dx_1 = \Theta_{\xi_R}(u)$$

$$(4.316)$$

In the very same way, the characteristic function of the quadrature is given by

$$\Theta_\zeta(0, v) = \Theta_{\xi_I}(v), \qquad (4.317)$$

These last two equations allow one to determine the PDFs of $\xi_R(t)$ and $\xi_I(t)$ through $\Theta_\xi(u, v)$. In fact

$$p_{\xi_R}(x_1) = \frac{1}{2\pi} \int_{-\infty}^{\infty} \Theta_{\xi_R}(u) \exp[-jux_1] du = \frac{1}{2\pi} \int_{-\infty}^{\infty} \Theta_\zeta(u, 0) \exp[-jux_1] du \qquad (4.318)$$

and

$$p_{\xi_I}(x_2) = \frac{1}{2\pi} \int_{-\infty}^{\infty} \Theta_{\xi_I}(v) \exp[-jvx_2] dv = \frac{1}{2\pi} \int_{-\infty}^{\infty} \Theta_\zeta(0, v) \exp[-jvx_2] dv \qquad (4.319)$$

As the next step, one can express the joint characteristic function $\Theta_\xi(u, v)$, given by eq. (4.315), in terms of the joint PDF $p_{A,\Psi}(a, \psi)$ of the magnitude and phase. In order to do so, the following change of variables can be used in eq. (4.315)

$$\begin{cases} x_1 = a \cdot \cos \psi \\ x_2 = a \cdot \sin \psi \end{cases} \qquad (4.320)$$

The Jacobian J of this transformation is just

$$J = \begin{vmatrix} \cos \psi & \sin \psi \\ -a \cdot \sin \psi & a \cdot \cos \psi \end{vmatrix} = a \qquad (4.321)$$

and, thus, the relationship between $P_{A,\Psi}(a, \psi)$ and $p_\xi(x_1, x_2)$ has the following form (see eq. (2.163) in section 2.6)

$$p_{A,\Psi}(a, \psi) = p_\xi(a \cos \psi, a \sin \psi)a \tag{4.322}$$

or

$$p_\xi(a \cos \psi, a \sin \psi) = \frac{p_{A,\Psi}(a, \psi)}{a} \tag{4.323}$$

After changing the variables in eq. (4.315), using eqs. (4.320) and (4.323), one obtains

$$
\begin{aligned}
\Theta_{\xi_R}(u) = \Theta_\xi(u, 0) &= \int_{-\infty}^\infty \int_{-\infty}^\infty p_\xi(x_1, x_2) \exp[jux_1] dx_1 dx_2 \\
&= \int_0^\infty \int_0^{2\pi} p_\xi(a \cos \psi, a \sin \psi) \exp[ju \cos \psi a]a \, da \, d\psi \tag{4.324} \\
&= \int_0^\infty \int_0^{2\pi} P_{A,\Psi}(a, \psi) \exp[jua \cos \psi] da \, d\psi
\end{aligned}
$$

The exponential term in the last equation can be expressed as a sum of Bessel functions of the first kind (eqs. (9.1.44–9.1.45) on page 361 in [29])

$$\exp[jua \cos \psi] = \sum_{n=-\infty}^\infty j^n J_n(ua) \exp[jn\psi] \tag{4.325}$$

so that eq. (4.324) becomes

$$
\begin{aligned}
\Theta_{\xi_R}(u) &= \int_0^\infty \int_0^{(2\pi)} \sum_{n=-\infty}^\infty j^n J_n(ua) \exp[jn\psi] p_{A,\Psi}(a, \psi) da \, d\psi \\
&= \sum_{n=-\infty}^\infty j^n \int_0^\infty J_n(ua) \int_0^{2\pi} \exp[jn(\omega_0 t + \varphi)] P_{A,\phi}(a, \varphi) d\varphi \, da \\
&= \sum_{n=-\infty}^\infty j^n \int_0^\infty J_n(ua) p_A(a) da \int_0^{2\pi} \exp[jn(\omega_0 t + \varphi)] p_{\phi|A}(\varphi \mid a) d\varphi \tag{4.326}
\end{aligned}
$$

Here, $p_{\phi|A}(\varphi \mid a)$, stands for PDF of the phase, $\phi(t)$, given a value of a for the magnitude

$$p_{\phi|A}(\varphi \mid a) = \frac{p_{A,\phi}(a, \varphi)}{p_A(a)} \tag{4.327}$$

One can obtain $\Theta_{\xi_R}(u)$ which is independent of time (i.e. $\xi_R(t)$ will be a stationary process) if and only if[17]

$$p_{\phi|A}(\varphi \mid a) = \frac{1}{2\pi} \tag{4.328}$$

[17]Recall that the expression for the total phase $\Psi(t)$ contains the term $\omega_0 t$. This time dependence is gone if eq. (4.328) holds.

This means that the distribution of the phase does not depend on the value of a, i.e. the magnitude and the phase are statistically independent[18]. Also, taking into account that

$$\int_0^{2\pi} \exp[jn(\omega_0 t + \varphi)]d\varphi = 0 \tag{4.329}$$

for $n > 0$, eq. (4.326) can be rewritten as

$$\Theta_{\xi_R}(u) = \int_0^\infty J_0(ua)p_A(a)da \tag{4.330}$$

or, taking the inverse Fourier transform

$$p_{\xi_R}(x_1) = \frac{1}{2\pi} \int_{-\infty}^\infty \Theta_{\xi_R}(u) \exp[-jx_1 u]du = \frac{1}{2\pi} \int_{-\infty}^\infty \int_0^\infty J_0(ua)p_A(a) \exp[-jx_1 u]da\,du. \tag{4.331}$$

In order to express $p_A(a)$ in terms of $p_{\xi_R}(x_1)$, it could be observed that $\Theta_{\xi_R}(u)$ is just the Hankel transform[19] (see [38]) of $p_A(a)/a$ and thus can be easily inverted to produce

$$\frac{p_A(a)}{a} = \int_0^\infty J_0(ua)\Theta_{\xi_R}(u)u\,du = \int_0^\infty uJ_0(ua) \int_{-\infty}^\infty p_{\xi_R}(x_1) \exp[jx_1 u]dx_1\,du \tag{4.332}$$

The transform pair given by eqs. (4.331) and (4.332) is known as the Blanc-Lapierre transform [19].

An alternative expression can be obtained by rewriting eq. (4.331) as

$$\begin{aligned} p_{\xi_R}(x_1) &= \frac{1}{2\pi} \int_{-\infty}^\infty \int_0^\infty J_0(ua)p_A(a) \exp[-jx_1 u]da\,du \\ &= \frac{1}{2\pi} \int_0^\infty p_A(a)da \int_{-\infty}^\infty J_0(ua) \exp[-jx_1 u]du \end{aligned} \tag{4.333}$$

Since the inner integral is equal to (equation in [39])

$$\int_{-\infty}^\infty J_0(ua) \exp[-jx_1 u]du = \begin{cases} \dfrac{2}{\sqrt{a^2 - x_1^2}} & \text{if } a \ge |x_1| \\ 0 & \text{if } a < |x_1| \end{cases} \tag{4.334}$$

one can write

$$p_{\xi_R}(x_1) = \frac{1}{\pi} \int_{|x_1|}^\infty \frac{p_a(a)}{\sqrt{a^2 - x_1^2}}da = \frac{1}{\pi} \int_0^\infty p_A(|x_1|\cosh a)da \tag{4.335}$$

---

[18]This condition can be relaxed if only wide sense stationarity is required.
[19]A pair of Hankel transforms is defined by $F(u) = \int_0^\infty f(x)J_0(xu)x\,dx$, $f(x) = \int_0^\infty F(u)J_0(xu)u\,du$ [38].

4.7.2.2 Kluyver Equation

Very often a narrow band random process $\xi(t)$ on a physical level is obtained as a sum of a number of narrow band processes with known characteristics. For example, multipath propagation of radio waves results in a weighted sum of randomly amplified and delayed replicas of the transmitted signal. A similar situation is encountered with sea radar when the reflected wave is a combination of reflections from numerous irregularities of the sea surface. A great number of similar examples can be found elsewhere. In such problems, it is beneficial to find a relationship between the PDF of each summand and the PDF of the net narrow band process. This problem is treated in this section.

Let a complex random variable ζ be a sum of N independent complex random variables ζ_n, $n = 1, 2, \ldots, N$:

$$\zeta = \xi + j\eta = \sum_{n=1}^{N} \zeta_n = \sum_{n=1}^{N} \xi_n + j \sum_{n=1}^{N} Y\eta_n \tag{4.336}$$

where $\xi_n = \mathrm{Re}(\zeta_n)$, $\eta_n = \mathrm{Im}(\zeta_n)$. It is assumed that the joint PDF $p_{\xi_n, \eta_n}(x_n, y_n)$ is known for each individual ζ_n. This can be recast in terms of a joint PDF of magnitude A_n and phase Ψ_n as in eq. (4.323)

$$\zeta_n = A_n \exp(j\Psi_n), \quad A_n = \sqrt{\xi_n^2 + \eta_n^2}, \quad \Psi_n = \mathrm{atan}\frac{\eta_n}{\xi_n} \tag{4.337}$$

$$p_{A_n, \Psi_n}(a_n, \varphi_n) = a_n p_{\xi_n, \eta_n}(a_n \cos\varphi_n, a_n \sin\varphi_n) = \frac{1}{2\pi} p_{A_n}(a_n) \tag{4.338}$$

The process $\zeta(t)$ itself can also be described by the joint PDF of its real and imaginary parts $p_\zeta(x, y)$ or, equivalently, by its characteristic function

$$\Theta_\zeta(u, v) = \int_{-\infty}^{\infty} \int_{-\infty}^{\infty} p_\zeta(x, y) \exp[j(ux + vy)]dx\,dy \tag{4.339}$$

Since ζ is a sum of N independent random variables its characteristic function is a product of the characteristic functions of each single term [18], i.e.

$$\Theta_\xi(u) = \prod_{n=1}^{N} \Theta_{\xi_n}(u) = \prod_{n=1}^{N} \int_0^\infty J_0(ua_n)p_{A_n}(a_n)da_n = \prod_{n=1}^{N} \langle J_0(ua_n) \rangle \tag{4.340}$$

A similar equation holds for $\Theta_\eta(u)$. In turn, according to the Blanc-Lapierre transformation (eq 4.332), $p_A(a)$ can be expressed in terms of $\Theta_\xi(u)$ as

$$p_A(a) = a \int_0^\infty J_0(ua)\Theta_X(u)u\,du = a \int_0^\infty J_0(ua) \prod_{n=1}^{N} \langle J_0(ua_n) \rangle u\,du$$

$$= a \int_0^\infty J_0(ua) \left[\prod_{n=1}^{N} \int_0^\infty J_0(ua_n)p_{A_n}(a_n)da_n \right] u\,du \tag{4.341}$$

This equation relates the distribution of the magnitude of the sum through distributions of the magnitude of each individual component.

In practice, it is the value of magnitude squared, $I = a^2$, which is available for measurement. In this case, the PDF of the intensity I can be easily found to be

$$p_i(I) = \frac{p_A(\sqrt{I})}{2\sqrt{I}} = \frac{\sqrt{I}\int_0^\infty J_0(u\sqrt{I})\left[\prod_{n=1}^N \int_0^\infty J_0(ua_n)p_{A_n}(a_n)da_n\right]u\,du}{2\sqrt{I}}$$

$$= \frac{1}{2}\int_0^\infty J_0(u\sqrt{I})\prod_{n=1}^N \langle J_0(ua_n)\rangle u\,du \qquad (4.342)$$

The last equation is known as the Kluyver equation.

4.7.2.3 Relations Between Moments of $p_{A_n}(a_n)$ and $p_i(I)$

The Kluyver equation (4.342) can rarely be used to produce analytical results. However, a useful relation between the moments of the sum and those of its components can be analytically obtained. In order to do so, one can make use of the moment generating function $\Phi(\lambda)$ as discussed in Section 2.1.1.

$$m_{kI} = (-1)^N\frac{\partial^N}{\partial\lambda^N}\Phi(\lambda)\bigg|_{\lambda=0} \qquad (4.343)$$

A Taylor-type expansion for $\Phi(\lambda)$ can be obtained directly from the Kluyver equation (4.342)

$$\Phi(\lambda) = \int_0^\infty e^{-\lambda I}p_i(I)dI = \int_0^\infty e^{-\lambda I}\frac{1}{2}\int_0^\infty J_0(u\sqrt{I})\prod_{n=1}^N \langle J_0(ua_n)\rangle u\,du\,dI$$

$$= \int_0^\infty \prod_{n=1}^N \langle J_0(ua_n)\rangle u\,du\,\frac{1}{2}\int_0^\infty e^{-\lambda I}J_0(u\sqrt{I})dI \qquad (4.344)$$

The first step is to calculate the inner integral in eq. (4.344). This can be achieved by using eq. (6.614-1) on page 709 in [39]

$$\int_0^\infty e^{-\alpha x}J_\nu(\beta\sqrt{x})dx = \frac{\beta}{4}\sqrt{\frac{\pi}{\alpha^3}}\exp\left(-\frac{\beta^2}{8\alpha}\right)\times\left[I_{\frac{\nu-1}{2}}\left(\frac{\beta^2}{8\alpha}\right) - I_{\frac{\nu+1}{2}}\left(\frac{\beta^2}{8\alpha}\right)\right] \qquad (4.345)$$

and setting $\alpha = \lambda$, $\beta = u$, $\nu = 0$, $x = I$, which implies that

$$\frac{\beta^2}{8\alpha} = \frac{u^2}{8\lambda}, \frac{\nu-1}{2} = \frac{1}{2}, \frac{\nu+1}{2} = \frac{1}{2} \qquad (4.346)$$

and thus

$$\int_0^\infty e^{-\lambda I}J_0(u\sqrt{I})dI = \frac{u}{4}\sqrt{\frac{\pi}{\lambda^3}}\exp\left(-\frac{u^2}{8\lambda}\right)\times\left[I_{-\frac{1}{2}}\left(\frac{u^2}{y\lambda}\right) - I_{\frac{1}{2}}\left(\frac{u^2}{8\lambda}\right)\right] \qquad (4.347)$$

This last equation can be further simplified by observing that (using 10.2.13–14 on page 443 in [29])

$$I_{1/2}(z) = \sqrt{\frac{2z}{\pi}\frac{\sinh z}{z}} = \sqrt{\frac{2}{\pi z}}\sinh z \tag{4.348}$$

$$I_{-1/2}(z) = \sqrt{\frac{2z}{\pi}\frac{\cosh z}{z}} = \sqrt{\frac{2}{\pi z}}\cosh z \tag{4.349}$$

and thus

$$I_{-\frac{1}{2}}\left(\frac{u^2}{8\lambda}\right) - I_{\frac{1}{2}}\left(\frac{u^2}{8\lambda}\right) = \sqrt{\frac{16\lambda}{\pi u^2}}\exp\left(-\frac{u^2}{8\lambda}\right) \tag{4.350}$$

Taking eq. (4.350) into account, eq. (4.347) becomes

$$\int_0^\infty e^{-\lambda I}J_0(u\sqrt{I})dI = \frac{u}{4}\sqrt{\frac{\pi}{\lambda^3}}\exp\left(-\frac{u^2}{8\lambda}\right) \times \sqrt{\frac{16\lambda}{\pi u^2}}\exp\left(-\frac{u^2}{8\lambda}\right) = \frac{1}{\lambda}\exp\left(-\frac{u^2}{4\lambda}\right) \tag{4.351}$$

Finally, the expression for $\Phi(\lambda)$ becomes

$$\Phi(\lambda) = \frac{1}{2}\int_0^\infty \prod_{n=1}^N \langle J_0(ua_n)\rangle u\,du\frac{1}{\lambda}\exp\left(-\frac{u^2}{4\lambda}\right) \tag{4.352}$$

As the next step, let us use the following substitution of variables in eq. (4.352)

$$x = u^2/4\lambda, u = \sqrt{4\lambda x}, \quad du = \sqrt{\lambda}\frac{dx}{x} \tag{4.353}$$

so that eq. (4.352) becomes

$$\Phi(\lambda) = \frac{1}{2}\int_0^\infty \prod_{n=1}^N \langle J_0(\sqrt{4\lambda x}a_n)\rangle \exp(-x)dx \tag{4.354}$$

The last integral cannot be calculated analytically in closed form, therefore one can use the Taylor expansion of $J_0(\sqrt{4\lambda x}a_n)$ with respect to variable λ. Using eq 9.1.10 in [29]

$$J_\nu(z) = \left(\frac{z}{2}\right)^\nu \sum_{k=0}^\infty \frac{(-z^2/4)^k}{k!\Gamma(\nu+k+1)} \tag{4.355}$$

with $\nu = 0$ and $z = \sqrt{4\lambda x}a_n$ one can obtain

$$J_0(\sqrt{4\lambda x}a_n) = \sum_{k_n=0}^\infty \frac{(-1)^{k_n}\lambda^{k_n}x^{k_n}a_2^{2k_n}}{k_n!k_n!}, \tag{4.356}$$

and, thus

$$\langle J_0(\sqrt{4\lambda x}a_n)\rangle = \sum_{k_n=0}^{\infty} \frac{(-1)^{k_n}\lambda^{k_n}x^{k_n}\langle a_n^{2k_n}\rangle}{k_n!k_n!} \tag{4.357}$$

Furthermore

$$\prod_{n=1}^{N}\langle J_0(\sqrt{4\lambda x}a_n)\rangle = \prod_{n=1}^{N}\sum_{k_n=0}^{\infty} \frac{(-1)^{k_n}\lambda^{k_n}x^{k_n}\langle a_n^{2k_n}\rangle}{k_n!k_n!}$$

$$= \sum_{k_N=0}^{\infty}\sum_{k_{N-1}=0}^{\infty}\cdots\sum_{k_1=0}^{\infty} \frac{(-1)^{m}\lambda^{m}x^{m}\langle a_{N-1}^{2k_N}\rangle\langle a_{N-1}^{2k_{N-1}}\rangle\cdots\langle a_1^{2k_1}\rangle}{k_N!k_N!k_{N-1}!K_N\cdots k_1!k_1!}$$

$$= \sum_{m=0}^{\infty}(-1)^{m}c_m\lambda^{m}x^{m} \tag{4.358}$$

where

$$m = \sum_{n=1}^{N}k_n \tag{4.359}$$

and

$$c_m = \sum_{k_N=0}^{\infty}\sum_{k_{N-1}=0}^{\infty}\cdots\sum_{k_1=0}^{\infty} \frac{\langle a_{N-1}^{2k_N}\rangle\langle a_{N-1}^{2k_{N-1}}\rangle\cdots\langle a_1^{2k_1}\rangle}{k_N!k_N!k_{N-1}!k_{N-1}\cdots k_1!k_1!}\Bigg|_{\sum_{n=1}^{N}k_n=m} \tag{4.360}$$

Finally, integrating eq. (4.354) with respect to x and taking advantage of eq. (4.359) one obtains

$$\Phi(\lambda) = \int_0^{\infty}\left(\sum_{m=0}^{\infty}(-1)^{m}c_m\lambda^{m}x^{m}\right)\exp(-x)dx = \sum_{m=0}^{\infty}(-1)^{m}c_m\lambda^{m}m! \tag{4.361}$$

Comparing this to eq. (4.358), the following expression can be derived

$$m_{kI} = k!k!c_k = k!k!\sum_{k_N=0}^{\infty}\sum_{k_{N-1}=0}^{\infty}\cdots\sum_{k_1=0}^{\infty} \frac{\langle a_{N-1}^{2k_N}\rangle\langle a_{N-1}^{2k_{N-1}}\rangle\cdots\langle a_1^{2k_1}\rangle}{k_N!k_N!k_{N-1}!k_{N-1}!\cdots k_1!k_1!}\Bigg|_{\sum_{n=1}^{N}k_n=k} \tag{4.362}$$

4.7.2.4 The Gram–Charlier Series for $p_{\xi_R}(x)$ and $p_i(I)$

As has been mentioned above, it is the distribution of the intensity $p_i(I)$ which is usually available for measurement. At the same time, it is important for numerical simulations to know and reproduce the statistic of the field amplitude $p_{\xi_R}(x)$. The exact analytical form, given by eqs. (4.330)–(4.331) is rarely achievable (however, a number of results can be found in [19]). Important relations like (4.362) give insight into how the distribution behaves but do not really allow its numerical simulation. In this section, useful results allowing the reconstruction of the PDF $p_{\xi_R}(x)$ from the PDF of the intensity $p_i(I)$ are derived.

As the first step, the Blanc-Lapierre transformation (4.331) can be rewritten in terms of $p_i(I)$, using the fact that

$$p_i(I) = \frac{p_A(\sqrt{I})}{2\sqrt{I}} \Rightarrow p_A(\sqrt{I}) = 2\sqrt{I}p_i(I) \tag{4.363}$$

and thus

$$
\begin{aligned}
p_{\xi_R}(x) &= \frac{1}{2\pi} \int_{-\infty}^{\infty} \int_0^{\infty} J_0(ua)p_a(a)da \exp[-jxu]du \\
&= \frac{1}{2\pi} \int_{-\infty}^{\infty} \int_0^{\infty} J_0(u\sqrt{I})p_A(\sqrt{I}) \exp[-jxu]du \, d\sqrt{I} \\
&= \frac{1}{2\pi} \int_{-\infty}^{\infty} \int_0^{\infty} J_0(u\sqrt{I})2\sqrt{I}p_i(I)\frac{dI}{2\sqrt{I}}\exp[-jxu]du \\
&= \frac{1}{2\pi} \int_{-\infty}^{\infty} \exp[-jxu]du \int_0^{\infty} J_0(u\sqrt{I})p_i(I)dI
\end{aligned} \tag{4.364}
$$

Let now the PDF $p_i(I)$ be represented by its expansion through the Laguerre polynomials $L_k(\beta I)$ as discussed in section 2.2, i.e.

$$p_i(I) = \beta \exp[-\beta I] \sum_{k=0}^{\infty} \alpha_k L_k(\beta I) \tag{4.365}$$

where

$$\alpha_k = \int_0^{\infty} p_i(I)L_k(\beta I)dI \tag{4.366}$$

In this case, eq. (4.364) becomes

$$
\begin{aligned}
p_{\xi_R}(x) &= \frac{1}{2\pi} \int_{-\infty}^{\infty} \exp[-jxu]du \int_0^{\infty} J_0(u\sqrt{I})\beta \exp[-\beta I] \sum_{k=0}^{\infty} \alpha_k L_k(\beta I)dI \\
&= \frac{\beta}{2\pi} \sum_{k=0}^{\infty} \alpha_k \int_{-\infty}^{\infty} \exp[-jxu]du \int_0^{\infty} J_0(u\sqrt{I})L_k(\beta I) \exp[-\beta I]dI \\
&= \frac{\beta}{2\pi} \sum_{k=0}^{\infty} \alpha_k \int_{-\infty}^{\infty} F_k(u) \exp[-jxu]du
\end{aligned} \tag{4.367}
$$

where

$$F_k(u) = \int_0^{\infty} J_0(u\sqrt{I})L_k(\beta I) \exp[-\beta I]dI \tag{4.368}$$

In order to simplify eq. (4.368), one can use the following integral (2.19.12-7 on page 474 in [40])

$$\int_0^\infty x^{\lambda/2} e^{-px} \left\{ \begin{array}{l} J_\lambda(b\sqrt{x}) \\ I_\lambda(b\sqrt{x}) \end{array} \right\} L_n^\lambda(cx) dx = J(p,c) \tag{4.369}$$

where

$$J(p,c) = \left(\frac{b}{2}\right)^\lambda \frac{(p-c)^n}{p^{\lambda+n+1}} \exp\left(-\frac{b^2}{4p}\right) L_n^\lambda\left(\pm\frac{b^2 c}{4pc - 4p^2}\right), \qquad p \neq c \tag{4.370}$$

$$J(c,c) = \frac{(\pm 1)^n}{n! c^{\lambda+n+1}} \left(\frac{b}{2}\right)^{\lambda+2n} \exp\left(\mp\frac{b^2}{4c}\right) \tag{4.371}$$

Using the last equation with $\lambda = 0$, $p = c = \beta$, $n = k$ and $b = u$, one can obtain

$$F_k(u) = \frac{1}{k! \beta^{k+1}} \left(\frac{u}{2}\right)^{2k} \exp\left(-\frac{u^2}{4\beta}\right) \tag{4.372}$$

Rewriting eq. (4.367) by taking into account eq. (4.372), one can obtain with $u = I$

$$\begin{aligned} p_{\xi_R}(x) &= \frac{\beta}{2\pi} \sum_{k=0}^\infty \alpha_k \int_{-\infty}^\infty F_k(u) \exp[-jxu] du \\ &= \frac{\beta}{2\pi} \sum_{k=0}^\infty \alpha_k \int_{-\infty}^\infty \frac{1}{k! \beta^{k+1}} \left(\frac{u}{2}\right)^{2k} \exp\left(-\frac{u^2}{4\beta}\right) \exp[-jxu] du \\ &= \frac{\sqrt{\beta}}{\pi} \sum_{k=0}^\infty \frac{\alpha_k}{k!} \int_{-\infty}^\infty t^{2k} \exp(-t^2) \exp[-j(2\sqrt{\beta}x)t] dt = \frac{\sqrt{\beta}}{\pi} \sum_{k=0}^\infty \frac{\alpha_k}{k!} G_k(x) \end{aligned} \tag{4.373}$$

where the following notation is used

$$G_k(x) = \int_{-\infty}^\infty t^{2k} \exp(-t^2) \exp[-j(2\sqrt{\beta}x)t] dt = \exp[-\beta x^2] \int_{-\infty}^\infty t^{2k} \exp[-(t + j\sqrt{\beta}x)^2] dt \tag{4.374}$$

Comparing the standard integral 3.462-4 on page 338 in [39]

$$\int_{-\infty}^\infty x^n \exp[-(x-\beta)^2] dx - (2j)^{-n} \sqrt{\pi} II_n(j\beta) \tag{4.375}$$

one can obtain

$$G_k(x) = \exp[-\beta x^2] \int_{-\infty}^\infty t^{2k} \exp[-(t + j\sqrt{\beta}x)^2] dt = 2^{-2k}(-1)^k \exp[-\beta x^2] \sqrt{\pi} H_{2k}(\sqrt{\beta}x) \tag{4.376}$$

and thus

$$p_{\xi_R}(x) = \frac{\sqrt{\beta}}{\pi} \sum_{k=0}^{\infty} \frac{\alpha_k}{k!} G_k(x) = \frac{\sqrt{\beta}}{\sqrt{\pi}} \exp[-\beta x^2] \sum_{k=0}^{\infty} \frac{(-1)^k \alpha_k}{2^{2k} k!} H_{2k}(\sqrt{\beta} x) \qquad (4.377)$$

Here $H_n(x)$ is the Hermitian polynomial of order n as discussed in section 2.2. The last expression should not come as a surprise. If $p_i(I)$ is an exponential distribution (χ^2 of two degrees of freedom) then $\alpha_0 = 1$ and $\alpha_k = 0$ for $k > 0$. This reduces eq. (4.377) to the Gaussian PDF as it must. At the same time, eq. (4.377) is nothing but the Gram–Charlier expansion of the PDF $p_{\xi_R}(x_1)$. The odd order terms are missing due to the fact that $p_{\xi_R}(x_1)$ is a symmetrical PDF.

Treatment of non-stationary processes becomes very complicated since independence between the phase and the magnitude does not hold. It is possible to obtain a number of useful results for the Gaussian process using direct calculations of the related statistics. This approach is described in some detail in section 4.7.3 below. A number of examples for the non-Gaussian case is considered in section 4.7.4.

4.7.3 Gaussian Narrow Band Process

4.7.3.1 First Order Statistics

Let $\xi_I(t)$ and $\xi_Q(t)$ be two independent Gaussian processes with equal variance σ^2 and different mean values m_I and m_Q respectively. These can be considered as components of a complex random process

$$\xi(t) = \xi_I(t) + j\xi_Q(t) = A(t) \exp[j\Psi(t)] \qquad (4.378)$$

The goal in this section is to derive first order statistics of the envelope $A(t)$ and phase $\Psi(t)$ of this complex Gaussian process.

Since in-phase and quadrature components are independent, their joint PDF is a product of the marginal PDF of each component, i.e.

$$p_{\xi_I \xi_Q}(x, y) = p_{\xi_I}(x) p_{\xi_Q}(y) = \frac{1}{2\pi\sigma^2} \exp\left[-\frac{(x - m_I)^2 + (y - m_Q)^2}{2\sigma^2} \right] \qquad (4.379)$$

Making a transformation of variables according to

$$\begin{aligned} x &= A \cos \phi \\ y &= A \sin \phi \end{aligned} \qquad (4.380)$$

the Jacobian of this transformation is

$$J = \begin{vmatrix} \dfrac{\partial x}{\partial A} & \dfrac{\partial y}{\partial A} \\ \dfrac{\partial x}{\partial \phi} & \dfrac{\partial y}{\partial \phi} \end{vmatrix} = \begin{vmatrix} \cos \phi & \sin \phi \\ -A \sin \phi & A \cos \phi \end{vmatrix} = A \qquad (4.381)$$

and, thus, the joint PDF of the envelope and phase is

$$
\begin{aligned}
p_{A,\phi}(A, \phi) &= p_{\xi_I \xi_Q}(A \cos \phi, A \sin \phi)A \\
&= \frac{A}{2\pi\sigma^2} \exp\left[-\frac{(A \cos \phi - m_I)^2 + (A \sin \phi - m_Q)^2}{2\sigma^2}\right] \\
&\times \frac{A}{2\pi\sigma^2} \exp\left[-\frac{A^2 + m_I^2 + m_Q^2 - 2A(m_I \cos \phi + m_Q \sin \phi)}{2\sigma^2}\right]
\end{aligned}
\tag{4.382}
$$

Further simplification can be achieved by introducing two new parameters

$$
m = \sqrt{m_I^2 + m_Q^2}, \qquad \psi = a \tan \frac{m_Q}{m_I}
\tag{4.383}
$$

and thus

$$
m_i = m \cos \psi, \qquad m_Q = m \sin \psi
\tag{4.384}
$$

As a result, eq. (4.382) can be rewritten as

$$
\begin{aligned}
p_{A,\phi}(A, \phi) &= \frac{A}{2\pi\sigma^2} \exp\left[-\frac{A^2 + m^2 - 2Am(\cos \psi \cos \phi + \sin \psi \sin \phi)}{2\sigma^2}\right] \\
&= \frac{A}{2\pi\sigma^2} \exp\left[\frac{A^2 + m^2 - 2Am \cos(\phi - \psi)}{2\sigma^2}\right]
\end{aligned}
\tag{4.385}
$$

This in hand, one can calculate the PDF of the amplitude by integrating expression (4.385) over ϕ:

$$
\begin{aligned}
p_A(A) &= \int_{-\pi}^{\pi} p_{A,\phi}(A, \phi)d\phi = \int_{-\pi}^{\pi} \frac{A}{2\pi\sigma^2} \exp\left[-\frac{A^2 + m^2 - 2Am \cos(\phi - \psi)}{2\sigma^2}\right]d\phi \\
&= \frac{A}{2\pi\sigma^2} \exp\left[-\frac{A^2 + m^2}{2\sigma^2}\right] \int_{-\pi}^{\pi} \exp\left[\frac{Am \cos(\phi - \psi)}{\sigma^2}\right]d\phi = \frac{A}{\sigma^2} \exp\left[-\frac{A^2 + m^2}{2\sigma^2}\right] I_0\left(\frac{mA}{\sigma^2}\right)
\end{aligned}
\tag{4.386}
$$

Here the following standard integral has been used

$$
\int_{-\pi}^{\pi} \exp\left[\frac{Am \cos(\phi - \psi)}{\sigma^2}\right]d\phi = \int_{-\pi}^{\pi} \exp\left[\frac{Am \cos(\phi)}{\sigma^2}\right]d\phi = 2\pi I_0\left(\frac{Am}{\sigma^2}\right)
\tag{4.387}
$$

Thus, the distribution of the envelope is Rician. In the very same manner, the distribution of the phase can be found to be

$$
p_\phi(\phi) = \int_{-\pi}^{\pi} p_{A,\phi}(A, \phi)dA = \int_0^\infty \frac{A}{2\pi\sigma^2} \exp\left[-\frac{A^2 + m^2 - 2Am \cos(\phi - \psi)}{2\sigma^2}\right]dA
\tag{4.388}
$$

Using substitutions

$$z = \frac{A - m\cos(\phi - \psi)}{\sigma}, \qquad dA = \sigma dz, \qquad A = \sigma z + m\cos(\phi - \psi) \qquad (4.389)$$

eq. (4.388) becomes

$$p_\phi(\phi) = \int_{-m\cos(\phi-\psi)}^{\infty} \frac{\sigma z + m\cos(\phi - \psi)}{2\pi\sigma^2} \exp\left[-\frac{(\sigma z)^2 + m^2 + m^2\cos^2(\phi - \psi)}{2\sigma^2}\right] \sigma dz$$

$$= \frac{1}{2\pi} \exp\left[-\frac{m^2}{2\sigma^2}\right] + \frac{m\cos(\phi - \psi)}{\sigma\sqrt{2\pi}} \Phi\left(\frac{m\cos(\phi - \psi)}{\sigma}\right) \exp\left[-\frac{m^2\sin^2(\phi - \psi)}{2\sigma^2}\right]$$

$$(4.390)$$

where $\Phi(s)$ is the CDF of the standard Gaussian PDF

$$\Phi(s) = \int_{-\infty}^{s} \frac{1}{\sqrt{2\pi}} \exp\left[-\frac{x^2}{2}\right] dx \qquad (4.391)$$

For a particular case of $m = m_I = m_Q = 0$

$$p_A(A) = \frac{A}{\sigma^2} \exp\left[-\frac{A^2}{2\sigma^2}\right] \qquad (4.392)$$

$$p_\phi(\phi) = \frac{1}{2\pi} \qquad (4.393)$$

4.7.3.2 Correlation Function of the In-phase and Quadrature Components

Second order statistics of the in-phase and quadrature components can be easily found using the fact that

$$S_I(\omega) = |H_I(\omega)|^2 \frac{N_0}{2}, \qquad S_Q(\omega) = |H_Q(\omega)|^2 \frac{N_0}{2} \qquad (4.394)$$

where $S_I(\omega)$ and $S_Q(\omega)$ are power spectral densities (PSD) of the in-phase and quadrature components. According to the Wiener–Khinchin theorem, we have then

$$R_I(\tau) = \frac{1}{2\pi} \int_{-\infty}^{\infty} S_I(\omega) \exp[j\omega\tau] d\omega \qquad (4.395)$$

Alternatively, the correlation function can be obtained through factorization, as described below. In general, the transfer function of a filter can be represented as a ratio of two polynomials

$$H(s) = \frac{B(s)}{A(s)} = \frac{b_n s^n + b_{n-1} s^{n-1} + \cdots + b_0}{a_n s^n + a_{n-1} s^{n-1} + \cdots + a_0} \qquad (4.396)$$

One can expand $H(s)$ using the method of partial fraction expansion:

$$H(s) = \frac{a_1}{s - p_1} + \frac{a_2}{s - p_2} + \cdots + \frac{a_n}{s - p_n} + K(s) \qquad (4.397)$$

where a_k are the residues, p_k are the poles and $K(s)$ is the remaining direct term.

Assuming that $K(s) = 0$, one can see that the time-domain response of the filter to an impulse is given by:

$$h(t) = \int_{-\infty}^{\infty} H(\omega) \exp[-j\omega t] d\omega = \int_{-\infty}^{\infty} \left(\frac{a_1}{s - p_1} + \cdots + \frac{a_n}{s - p_n} \right) \exp[-j\omega t] d\omega = \sum_{k=1}^{n} a_k \exp[p_k t]$$

$$(4.398)$$

Now that the expression for $h(t)$ is factorized, one can determine the autocorrelation function:

$$R(\tau) = \int_0^{\infty} h(t) h(t + \tau) dt = \int_0^{\infty} \left\{ \sum_{k=1}^{n} \sum_{l=1}^{n} a_k a_l \exp[p_k \tau] \exp[(p_k + p_l)t] \right\} dt$$

$$\sum_{k=1}^{n} \sum_{l=1}^{n} a_k a_l \exp[p_k \tau] \int_0^{\infty} \exp[(p_k + p_l)t] dt = \sum_{k=1}^{n} \sum_{l=1}^{n} a_k a_l \frac{\exp[p_k \tau]}{p_k + p_l}$$

$$= \sum_{k=1}^{n} a_k \exp[p_k t] \sum_{l=1}^{n} \frac{a_l}{p_k + p_l}$$

$$(4.399)$$

Finally, the expression for $R(\tau)$ has the following form

$$R(\tau) = \sum_{k=1}^{n} c_k \exp[p_k \tau] \qquad (4.400)$$

where coefficients c_k are given by:

$$c_k = a_k \sum_{l=1}^{n} \frac{a_l}{p_k + p_l} \qquad (4.401)$$

4.7.3.3 Second Order Statistics of the Envelope

In order to find second order statistics of the envelope and the phase, one ought to derive the fourth order joint probability density function of $\xi_I(t)$, $\xi_I(t + \tau)$, $\xi_Q(t)$ and $\xi_Q(t + \tau)$. Since $\xi_I(t)$ and $\xi_Q(t + \tau)$ are independent in any moment of time, one obtains

$$p_{\xi_I \xi_Q \xi_{I\tau} \xi_{Q\tau}}(x, y, x_\tau, y_\tau) = p_{\xi_I \xi_\tau}(x, x_\tau) p_{\xi_Q \xi_{Q\tau}}(y, y_\tau)$$

$$= \frac{1}{2\pi\sigma^2 \sqrt{1 - \rho^2(\tau)}} \exp\left[-\frac{x^2 - 2\rho(\tau)xx_\tau + x_\tau^2}{2\sigma^2(1 - \rho^2(\tau))} \right]$$

$$\times \frac{1}{2\pi\sigma^2 \sqrt{1 - \rho^2(\tau)}} \exp\left[-\frac{y^2 - 2\rho(\tau)yy_\tau + y_\tau^2}{2\sigma^2(1 - \rho^2(\tau))} \right] \qquad (4.402)$$

or

$$p_{\xi_I \xi_Q \xi_{I\tau} \xi_{Q\tau}}(x,y,x_\tau,y_\tau) = \frac{1}{4\pi^2 \sigma^4(1 - \rho^2(\tau))} \exp\left[-\frac{x^2 + y^2 - 2\rho(\tau)(xx_\tau + yy_\tau) + x_\tau^2 + y_\tau^2}{2\sigma^2(1 - \rho^2(\tau))} \right]$$

(4.403)

Here, $\rho(\tau)$ is the normalized correlation function, such that $\rho(0) = 1$.

Changing variables according to

$$x = A\cos\phi, \qquad y = A\sin\phi$$
$$x_\tau = A_\tau \cos\phi_\tau, \quad y_\tau = A_\tau \sin\phi_\tau$$

(4.404)

and noticing that the Jacobian of the transformation (4.404) is

$$J = \begin{vmatrix} \cos\phi & \sin\phi & 0 & 0 \\ -A\sin\phi & A\cos\phi & 0 & 0 \\ 0 & 0 & \cos\phi_\tau & \sin\phi_\tau \\ 0 & 0 & -A_\tau\sin\phi_\tau & A_\tau\cos\phi_\tau \end{vmatrix} = AA_\tau$$

(4.405)

the following joint PDF of magnitude and phase in two separate moments of time can be obtained

$$p_{AA_\tau\phi\phi_\tau}(A, A_\tau, \phi, \phi_\tau)$$
$$= \frac{AA_\tau}{4\pi^2\sigma^4(1 - \rho^2(\tau))} \exp\left\{ -\frac{1}{2\sigma^2(1 - \rho^2(\tau))} [A^2 + A_\tau^2 - 2\rho(\tau)AA_\tau\cos(\phi_\tau - \phi)] \right\}$$

(4.406)

A joint probability density function of amplitude in two separate moments of time can be calculated by integrating (4.406) over phase variables, i.e.

$$p_{AA_\tau}(A, A_\tau) = \int_{-\pi}^{\pi}\int_{-\pi}^{\pi} p_{AA_\tau\phi\phi_\tau}(A, A_\tau, \phi, \phi_\tau) d\phi\, d\phi_\tau$$

$$= \frac{AA_\tau}{4\pi^2\sigma^4(1 - \rho^2(\tau))} \exp\left\{ -\frac{A^2 + A_\tau^2}{2\sigma^2(1 - \rho^2(\tau))} \right\}$$

$$\times \int_{-\pi}^{\pi}\int_{-\pi}^{\pi} \exp\left\{ \frac{\rho(\tau)AA_\tau\cos(\phi_\tau - \phi)}{\sigma^2(1 - \rho^2(\tau))} \right\} d\phi\, d\phi_\tau$$

$$\frac{AA_\tau}{\sigma^4(1 - \rho^2(\tau))} \exp\left\{ -\frac{A^2 + A_\tau^2}{2\sigma^2(1 - \rho^2(\tau))} \right\} I_0\left[\frac{\rho(\tau)}{1 - \rho^2(\tau)} \frac{AA_\tau}{\sigma^2} \right]$$

(4.407)

Here, once again, the integral (4.387) has been used. The last expression allows one to calculate the correlation function of the envelope. The first step is to calculate an average value of the magnitude. Since it is already known that the marginal PDF is given by the

Rayleigh distribution (4.392), the mean and variance can be easily found to be

$$m_A = \int_0^\infty A p_A(A) dA = \int_0^\infty \frac{A^2}{\sigma^2} \exp\left[-\frac{A^2}{2\sigma^2}\right] dA = \sqrt{\frac{\pi}{2}}\sigma \tag{4.408}$$

$$\sigma_A^2 = \int_0^\infty (A - m_A)^2 p_A(A) dA = \int_0^\infty \left(A - \sqrt{\frac{\pi}{2}}\sigma\right)^2 \frac{A}{\sigma^2} \exp\left[-\frac{A^2}{2\sigma^2}\right] dA = \left(2 - \frac{\pi}{2}\right)\sigma^2 \tag{4.409}$$

Further simplification can be achieved by normalizing the magnitude variables by variance

$$z_1 = \frac{A}{\sigma}, \qquad z_2 = \frac{A_\tau}{\sigma}, \qquad J = \sigma^2 \tag{4.410}$$

and, thus

$$p_{AA_\tau}(z_1, z_2) = \frac{z_1 z_2}{1 - \rho^2(\tau)} \exp\left\{-\frac{z_1^2 + z_2^2}{2(1 - \rho^2(\tau))}\right\} I_0\left[\frac{\rho(\tau)}{1 - \rho^2(\tau)} z_1 z_2\right] \tag{4.411}$$

Using a Bessel function expansion in terms of the Laguerre polynomial

$$I_0(a z_1 z_2) = \sum_{n=0}^\infty L_n(z_1^2/2) L_n(z_2^2/2) a^n \tag{4.412}$$

the expression (4.411) can be transformed into

$$p_{AA_\tau}(z_1, z_2) = z_1 z_2 \exp\left\{-\frac{z_1^2 + z_2^2}{2}\right\} \sum_{n=0}^\infty L_n\left(\frac{z_1^2}{2}\right) L_n\left(\frac{z_1^2}{2}\right) \frac{\rho^{2n}}{n! n!} \tag{4.413}$$

The expansion (4.413) can be now used to calculate the second moment of the envelope

$$\langle AA_\tau \rangle = \sigma^2 \langle z_1 z_2 \rangle = \sigma^2 \int_{-\infty}^\infty \int_{-\infty}^\infty z_1 z_2 p_{AA_\tau}(z_1, z_2) dz_1 \, dz_2$$

$$= \sigma^2 \sum_{n=0}^\infty \frac{\rho^{2n}}{n! n!} \int_{-\infty}^\infty \int_{-\infty}^\infty z_1 z_2 \exp\left\{-\frac{z_1^2 + z_2^2}{2}\right\} L_n\left(\frac{z_1^2}{2}\right) L_n\left(\frac{z_1^2}{2}\right) dz_1 \, dz_2 \tag{4.414}$$

$$= \sigma^2 \sum_{n=0}^\infty \frac{\rho^{2n}}{n! n!} \left[\int_{-\infty}^\infty z_1 \exp\left\{-\frac{z_1^2}{2}\right\} L_n\left(\frac{z_1^2}{2}\right) dz_1\right]^2$$

Finally

$$C_{AA}(\tau) = \langle AA_\tau \rangle - m_A^2 = \frac{\pi \sigma^2}{2} \sum_{n=1}^\infty \left(\frac{1}{2n!!}\right)^2 \rho^{2n}(\tau) = \sigma_A^2 \frac{\pi}{4 - \pi} \sum_{n=1}^\infty \left(\frac{1}{2n!!}\right)^2 \rho^{2n}(\tau) \tag{4.415}$$

If only two terms are taken into account then

$$C_{AA}(\tau) = \sigma_A^2 (0.921 \rho^2(\tau) + 0.058 \rho^4(\tau) + \cdots) \approx \sigma_A^2 \rho^2(\tau) \tag{4.416}$$

Introducing a new variable $\Delta \phi = \phi_\tau - \phi$ and integrating with reference to eq. (4.406) over A, A_τ and ϕ, we can obtain the expression for the PDF of the phase difference in two moments of time separated by τ

$$p_{\Delta\phi}(\Delta\phi) = \frac{1 - \rho^2(\tau)}{2\pi} \left[\frac{1}{1 - x^2} + x \frac{\frac{\pi}{2} + a \sin x}{(1 - x^2)^{3/2}} \right], \qquad x = \rho(\tau) \cos \Delta\phi \tag{4.417}$$

4.7.3.4   Level Crossing Rate

The level crossing rate (LCR) of the random process is defined as the number of crossings of a fixed threshold R per unit time in the positive direction (i.e. from a level below R to a level above R). In order to find the LCR, one needs to calculate the joint probability density of the random process and its derivative $p_{\xi\dot{\xi}}(x, \dot{x})$. With this in hand, the LCR can be calculated as

$$LCR(R) = \int_0^\infty \dot{x} p_{\xi\dot{\xi}}(x, \dot{x}) \mathrm{d}\dot{x} \tag{4.418}$$

For a random process with Gaussian components

$$p_{A\dot{A}}(A, \dot{A}) = p_A(A) p_{\dot{A}}(\dot{A}) \tag{4.419}$$

where

$$p_A(A) = \frac{A}{\sigma^2} \exp\left(-\frac{A^2 + m^2}{2\sigma^2}\right) I_0\left(\frac{mA}{\sigma^2}\right) \tag{4.420}$$

and

$$p_{\dot{A}}(\dot{A}) = \frac{1}{\sigma\sqrt{2\pi(-\rho_0'')}} \exp\left[-\frac{\dot{A}^2}{2\sigma^2(-\rho_0'')}\right] \tag{4.421}$$

The following short notation is also used

$$-\rho_0'' = -\frac{\partial^2}{\partial \tau^2} \rho(\tau)\Big|_{\tau=0} \tag{4.422}$$

Thus, the LCR can be calculated as

$$LCR(R) = \int_0^\infty p_A(R)p_{\dot{A}}(\dot{A})\dot{A}\,d\dot{A} = p_A(R)\int_0^\infty p_{\dot{A}}(\dot{A})\dot{A}\,d\dot{A}$$

$$= p_A(R)\int_0^\infty \frac{\dot{A}}{\sigma\sqrt{2\pi(-\rho_0'')}}\exp\left[-\frac{\dot{A}^2}{2\sigma^2(-\rho_0'')}\right]d\dot{A} \qquad (4.423)$$

$$= \frac{p_A(R)}{\sigma\sqrt{2\pi(-\rho_0'')}}\int_0^\infty \sigma^2(-\rho_0'')\exp\left[-\frac{\dot{A}^2}{2\sigma^2(-\rho_0'')}\right]d\frac{\dot{A}^2}{2\sigma^2(-\rho_0'')}$$

$$\frac{p_a(R)\sigma^2(-\rho_0'')}{\sigma\sqrt{2\pi(-\rho_0'')}}\int_0^\infty \exp[-t]dt = p_A(R)\sigma\sqrt{\frac{-\rho_0''}{2\pi}} = \frac{R}{\sigma}\exp\left[-\frac{R^2+m^2}{2\sigma^2}\right]I_0\left(\frac{mR}{\sigma^2}\right)\sqrt{\frac{-\rho_0''}{2\pi}}$$

In the particular case of $m = 0$, one can obtain

$$LCR(R) = \frac{R}{\sigma}\exp\left[-\frac{R^2}{2\sigma^2}\right]\sqrt{\frac{-\rho_0''}{2\pi}} \qquad (4.424)$$

In many cases, it is easier to calculate the coefficient $-\rho_0''$ through the power spectral density. Indeed, using properties of the Fourier transform, one can obtain

$$-\rho_0'' = \frac{\partial^2}{\partial\tau^2}\rho(\tau)\bigg|_{\tau=0} = \frac{1}{2\pi\sigma^2}\int_0^\infty \omega^2 S(\omega)\,d\omega \qquad (4.425)$$

4.7.4 Examples of Non-Gaussian Narrow Band Random Processes

4.7.4.1 K Distribution

The K distribution

$$p_A(a) = \frac{2b}{\Gamma(\nu)}\left(\frac{ba}{2}\right)^\nu K_{\nu-1}(ba), \quad \nu > 0, x \geq 0 \qquad (4.426)$$

is a rare case when the Blanc-Lapierre transformation can be calculated analytically. In order to do so, the characteristic function of the PDF (4.426) using the Bessel transform [19,38] can be found first

$$\theta_{\xi_R}(s) = \int_0^\infty p_A(a)J_0(sa)da = \frac{b^{\nu+1}}{2^{\nu-1}\Gamma(\nu)}\int_0^\infty A^\nu K_{\nu-1}(ba)J_0(sa)da \qquad (4.427)$$

It can be calculated using the standard integral (6.576–7 on page 694 in [39])

$$\int_0^\infty x^{\eta+\mu+1}J_\mu(ax)K_\eta(bx)dx = 2^{\mu+\eta}a^\mu b^\eta \frac{\Gamma(\mu+\eta+1)}{(a^2+b^2)^{\mu+\eta+1}} \qquad (4.428)$$

where the following should be used:

$$\mu = 0, \qquad a = s, \qquad x = A, \qquad \eta = \nu - 1 \tag{4.429}$$

This yields

$$\theta_{\xi_R}(s) = \frac{b^{\nu+1}}{2^{\nu-1}\Gamma(\nu)} 2^{\nu-1} b^{\nu-1} \frac{\Gamma(\nu)}{(a^2+b^2)^\nu} = \frac{b^{2\nu}}{(s^2+b^2)^\nu} \tag{4.430}$$

The PDF $p_x(x)$ itself can be found by using an inverse Fourier transform of its characteristic function (4.430) [40]:

$$
\begin{aligned}
p_{\xi_R}(x) &= \frac{1}{2\pi} \int_{-\infty}^{\infty} \theta_{\xi_R}(s) \exp(-isx) ds = \frac{b^{2\nu}}{2\pi} \int_{-\infty}^{\infty} \frac{\exp(-isx)}{(s^2+b^2)^\nu} ds \\
&= \int_{-\infty}^{\infty} \frac{\exp(-isx)}{(is+b)^\nu(-is+b)^\nu} ds \\
&= \frac{b^{2\nu}}{2\pi} \int_{-\infty}^{\infty} (is+b)^{-2\frac{\nu}{2}}(-is+b)^{-2\frac{\nu}{2}} \exp(-isx) ds
\end{aligned}
\tag{4.431}
$$

The last integral is a table integral (3.384–9 on page 321 in [39]):

$$\int_{-\infty}^{\infty} (\beta + jx)^{-2\mu}(\beta - jx)^{-2\mu} e^{-jpx} dx = 2\pi(2\beta)^{-2\mu} \frac{|p|^{2\mu-1}}{\Gamma(2\mu)} W_{0,\frac{1}{2}-2\mu}(2\beta|p|) \tag{4.432}$$

Thus, setting

$$\mu = \frac{\nu}{2}, \qquad \beta = b, \quad p \rightarrow x, \qquad x \rightarrow s \tag{4.433}$$

the expression for the PDF becomes

$$p_{\xi_R}(x) = \frac{b^\nu}{2\pi} 2\pi(2b)^{-\nu} \frac{|x|^{\nu-1}}{\Gamma(\nu)} W_{0,\frac{1}{2}-\nu}(2b|x|) = \frac{2^{-2\nu+1}b}{\Gamma(\nu)} (2b|x|)^{\nu-1} W_{0,\frac{1}{2}-\nu}(2b|x|) \tag{4.434}$$

This equation can be further simplified using the following expression for the Whittaker function $W_{0,1/2-\nu}(°)$, first in terms of the Kummer function $U(a,b,z)$ and then in terms of the modified Bessel function K of the second kind (eq. 13.1–33 on page 505 and eq. 13.6–21 in Table 3.6 on page 510 in [29]):

$$W_{0,\frac{1}{2}-\nu}(z) = \exp\left(-\frac{z}{2}\right) z^{-\nu} U(1-\nu, 2-2\nu, z) = \frac{z^{1/2}}{\sqrt{\pi}} K_{\nu-\frac{1}{2}}\left(\frac{z}{2}\right) \tag{4.435}$$

and thus

$$p_{\xi_R}(x) = \frac{2^{-\nu+\frac{1}{2}}b}{\Gamma(\nu)\sqrt{\pi}} (b|x|)^{\nu-\frac{1}{2}} K_{\nu-\frac{1}{2}}(b|x|) \tag{4.436}$$

Therefore, the distribution of instantaneous values corresponding to the K distribution of the magnitude is again expressed in terms of the modified Bessel function K.

4.7.4.2 Gamma Distribution

Let the PDF of the intensity be the Gamma distribution

$$p_i(I) = \frac{\beta^{\alpha+1} I^\alpha \exp(-\beta I)}{\Gamma(\alpha+1)} \qquad (4.437)$$

Coefficients α_m of the expansion (4.365) can be found as

$$\alpha_m = \int_0^\infty p_i(I) L_m(\beta I) dI = \int_0^\infty \frac{\beta^{\alpha+1} I^\alpha \exp(-\beta I)}{\Gamma(\alpha+1)} L_m(\beta I) dI$$

$$= \frac{1}{\Gamma(\alpha+1)} \int_0^\infty (\beta I)^\alpha \exp[-\beta I] L_m(\beta I) d\beta I = \frac{1}{\Gamma(\alpha+1)} \int_0^\infty x^\alpha \exp[-x] L_m(x) dx \qquad (4.438)$$

The last integral can be calculated using the table integral 7.414–11 on page 845 in [39]

$$\int_0^\infty x^{\gamma-1} \exp[-x] L_m^\mu(x) dx = \frac{\Gamma(\gamma(\Gamma(1+\mu+m-\gamma))}{m!\Gamma(1+\mu-\gamma)}, \qquad \mathrm{Re}\{\gamma\} > 0 \qquad (4.439)$$

with $\gamma = \alpha + 1$, and $\mu = 0$:

$$\alpha_m = \frac{1}{\Gamma(\alpha+1)} \int_0^\infty x^\alpha \exp[-x] L_m(x) dx = \frac{1}{\Gamma(\alpha+1)} \frac{\Gamma(\alpha+1)\Gamma(1+m-\alpha-1)}{m!\Gamma(1-\alpha-1)} = \frac{\Gamma(m-\alpha)}{m!\Gamma(-\alpha)} \qquad (4.440)$$

Substituting eq. (4.440) into eq. (4.377) results in the following expression for the PDF of the amplitude of the distribution

$$p_{\xi_R}(x) = \frac{\sqrt{\beta}}{\sqrt{\pi}} \exp[-\beta x^2] \sum_{k=0}^\infty \frac{(-1)^k \Gamma(k-\alpha)}{2^{2k} K! K! \Gamma(-\alpha)} H_{2k}(\sqrt{\beta} x) \qquad (4.441)$$

A few examples showing the accuracy of the approximation of the Gamma PDF can be found in Fig. 4.8. Since the coefficients of the expansion (4.441) decay as fast as α_k, one can expect that the approximation (4.441) converges as fast to the exact PDF $p_{\xi_R}(x)$ as the corresponding approximation of $p_i(I)$ converges to the exact Gamma PDF.

4.7.4.3 Log-Normal Distribution

The log-normal distribution is also frequently used in modeling of random EM fields. In this case

$$p_i(I) = \frac{1}{I\sigma\sqrt{2\pi}} \exp\left[-\frac{(\ln(I) - a)^2}{2\sigma^2}\right] \qquad (4.442)$$

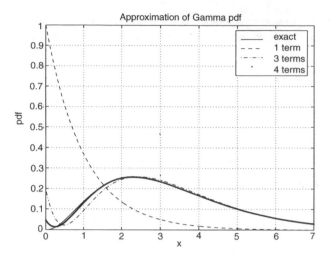

Figure 4.8 Approximation of the Gamma distribution by its Gram–Charlier series.

with moments given by

$$m_{Ln} = \exp\left(an + \frac{n^2 \sigma^2}{2} \right) \tag{4.443}$$

It is impossible to obtain an expression similar to eq. (4.440) in this case. However, the fact that the moments of this distribution are analytically known can be used as follows.

Let $L_k(\beta I)$ be rewritten as a power series of (βI):

$$L_k(\beta I) = \sum_{l=0}^{k} a_l^{(k)} (\beta I)^l \tag{4.444}$$

where the coefficients $a_l^{(k)}$ can be recursively calculated using eq. 22.3.9 on page 775 in [29]

$$a_l^{(k)} = (-1)^l \binom{k}{k-l} \frac{1}{l!} = (-1)^l \frac{k!}{(k-l)!} \frac{1}{l!l!} \tag{4.445}$$

Plugging eq. (4.444) into eq. (4.366), one can obtain a recursive algorithm for the calculation of coefficients in expansion (4.377)

$$\alpha_k = \int_0^\infty p_i(I) \left(\sum_{l=0}^{k} a_l^{(k)} (\beta I)^l \right) dI = \sum_{l=0}^{k} a_l^{(k)} \beta^l m_l = \sum_{l=0}^{k} a_l^{(k)} \beta^l \exp\left(al + \frac{l^2 \sigma^2}{2} \right) \tag{4.446}$$

thus leading to the final expression for $p_{\xi_R}(x)$ in the form

$$p_{\xi_R}(x) = \frac{\sqrt{\beta}}{\sqrt{\pi}} \exp[-\beta x^2] \sum_{k=0}^{\infty} \frac{1}{2^{2k}} H_{2k}(\sqrt{\beta} x) \sum_{l=0}^{k} \frac{(-1)^{k-l}}{(k-l)!l!l!} \beta^l \exp\left(al + \frac{l^2 \sigma^2}{2} \right) \tag{4.447}$$

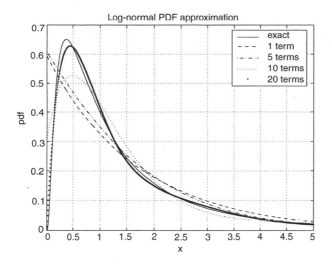

Figure 4.9 Approximation of log-normal PDF by its Gram–Charlier series.

A few examples showing the accuracy of approximation of the log-normal PDF can be found in Fig. 4.9.

4.7.4.4 A Narrow Band Process with Nakagami Distributed Envelope

In this case, the distribution of the envelope is described by the Nakagami PDF

$$p_A(a) = \frac{2}{\Gamma(m)} \left(\frac{m}{\Omega}\right)^m a^{2m-1} \exp\left(-\frac{ma^2}{\Omega}\right), \qquad m \geq \frac{1}{2}, a \geq 0 \tag{4.448}$$

The characteristic function can be found by first finding the characteristic function of the PDF (4.448) using the Bessel transform [19,38]

$$\theta_{\xi_R}(s) = \int_0^\infty p_A(a) J_0(sa) da = \frac{2}{\Gamma(m)} \left(\frac{m}{\Omega}\right)^m \int_0^\infty a^{2m-1} \exp\left(-\frac{ma^2}{\Omega}\right) J_0(sa) da \tag{4.449}$$

This integral can be calculated using standard integral (6.631–1 on page 698 in [39])

$$\int_0^\infty x^\mu \exp(-\alpha x^2) J_\nu(\beta x) dx = \frac{\beta^\nu \Gamma\left(\frac{\nu+\mu+1}{2}\right)}{2^{\nu+1} \alpha^{\frac{\nu+\mu+1}{2}} \Gamma(\nu+1)} {}_1F_1\left(\frac{\nu+\mu+1}{2}; \nu+1; \frac{-\beta^2}{4\alpha}\right) \tag{4.450}$$

by setting

$$\nu = 0, \qquad \beta = s, \qquad \alpha = \frac{m}{\Omega} \quad \text{and} \quad \mu = 2m - 1 \tag{4.451}$$

Here $_1F_1(a; b; z)$ is the so-called confluential hypergeometric function[20] [39]. As a result

$$\theta_{\xi_R}(s) = {}_1F_1\left(m; 1; -\frac{s^2\Omega}{4m}\right) \tag{4.452}$$

The distribution of the in-phase component can be found through the Fourier transform of the characteristic function, i.e.

$$p_{\xi_R}(x) = \frac{1}{2\pi}\int_{-\infty}^{\infty} {}_1F_1\left(m; 1; -\frac{s^2\Omega}{4m}\right)\exp(-isx)ds = \frac{1}{\pi}\int_0^{\infty} {}_1F_1\left(m; 1; -\frac{s^2\Omega}{4m}\right)\cos(sx)dx$$

$$\tag{4.453}$$

Using the substitution $z = \frac{1}{2}\sqrt{\frac{m}{\Omega}}s$, the last integral is converted to

$$p_{\xi_R}(x) = \frac{2}{\pi}\sqrt{\frac{m}{\Omega}}\int_0^{\infty} {}_1F_1(m; 1; -z^2)\cos\left(2x\sqrt{\frac{m}{\Omega}}z\right)dz \tag{4.454}$$

which, in turn, can be calculated with the help of the standard integral 7.642 on page 822 in [39]

$$\int_0^{\infty}\cos(2xy){}_1F_1(a; c; -x^2)dx = \frac{\sqrt{\pi}\,\Gamma(c)}{2\,\Gamma(a)}y^{2a-1}\exp(-y^2)\Psi\left(c - \frac{1}{2}, a + \frac{1}{2}; y^2\right) \tag{4.455}$$

by setting

$$c = 1, \qquad y = x\sqrt{\frac{m}{\Omega}} \quad \text{and} \quad a = m \tag{4.456}$$

Thus

$$p_{\xi_R}(|x|) = \left(\frac{m}{\Omega}\right)^m \frac{1}{\Gamma(m)\sqrt{\pi}}|x|^{2m-1}\exp\left(-\frac{m}{\Omega}x^2\right)\Psi\left(\frac{1}{2}, m + \frac{1}{2}; \frac{x^2 m}{\Omega}\right) \tag{4.457}$$

Here $\Psi(a, b; z)$ is the second confluent hypergeometric function, also known as the Kummer $U(a, b; z)$ function [39]. Since the Kummer function $\Psi(a, b; z)$ can also be expressed in terms of the Whittaker function $W_{\alpha,\beta}(z)$ as

$$\Psi\left(\frac{1}{2} + \mu - \kappa, 1 + 2\mu; z\right) = z^{-(\frac{1}{2}+\mu)}\exp\left(\frac{z}{2}\right)W_{\kappa,\mu}(z) \tag{4.458}$$

Equation (4.457) can be rewritten in the form suggested in [41]

$$p_{\xi_R}(|x|) = \frac{1}{\Gamma(m)\sqrt{\pi}}\left(\frac{m}{\Omega}\right)^{\frac{2m-1}{4}}|x|^{m-\frac{3}{2}}\exp\left(-\frac{m}{2\Omega}x^2\right)W_{\frac{2m-1}{4},\frac{2m-1}{4}}\left(\frac{mx^2}{\Omega}\right) \tag{4.459}$$

---

[20]The same function is known as the Kummer M function $M(a, b; z)$ or $\Phi(a, b; z)$ [39].

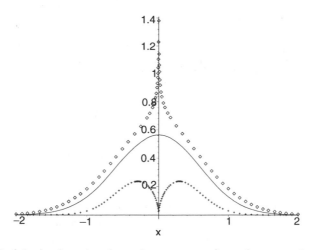

Figure 4.10 PDF of the in-phase (quadrature) component of a stationary random process with the Nakagami distribution for different values of the severity parameter m.

The shape of the PDF of the in-phase component is shown in Fig. 4.10 for different values of parameter m.

Alternatively, one can obtain a series expansion for $p_{\xi_R}(x)$ in terms of Hermitian polynomials by following the procedure established in section 2.2. Indeed, using eqs. (2.65) and (4.366)

$$p_i(I) = \frac{1}{\Gamma(m)} I^{m-1} \exp\left(-\frac{I}{\Omega}\right) = \frac{1}{\Omega} \exp\left(-\frac{I}{\Omega}\right)\left[1 + \sum_{k=1}^{\infty} \alpha_k L_k\left(\frac{x}{\Omega}\right)\right] \qquad (4.460)$$

with

$$\alpha_k = \int_0^\infty p_i(I) L_k\left(\frac{I}{\Omega}\right) dI = \frac{1}{\Gamma(m)} \int_0^\infty I^{m-1} \exp\left(-\frac{I}{\Omega}\right) L_k\left(\frac{I}{\Omega}\right) dI = \frac{\Gamma(1+k-m)}{\Gamma(1-m)k!} \Omega^m \qquad (4.461)$$

one obtains that the series (4.377) for the PDF of the in-phase component becomes

$$p_{\xi_R}(x) = \frac{1}{\sqrt{\pi \Omega}} \exp\left[-\frac{x^2}{\Omega}\right] \sum_{k=0}^{\infty} \frac{(-1)^k \Gamma(1+k-m)}{2^{2k} k! \Gamma(1-m)k!} H_{2k}\left(\frac{x}{\sqrt{\Omega}}\right) \qquad (4.462)$$

For integer values of $m = n$, $n = 1, 2, \ldots$, expressions (4.452) and (4.457) can be further simplified since

$$_1F_1\left(n; 1; -\frac{s^2\Omega}{4n}\right) = \exp\left(-\frac{s^2\Omega}{4n}\right) L_{n-1}\left(\frac{s^2\Omega}{4n}\right) = \exp\left(-\frac{s^2\Omega}{4n}\right)\left(1 + \sum_{k=1}^{n-1} c_k \left(\frac{s^2\Omega}{4n}\right)^k\right) \qquad (4.463)$$

and, thus

$$
\begin{aligned}
p_{\xi_R}(x) &= \frac{1}{2\pi} \int_{-\infty}^{\infty} \exp\left(-\frac{s^2\Omega}{4n}\right) \left[1 + \sum_{k=1}^{n-1} c_k \left(\frac{s^2\Omega}{4n}\right)^k\right] \exp(-isx)\,dx \\
&= \frac{1}{2\pi} \int_{-\infty}^{\infty} \left(-\frac{s^2\Omega}{4n}\right) \exp(-isx)\,ds + \sum_{k=1}^{n-1} c_k \frac{1}{2\pi} \int_{-\infty}^{\infty} \left(\frac{s^2\Omega}{4n}\right)^k \exp\left(-\frac{s^2\Omega}{4n}\right) \exp(-isx)\,ds \\
&= \frac{\exp\left(-\dfrac{x^2}{\Omega}\right)}{\sqrt{\pi\Omega}} \left[1 + \sum_{k=1}^{n-1} c_k \frac{(-1)^k}{2^{2n}} \exp\left(-\frac{x^2}{\Omega}\right) H_{2n}\left(\frac{x}{\sqrt{\Omega}}\right)\right]
\end{aligned}
\tag{4.464}
$$

Knowledge of the characteristic function (4.452) of the in-phase process also allows one to derive an expression for the PDF of the envelope of the mixture of the narrow band process with the Nakagami envelope and a narrow band Gaussian noise $\xi(t)$ with the variance $\sigma^2 = N_0\Delta f$. Indeed, since in this case

$$
\psi_R(t) = \xi_R(t) + \xi(t)
\tag{4.465}
$$

and, using independence of the noise and the signal

$$
\Theta_\psi(js) = \Theta_{\xi_R}(js)\Theta_\xi(js) = {}_1F_1\left(m; 1; -\frac{s^2\Omega}{4m}\right) \exp\left(-\frac{\sigma^2 s^2}{2}\right)
\tag{4.466}
$$

Taking the inverse Bessel transform (4.332), one obtains the PDF $p_\psi(a)$ of the envelope of the process $\psi(t)$

$$
p_{A_\psi}(a) = a \int_0^\infty \Theta_\psi(js) J_0(as) s\,ds = a \int_0^\infty {}_1F_1\left(m; 1; -\frac{s^2\Omega}{4m}\right) \exp\left(-\frac{\sigma^2 s^2}{2}\right) J_0(as) s\,ds
\tag{4.467}
$$

This integral can be expressed in terms of Meijer functions [39] using the table integral (7.663-1 on page 826 of [39])

$$
\int_0^\infty x^{2\zeta} {}_1F_1(a; b; -\lambda x^{2n}) J_\nu(xy)\,dx = \frac{2^{2\zeta}}{\Gamma(a)y^{2\zeta+1}} G_{23}^{21}\left(\frac{y^2}{4\lambda}\middle|\begin{matrix}1,b \\ \frac{1}{2}+\zeta+\frac{1}{2}\nu, a, \frac{1}{2}+\zeta-\frac{1}{2}\nu\end{matrix}\right)
\tag{4.468}
$$

if the exponent $\exp[-\sigma^2 s^2/2]$ is represented by its Taylor series. Setting

$$
a = m, \qquad b = 1, \qquad \lambda = \frac{\Omega}{4m}, \qquad \nu = 0, \qquad \zeta = n + \frac{1}{2}
\tag{4.469}
$$

one obtains

$$
\int_0^\infty s^{2n+1} {}_1F_1\left(m; b; -\frac{s^2\Omega}{4m}\right) J_0(as)\,dx = \frac{2^{2n}}{\Gamma(m)a^{2n+1}} G_{23}^{21}\left(\frac{ma^2}{\Omega}\middle|\begin{matrix}1,1 \\ n+1, m, n+1\end{matrix}\right)
\tag{4.470}
$$

and, thus

$$p_{A_\psi}(a) = a \sum_{k=0}^{\infty} (-1)^k \frac{\sigma^{2k}}{2^k k!} \int_0^{\infty} {}_1F_1\left(m; 1; -\frac{s^2\Omega}{4m}\right) s^{2k+1} J_0(as) s \, ds$$

$$= \frac{1}{\Gamma(m)} \sum_{k=0}^{\infty} (-1)^k \frac{\sigma^{2k}}{k!} \frac{2^k}{a^{2k}} G_{23}^{21}\left(\frac{ma^2}{\Omega}\bigg|_{k+1,m,k+1}^{1,1}\right) \qquad (4.471)$$

Although this equation is convenient in calculations of the PDF itself[21] it is not useful in analysis.

4.8   SPHERICALLY INVARIANT PROCESSES

4.8.1   Definitions

A formal definition of a Spherically Invariant Vector (SIRV) is as follows: a random vector $\eta = [\eta_1, \eta_2, \ldots, \eta_N]^T$ is called a spherically invariant vector if its PDF is uniquely defined by the specification of a mean vector m_η, a covariance matrix $C_{\eta\eta}$, and the characteristic PDF $p_\mu(s)$. It is also known as an elliptically contoured distribution [42–68]. This definition does not provide any specific details about the general form of the PDF of a SIRV. However, the answer is given by the following.

Representation theorem [42]:   If a random vector η is a SIRV, then there exists a non-negative random variable μ such that the PDF of the random vector η conditioned on μ is a multivariate Gaussian PDF:

$$\eta = \mu\xi \qquad (4.472)$$

A Spherically Invariant Random Process (SIRP) $\eta(t)$ is a random process such that every random vector n obtained by means of sampling of the random process $\eta(t)$ is a SIRV.

It is important to distinguish between a SIRP and a compound process, obtained as a product of two random processes

$$\eta(t) = \mu(t)\xi(t) \qquad (4.473)$$

which will be extensively used in the following derivations. It is clear from the definition of the SIRV that any particular realization of a SIRV is a Gaussian process with variance changing from one realization to another. Thus, while the process $\eta(t)$ is non-Gaussian and stationary, it is not an ergodic process, since the distribution of the variance cannot be estimated by observing a single realization of the process. On the contrary, a compound process defined by a similar equation, eq. (4.473), defines an ergodic process. However, the structure of the compound process is more complicated since every realization is not Gaussian and it will not be possible to use general expressions for the joint probability density of an arbitrary order. Nevertheless, it will be shown that a significant gain can be obtained from the

---

[21]The Meijer functions are implemented in such products as Maple© or Mathematica©

representation of a compound process as a product of a Gaussian process and an exponentially correlated Markov process.

4.8.2 Properties

4.8.2.1 Joint PDF of a SIRV

It is often convenient to assume that the modulating random variable μ is independent on the underlying zero mean Gaussian vector ξ. In this case, it is relatively easy to obtain the joint PDF $p_\eta(x)$ of the vector η in terms of the covariance matrix $C_{\xi\xi}$ of the Gaussian vector ξ and the characteristic PDF $p_\mu(s)$ of the modulating variable μ. Indeed, the PDF $p_\xi(x)$ is given by

$$p_\xi(x) = \frac{1}{(2\pi)^{N/2}\sqrt{|C_{\xi\xi}|}} \exp\left[-\frac{x^\mathrm{T}C_{\xi\xi}^{-1}x}{2}\right] = \frac{1}{(2\pi)^{N/2}\sqrt{|C_{\xi\xi}|}} \exp\left[-\frac{p}{2}\right] = p_\xi(p), p = x^\mathrm{T}C_{\xi\xi}^{-1}x$$

$$(4.474)$$

Here p is a scalar parameter, defined as a quadratic form of the vector x. Equation (4.474) indicates a well-known property of a Gaussian vector. The conditional PDF is thus given by

$$p_{\eta|\mu}(y \mid s) = \frac{1}{(2\pi)^{N/2}\sqrt{|C_{\xi\xi}|}} s^{-N} \exp\left[-\frac{x^\mathrm{T}C_{\xi\xi}^{-1}x}{2s^2}\right] = \frac{s^{-N}}{(2\pi)^{N/2}\sqrt{|x_{\xi\xi}|}} \exp\left[-\frac{p}{ds^2}\right] \quad (4.475)$$

since multiplication by a fixed number scales the covariance matrix by a factor s^{-2} (see section 2.3). Finally, the unconditional PDF $p_\eta(x)$ can be obtained by integration of the PDF (4.475) over the range of μ and the weight $p_\mu(s)$:

$$p_\eta(y, N) = \int_0^\infty p_{\eta|\mu}(y \mid s)p_\mu(s)\mathrm{d}s = \frac{1}{(2\pi)^{N/2}\sqrt{|C_{\xi\xi}|}} \int_0^\infty s^{-N} \exp\left[-\frac{p}{2s^2}\right]p_\mu(s)\mathrm{d}s$$

$$= \frac{1}{(2\pi)^{N/2}\sqrt{|C_{\xi\xi}|}} h_N(p) \tag{4.476}$$

where

$$h_N(p) = \int_0^\infty s^{-N} \exp\left[-\frac{p}{2s^2}\right]p_\mu(s)\mathrm{d}s \tag{4.477}$$

Equations (4.476)–(4.477) show that the PDF of a SIRV is a function of a quadratic form p, as is the Gaussian distribution. However, the form of this dependence is more complicated than a simple exponent. Thus, a SIRV can be considered as a natural extension of a Gaussian random vector. It is shown in a number of publications [47–58] that a SIRV inherits a number of properties of a Gaussian vector such as invariance under a deterministic linear transformation, properties of an optimal estimator, etc.

It follows from eq. (4.472) that the covariance matrices M_η of a SIRV and the covariance matrix of the underlying Gaussian process M_ξ are

$$M_\eta = E\{\mu^2\}M_\xi = m_{2\mu}M_\xi \qquad (4.478)$$

Without loss of generality, it is possible to assume that the second moment $m_{2\mu}$ is unity: $m_{2\mu} = 1$. In this case, the correlation matrix of the SIRV is the same as that of the underlying Gaussian process.

Setting $N = 1$ in eq. (4.476), the marginal PDF of the resulting process $p_\eta(y)$ is related to the marginal PDF $p_\mu(s)$ of the modulation process as

$$p_\eta(y) = \frac{1}{\sqrt{2\pi}\sigma} \int_0^\infty \exp\left[-\frac{y^2}{2s^2\sigma^2}\right] p_\mu(s)\frac{ds}{s} \qquad (4.479)$$

Taking the Fourier transform of both sides, the following equation for the characteristic function of the output process can be easily derived

$$\begin{aligned}
\Theta_\eta(j\omega) &= \int_{-\infty}^\infty p_\eta(y)\exp(j\omega y)dy = \frac{1}{\sqrt{2\pi}\sigma}\int_0^\infty p_\mu(s)\frac{ds}{s}\int_{-\infty}^\infty \exp\left[-\frac{y^2}{2s^2\sigma^2}\right]\exp(j\omega y)dy \\
&= \int_0^\infty p_\mu(s)\exp\left(-\frac{s^2\sigma^2\omega^2}{2}\right)ds = \int_0^\infty p_\mu(s)\left[\sum_{k=0}^\infty (-1)^k\frac{s^{2k}\sigma^{2k}\omega^{2k}}{2^k k!}\right]ds \\
&= 1 + \sum_{k=1}^\infty (-1)^k\frac{\sigma^{2k}\omega^{2k}}{2^k k!}m_{2k}^\mu \qquad (4.480)
\end{aligned}$$

Comparing this expansion with the Taylor series of the characteristic function, $\Theta_\eta(j\omega)$, the following relations between the moments of the modulating PDF and the resulting process can be easily deduced

$$\Theta_\eta(j\omega) = \sum_{l=0}^\infty \frac{(-j)^l m_l^\eta}{l!}\omega^l = 1 + \sum_{k=1}^\infty (-1)^k\frac{\sigma^{2k}\omega^{2k}}{d^k k!}m_{2k}^\mu \qquad (4.481)$$

or, equivalently

$$m_{2k-1}^\eta = 0, m_{2k}^\eta = \frac{(2k)!\sigma^{2k}}{2^k k!}m_{2k}^\mu = m_{2k}^\xi m_{2k}^\mu \qquad (4.482)$$

The last equation can be of course directly obtained by manipulating moment brackets, as in section 2.8. Interestingly, only even moments of the modulating PDF matter, since the odd moments are cancelled by zero odd moments of the Gaussian distribution.

4.8.2.2 Narrow Band SIRVs

If the underlying Gaussian process $\xi(t)$ in eq. (4.472) is a narrow band process, then the resulting SIRV (or compound process) will also be a narrow band process. In turn, the real

Gaussian process $\xi(t)$ can be considered as a real part of a complex process

$$\zeta(t) = \xi_R(t) + j\xi_I(t), \xi(t) = \mathrm{Re}\{\zeta(t)\} \tag{4.483}$$

where the in-phase component $\xi_R(t)$ and the quadrature component $\xi_I(t)$ are defined using the Hilbert transformation, as described in section 4.7.1. As a result, the process $\eta(t)$ can be thought of as the real part of a complex process

$$\iota(t) = \mu(t)\zeta(t) = \mu(t)\xi_R(t) + j\mu(t)\xi_I(t) \tag{4.484}$$

with the envelope defined as

$$A_\iota(t) = |\iota(t)| = \mu(t)|\zeta(t)| = \mu(t)A_\xi(t) \tag{4.485}$$

Thus, the envelope of the resulting process is an envelope of a complex Gaussian process modulated by a random variable μ or a random process $\mu(t)$. Since the envelope of a narrow band Gaussian process is the Rayleigh process

$$p_{A_\xi}(a) = \frac{a}{\sigma^2}\exp\left(-\frac{a^2}{2\sigma^2}\right) \tag{4.486}$$

the PDF of the envelope $p_{A_\eta}(a)$ of the process $\eta(t)$ can be expressed as

$$p_{A_\eta}(a) = \int_0^\infty p_\mu(s)\frac{a}{\sigma^2}\exp\left(-\frac{a^2}{2s^2\sigma^2}\right)\frac{ds}{s^2} \tag{4.487}$$

4.8.3 Examples

A number of analytically tractable examples are summarized in Table 4.1.

Table 4.1

Marginal PDF	$p_\eta(y)$	$p_\mu(s), s \geq 0$	$h_{2N}(p)$
Gaussian	$\dfrac{\exp\left(-\frac{y^2}{2\sigma^2}\right)}{\sqrt{2\pi\sigma^2}}$	$\delta(s-1)$	$\exp\left(-\frac{p}{2}\right)$
Two-side Laplace	$\frac{b}{2}\exp(-b|x|)$	$b^2 s\exp\left(-\frac{b^2 s^2}{2}\right)$	$\dfrac{b^{2N}}{(b\sqrt{p})^{N-1}}k_{N-1}(b\sqrt{p})$
Cauchy	$\dfrac{b}{\pi(b^2+x^2)}$	$\dfrac{b^2}{s^3}\exp\left(-\frac{b^2}{2s^2}\right)$	$\dfrac{2^N b\Gamma\left(N+\frac{1}{2}\right)}{\sqrt{\pi}(b^2+p)^{N+\frac{1}{2}}}$
K-distribution	$\dfrac{2b}{\Gamma(\nu)}\left(\frac{bx}{2}\right)^\nu K_{\nu-1}(bx)$	No analytical expression	$\dfrac{b^{2N}(b\sqrt{p})^{\nu-N}}{\Gamma(\nu)2^{\nu-1}}K_{N-\nu}(b\sqrt{p})$

REFERENCES

1. W. Feller, *An Introduction to Probability Theory and Its Applications*, New York: Wiley, 1950–1966.
2. E. Dynkin, *Theory of Markov Processes*, Englewood Cliffs, NJ: Prentice-Hall, 1961.
3. J. Doob, *Stochastic Processes*, New York: Wiley, 1990.
4. J. Proakis, *Digital Communications*, Boston: McGraw-Hill, 2001.
5. L. Zadeh, *System Theory*, New York: McGraw-Hill, 1969.
6. R. Dorf (Ed.), *The Engineering Handbook*, CRC Press, 1995.
7. A. Papoulis, *The Fourier Integral and Its Applications*, New York: McGraw-Hill, 1962.
8. G. Box, G. Jenkins, and G. Reinsel, *Time Series Analysis: Forecasting and Control*, Englewood Cliffs, NJ: Prentice-Hall, 1994.
9. A. Elwalid, D. Heyman, T. Lakshman, D. Mitra, and A. Weiss, Fundamental Bounds and Approximations for ATM Multiplexers with Applications to Video Teleconferencing, *IEEE Journal on Selected Areas of Communications*, vol. **13**, no. 6, 1995, pp. 1004–1016.
10. M. Zorzi, Some Results on Error Control for Burst-error Channels Under Delay Constraints, *IEEE Transactions Vehicular Technology*, vol. **50**, no. 1, 2001, pp. 12–24.
11. K. Baddour and N. Beaulieu, Autoregressive models for fading channel simulation, *Proceedings of GLOBECOM-2001*, November, 2001, vol. **2**, pp. 1187–1192.
12. P. Rao, D. Johnson, and D. Becker, Generation and analysis of non-Gaussian Markov time series Signal Processing, *IEEE Transactions on Acoustics, Speech, and Signal Processing*, vol. **40**, no. 4, April 1992, pp. 845–856.
13. A. Oppenheim and R. Shafer, *Discrete-time Signal Processing*, Upper Saddle River: Prentice Hall, 1999.
14. M. Schetzen, *The Volterra and Wiener Theories of Nonlinear Systems*, New York: Wiley, 1980.
15. A.N. Malakhov, *Cumulant Analysis of Non-Gaussian Random Processes and Their Transformations*, Moscow: Sovetskoe Radio, 1978 (in Russian).
16. J.K. Ord, *Families of Frequency Distributions*, London: Griffin, 1972.
17. P. Bello, Characterization of Randomly Time-Variant Linear Channels, *IEEE Transactions Communications*, vol. **11**, no. 4, December, 1963, pp. 360–393.
18. A. Papoulis, *Probability, Random Variables, and Stochastic Processes*, New York: McGraw-Hill, 1984.
19. S.M. Rytov, Yu. A. Kravtsov, and V.I. Tatarskii, *Principles of Statistical Radiophysics*, New York: Springer-Verlag, 1987–1989.
20. V. Tikhonov and V. Khimenko, *Excursions of Trajectories of Random Processes*, Moscow: Nauka, 1987 (In Russian).
21. S.O. Rice, Mathematical Analysis of Random Noise, *Bell Systems Technical Journal*, vol. **24**, January, 1945, pp. 46–156.
22. S.O. Rice, Statistical Properties of a Sine Wave Plus Random Noise, *Bell Systems Technical Journal*, vol. **27**, January, 1948, pp. 109–157.
23. M. Patzold, *Mobile Fading Channels*, John Wiley & Sons, 2002.
24. J. Lai and N.B. Nandayam, Minimum Duration Outages in Rayleigh fading Channels, *IEEE Transactions on Communications*, vol. **49**, October, 2001, pp. 1755–1761.
25. W. Lindsey and M. Simon, *Telecommunication Systems Engineering*, Englewood Cliffs, NJ: Prentice-Hall, 1973.
26. M.D. Yacoub, M.V. Barbin, M.S. de Castro, and J.E. Vargas Bautista, Level Crossing Rate of Nakagami-*m* Fading Signals: Field Trials and Validation, *Electronics Letters*, vol. **36**, no. 4, 2000, pp. 355–357.
27. J. Lai and N.B. Nandayam, Packet Error Rate for Burst-Error-Correcting Codes in Rayleigh Fading Channels, *Proceedings of IEEE VTC*, 1998, pp. 1568–1572.
28. C. Nikias and A. Petropulu, *Higher-Order Spectra Analysis*, New Jersey: Prentice-Hall, 1993.

29. M. Abramowitz and I. Stegun, *Handbook of Mathematical Functions*, New York: Dover, 1970.

30. W. Jakes, *Microwave Mobile Communications*, New York: Wiley, 1974.

31. M. Nakagami, The *m*-Distribution—A General Formula of Intensity Distribution of rapid Fading, In: W.C. Hoffman, Ed., *Statistical Methods in Radio Wave Propagation*, New York: Pergamon, 1960, pp. 3–36.

32. Y. Goldstein and V.Ya.Kontorovich, Influence of the Statistical Properties of Radio Channel on Error Grouping, *Telecommunications and Radio Engineering*, vol. **22**, no. 8, 1968, pp. 6–8.

33. V.Ya. Kontorovich, Questions of Noise Immunity of Communication Systems Due to Gaussian and Non-Gaussian Interference, *Ph.D. Thesis*, Leningrad Bonch-Bruevich Telecommunications Institute 1968, (In Russian).

34. F. Ramos, V. Kontorovich, and M. Lara-Barron, Simulation of Nakagami Fading with Given Autocovariance Function, *IEEE Intern. Symp. on Intellegent Sign. Proc. and Commun. Systems*, ISPACS 2000, Hawaii, November, 5–8, 2000, pp. 561–564.

35. Cyril-Daniel Iskander and P.T. Mathiopolous, Analytical Level-Crossing Rates and Average Fade Durations for Diversity techniques in Nakagami Fading Channels, *IEEE Transactions on Communications*, vol. **50**, no. 8, August, 2002, pp. 1301–1309.

36. R. Ertel, P. Cardieri, K. Sowerby, T.S. Rappaport, and J. Reed, Overview of Spatial Channel Models for Antenna Array Communication Systems, *IEEE Transactions on Personal Communications*, vol. **5**, no. 1, 1998, pp. 10–22.

37. V. Tikhonov, On a Markov Nature of the Envelope of the Narrowband Random Process, *Radiotechnika and Elekronika*, vol. **6**, no. 7, 1961 (In Russian).

38. A.H. Zemanian, *Generalized Integral Transformations*, New York: Dover, 1987.

39. I.S. Gradshteyn and I.M. Ryzhik, *Table of Integrals, Series, and Products*, New York: Academic Press, 1980.

40. A.P. Prudnikov, Yu. A. Brychkov, and O.I. Marichev, *Integrals and Series*, New York: Gordon and Breach Science Publishers, 1986.

41. D. Klovsky, V. Kontorovich, and S. Shirokov, *Models of Continuous Communication Channels Based on Stochastic Differential Equations*, Moscow: Radio i sviaz, 1984, (in Russian).

42. K. Yao, A Representation Theorem and its Applications to Spherically-Invariant Random Processes, *IEEE Transactions on Information Theory*, vol. **19**, no. 5, 1973, pp. 600–608.

43. A. Abdi, H. Barger and M. Kaveh, Signal Modeling in Wireless Fading Channels Using Spherically Invariant Processes, *Proceedings of ICASSP* vol. **5**, 2000, 2997–3000.

44. G. Jacovitti and A. Neri, Special Techniques for The Estimation of the Acf of Spherically Invariant Random Processes, *Spectrum Estimation and Modeling*, Fourth Annual ASSP Workshop, 3–5 Aug, 1988, pp. 379–382.

45. L. Izzo, L. Paura, and M. Tanda, Performance of a Square-Law Combiner for Reception of Nakagami Fading Orthogonal Signals in Spherically Invariant Noise, *Proceedings of IEEE Symposium Information Theory*, June-1 July, 1994, p. 91.

46. L. Izzo and M. Tanda, Diversity Reception of Fading Signals in Spherically Invariant Noise, *IEE Proceedings on Communications*, vol. **145**, no. 4, 1998, pp. 272–276.

47. Y. Kung, A Representation Theorem and its Applications to Spherically-Invariant Random Processes, *IEEE Transactions on Information Theory*, vol. **19**, no. 5, September, 1973, pp. 600–608.

48. M. Rangaswamy, D. Weiner, and A. Ozturk, Non-Gaussian Random Vector Identification Using Spherically Invariant Random Processes, *IEEE Transactions on Aerospace and Electronic Systems*, vol. **29**, no. 1, 1993, pp. 111–124.

49. L. Izzo and M. Tanda, Array Detection of Random Signals Embedded in an Additive Mixture of Spherically Invariant and Gaussian Noise, *IEEE Transactions on Communications*, vol. **49**, no. 10, 2001, pp. 1723–1726.

50. B. Picinbono, Spherically Invariant and Compound Gaussian Stochastic Processes, *IEEE Transactions on Information Theory*, vol. **16**, no. 1, 1970, pp. 77–79.

51. G. Wise and N. Gallagher, Jr., On Spherically Invariant Random Processes, *IEEE Transactions on Information Theory*, vol. **24**, no. 1, 1978, pp. 118–120.

52. L. Hoi and A. Cambanis, On the Rate Distortion Functions of Spherically Invariant Vectors and Sequences, *IEEE Transactions on Information Theory*, vol. **24**, no. 3, 1978, pp. 367–373.

53. M. Rupp, The Behavior of LMS and NLMS Algorithms in the Presence of Spherically Invariant Processes, *IEEE Transactions on Signal Processing*, vol. **41**, no. 3, 1993, pp. 1149–1160.

54. M. Rupp and R. Frenzel, Analysis of LMS and NLMS Algorithms with Delayed Coefficient Update Under the Presence of Spherically Invariant Processes, *IEEE Transactions on Signal Processing*, vol. **42**, no. 3, 1994, pp. 668–672.

55. T. Barnard and D. Weiner, Non-Gaussian Clutter Modeling with Generalized Spherically Invariant Random Vectors, *IEEE Transactions on Signal Processing*, vol. **44**, no. 10, 1996, pp. 2384–2390.

56. E. Conte, M. Lops, and G. Ricci, Adaptive Matched Filter Detection in Spherically Invariant Noise, *IEEE Signal Processing Letters*, Volume: **3** Issue: 8, August, 1996, pp. 248–250.

57. E. Conte, A. De Maio, and G. Ricci, Recursive Estimation of the Covariance Matrix of a Compound-Gaussian Process and Its Application to Adaptive CFAR Detection, *IEEE Transactions on Signal Processing*, vol. **50**, no. 8, 2002, pp. 1908–1915.

58. M. Rangaswamy and J. Michels, A Parametric Multichannel Detection Algorithm for Correlated Non-Gaussian Random Processes, *Proceedings of IEEE National Radar Conference*, 1997, 13–15 May, 1997, pp. 349–354.

59. V. Aalo and J. Zhang, Average Error Probability of Optimum Combining With a Co-Channel Interferer In Nakagami Fading, *Proceedings of Wireless Communications and Networking Conference*, 2000, vol. **1**, 2000, pp. 376–381.

60. J. Roberts, Joint Phase and Envelope Densities of a Higher Order, *IEE Proceedings of Radar, Sonar and Navigation*, vol. **142**, no. 3, 1995, pp. 123–129.

61. E. Conte, M. Longo, and M. Lops, Modeling and Simulation of Non-Rayleigh Radar Clutter, *IEE Proceedings of Radar and Signal Processing F*, vol. **138**, no. 2, 1991, pp. 121–130.

62. A. Gualtierotti, A Likelihood Ratio Formula for Spherically Invariant Processes, *IEEE Transactions on Information Theory*, vol. **22**, no. 5, 1976, pp. 610–610.

63. M. Rangaswamy, D. Weiner, and A. Ozturk, Computer Generation of Correlated Non-Gaussian Radar Clutter, *IEEE Transactions on Aerospace and Electronic Systems*, vol. **31**, no. 1, 1995, pp. 106–116.

64. M. Rangaswamy, J. Michels, and D. Weiner, Multichannel Detection for Correlated Non-Gaussian Random Processes Based on Innovations, *IEEE Transactions on Signal Processing*, vol. **43**, no. 8, 1995, pp. 1915–1922.

65. L. Izzo and M. Tanda, Asymptotically Optimum Diversity Detection In Correlated non-Gaussian Noise, *IEEE Transactions on Communications*, vol. **44**, no. 5, 1996, pp. 542–545.

66. K. Gerlach, Spatially Distributed Target Detection in Non-Gaussian Clutter, *IEEE Transactions* vol. **35**, no. 3, 1999, pp. 926–934.

67. I. Blake and J. Thomas, On a Class of Processes Arising in Linear Estimation Theory, *IEEE Transactions on Information Theory*, vol. **14**, no. 1, 1968, pp. 12–16.

68. G. Gabor and Z. Gyorfi, On the Higher Order Distributions of Speech Signals, *IEEE Transactions on Acoustics, Speech, and Signal Processing*, vol. **36**, no. 4, 1988, pp. 602–603.

69. R. Vijayan and J.M. Holzman, Foundations for Level Crossing Analysis of Handoff Algorithms, *Proceedings of ICC*, 1993, pp. 935–939.

70. Ya. Fomin, *Excursions of Random Process*, Moscow, Sovetskoe Radio, 1980 (in Russian).

71. J. Galambos, *The Asymptotic Theory of Extreme Order Statistics*, New York: Wiley, 1978.

72. T.H. Lehman, A Statistical Theory of Electromagnetic Fields in Complex Cavities, *Interaction Note 494*, May, 1993.

73. R. Holland and R. St. John, Statistical Coupling of EM fields to Cables in an Overmode Cavity, *Proceedings of ACES*, 1996, pp. 877–887.

74. R. Holland and R. St. John, Statistical Description of Cable Current Response Inside a Leaky Enclosure, *Proceedings of ACES*, 1997.
75. R. Holland and R. St. John, Statistical Response of EM-Driven Cables Inside an Overmoded Enclosure, *Accepted* for *IEEE Transactions on EMC*.
76. R. St. John, Approximate Field Component Statistics of the Lehman, Overmoded Cavity Distribution, *Internet*.
77. R. Holland and R. St. John, Statistical Response of Enclosed Systems to HPM Environment, *Proceedings of ACE*, 1994, pp. 554–568.
78. R. Holland and R. St. John, *Statistical Electromagnetics*, Philadelphia: Taylor and Francis Scientific Publishers, in press.
79. D.A. Hill, Spatial Correlation Function for Fields in a Reverberation Chamber, *IEEE Transactions on EMC*, Vol. **37**, No. 1, 1995, p. 138.
80. J.G. Kostas and B. Boverie, Statistical Model for a Mode-Stirred Chamber, *IEEE Transactions on EMC*, Vol. **33**, No. 4, November, 1991, pp. 366–370.
81. A. Molisch, *Wideband Wireless Digital Communications*, Upper Saddle River, NY: Prentice-Hall, 2001.

5

Markov Processes and Their Description

5.1 DEFINITIONS

Let $\xi(t)$ be a random process which depends on only one parameter—time[1]. Conditional on the range of values which the random process assumes and the nature of the parameter t, one can distinguish five major types of random process:

1. A discrete random sequence $\xi_m = \xi(t_m)$ or a random chain (discrete process in discrete time). In this case, the random process attains a fixed and countable set of values x_k, $k = 1, \ldots, K$, and the time parameter is discrete as well, i.e. t belongs to a discrete countable set t_m, $m = 1, \ldots, M$. It is important to note that both K and M can be infinite. Processes of this type are often encountered in practice: a random tossing of an object, random telegraph signals and remote sensing are just a few examples. Random discrete sequences are often obtained as a result of sampling and quantization of a continuous random process. Most of the modern communication systems use such signals and thus can be described by random chains.

2. A random sequence ξ_m (continuous process in discrete time). This kind of process is different from the discrete chain only in the range of attainable values. In the case of the random sequence it is a continuum of values rather than a discrete set as in the former case. Quite often a random sequence is obtained as the result of sampling of a continuous random process.

3. A discrete random process or continuous time chain (discrete process with continuous time). In this case the random process $\xi(t)$ assumes values from a countable set x_k, $k = 1, \ldots,$ K, while the parameter t continuously changes from $t = t_0$ to $t_0 + T$. It is also common to assume that $t_0 = 0$. The length of a serving queue and the number of users of a computer network are two examples of processes of this type. Often, a continuous time chain can be considered as an output of a quantizer driven by a continuous random process.

4. A continuous random process $\xi(t)$. In this case, the random process attains a value from a continuum of values with the time parameters also changing continuously. It is also often

[1]Among the other possible choices the spatial parameter is often used to model space–time variation of signals.

Stochastic Methods and Their Applications to Communications.
S. Primak, V. Kontorovich, V. Lyandres
© 2004 John Wiley & Sons, Ltd   ISBN: 0-470-84741-7

assumed that the trajectory of the continuous process does not experience large vertical jumps. Noise is one of quite a few examples of this kind of random process.

5. Mixed random processes. In this case, a random process is a mixture of the different types listed above. In particular, continuous processes with jumps can be considered as mixed. In this case, the random process resembles a continuous random process most of the time. However, from time to time, a rapid variation in the value (jump) appears, often due to changes in the system structure. Noise in communication systems using power lines and channels with impulsive and continuous noise are systems whose random structure can be described as a mixed process. Often, it is possible to describe a mixed process using a vector random process $\xi(t) = [\lambda(t), \theta(t)]^{\mathrm{T}}$, where $\lambda(t)$ is a continuous random process (type 4) and $\theta(t)$ is a discrete chain[2].

If, in addition, a process poses a Markov property as defined below, one can distinguish Markov chains, Markov sequences, continuous time Markov chains, continuous Markov processes and mixed Markov processes. A few typical trajectories are shown in Figure 5.1.

The classification can be extended to a case of a random vector $\xi(t) = [\xi_1(t), \xi_2(t), \ldots, \xi_r(t)]^{\mathrm{T}}$. In this case the vector process is called r-dimensional. In general, each component belongs to a different type of process and thus classification should be done on a component basis.

Extensive literature exists on the theory of Markov processes and their applications [1–6 to name a few]. Only a brief review is included in this book to facilitate further considerations.

5.1.1   Markov Chains

Let a physical system which has K possible states $\vartheta^{(1)}, \vartheta^{(2)}, \ldots, \vartheta^{(K)}$ be considered. The system can change its state randomly, depending on the external factors. These changes appear in a sudden manner at time instants $t_1 < t_2 < t_3 < \ldots$, i.e. transitions $\theta_1 \rightarrow \theta_2 \rightarrow \theta_3 \rightarrow \ldots$ take place only at these specified moments of time. Here $\theta_m = \theta(t_m)$ stands for the state of the system after $m - 1$ transitions, and $\theta_1 = \theta(t_1)$ is the initial state of the system (see Fig. 5.1(a)). Of course all θ_k are chosen from a finite set $\{\vartheta^{(l)}\}, l = 1, \ldots, K$.

A complete description of such a system can be achieved if the joint probability $P(\theta_1, \theta_2, \ldots, \theta_m)$ can be calculated for an arbitrary m. Using the Bayes formula one can write

$$P(\theta_{k_1}, \theta_{k_2}, \ldots, \theta_{k_m}) = P(\theta_1 = \vartheta_{k_1}, \theta_2 = \vartheta_{k_2}, \ldots, \theta_m = \vartheta_{k_m})$$
$$= P(\theta_1 = \vartheta_{k_1}, \theta_2 = \vartheta_{k_2}, \ldots, \theta_{m-1} = \vartheta_{k_{m-1}})$$
$$\times P(\theta_m = \vartheta_{k_m} \mid \theta_1 = \vartheta_{k_1}, \theta_2 = \vartheta_{k_2}, \ldots, \theta_{m-1} = \vartheta_{k_{m-1}}) \quad (5.1)$$

---

[2]The notation $\xi(t)$ for a continuous random process and $\theta(t)$ for a discrete random process or chain is reserved for the presentation in this chapter.

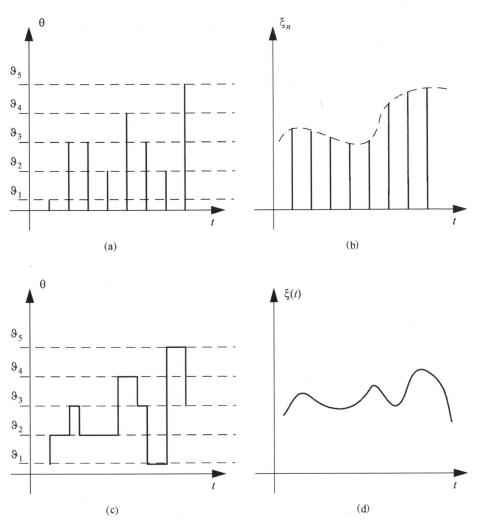

Figure 5.1 Classification of Markov processes: (a) Markov chain; (b) Markov sequence; (c) continuous time Markov chain; (d) continuous Markov process.

where ϑ_{k_μ} is one of the possible states $\vartheta^{(\mu)}$. In the very same manner, one can further expand $P(\theta_1 = \vartheta_{k_1}, \theta_2 = \vartheta_{k_2}, \ldots, \theta_{m-1} = \vartheta_{k_{m-1}})$ in terms of conditional probabilities to obtain

$$P(\theta_{k_1}, \theta_{k_2}, \ldots, \theta_{k_m}) = P_1(\theta_1) \prod_{\mu=2}^{m} \pi_\mu(\theta_\mu \mid \theta_1, \ldots, \theta_{\mu-1}) \qquad (5.2)$$

where $P_1(\theta_1)$ is the probability of the initial state $\theta_1 = \theta(t_1)$ and

$$\pi_\mu(\vartheta_{k_\mu} \mid \vartheta_{k_1}, \ldots, \vartheta_{k_{\mu-1}})$$
$$= P(\theta_\mu = \vartheta_{k_\mu} \mid \theta_1 = \vartheta_{k_1}, \theta_2 = \vartheta_{k_2}, \ldots, \theta_{\mu-1} = \vartheta_{k_{\mu-1}}), \qquad 2 \leq \mu \leq m \qquad (5.3)$$

is the conditional probability to reach state ϑ_{k_μ} at the time instant t_μ if the previous states are $\theta_1, \theta_2, \ldots, \theta_{\mu-1}$. The probability $\pi_\mu(\vartheta_{k_\mu} \mid \vartheta_{k_1}, \ldots, \vartheta_{k_{\mu-1}})$ is also known as the transitional probability.

Thus, in order to calculate the joint probabilities $P(\theta_1, \theta_2, \ldots, \theta_m)$, it is sufficient to know the initial state of the system and all the transitional probabilities π_μ, or the methodology of their calculation. There are a few specific cases when such calculations are relatively simple:

1. Changes in all states are independent. In this case, the transition to the next state does not depend on the previous state of the system, i.e.

$$\pi_m(\theta_m \mid \theta_1, \ldots, \theta_{m-1}) = P_m(\theta_m) \tag{5.4}$$

and eq. (5.2) is significantly simplified to become

$$P(\theta_1, \theta_2, \ldots, \theta_m) = \prod_{\mu=1}^{m} P_\mu(\theta_\mu) \tag{5.5}$$

As can be seen, this case coincides with the case of independent trials [4,5]. It is also a degenerate case of a Markov chain considered below.

2. A more general model can be obtained if the transitional probabilities depend only on the current state $\theta_{m-1} = \theta(t_{m-1})$ of the system and do not depend on its past states at the time instances $t_1, t_2, \ldots, t_{m-2}$. This property is general to all types of Markov process. For this reason Markov processes are often called processes without memory.[3]

    Mathematically speaking, the following equality holds for any index m

$$\pi_m(\theta_m \mid \theta_1, \ldots, \theta_{m-1}) = \pi_m(\theta_m \mid \theta_{m-1}) \tag{5.6}$$

Conditional probabilities $\pi_m(\theta_m \mid \theta_{m-1})$ are also called transition probabilities. Equation (5.6) defines the so-called Markov property of a random chain (process).

Using the chain rule, one can easily transform eq. (5.2) into

$$P(\theta_1, \theta_2, \ldots, \theta_m) = P_1(\theta_1) \prod_{\mu=2}^{m} \pi_\mu(\theta_\mu \mid \theta_{\mu-1}) \tag{5.7}$$

As a result, the complete description of a Markov chain can be achieved by providing the probability of the initial state $P_1(\theta_1)$ and the transitional probabilities $\pi_\mu(\theta_\mu \mid \theta_{\mu-1})$.

It is important to emphasize that eq. (5.7) holds true for the case of two moments of time ($m = 2$) for any random process. However, for the case of three or more moments, this equation does not describe a general situation, i.e. not all random processes are Markov.

---

[3]In the case (2), the process under consideration does not remember its past only, in contrast to a process considered in the case (1) which does not remember its past or its present.

3. In a similar manner, one could define a Markov process of order $n(n \geq 1)$: if the transitional PDF depends only on n previous states (including the current state) of the system, and does not depend on a more distant past. Mathematically speaking

$$\pi_m(\theta_m \mid \theta_1, \ldots, \theta_{m-1}) = \begin{cases} \pi_m(\theta_m \mid \theta_{m-n}, \ldots, \theta_{m-1}), & m > n \\ \pi_m(\theta_m \mid \theta_1, \ldots, \theta_{m-1}), & 2 \leq m \leq n \end{cases} \tag{5.8}$$

A regular Markov chain, defined in Section 2, corresponds to the case $n = 1$.

A Markov chain of order n can be reduced to a simple vector Markov chain for a vector of dimension n. In order to do so, one can consider n sequential states of the system as components of a single n-dimensional vector $\boldsymbol{\theta}$ defined as

$$\boldsymbol{\theta}_m = \{\theta_{i,m}, i = 1, \ldots, n\} = \{\theta_{m-n+1}, \ldots, \theta_m\}, \quad \theta_{i,m} = \theta_{m-n+i} \tag{5.9}$$

On the previous time step the vector $\boldsymbol{\theta}_{m-1}$ is expressed as

$$\{\theta_{i,m-1}, i = 1, \ldots, n\} = \{\theta_{m-n}, \ldots, \theta_{m-1}\} \tag{5.10}$$

and the following relations between components of the vector are satisfied

$$\theta_{i,m} = \theta_{i+1,m-1} = \theta_{m-n+i}, \quad \theta_{1,m-1} = \theta_{m-n}, \quad n < m \tag{5.11}$$

Thus, a Markov chain of the order n is a particular case of a simple Markov vector of dimension n. In this case, the transitional PDF is defined as

$$\pi_m(\boldsymbol{\theta}_m \mid \boldsymbol{\theta}_{m-1}) = \pi_m(\theta_{n,m} \mid \theta_{1,m-1}, \ldots, \theta_{n,m-1}) \prod_{i=1}^{n-1} \delta(\theta_{i,m}, \theta_{i+1,m-1}) \tag{5.12}$$

where $\delta(\mu, \nu) = \delta_{\mu\nu}$ is the Kronecker symbol. This relation follows from the definition (5.11) of the components of the vector $\boldsymbol{\theta}_m$.

It is often necessary to evaluate the probabilities $P_m(\boldsymbol{\theta}_m)$ and $P_{m-1}(\boldsymbol{\theta}_{m-1})$ of the system state at two sequential moments of time. For a simple Markov chain, the necessary relation can be obtained directly from eq. (5.7)

$$P_m(\boldsymbol{\theta}_m) = \sum_{\theta_{m-1}=\vartheta_1}^{\vartheta_k} \pi_m(\boldsymbol{\theta}_m \mid \boldsymbol{\theta}_{m-1}) P_{m-1}(\boldsymbol{\theta}_{m-1}), \quad m = 2, \ldots, M \tag{5.13}$$

Indeed, it follows from eq. (5.7) that

$$P(\boldsymbol{\theta}_1, \ldots, \boldsymbol{\theta}_m) = \pi_m(\boldsymbol{\theta}_m \mid \boldsymbol{\theta}_{m-1}) P(\boldsymbol{\theta}_1, \ldots, \boldsymbol{\theta}_{m-1}) \tag{5.14}$$

Adding both sides of eq. (5.14) over $\boldsymbol{\theta}_1, \ldots, \boldsymbol{\theta}_{m-1}$, one immediately obtains eq. (5.13).

If a simple vector Markov chain of dimension r is considered, $\boldsymbol{\theta}_m = \{\theta_{i,m}, i = 1, \ldots, r\}$, then the recurrent relation (similar to eq. (5.13)) has the following form

$$P_m(\theta_{1,m}, \ldots, \theta_{r,m})$$

$$= \sum_{\theta_{1,m-1}=\vartheta_{1,1}}^{\vartheta_{1,K_1}} \cdots \sum_{\theta_{r,m-1}=\vartheta_{r,1}}^{\vartheta_{r,K_r}} \pi_m(\theta_{1,m}, \ldots, \theta_{r,m} \mid \theta_{1,m-1}, \ldots, \theta_{r,m-1}) P(\theta_{1,m-1}, \ldots, \theta_{r,m-1})$$

$$(5.15)$$

or

$$P_m(\boldsymbol{\theta}_m) = \sum_{\theta_{m-1}=\vartheta} \pi_m(\boldsymbol{\theta}_m \mid \boldsymbol{\theta}_{m-1}) P_{m-1}(\boldsymbol{\theta}_{m-1}) \tag{5.16}$$

and the summation is conducted over all possible states $\boldsymbol{\vartheta} = \{\vartheta_{i,k_i}, k_i = 1, \ldots, K_i\}$ of each component $\theta_{i,m}, i = 1, \ldots, r$. Here K_i is the number of states of the i-th component.

For a Markov chain of order n, one can write an equation similar to eq. (5.7), based on eqs. (5.1) and (5.8)

$$P_m(\boldsymbol{\theta}_{m-n+1}, \ldots, \boldsymbol{\theta}_m) = \sum_{\theta_{m-n}=\vartheta_1}^{\vartheta_K} \pi_m(\boldsymbol{\theta}_m \mid \boldsymbol{\theta}_{m-n}, \ldots, \boldsymbol{\theta}_{m-1}) \times P_{m-1}(\boldsymbol{\theta}_{m-n}, \ldots, \boldsymbol{\theta}_{m-1}) \tag{5.17}$$

where $m > n$. At the initial steps $2 \leq m \leq n$, we have to set $n = m - 1$. In this case the following recurrent relation holds

$$P_m(\theta_2, \ldots, \theta_m) = \sum_{\theta_m=\vartheta_1}^{\vartheta_K} \pi_m(\theta_m \mid \theta_1, \ldots, \theta_{m-1}) \times P_{m-1}(\theta_{m-n}, \ldots, \theta_{m-1}) \tag{5.18}$$

Expressions (5.17) and (5.18) relate joint probabilities of the last n states of the discrete random variable $\{\theta_m\}$.

It can also be noted that the same relations can be obtained from the recurrent relation (5.16) for the probabilities of states of the Markov chain of order n if one sets $r = n$ and uses eqs. (5.10) and (5.12). In this case the state space of each component is identical and thus

$$P_m(\boldsymbol{\theta}_m) = \sum_{\theta_{1,m-1}=\vartheta_1}^{\vartheta_K} \pi_m(\theta_{n,m} \mid \theta_{1,m-1}, \theta_{1,m}, \ldots, \theta_{n-1,m}) \times P_{m-1}(\theta_{1,m-1}, \theta_{1,m}, \ldots, \theta_{n-1,m})$$

$$= \sum_{\theta_{m-n}=\vartheta_1}^{\vartheta_K} \pi_m(\theta_m \mid \theta_{m-n}, \theta_{m-n+1}, \ldots, \theta_{m-1}) \times P_{m-1}(\theta_{m-n}, \theta_{m-n+1}, \ldots, \theta_{m-1})$$

$$(5.19)$$

where $n < m(m \geq 2)$. The obtained relation coincides with eq. (5.17).

As an example of the application of a simple Markov chain, one can consider the following problem. Let the initial state of a system be known at the initial moment of time t_1

and let the transition probabilities be specified as

$$p_k(m) = P(\theta_m = \vartheta_k), \quad \pi_{jk}(\mu, m) = P\{\theta_m = \vartheta_k \mid \theta_\mu = \vartheta_j\}, \quad j, k = 1, \ldots, K, 1 \le \mu \le m$$
$$(5.20)$$

The interpretation of this notation is as follows: the probability $p_k(m)$ describes the (unconditional) probability of finding the system at the state ϑ_k at the $(m - 1)$-th step (i.e. at the moment of time $t = t_m$); the probability $p_k(0) = p_k^0$ describes the initial distribution of the states of the system. Conditional probabilities $\pi_{jk}(\mu, m)$ define the probability of the system to reside at the state ϑ_k at the moment of time t_m, given that the system has resided at the state ϑ_μ at the moment of time $t = t_\mu < t_m$. This probability is often referred to as the transitional probability from the state ϑ_j to the state ϑ_k over $(m - \mu)$ steps. It is required to find the probabilities of states of the system at the moment of time $t_m > t_1$, and to find their limit when m approaches infinity.

It is clear that all the quantities defined by eq. (5.20) are non-negative and satisfy the following normalization conditions

$$\sum_{k=1}^{K} p_k(m) = 1, \quad p_k(m) \ge 0, \quad m = 1, \ldots, M \tag{5.21}$$

$$\sum_{k=1}^{K} \pi_{jk}(\mu, m) = 1, \quad \pi_{jk}(\mu, m) \ge 0, \quad j = 1, \ldots, K \tag{5.22}$$

Using the Bayes theorem one can immediately write

$$p_k(m) = \sum_{j=1}^{K} p_j(\mu) \pi_{jk}(\mu, m), \quad k = 1, \ldots, K, \quad 1 \le \mu \le m \tag{5.23}$$

Transition of the system from the state ϑ_j to the state ϑ_k over a few steps can be completed in a number of ways. In other words, the system can reside in different intermediate states ϑ_i. Using the Bayes rule and the definition of a Markov process one deduces

$$\pi_{jk}(\mu, m) = \sum_{i=1}^{K} \pi_{ji}(\mu, n) \pi_{ik}(n, m), j, k = 1, \ldots, K, \quad \mu < n < m \tag{5.24}$$

This relation for a chain with a finite number of states was developed by A. Markov and is known as the Markov equation. It is a particular case of the more general Kolmogorov–Chapman–Smoluchowski equation considered below.

It is possible to rewrite eqs. (5.23) and (5.24) in a compact form if the following matrix notations are used:

$$\mathbf{\Pi}(\mu, m) = \| \pi_{jk}(\mu, m) \|, \quad \mathbf{P}(m) = \| p_k(m) \| \tag{5.25}$$

In this case eq. (5.31) becomes

$$P(m) = \mathbf{\Pi}^{\mathrm{T}}(\mu, m)P(\mu) \tag{5.26}$$

$$\mathbf{\Pi}(\mu, m) = \mathbf{\Pi}(\mu, n)\mathbf{\Pi}(n, m) \tag{5.27}$$

By design, the matrix $\mathbf{\Pi}(\mu, m)$ is a square matrix with non-negative elements. In addition, a sum of elements in any row totals unity. Such matrices are called Markov or stochastic matrices.

A distinction can be made between homogeneous and non-homogeneous Markov chains. A homogeneous chain can be characterized by the fact that its transitional matrix depends only on the difference of its arguments, i.e.

$$\mathbf{\Pi}(\mu, m) = \mathbf{\Pi}(m - \mu), \pi_{jk}(\mu, m) = \pi_{jk}(m - \mu), \quad m > \mu \tag{5.28}$$

All other chains are non-homogeneous. For a homogeneous Markov chain, one can consider a one-step transitional matrix, which does not depend on any time parameter

$$\mathbf{\Pi} = \mathbf{\Pi}(1), \qquad \pi_{jk} = \pi_{jk}(1) \tag{5.29}$$

There is a simple relation between the matrix $\mathbf{\Pi}$ and the transitional probability matrix $\mathbf{\Pi}(\mu, m) = \mathbf{\Pi}(m - \mu)$ between the μ-th and the m-th time step. Indeed, using property (5.27) one can sequentially obtain

$$\mathbf{\Pi}(m - \mu) = \mathbf{\Pi}(m - \mu - 1)\mathbf{\Pi}(1) = \mathbf{\Pi}(m - \mu - 2)\mathbf{\Pi}(1)\mathbf{\Pi}(1) = \mathbf{\Pi}^{m-\mu} \tag{5.30}$$

Thus, the k-step transitional matrix is just the one-step transitional matrix raised to the power k

$$\mathbf{\Pi}(k) = \mathbf{\Pi}^k \tag{5.31}$$

Furthermore, using a property of matrix transposition, it is possible to calculate probabilities $P(m + k)$ of states of the system after m steps if the initial probabilities $P(k)$ are known. Indeed, using eq. (5.30) in eq. (5.26) one readily concludes that

$$P(m + k) = \mathbf{\Pi}^{\mathrm{T}}(k, m)P(k) = (\mathbf{\Pi}^m)^{\mathrm{T}}P(k) = (\mathbf{\Pi}^{\mathrm{T}})^m P(k) \tag{5.32}$$

A class of homogeneous Markov chains can be further restricted to stationary chains. For a stationary chain, the probability $P(m)$ of the states at each step of the system does not depend on the index m, i.e. they are equal at each step. All other chains are non-stationary. It follows from the definition of a stationary chain that

$$P(m) = P(1) = P = \| p_k \| \tag{5.33}$$

This, combined with eq. (5.26), shows that

$$P = \mathbf{\Pi}^{\mathrm{T}}P \tag{5.34}$$

i.e. the vector of stationary probabilities is the eigenvector of the transitional matrix $\mathbf{\Pi}$ corresponding to the eigenvalue $\lambda = 1 \cdot n$.

It is important in many applications to decide if there is a limiting stationary distribution $P = \lim_{m \to \infty} P(m)$, and, if it exists, how to find it. The following theorems provide an answer to this question. A detailed proof can be found in the literature [7].

It is said that a state ϑ_k is reachable from a state ϑ_j, if, after a certain (finite) number of steps, there is a positive probability that the system can reach the state ϑ_k. In other words, there is a finite number m such that $\pi_{kj}(m) > 0$.

Theorem 1. For any Markov chain with a finite number of states and with all states reachable from all other states, i.e. for any j and k there is an integer m_{jk} such that

$$\pi_{jk}(m_{jk}) > 0 \cdot \tag{5.35}$$

there are final probabilities, such that

$$p_k = \lim_{M \to \infty} p_k(M) \tag{5.36}$$

These final probabilities do not depend on the initial probabilities $p_k(1)$ at the first step. A Markov chain for which the final probabilities exist is called an ergodic Markov chain.

Theorem 2. If the condition (5.35) is satisfied, then the final probabilities $p_k, k = 1, \ldots, K$ are the unique solution of the system of linear equations

$$p_k = \sum_{j=1}^{K} p_j \pi_{jk}, \quad k = 1, \ldots, K \tag{5.37}$$

subject to an additional condition (normalization)

$$\sum_{j=1}^{K} p_j = 1 \tag{5.38}$$

These final probabilities form a stationary probability vector (distribution) $P = \| p_k \|$. A Markov chain is stationary if its initial distribution $P(1)$ coincides with the final probabilities. It is easy to see that the matrix form of eq. (5.37) coincides with that of eq. (5.34).

It can be seen from eqs. (5.37) and (5.38) that there are $K + 1$ equations for determining only K unknown final probabilities. However, the first K equations in eq. (5.37) are linearly dependent. Indeed, by summing both sides of the equation over all possible values of the index k one obtains an identity, thus indicating that only $(K - 1)$ equations are really independent. As a result, the unknown probabilities p_k could be defined from $(K - 1)$ eq. (5.37) and an additional eq. (5.38). In order to illustrate this fact, the following two important examples are considered.

Example: A symmetric Markov chain with two states (Figure 5.2).

Let $\theta_m, m = 1, \ldots, M$ be a Markov chain with two possible states ϑ_1 and ϑ_2. In the initial state the probability of the state ϑ_1 is $p_1(0) = p_1^0 = \alpha$, while the probability of the second

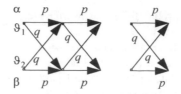

Figure 5.2   A symmetric Markov chain with two states.

state ϑ_2 is $p_2(0) = p_2^0 = \beta$, where $\alpha + \beta = 1$. One-step transitional probabilities are described by the following equalities: $\pi_{11} = \pi_{22} = p$ and $\pi_{12} = \pi_{21} = q$, where $p + q = 1$, $p > 0$. Since its one-step transitional matrix Π is symmetric, the corresponding Markov chain is called a symmetric Markov chain.

The goal is to determine the transitional probabilities $\pi_{jk}(m - 1)$ after $(m - 1)$ steps, probabilities $p_k(m)$ of each state on the m-th step, and the final probabilities p_k. One also requires to determine the reverse probabilities $\pi_{\overline{kj}}(m - 1) = P\{\theta_1 = \vartheta_j \mid \theta_m = \vartheta_k\}$. In particular, the probability $\pi_{\overline{11}}(m - 1) = P\{\theta_1 = \vartheta_1 \mid \theta_m = \vartheta_1\}$ is of interest. This symmetric Markov chain is homogeneous, since the transitional probabilities do not depend on the step m. Setting $\mu = 1$ and $n = 0$ in the Markov equation (5.27), one immediately obtains

$$\pi_{jk}(m - 1) = \sum_{i=1}^{2} \pi_{ji}\pi_{ik}(m - 2) \tag{5.39}$$

At the same time, the equation for the probabilities of the states on the m-th step can be found from the expression (5.26):

$$p_k(m) = \sum_{j=1}^{2} p_j(1)\pi_{jk}(m_1) \tag{5.40}$$

The probabilities $\pi_{jk}(m - 1)$ can be calculated using two methods: either by using eq. (5.31) and matrix calculus or by calculating these probabilities step by step, using eq. (5.39). The second technique is investigated here in more detail.

The normalization condition (5.22) for the case of $\mu = 0$ can be written as

$$\pi_{11}(m - 1) + \pi_{12}(m - 1) = 1$$
$$\pi_{22}(m - 1) + \pi_{21}(m - 1) = 1 \tag{5.41}$$

Since there is a complete symmetry in this problem, one can justify that $\pi_{12}(m - 1) = \pi_{21}(m - 1)$, and, consequently, $\pi_{11}(m - 1) = \pi_{22}(m - 1)$. Thus one needs to calculate only one of these probabilities in order to find all the others. For example, one can concentrate on calculating $\pi_{11}(m - 1)$. In order to do that, relation (5.22) can be exploited to produce

$$\pi_{11}(m) = \pi_{11}\pi_{11}(m - 1) + \pi_{12}\pi_{12}(m - 1) = p\pi_{11}(m - 1) + q[1 - \pi_{11}(m - 1)]$$
$$= (p - q)\pi_{11}(m - 1) + \frac{1 - (p - q)}{2} = \frac{1}{2} + \frac{(p - q)}{2}[2\pi_{11}(m - 1) - 1] \tag{5.42}$$

Since, for $m = 1$ the system is in the initial state, one deduces that

$$\pi_{11}(0) = \pi_{22}(0) = 1, \qquad \pi_{12}(0) = \pi_{21}(0) = 0 \tag{5.43}$$

Now it is possible to derive the expression for all the transitional probabilities using eqs. (5.41) and (5.42) in a sequential manner. Indeed, it follows from eqs. (5.42) and (5.43) that

$$\pi_{11}(1) = \frac{1}{2} + \frac{(p-q)}{2}[2\pi_{11}(0) - 1] = \frac{1}{2} + \frac{(p-q)}{2}$$

$$\pi_{11}(2) = \frac{1}{2} + \frac{(p-q)}{2}[2\pi_{11}(1) - 1] = \frac{1}{2} + \frac{(p-q)^2}{2} \tag{5.44}$$

$$\cdots$$

$$\pi_{11}(m) = \frac{1}{2} + \frac{(p-q)}{2}[2\pi_{11}(m-1) - 1] = \frac{1}{2} + \frac{(p-q)^m}{2}$$

Once $\pi_{11}(m)$ is found, the rest of the transitional probabilities can be found from eq. (5.41):

$$\pi_{22}(m) = \pi_{11}(m) = \frac{1}{2} + \frac{(p-q)^m}{2} \tag{5.45}$$

$$\pi_{12}(m) = \pi_{21}(m) = 1 - \pi_{11}(m) = \frac{1}{2} - \frac{(p-q)^m}{2} \tag{5.46}$$

Finally, probabilities of each state can be found through the probabilities in the initial state and the transitional probabilities (5.44)–(5.45)

$$p_1(m) = p_1(1)\pi_{11}(m-1) + p_2(1)\pi_{21}(m-1)$$

$$= \alpha\left[\frac{1}{2} + \frac{(p-q)^m}{2}\right] + \beta\left[\frac{1}{2} - \frac{(p-q)^m}{2}\right] = \frac{1}{2} + \frac{(\alpha - \beta)}{2}(p-q)^{m-1} \tag{5.47}$$

$$p_2(m) = p_2(1)\pi_{22}(m-1) + p_1(1)\pi_{12}(m-1)$$

$$= \beta\left[\frac{1}{2} + \frac{(p-q)^m}{2}\right] + \alpha\left[\frac{1}{2} - \frac{(p-q)^m}{2}\right] = \frac{1}{2} + \frac{(\beta - \alpha)}{2}(p-q)^{m-1} \tag{5.48}$$

Since $p > 0$ and all the states are reachable, then $(p - q) < 1$ and

$$\lim_{m\to\infty} p_1(m) = \lim_{m\to\infty} p_2(m) = \frac{1}{2} \tag{5.49}$$

This indicates that this chain is an ergodic Markov chain and the final probabilities p_1 and p_2 of the states can be found. The latter can be accomplished by solving a system of linear equations, obtained directly from eqs. (5.37) and (5.38)

$$p_1 = pp_1 + qp_2 \quad p_2 = qp_1 + pp_2 \qquad \text{or} \qquad p_1 + p_2 = 1 \tag{5.50}$$

The only solution of either of these systems is $p_1 = p_2 = 0.5$.

   Further analysis of expressions of probabilities of states shows that there is a transient process before a stationary (final) distribution of states is achieved. It also becomes obvious that the approach to stationarity has an exponential nature, since the variable part of $p_1(m)$ is proportional to $(p - q)^m$. On the other hand if $p = q = 0.5$, the final probabilities are immediately achieved and the process is stationary.

In order to find the reversed probabilities $\pi_{\overline{kj}}(m - 1) = P\{\theta_1 = \vartheta_j \mid \theta_m = \vartheta_k\}$, the Bayes theorem can be used. Indeed, according to this theorem

$$\pi_{\overline{kj}}(m - 1) = P\{\theta_1 = \vartheta_j \mid \theta_m = \vartheta_k\} = \frac{P\{\theta_1 = \vartheta_j, \theta_m = \vartheta_k\}}{p_k(m)} = \frac{\pi_{jk}(m)p_j(1)}{p_k(m)} \tag{5.51}$$

In particular, this leads to the following expression for the reversed transition probability $\pi_{\overline{11}}(m - 1)$

$$\pi_{\overline{11}}(m - 1) = \frac{\pi_{11}(m)p_1(1)}{p_1(m)} = \frac{\pi_{11}(m)p_1(1)}{p_1(m)} = \frac{\alpha + \alpha(p - q)^m}{1 + (\alpha - \beta)(p - q)^m} \tag{5.52}$$

This clearly shows that, in general, the reversed transitional probability is not equal to the forward transitional probability.

Example: A quasi-random telegraph signal. Let $\theta(t)$ be a random process with two states $\pm\vartheta_0$, which changes states only at the fixed moments of time $t_n = \Delta \pm nT_0$, where T_0 is a constant, $n = 0, 1, 2, \ldots$, and Δ is a random variable uniformly distributed on the interval $[0, T_0]$ and independent of $\theta(t)$. It is assumed that the conditions of the previous example also hold. In particular, $\vartheta_1 = \vartheta_0$ and $\vartheta_2 = -\vartheta_0$. Such a signal is known as a quasi-random telegraph signal. It is important to note that the change of state is only allowed in certain instants of time, in contrast to the example considered later in Section 5.1.3. It is required to find the correlation function and the power spectrum density of such a signal in the stationary case, i.e. assuming that $p_1 = p_2 = 0.5$.

The expectation $\langle\theta(t)\rangle$ of the process is equal to zero since

$$\langle\theta(t)\rangle = p_1\vartheta_0 + p_2(-\vartheta_0) = 0 \tag{5.53}$$

Thus, its covariance function $C(\tau)$ is just an average of the product of two values of the process at separate moments of time, i.e.

$$C(\tau) = \langle\theta(t)\theta(t + \tau)\rangle \tag{5.54}$$

The covariance function can be first evaluated assuming that the value of Δ is fixed. Let the interval of length τ contain m points where the process can change its value, as shown in Fig. 5.3. Since the signal can assume only two values $\pm\vartheta_0$, the product $\theta(0)\theta(\tau)$ assumes only two possible values as well. This value is ϑ_0^2 if both ends of the interval lie on the same side of the real axis, i.e. if either $\theta(0) = \theta(\tau) = \vartheta_0$ or $\theta(0) = \theta(\tau) = -\vartheta_0$. If the ends of these intervals lie on pulses of different polarity then $\theta(0)\theta(\tau) = -\vartheta_0^2$. Thus, one can write

$$C_m(\tau \mid \Delta) = \vartheta_0^2 P\{\theta(0) = \vartheta_0, \theta(\tau) = \vartheta_0\} + \vartheta_0^2 P\{\theta(0) = -\vartheta_0, \theta(\tau) = -\vartheta_0\}$$
$$-\vartheta_0^2 P\{\theta(0) = \vartheta_0, \theta(\tau) = -\vartheta_0\} - \vartheta_0^2 P\{\theta(0) = -\vartheta_0, \theta(\tau) = \vartheta_0\}$$
$$= \vartheta_0^2[p_1\pi_{11}(m) + p_2\pi_{22}(m)] - \vartheta_0^2[p_1\pi_{12}(m) + p_2\pi_{21}(m)] \tag{5.55}$$

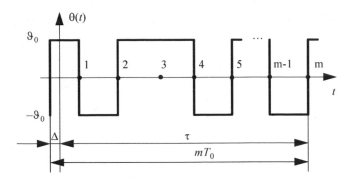

Figure 5.3   Quasi-random telegraph signal.

Since for a stationary case $p_1 = p_2 = 0.5$ and, as has been shown in the example above,

$$\pi_{11}(m) = \pi_{22}(m) = \frac{1 + (p-q)^m}{2}, \qquad \pi_{12}(m) = \pi_{21}(m) = \frac{1 - (p-q)^m}{2} \qquad (5.56)$$

then the expression for the covariance function becomes

$$C_m(\tau \mid \Delta) = \vartheta_0^2 (p-q)^m \qquad (5.57)$$

If the values of τ and m are such that $(m-1)T_0 \le \tau \le m T_0$ then there are two possibilities:

1. If $\Delta < m T_0 - \tau$, then the interval τ contains $(m-1)$ points where the system is able to change its state. As a result, the covariance function for these values of Δ and τ is

$$C_{m-1}(\tau \mid \Delta) = \vartheta_0^2 (p-q)^{m-1} \qquad (5.58)$$

2. If $\Delta > m T_0 - \tau$ then there are m points where the system is able to change its state. As a result, the correlation function for these values of Δ and τ is

$$C_m(\tau \mid \Delta) = \vartheta_0^2 (p-q)^m \qquad (5.59)$$

Since the random variable Δ is uniformly distributed on the interval $[0, T_0]$, its probability density function is a constant

$$p_\Delta(\Delta) = \frac{1}{T_0}, \qquad \Delta \in [0, T_0] \qquad (5.60)$$

and the unconditional covariance function $C(\tau)$ can be obtained by averaging the conditional correlation function $C(\tau \mid \Delta)$

$$C(\tau) = \int_0^{T_0} C(\tau \mid \Delta) p_\Delta(\Delta) d\Delta = \frac{1}{T_0} \left[\int_0^{mT_0-\tau} C_{m-1}(\tau \mid \Delta) d\Delta + \int_{mT_0-\tau}^{T_0} C_m(\tau \mid \Delta) d\Delta \right]$$

$$= \vartheta_0^2 \left(m - \frac{\tau}{T_0} \right) (p-q)^{m-1} + \vartheta_0^2 \left[\frac{\tau}{T_0} - (m-1) \right] (p-q)^m \tag{5.61}$$

Given that $C(\tau)$ is an even function of τ, the final expression for the correlation function takes the following form

$$C(\tau) = \vartheta_0^2 \left(m - \frac{|\tau|}{T_0} \right) (p-q)^{m-1} + \vartheta_0^2 \left[\frac{|\tau|}{T_0} - (m-1) \right] (p-q)^m \tag{5.62}$$

The power spectral density of the quasi-random telegraph signal can be then found by applying the Wiener–Khinchin theorem

$$S(\omega) = 2 \int_0^\infty C(\tau) \cos \omega\tau d\tau$$

$$= \sum_{m=0}^\infty 2\vartheta_0^2 (p-q)^{m-1} \int_{(m-1)T_0}^{mT_0} \left[\left(m - \frac{\tau}{T_0} \right) + \left[\frac{\tau}{T_0} - (m-1) \right] (p-q) \right] \cos \omega\tau d\tau$$

$$\times \frac{\vartheta_0^2 t_0 [1 - (p-q)^2]}{1 - 2(p-q)\cos \omega T_0 + (p-q)^2} \left(\frac{\sin\left(\frac{\omega T_0}{2}\right)}{\frac{\omega T_0}{2}} \right)^2 \tag{5.63}$$

For a particular case of $p = q = 0.5$ this expression can be further simplified to produce

$$k(\tau) = \begin{cases} \vartheta_0^2 \left(1 - \frac{|\tau|}{T_0} \right), & |\tau| < T_0 \\ 0, & |\tau| > T_0 \end{cases}, \quad S(\omega) = \vartheta_0^2 T_0 \left(\frac{\sin\left(\frac{\omega T_0}{2}\right)}{\frac{\omega T_0}{2}} \right)^2 \tag{5.64}$$

It is important to note that if Δ is deterministic and equal to zero, $\Delta = 0$, then the quasi-telegraph signal becomes non-stationary. The easiest way to show this is to consider a case of $p = q = 0.5$. For this signal the covariance function is[4]

$$C(t_1, t_2) = \langle \theta(t_1)\theta(t_2) \rangle = \vartheta_0^2 P\{\theta(t_1) = \vartheta_0, \theta(t_2) = \vartheta_0\} + \vartheta_0^2 P\{\theta(t_1) = -\vartheta_0, \theta(t_2) = -\vartheta_0\}$$

$$- \vartheta_0^2 P\{\theta(t_1) = \vartheta_0, \theta(t_2) = -\vartheta_0\} - \vartheta_0^2 P\{\theta(t_1) = -\vartheta_0, \theta(t_2) = \vartheta_0\}$$

$$= \begin{cases} \vartheta_0^2 (m-1)T_0 < t_1, & t_2 < mT_0 \\ 0, & \text{otherwise} \end{cases} \tag{5.65}$$

---

[4]This signal still has zero mean at any moment of time.

Thus, the convariance function depends separately on the values of t_1 and t_2 and the process $\theta(t)$ is non-stationary.

5.1.2 Markov Sequences

A Markov sequence is a sequence of random variables $\xi_m = \xi(t_m)$ which attains a continuum of values at discrete moments of time $t_1 < t_2 < \cdots < t_M$. A sampled continuous Markov random process is an example of a Markov sequence. A Markov sequence is a good way to describe coherent digital communication systems, since they change their state only at certain fixed moments of time. At the same time a number of possible states is so large that the description in terms of a discrete number of states is rather complicated.

A sequence of random variables ξ_m is called a Markov sequence if for any m the conditional distribution function (or, equivalently, the conditional probability density) satisfies the following conditions

$$P_m(x_m \mid x_1, \ldots, x_{m-1}) = P_m(x_m \mid x_{m-1}) \tag{5.66}$$

$$\pi_m(x_m \mid x_1, \ldots, x_{m-1}) = \pi_m(x_m \mid x_{m-1}) \tag{5.67}$$

These relations reflect the fact that, for a Markov sequence, conditional probabilities and distributions for the future moment of time t_m depend only on the state of the system at the current moment of time t_{m-1}, not on the past state of the system.

It follows from eq. (5.67) that the joint probability density function $p_M(x_1, \ldots, x_M)$ can be expressed in terms of one-step transitional probability densities $\pi_m(x_m \mid x_{m-1})$, $m = 2, \ldots, M$ and the probability of the initial state $p_1(x_1)$:

$$p_m(x_1, \ldots, x_m) = p_1(x_1) \prod_{\mu=2}^{m} \pi_\mu(x_\mu \mid x_{\mu-1}) \tag{5.68}$$

The last equation can also be considered as a definition of a Markov sequence, since it follows from eq. (5.68) that

$$\pi_m(x_m \mid x_1, \ldots, x_{m-1}) = \frac{p_m(x_1, \ldots, x_m)}{p_{m-1}(x_1, \ldots, x_{m-1})} = \pi_m(x_m \mid x_{m-1}) \tag{5.69}$$

The following is a brief summary of properties of a Markov sequence.

1. If the state of a Markov sequence is known at the current moment of time, its future state does not depend on any past state. This means that if the state x_n is known, then for any $m > n > \mu$ random variables x_m and x_μ are independent:

$$p(x_m, x_\mu \mid x_n) = \pi_{mn}(x_m \mid x_n)\pi_{\overline{\mu n}}(x_\mu \mid x_n) \tag{5.70}$$

Indeed, using eq. (5.67) one can write

$$p(x_m, x_\mu \mid x_n) = \frac{p(x_\mu, x_n, x_m)}{p(x_n)} = \frac{p(x_\mu, x_n)\pi_{mn}(x_m \mid x_n)}{p(x_n)} = \pi_{mn}(x_m \mid x_n)\pi_{\overline{\mu n}}(x_\mu \mid x_n) \tag{5.71}$$

A similar statement can be derived for a few moments in the past and in the future.

2. Any sub-sequence of a Markov sequence is also a Markov sequence, i.e. for a given t_m and moments of time $t_{m_1} < t_{m_2} < \cdots < t_{m_n}$ the following equation is valid:

$$\pi(x_{m_n} \mid x_{m_1}, \ldots, x_{m_{n-1}}) = \pi(x_{m_n} \mid x_{m_{n-1}}) \tag{5.72}$$

3. A Markov sequence remains Markov if the time direction is reversed, i.e.

$$\pi(x_m \mid x_{m+1}, \ldots, x_{m+n}) = \pi(x_m \mid x_{m+1}) \tag{5.73}$$

This can be proven by expanding eq. (5.73) using the Bayes formula and the property (5.67)

$$\pi(x_m \mid x_{m+1}, \ldots, x_{m+n}) = \frac{p(x_m, x_{m+1}, \ldots, x_{m+n})}{p(x_{m+1}, \ldots, x_{m+n})}$$

$$= \frac{p(x_m, x_{m+1})}{p(x_{m+1})} \times \frac{p(x_{m+2}, \ldots, x_{m+n} \mid x_{m+1})}{p(x_{m+2}, \ldots, x_{m+n} \mid x_{m+1})} = \pi(x_m \mid x_{m+1}) \tag{5.74}$$

4. The transitional probability density satisfies the following equation

$$\pi(x_m \mid x_\mu) = \int_{-\infty}^{\infty} \pi(x_m \mid x_n)\pi(x_n \mid x_\mu)\mathrm{d}x_n, \quad m > n > \mu \tag{5.75}$$

which is a particular case of the Kolmogorov–Chapman equation. Indeed, for any Markov sequence, the joint probability density at three moments of time can be expressed as

$$p(x_\mu, x_n, x_m) = p(x_\mu)\pi(x_n \mid x_\mu)\pi(x_m \mid x_n) \tag{5.76}$$

Integrating the latter over x_n one obtains

$$p(x_\mu, x_m) = \int_{-\infty}^{\infty} p(x_\mu, x_n, x_m)\mathrm{d}x_n = p(x_\mu)\int_{-\infty}^{\infty} \pi(x_n \mid x_\mu)\pi(x_m \mid x_n)\mathrm{d}x_n \tag{5.77}$$

which is equivalent to eq. (5.75).

A Markov sequence is called a homogeneous Markov sequence if the transitional probability density $\pi_m(x_m \mid x_{m-1})$ does not depend on m. A Markov sequence is called a stationary Markov sequence if it is a homogeneous sequence and the probabilities of states do not depend on the instant of time

$$p_m(x_m) = p(x_m) = \text{const.} \tag{5.78}$$

Similar to a Markov chain of n-th order, one can consider a Markov sequence of order $n \geq 1$, defined as

$$\pi_m(x_m \mid x_1, \ldots, x_{m-1}) = \begin{cases} \pi_m(x_m \mid x_{m-n}, \ldots, x_{m-1}), & m > n \\ \pi_m(x_m \mid x_1, \ldots, x_{m-1}), & 2 \leq m \leq n \end{cases} \tag{5.79}$$

For $n = 1$ the Markov chain of order n coincides with a simple Markov chain. For $m \leq n$ one has to formally assume $n = m - 1$ $(m \geq 2)$.

It is possible to introduce an n-dimensional random vector

$$\boldsymbol{\xi}_m = \{\xi_{i,m}, i = 1, \ldots, n\} \tag{5.80}$$

with components on two sequential steps defined as

$$\xi_{i,m} = \xi_{i+1,m-1} = \xi_{m-n+1}, \xi_{1,m-1} = \xi_{m-n}(n < m) \tag{5.81}$$

It is easy to see that this new vector is a simple Markov vector. Its transitional probability density can be expressed as

$$\pi_m(\boldsymbol{X}_m \mid \boldsymbol{X}_{m-1}) = \pi_m(x_{n,m} \mid x_{1,m-1}, \ldots, x_{n,m-1}) \prod_{i=1}^{n-1} \delta(x_{i,m} - x_{i+1,m-1}) \tag{5.82}$$

The probability density $p_m(x_{m-n+1}, \ldots, x_m)$ obeys the following recurrence equations, similar to eq. (5.13):

$$p_m(x_{m-n+1}, \ldots, x_m) = \int_{-\infty}^{\infty} \pi_m(x_m \mid x_{m-n}, \ldots, x_{m-1}) p_{m-1}(x_{m-n}, \ldots, x_{m-1}) dx_{m-n} \tag{5.83}$$

where $m > n$. On the steps $2 \leq m \leq n$ the following equality holds for a Markov vector

$$p_m(x_2, \ldots, x_m) = \int_{-\infty}^{\infty} \pi_m(x_m \mid x_1, \ldots, x_{m-1}) p_{m-1}(x_1, \ldots, x_{m-1}) dx_1 \tag{5.84}$$

In the case of a simple scalar Markov chain eq. (5.84) becomes

$$p_m(x_m) = \int_{-\infty}^{\infty} \pi_m(x_m \mid x_{m-1}) p_{m-1}(x_{m-1}) dx_{m-1} \tag{5.85}$$

Similarly, for a vector case eq. (5.84) becomes

$$p_m(\boldsymbol{X}_m) = \int_{-\infty}^{\infty} \pi_m(\boldsymbol{X}_m \mid \boldsymbol{X}_{m-1}) p_{m-1}(\boldsymbol{X}_{m-1}) d\boldsymbol{X}_{m-1} \tag{5.86}$$

where $m = 2, \ldots, M$ and the integration is performed over all the components $x_{i,m-1}(i = 1, \ldots, r)$. Proofs of eqs. (5.82)–(5.86) are similar to those for the case of a Markov chain.

Example. Let independent random variables $\lambda_1, \lambda_2, \ldots, \lambda_m, \ldots$ have probability densities $p_m(x)$, respectively. It is possible to prove that the following sequence

$$\xi_1 = \lambda_1, \quad \xi_2 = \lambda_1 + \lambda_2 \ldots, \quad \xi_m = \lambda_1 + \lambda_2 + \cdots + \lambda_m, \ldots \tag{5.87}$$

is a Markov sequence.

Since differences $\xi_{k+1} - \xi_k = \lambda_k$ are independent, one can obtain an expression of joint probability density function of the first m random variables ξ_k as

$$p(x_1, \ldots, x_m) = p_1(x_1)p_2(x_2 - x_1) \ldots p_m(x_m - x_{m-1}) \tag{5.88}$$

Using this relation, the expression for transitional probability becomes

$$\pi_m(x_m \mid x_1, \ldots, x_{m-1}) = \frac{p(x_1, \ldots, x_m)}{p(x_1, \ldots, x_{m-1})} = p_m(x_m - x_{m-1}) \tag{5.89}$$

Since this transitional density does not depend on $x_1, \ldots, x_{m-2}$ the considered sequence is a Markov sequence.

If, in addition, all the random variables λ_m have zero mean, $\langle \lambda_m \rangle = 0$, then $\langle \xi_m \rangle = 0$. Furthermore, given that $\xi_m = \lambda_m + \xi_{m-1}$, then

$$\langle \xi_m \mid \xi_{m-1} \rangle = \langle \lambda_m \mid \xi_{m-1} \rangle + \langle \xi_{m-1} \mid \xi_{m-1} \rangle = \xi_{m-1} \tag{5.90}$$

since λ_m does not depend on ξ_{m-1} and $\langle \lambda_m \rangle = 0$. Finally, using a Markov property of the sequence it is possible to conclude that

$$\langle x_m \mid x_1, \ldots, x_{m-1} \rangle = x_{m-1} \tag{5.91}$$

A sequence of random variables which has the property described by eq. (5.91) is called a martingale [8].

Example. Let $\xi_1, \ldots, \xi_M$ be a sequence of independent random variables, described by the probability density function $p_m(x_m), m = 1, \ldots, M$. A new sequence $\{\lambda_m, m = 1, \ldots, M\}$ is defined such that it satisfies the following equation

$$\sum_{\mu=m-n}^{m} a_{m-\mu}\lambda_\mu = \xi_m, \quad m > n \tag{5.92}$$

where $\{a_{m-\mu}\}$ are known coefficients, $a_0 \neq 0$ and $a_n \neq 0$.

It is clear that the sequence λ_m is a sequence of independent variables if $n = 0$ and Markov for $n = 1$. For $n > 1$ one has a case of a Markov chain of order n, which is described by the transitional probability density

$$\pi_m(x_m \mid x_{m-n}, \ldots, x_{m-1} = a_0 p_m \left(\sum_{\mu=m-n}^{m} a_{m-\mu} x_\mu \right) \tag{5.93}$$

A conditional expectation is then

$$\langle \lambda_m \mid \lambda_1, \ldots, \lambda_{m-1} \rangle = \langle \lambda_m \mid \lambda_{m-n}, \ldots, \lambda_{m-1} \rangle = a_0^{-1} \left[\langle x_m \rangle - \sum_{\mu=m-n}^{m} a_{m-\mu}\lambda_\mu \right] \tag{5.94}$$

It can be seen from this equation that the conditional mean depends on n previous values of the process $\lambda_{m-n}, \ldots, \lambda_{m-1}$.

5.1.3 A Discrete Markov Process

Let process $\theta(t)$ attain values from a finite set $\vartheta_1, \ldots, \vartheta_K$ and transitions from state to a state appear at random moments of time. The transitional probability

$$\pi_{ij}(t_0, t) = P\{\theta(t) = \vartheta_j \mid (\theta(t_0) = \vartheta_i)\}, \quad t > t_0 \tag{5.95}$$

describes the probability that the system is in state ϑ_j at the moment of time t, given that the system has been in the state ϑ_i at the moment of time t_0. It is clear that

$$\sum_{j=1}^{K} \pi_{ij}(t_0, t) = 1, \quad \pi_{ij}(t_0, t) \geq 0 \tag{5.96}$$

and

$$\pi_{ij}(t_0, t_0) = \delta_{ij} = \begin{cases} 1, & i = j \\ 0, & i \neq j \end{cases} \tag{5.97}$$

The main problem in studying Markov processes is to calculate transitional probabilities and unconditional probabilities of the states given the initial distribution of the states and one-step transitional probability densities.

The same considerations used to derive eq. (5.24) can be used to obtain the Kolmogorov–Chapman equation in the following form

$$\pi_{ij}(t_0, t + \Delta t) = \sum_{k=1}^{K} \pi_{ik}(t_0, t)\pi_{kj}(t, t + \Delta t), \quad t > t_0, \quad \Delta t > 0 \tag{5.98}$$

In the case of discrete Markov processes and a small time interval Δt, transitional probabilities can be written as

$$\begin{aligned} \pi_{kk}(t, t + \Delta t) &= P\{\theta(t + \Delta t) = \vartheta_k \mid \theta(t) = \vartheta_k\} = 1 - a_{kk}(t)\Delta t + o(\Delta t) \\ \pi_{kj}(t, t + \Delta t) &= P\{\theta(t + \Delta t) = \vartheta_j \mid \theta(t) = \vartheta_k\} = a_{kj}(t)\Delta t + o(\Delta t) \end{aligned} \tag{5.99}$$

where $o(\Delta t)$ are terms of the higher order with respect to Δt. It can be seen from eq. (5.99) that it is a characteristic property of a Markov process that for small increments of time Δt, the probability of π_{kk} remaining in the current state is much higher than the probability of π_{kj} changing state.

The first of the two equations (5.99) is consistent with eq. (5.97) and reflects two facts: (1) for $\Delta t = 0$ the system is certainly in the state ϑ_k; (2) the probability of a transition from the state ϑ_k to any other state in the future depends on the moment of time to be considered, and, for a small interval Δt, these probabilities are proportional to the duration of this time interval.

Since the probability of transition from one state to another is non-negative, then $a_{kj}(t) \geq 0$. Furthermore, it follows from eq. (5.96) that the following normalization condition must also be satisfied

$$a_{kk}(t) = \sum_{j \neq k} a_{kj}(t) \geq 0, \quad a_{kj}(t) \geq 0 \tag{5.100}$$

Substituting eq. (5.99) into the right-hand side of eq. (5.98) one obtains a system of differential equations for transitional probabilities

$$\frac{\partial}{\partial t}\pi_{ij}(t_0, t) = -a_{jj}(t)\pi_{ij}(t_0, t) + \sum_{k \neq j} a_{kj}(t)\pi_{ik}(t_0, t) \tag{5.101}$$

where $a_{kj}(t)$ satisfies the condition (5.100).

The solution of this system of differential equations with initial conditions (5.97) provides the dependence of the transitional probabilities on time t. If the number of possible states of the system is finite, then for any continuous functions $a_{ij}(t)$, satisfying the conditions (5.100), the system of differential equations (5.101) has a unique non-negative solution which defines a discrete Markov process.

Similarly to a Markov sequence, a Markov discrete process remains Markov when the direction of time is reversed. It can be shown that in addition to the system of equations (5.101), an alternative system of equations can be written. In order to derive this new system of equations one ought to consider a point in time which is close to the initial moment of time t_0 and rewrite eq. (5.98) in the following form

$$\pi_{ij}(t_0, t) = \sum_k \pi_{ik}(t_0, t_0 + \Delta t)\pi_{kj}(t_0 + \Delta t, t), \quad t > t_0 + \Delta t > t_0 \tag{5.102}$$

The next step is to use the approximation of probabilities for a small time interval Δt, similar to eq. (5.99)

$$\pi_{ii}(t_0, t_0 + \Delta t) = 1 - a_{ii}(t_0)\Delta t + o(\Delta t) \tag{5.103}$$

Substituting approximation (5.103) into eq. (5.102) and taking a limit for $\Delta t \to 0$ one obtains the so-called reversed[5] Kolmogorov equation for a discrete Markov process

$$\frac{\partial}{\partial t_0}\pi_{ij}(t_0, t) = -a_{ii}(t_0)\pi_{ij}(t_0, t) - \sum_{k \neq j} a_{ik}(t_0)\pi_{kj}(t_0, t), \quad t > t_0 \tag{5.104}$$

Equation (5.101) is satisfied not only by transitional probabilities $\pi_{ij}(t_0, t)$ but also by unconditional probabilities $p_j(t)$ of states at any moment of time t. Indeed, once the initial probabilities $p_j(0) = p_j(t_0)$ are known then, for a discrete Markov process, the following equations are satisfied

$$p_j(t) = \sum_i p_i(t_0)\pi_{ij}(t_0, t), p_j(t + \Delta t) = \sum_i p_i(t)\pi_{ij}(t, t + \Delta t) \tag{5.105}$$

After multiplication of both sides of eq. (5.101) by $p_i(t_0)$ and summing over all possible states one obtains

$$\frac{\partial}{\partial t}p_j(t) = -a_{jj}(t)p_j(t) + \sum_{k \neq j} a_{kj}(t)p_k(t) \tag{5.106}$$

---

[5]Equation (5.101) is called the "forward" Kolmogorov equation.

A unique solution of eq. (5.106) is obtained using the initial conditions

$$p_j(t) = p_j(t_0) = p_j^0 \qquad (5.107)$$

A discrete Markov process is called a homogeneous Markov process if its transitional probabilities $\pi_{ij}(t_0, t)$ depend only on time difference $\tau = t - t_0$, i.e.

$$\pi_{ij}(t_0, t) = \pi_{ij}(\tau), \qquad \tau = t - t_0 \qquad (5.108)$$

It follows from eq. (5.106) that, for a homogeneous Markov process, coefficients $a_{kj}(t) = a_{kj}$ are constants (as functions of time) and eqs. (5.99)–(5.103) are simplified to a system of linear differential equations with constant coefficients

$$\frac{d}{d\tau}\pi_{ij}(\tau) = -a_{jj}\pi_{ij}(\tau) + \sum_{k \neq j} a_{kj}\pi_{ik}(\tau) \qquad (5.109)$$

$$\pi_{ij}(0) = \delta_{ij} \qquad (5.110)$$

A homogeneous discrete Markov process is called ergodic if there is a limit

$$\lim_{\tau \to \infty} \pi_{ij}(\tau) = p_j \qquad (5.111)$$

which does not depend on the initial state ϑ_i. Probabilities p_j define an equilibrium, or final, distribution of the states of the system. These probabilities can be defined from algebraic equations, which are the limiting case of eq. (5.109)

$$-a_{jj}p_j + \sum_{k \neq j} a_{kj}p_k = 0, \qquad \sum_{k=1}^{K} p_k = 1 \qquad (5.112)$$

The concept of the final probabilities is illustrated by the following examples.

Example: A discrete binary Markov process (signal). Let $\theta(t)$ be a discrete Markov process with two possible states $\vartheta_1 = 1$ and $\vartheta_2 = -1$. The probability of transition from the state $\vartheta_1 = 1$ to the state $\vartheta_2 = -1$ for a short time interval Δt is equal to $\lambda \Delta t$ and the probability of the transition from the state $\vartheta_2 = -1$ to the state $\vartheta_1 = 1$ is equal to $\mu \Delta t$. Initially, the distribution of these two states is described by the following probabilities $\alpha = P\{\theta(t_0) = \vartheta_1\}$ and $\beta = P\{\theta(t_0) = \vartheta_2\} = 1 - \alpha$. It is required to determine the probabilities of transitions $\pi_{ij}(t_0, t)$, the final stationary probabilities of each state, as well as the mean values of the process $\theta(t)$ and its correlation function.

It is worth noting that if a binary Markov process $\Phi(t)$ has two states $\vartheta_1 = \varphi_1$ and $\vartheta_2 = \varphi_2$, then it could be represented using a linear transformation of a binary process $\theta(t)$ with two states ± 1:

$$\Phi(t) = \frac{\varphi_1 + \varphi_2}{2} + \frac{\varphi_1 - \varphi_2}{2}\theta(t) \qquad (5.113)$$

Thus, the statistical properties of the process $\Phi(t)$ can be easily expressed in terms of statistics of the process $\theta(t)$.

It follows from the definition of $\theta(t)$ that the following are the values of the constants in eq. (5.109): $a_{12} = \lambda$, $a_{22} = \mu$. Since all the coefficients a_{ij} are constants, the process $\theta(t)$ is homogeneous[6]. The next step is to rewrite eq. (5.109) as

$$
\left.
\begin{aligned}
\frac{\partial}{\partial t} \pi_{i1}(t_0, t) &= -\lambda \pi_{i1}(t_0, t) + \mu \pi_{i2}(t_0, t) \\
\frac{\partial}{\partial t} \pi_{i2}(t_0, t) &= -\mu \pi_{i2}(t_0, t) + \lambda \pi_{i1}(t_0, t)
\end{aligned}
\right\}, \quad i = 1, 2 \tag{5.114}
$$

Using property (5.96) one obtains $\pi_{12}(t_0, t) = 1 - \pi_{11}(t_0, t)$, which, in turn, implies that the first of the eqs. (5.114) can be rewritten as

$$
\frac{\partial}{\partial t} \pi_{11}(t_0, t) = -(\lambda + \mu)\pi_{11}(t_0, t) + \mu, \quad t > t_0 \tag{5.115}
$$

A general solution of this equation which satisfies the condition $\pi_{11}(t_0, t_0)$ is given by

$$
\pi_{11}(t_0, t) = \mu \int_{t_0}^{t} \exp[-(\lambda + \mu)(t - s)]ds + \exp[-(\lambda + \mu)(t - t_0)]
$$

$$
= \frac{\mu}{\lambda + \mu} + \frac{\lambda}{\lambda + \mu} \exp[-(\lambda + \mu)\tau] \tag{5.116}
$$

In the very same way a solution can be obtained for all transitional probabilities. Here, only the final results are presented:

$$
\pi_{11}(\tau) = \frac{\mu}{\lambda + \mu} + \frac{\lambda}{\lambda + \mu} \exp[-(\lambda + \mu)\tau], \pi_{12}(\tau) = \frac{\lambda}{\lambda + \mu}[1 - \exp[-(\lambda + \mu)\tau]] \tag{5.117}
$$

$$
\pi_{22}(\tau) = \frac{\lambda}{\lambda + \mu} + \frac{\mu}{\lambda + \mu} \exp[-(\lambda + \mu)\tau], \pi_{21}(\tau) = \frac{\mu}{\lambda + \mu}[1 - \exp[-(\lambda + \mu)\tau]] \tag{5.118}
$$

These equations indicate that the process considered is homogeneous since all the transitional probabilities depend only on the time difference $\tau = t - t_0$. In addition, this process is ergodic since the transitional probabilities approach a limit at $t \to \infty$ thus defining stationary probabilities

$$
p_1 = \frac{\mu}{\lambda + \mu}, \qquad p_2 = \frac{\lambda}{\lambda + \mu} \tag{5.119}
$$

[6]See derivations below.

Given an initial distribution of the probabilities and the transitional probabilities it is possible to find unconditional probabilities of each state:

$$p_1(t_0 + s) = \alpha\pi_{11}(s) + (1 - \alpha)\pi_{21}(s) = \frac{\mu}{\lambda + \mu} + \left(\alpha - \frac{\mu}{\lambda + \mu}\right)\exp[-(\lambda + \mu)s]$$

$$p_2(t_0 + s) = \alpha\pi_{22}(s) + (1 - \alpha)\pi_{12}(s) = \frac{\lambda}{\lambda + \mu} - \left(\alpha - \frac{\mu}{\lambda + \mu}\right)\exp[-(\lambda + \mu)s]$$

(5.120)

The average $\langle\theta(t)\rangle$ of the process $\theta(t)$ can be evaluated directly from the definition of average and the expression (5.120) for probabilities of the states at any moment of time

$$m_\theta(t) = \langle\theta(t)\rangle = 1 \cdot p_1(t) + (-1) \cdot p_2(t) = \alpha\pi_{11}(\tau) + (1 - \alpha)\pi_{21}(\tau)$$

$$- \alpha\pi_{22}(\tau) - (1 - \alpha)\pi_{12}(\tau) = \frac{\mu - \lambda}{\lambda + \mu} + 2\left(\alpha - \frac{\mu}{\lambda + \mu}\right)\exp[-(\lambda + \mu)\tau] \qquad (5.121)$$

Here $\tau - t - t_0$

The next step is to find the covariance function

$$C_\theta(s, \tau) - \langle\theta(t_0 + s)\theta(t_0 + s + \tau)\rangle - \langle\theta(t_0 + s)\rangle\langle\theta(t_0 + s + \tau)\rangle, \quad s > 0, \tau > 0 \quad (5.122)$$

Once again, using a definition of the average one obtains

$$\langle\theta(t_0 + s)\theta(t_0 + s + \tau)\rangle = 1 \cdot p_1(t_0 + s)\pi_{11}(\tau) + p_2(t_0 + s)\pi_{22}(\tau)$$

$$+ (-1) \cdot p_1(t_0 + s)\pi_{12}(\tau) - p_2(t_0 + s)\pi_{21}(\tau), \quad s > 0, \tau > 0$$

(5.123)

Finally, using eqs. (5.117), (5.118) and (5.123) the expression for the covariance function is reduced to

$$C_\theta(s, \tau) = \left\{ \frac{4\mu\lambda}{(\lambda + \mu)^2} - 4\left[\frac{\mu - \lambda}{\lambda + \mu} + \left(\alpha - \frac{\mu}{\lambda + \mu}\right)\exp[-(\lambda + \mu)s]\right] \right.$$

$$\left. \times \left(\alpha - \frac{\mu}{\lambda + \mu}\right)\exp[-(\lambda + \mu)s] \right\}\exp[-(\lambda + \mu)\tau]$$

(5.124)

If the process starts from the stationary distribution of the states, i.e. $\alpha = p_1 = \mu/(\lambda + \mu)$, then the process is a stationary process from the initial moment of time t_0. Its mean does not depend on time

$$m_\theta(t) = \frac{\mu - \lambda}{\lambda + \mu} = \text{const.} \qquad (5.125)$$

and its covariance function depends only on τ

$$C_\theta(\tau) = \frac{4\lambda\mu}{(\lambda + \mu)^2}\exp[-(\lambda + \mu)\tau], \quad \tau > 0 \qquad (5.126)$$

Finally, using symmetry of the covariance function of a stationary process one obtains

$$C_\theta(\tau) = \frac{4\lambda\mu}{(\lambda+\mu)^2}\exp[-(\lambda+\mu)|\tau|] \tag{5.127}$$

If $\lambda = \mu$ the binary Markov process is called a symmetric binary Markov process or a random telegraph signal. Using this condition, eqs. (5.117)–(5.127) can be significantly simplified to produce

$$\pi_{11}(\tau) = \pi_{22}(\tau) = \frac{1}{2}(1 + e^{-2\lambda\tau}),$$

$$p_1 = p_2 = \frac{1}{2}, m_\theta = 0, K_\theta(\tau) = \exp[-2\lambda\tau] \tag{5.128}$$

$$\pi_{12}(\tau) = \pi_{21}(\tau) = \frac{1}{2}(1 - e^{-2\lambda\tau})$$

5.1.4   Continuous Markov Processes

In this section a continuous Markov process is the subject of discussion. In this case the process can attain a continuum of values and is a function of a continuous parameter—time t. Let $\xi(t_0), \xi(t_1), \ldots, \xi(t_{m-1})$, and $\xi(t_m)$ be values of the random process $\xi(t)$ at the moments of time $t_0 < t_1 < \cdots < t_{m-1} < t_m$. The process $\xi(t)$ is called a continuous Markov process if its conditional transitional probability density

$$\pi_m(x_m, t_m \mid x_{m-1}, t_{m-1}; x_{m-2}, t_{m-2}; \ldots; x_0, t_0) = \frac{p_{m+1}(x_0, x_1, \ldots, x_m; t_0, t_1, \ldots, t_m)}{p_m(x_0, x_1, \ldots, x_{m-1}; t_0, t_1, \ldots, t_{m-1})} \tag{5.129}$$

depends only on the values of the process at the last moment of time $\xi(t_{m-1})$, i.e.

$$\pi_m(x_m, t_m \mid x_{m-1}, t_{m-1}; x_{m-2}, t_{m-2}; \ldots; x_0, t_0) = \pi_m(x_m, t_m \mid x_{m-1}, t_{m-1}), \quad m > 1 \tag{5.130}$$

As usual

$$p_m(x_0, x_1, \ldots, x_{m-1}; t_0, t_1, \ldots, t_{m-1}) = \frac{\partial^m}{\partial x_{m-1} \ldots \partial x_0} \text{Prob}\{\xi(t_{m-1}) < x_{m-1}, \ldots, \xi(t_0) < x_0\} \tag{5.131}$$

In other words, the future of a Markov process does not depend on its past if its value at the present is known.

It follows from eqs. (5.129) and (5.130) that

$$p_{m+1}(x_0, x_1, \ldots, x_m; t_0, t_1, \ldots, t_m) = \pi_m(x_m, t_m \mid x_{m-1}, t_{m-1})p_m(x_0, x_1, \ldots, x_{m-1}; t_0, t_1, \ldots, t_{m-1}) \tag{5.132}$$

Integration of the last equation over x_0 produces

$$p_m(x_1, \ldots, x_m; t_1, \ldots, t_m) = \pi_m(x_m, t_m \mid x_{m-1}, t_{m-1})p_{m-1}(x_1, \ldots, x_{m-1}; t_1, \ldots, t_{m-1}) \quad (5.133)$$

This indicates that for any $m > 1$, the expression for the joint probability function contains the same transitional probability density $\pi(x, t \mid x', t'), t > t'$, (from the state x' to the state x). Sequential application of integration to eq. (5.130) leads to the following important representation of a joint probability density function, considered at any $m + 1$ moments of time $t_m < t_{m-1} < \cdots < t_1 < t_0$

$$p_{m+1}(x_0, x_1, \ldots, x_m; t_0, t_1, \ldots, t_m) = \prod_{k=0}^{m-1} \pi(x_{m-k}, t_{m-k} \mid x_{m-k-1}, t_{m-k-1})p(x_0, t_0) \quad (5.134)$$

In particular

$$p_2(x_0, x_1; t_0, t_1) = \pi(x_1, t_1 \mid x_0, t_0)p(x_0, t_0) \quad (5.135)$$

This shows that a complete description of a continuous Markov process can be achieved if its marginal and transitional probability densities are known. This explains why Markov processes are so important in applications—a limited amount of information (only second order joint probability densities) is needed to obtain a complete description.

Using expression (5.130) and the Bayes formula one can show that a continuous valued Markov process remains Markov if time is reversed. In other words

$$\pi(x_0, t_0 \mid x_1, t_1; x_2, t_2; \ldots; x_m, t_m) = \pi(x_0, t_0 \mid x_1, t_1), \quad t_0 < t_1 < \cdots < t_m \quad (5.136)$$

Using the definition of the transitional probability density $\pi(x, t \mid x_0, t_0)$ one can obtain the following properties of the transitional density.

1. The transitional probability density is non-negative and is normalized to one after integration over the destination state x:

$$\pi(x, t \mid x_0, t_0) \geq 0 \quad \text{and} \quad \int_{-\infty}^{\infty} \pi(x, t \mid x_0, t_0)dx = 1 \quad (5.137)$$

2. The transitional probability density reduces to a delta function for coinciding moments of time

$$\pi(x, t \mid x_0, t) = \delta(x - x_0) \quad (5.138)$$

3. The transitional probability density satisfies the Kolmogorov–Chapman equation [1,2]

$$\pi(x, t \mid x_0, t_0) = \int_{-\infty}^{\infty} \pi(x, t \mid x', t')\pi(x', t' \mid x_0, t_0)d\lambda' \quad (5.139)$$

This equation can be easily derived using eq. (5.134) for $m = 2$

$$p_3(x_0, x_1, x; t_0, t_1, t) = \pi(x, t \mid x_1, t_1)\pi(x_1, t_1 \mid x_0, t_0)p(x_0, t_0) \quad (5.140)$$

Integration over x_1 produces the required relation (5.139).

A continuous Markov process is called a homogeneous Markov process if its transitional probability density depends only on a difference of time instances, i.e.

$$\pi(x,t \mid x_1,t_1) = \pi(x,x_1 \mid t - t_1) = \pi(x,x_1 \mid \tau), \quad \tau = t - t_1 \tag{5.141}$$

Furthermore, if $\pi(x,x_1 \mid \tau)$ approaches a limit $p_{st}(x)$ when $\tau \to \infty$

$$\lim_{\tau \to \infty} \pi(x,x_1 \mid \tau) = p_{st}(\lambda) \tag{5.142}$$

and this limit is independent of the initial state x' then this Markov process is called ergodic.

It is worth noting that the Kolmogorov–Chapman equation in the form (5.139) describes all types of Markov process considered above if the transitional density $\pi(x,t \mid x',t')$ can be a generalized function and the parameter t allows change on a discrete or continuous set. For example, for a Markov chain one can represent the transitional density as

$$\pi(x,t \mid x',t') = \pi(x,k\Delta t \mid x',(k-1)\Delta t) = \sum_{m=1}^{M}\sum_{n=1}^{M} P_{mn}\delta(x - x_m)\delta(x - x_n) \tag{5.143}$$

5.1.5 Differential Form of the Kolmogorov–Chapman Equation

The Kolmogorov–Chapman equation (5.139) is an integral equation and under some conditions it can be converted to a differential equation with a more revealing structure. The following section follows closely Section 3.4 in [1]. It is assumed for now that for all $\varepsilon > 0$

1. The following limit exists uniformly in x,x', and t such that $|x - x'| > \varepsilon$

$$\lim_{\Delta t \to 0} \frac{\pi(x,t + \Delta t; x',t)}{\Delta t} = W(x \mid x',t) \geq 0 \tag{5.144}$$

2. The following two limits are uniform in x', ε, and t

$$\lim_{\Delta t \to 0} \frac{1}{\Delta t} \int_{|x-x'|<\varepsilon} (x - x')\pi(x,t + \Delta t; x',t)dx = K_1(x,t) + O(\varepsilon) \tag{5.145}$$

$$\lim_{\Delta t \to 0} \frac{1}{\Delta t} \int_{|x-x'|<\varepsilon} (x - x')^2\pi(x,t + \Delta t; x',t)dx = K_2(x,t) + O(\varepsilon) \tag{5.146}$$

The condition (5.144) allows one to classify two types of continuous valued Markov process. If the limit is equal to zero, i.e.

$$\lim_{\Delta t \to 0} \frac{\pi(x,t + \Delta t; x',t)}{\Delta t} = 0 \tag{5.147}$$

this implies that there is no possibility that a discontinuity of size $x - x'$ would appear at the moment of time t. Otherwise, the quantity

$$P_{jump} = \lim_{\Delta t \to 0} \int_{|x-x'|>\varepsilon} \frac{\pi(x, t + \Delta t; x', t)}{\Delta t} dx = \int_{|x-x'|>\varepsilon} W(x \mid x', t) dx \qquad (5.148)$$

is the probability that a jump in magnitude of at least ε may appear at the moment of time t. The former is called a diffusion Markov process and the latter is a continuous process with jumps.

If a random process $\xi(t)$ is treated as a coordinate of some imaginary particle, the quantity $K_1(x, t)$ describes the average value of the velocity of the process at the moment of time t. Large deviations of the particle position are excluded in eq. (5.145) to avoid contributions from jumps: the uniformity in ε would insure that arbitrary small deviations can be considered, and, thus, contributions of larger velocities can be neglected. Using mechanical analogy the coefficient $K_1(x, t)$ is often called the drift of the Markov process and it is attributed to an external force (potential). Finally, the coefficient $K_2(x, t)$ describes the average energy of fluctuations of the particle around a fixed position x. Such deviations could be caused by external noise and in general do not pull the particle in a certain direction as an external force. It is common to call $K_2(x)$ the diffusion to reflect the fact that a particle can randomly change its position without external force. Thus, the motion of the particle can be thought of as a superposition of the drift, diffusion, and jumps.

Let $f(z)$ be a deterministic function which is at least twice differentiable and $\xi(t)$ be a continuous Markov process. In this case the rate of change in the average value of $f(\xi)$, given that $\xi(t_0) = y$, is defined by the following equation

$$\frac{\partial}{\partial t} \langle f(\xi) \mid y \rangle = \frac{\partial}{\partial t} \int_{-\infty}^{\infty} f(x) \pi(x, t \mid y, t_0) dx = \lim_{\Delta t \to 0} \frac{\int_{-\infty}^{\infty} f(x) [\pi(x, t + \Delta t \mid y, t_0) - \pi(x, t \mid y, t_0)] dx}{\Delta t}$$

$$= \lim_{\Delta t \to 0} \frac{\int_{-\infty}^{\infty} f(n) \int_{-\infty}^{\infty} [\pi(x, t + \Delta t \mid z, t) \pi(z, t \mid y, t_0) dz] dx - \int_{-\infty}^{\infty} f(z) \pi(z, t \mid y, t_0) dz}{\Delta t}$$

$$(5.149)$$

If $\varepsilon > 0$ is fixed then the region of integration can be divided into two intervals: $|x - z| \geq \varepsilon$ and $|x - z| < \varepsilon$, while the function $f(x)$ can be represented by its Taylor series

$$f(x) = f(z) + (x - z)\frac{d}{dz}f(z) + \frac{(x - z)^2}{2}\frac{d^2}{dz^2}f(z) + |x - z|^2 R(x, z) \qquad (5.150)$$

where $\lim_{|x-z| \to 0} |R(x, z)| = 0$. As a result, the Kolmogorov–Chapman equation (5.149) can be written as [2]

$$\frac{\partial}{\partial t} \langle f(\xi) \mid y \rangle = \lim_{\Delta t \to 0} \left\{ \int_{|x-z|<\varepsilon} \int \left[(x - z)\frac{\partial f}{\partial z} + \frac{1}{2}(x - z)^2 \frac{\partial^2 f}{\partial z^2} \right] \pi(x, t + \Delta t \mid z, t) \pi(z, t \mid y, t') dx dz \right.$$

$$\left. + P(x, z, t) + Q(x, z, t) + S(n, z, t) \right\} \qquad (5.151)$$

where

$$P(x, z, t) = \int_{|x-z|<\varepsilon} \int |x - z|^2 R(x, z)\pi(x, t + \Delta t \mid z, t)\pi(z, t \mid y, t')\mathrm{d}x\mathrm{d}z \qquad (5.152)$$

$$Q(x, z, f) = \int_{|x-z|\geq\varepsilon} \int f(x)\pi(x, t + \Delta t \mid z, t)\pi(z, t \mid y, t')\mathrm{d}x\mathrm{d}z \qquad (5.153)$$

$$S(x, z, t) = \int_{|x-z|<\varepsilon} \int f(z)\pi(x, t + \Delta t \mid z, t)\pi(z, t \mid y, t')\mathrm{d}x\mathrm{d}z$$

$$- \int\int f(z)\pi(x, t + \Delta t \mid z, t)\pi(z, t \mid y, t')\mathrm{d}x\mathrm{d}z \qquad (5.154)$$

Assuming uniform convergence allows one to interchange the limit operation with integration, thus the first line of eq. (5.151) can be rewritten with the help of eqs. (5.145) and (5.146) as

$$\int \left[a(z)\frac{\partial f}{\partial z} + \frac{1}{2}B(z)\frac{\partial^2 f}{\partial z^2} \right]\pi(z, t \mid y, t')\mathrm{d}z + O(\varepsilon) \qquad (5.155)$$

The remainder term (5.152) vanishes since the following estimate is valid

$$\left| \frac{1}{\Delta t}\int_{|x-z|<\varepsilon} |x - z|^2 R(x, z)\pi(x, t + \Delta t \mid z, t) \right| \leq \left| \frac{1}{\Delta t}\int_{|x-z|<\varepsilon} |x - z|^2 \pi(x, t + \Delta t \mid z, t) \right|$$

$$\max|R(x, z)| \rightarrow [B(z, t) + O(\varepsilon)]\max|R(x, z)| \rightarrow 0 \qquad (5.156)$$

It is easy to see that the remaining terms are simply

$$Q + S = \int_{|x-z|\geq\varepsilon} \int f(x)[W(z \mid x, t)\pi(x, t \mid y, t') - W(x \mid z, t)\pi(z, t \mid y, t')]\mathrm{d}x\mathrm{d}z \qquad (5.157)$$

Finally, taking the limit of both sides of eq. (5.149) when $\varepsilon \rightarrow 0$ one obtains

$$\frac{\partial}{\partial t}\int_{-\infty}^{\infty} f(z)p(z, t \mid y, t') = \int_{-\infty}^{\infty} \left[A(z)\frac{\partial f}{\partial z} + \frac{1}{2}B(z)\frac{\partial^2 f}{\partial z^2} \right]\pi(z, t \mid y, t')\mathrm{d}z$$

$$+ \int_{-\infty}^{\infty} f(z)\left\{ (P)\int_{-\infty}^{\infty} f(x)[W(z \mid x, t)\pi(x, t \mid y, t') - W(x \mid z, t)\pi(z, t \mid y, t')]\mathrm{d}x \right\}\mathrm{d}z$$

$$(5.158)$$

where $(P) \int$ stands for the principal value of the integral, i.e.

$$(P)\int_{-\infty}^{\infty} F(x, z)\mathrm{d}x = \lim_{\varepsilon\to 0}\int_{|x-z|>\varepsilon} F(x, z)\mathrm{d}x \qquad (5.159)$$

The necessity of considering the principal value stems from the fact that $w(x \mid z, t)$ is defined only for $x \neq z$, thus leaving a possibility for $W(x \mid x, t)$ to be infinite.

Since equality (5.158) holds for an arbitrary double differentiable function $f(z)$, one can hope to find some relations between the integrands on both sides of eq. (5.158) subject to certain boundary conditions. This can be accomplished through integrating by parts, which results in the following integro-differential equation, referred to as a differential Chapman–Kolmogorov equation [1]

$$\frac{\partial}{\partial p}\pi(z,t \mid y,t') = -\frac{\partial}{\partial z}[A(z,t)\pi(z,t \mid y,t')] + \frac{1}{2}\frac{\partial^2}{\partial z^2}[(z,t)\pi(z,t \mid y,t')]$$

$$+ \int_{-\infty}^{\infty} f(x)[W(z \mid x,t)\pi(x,t \mid y,t') - W(x \mid z,t)\pi(z,t \mid y,t')]dx \quad (5.160)$$

This equation must be supplemented by the following initial condition

$$\pi(z,t \mid y,t') = \delta(z-y) \quad (5.161)$$

and a set of boundary conditions which will be considered later. There are also some restrictions on the form of coefficients[7] $A(z)$ and $B(z)$. Some further classification of Markov processes can be achieved based on the representation (5.160): if $W(x \mid y) = 0$ then the process has a continuous trajectory and is called a diffusion Markov process, otherwise it is a process with jumps. For example, if $A(z) = B(z) = 0$ the trajectory of the process contains only jumps which appear at random according to a certain distribution. In the case of the diffusion process the differential Kolmogorov–Chapman equation is known as the Fokker–Planck equation, while for the case of process with jumps it is known as the Master equation. More details can be found in [1,2] and in the following sections.

5.2 SOME IMPORTANT MARKOV RANDOM PROCESSES

5.2.1 One-Dimensional Random Walk

The theory of Markov processes allows a great number of applied problems to be solved, due to the relative ease of finding a joint probability density of any order. In this section the problem of a one-dimensional random walk is considered as an example.

Let a particle be moving along θ axis, which is represented by a homogeneous Markov chain with an infinite number of states. The particle may change its position only at fixed moments of time $t_m = m\Delta t$. A transition is allowed only to the neighbouring state or the particle may remain in the current state. These transitions are described by the following probabilities

$$\pi_{j,j+1}(1) = p, \quad \pi_{j,j}(1) = 1-p-q, \quad \text{and} \quad \pi_{j,j-1} = q \quad (5.162)$$

Here, of course, $p + q \leq 1$. The corresponding geometry is shown in Fig. 5.5. For simplicity it is also assumed that $\Delta t = 1$ (Figs. 5.4–5.6).

---

[7]In future, alternative notation is used for $A(x)$ and $B(x)$: $K_1(x) = A(x)$ and $K_2(x) = B(x)$.

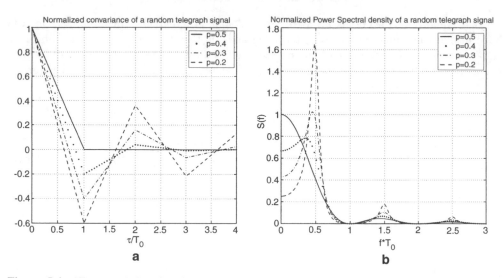

Figure 5.4   The correlation function (a) and the power spectral density (b) of the quasi-random telegraph signal.

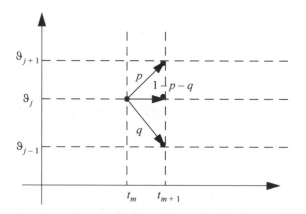

Figure 5.5   A one-dimensional random walk of a particle.

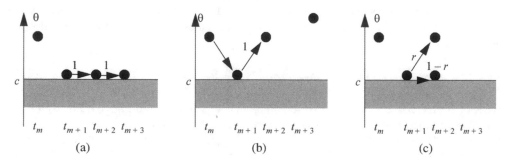

Figure 5.6   Different boundary conditions on a wall: (a) absorbing; (b) reflecting; (c) elastic.

At the initial moment of time $t = t_0 = 0$ the particle resides at $\vartheta = \vartheta_0$. At the next moment of time t_1 the position of the particle changes by a random increment ξ_1 with three possible values -1, 0, and 1 with the following probabilities

$$P(\xi_1 - 1) - p, \quad P(\xi_1 = 0) = 1 - p - q, \quad \text{and} \quad P(\xi_1 = -1) = q \tag{5.163}$$

At the moment of time $t_m = m$ the position of the particle is then given by

$$\vartheta_m = \vartheta_0 + \sum_{i=1}^{m} \xi_i \tag{5.164}$$

The particle under consideration can travel along the whole real axis $\vartheta \in (-\infty, \infty)$, and can be confined to a semi-infinite interval $\vartheta \in [c, \infty)$ ($\vartheta \in (-\infty, c]$) or to a finite interval $\vartheta \in [d, c]$. The first case is known as an unrestricted random walk, while the other two cases are known as a random walk with one or two walls at $\vartheta = c$ and $\vartheta = d$ respectively. Depending on the boundary conditions imposed on the wall one may distinguish between an absorbing wall, a reflecting wall, or an elastic wall. If a particle reaches an absorbing wall it remains on the wall forever. For a reflecting wall the position of the particle can be only decreased (increased) during the next step. In the case of an elastic wall the coordinate of the particle decreases (increases) with probability r or remains the same with probability $1 - r$. These three possibilities are shown in Fig. 5.6.

5.2.1.1   Unrestricted Random Walk

In the case of an unrestricted random walk the position ϑ of a particle can be anywhere on the real axis, i.e. it may vary from $-\infty$ to ∞. For simplicity it can be assumed that the initial position of the particle is at $\vartheta_0 = 0$. Then, after m steps the position is given by

$$\vartheta_m = \sum_{i=1}^{m} \xi_i \tag{5.165}$$

the random variable ϑ_m takes integer values from a finite interval $[-m, m]$. In order to arrive at the point with coordinate k, the particle must experience $m_1 \geq 0$ positive steps, $m_2 \geq 0$ negative steps, and $m_0 \geq 0$ zero steps such that

$$k = m_1 - m_2 \quad \text{and} \quad m_0 = m - (m_1 + m_2) \tag{5.166}$$

Thus, the probability for the particle to reside at $\theta_m = k$ is given by a trinomial law

$$P(\theta_m = k) = \sum \frac{m!}{m_0! m_1! m_2!} p^{m_1} (1 - p - q)^{m_0} q^{m_2} \tag{5.167}$$

where summation is taken over all possible indices m_1, m_2 and m_0 satisfying the restriction (5.166).

Let μ and σ^2 stand for the mean and the variance of the random variable ξ_i described above. It is clear that

$$\mu = 1 \cdot p + (-1 \cdot q) = p - q, \sigma^2 = (1)^2 p + (-1)^2 p + (-1)^2 q - \mu^2 = p + q - (p - q)^2$$

$$(5.168)$$

The process ϑ is a Markov process since increments ξ_i are assumed to be independent. The average and variance of the random variable ϑ_m are given by

$$\langle \vartheta_m \rangle = \left\langle \sum_{i=1}^{m} \xi_i \right\rangle = m\mu = m(p - q)$$

$$(5.169)$$

$$\sigma_m^2 = \langle (\vartheta_m - m\mu)^2 \rangle = m\sigma^2$$

The next step is to calculate the probability $P(j \leq \vartheta_m \leq k)$ of the event that the particle resides in one of the coordinates $j, j + 1, \ldots, k, j < k$. Since this probability is a sum of probabilities of each state, an approximation can be obtained from central limit theorem (CLT). According to CLT, the random variable θ_m is approximately Gaussian for a large number m, with the mean $m\mu$ and the variance $m\sigma^2$. Thus

$$P(j \leq \theta_m \leq k) = \frac{1}{\sqrt{2\pi m\sigma^2}} \int_j^k \exp\left[-\frac{(x - m\mu)^2}{2m\sigma^2}\right] dx = \Phi\left(\frac{k - m\mu}{\sigma\sqrt{m}}\right) - \Phi\left(\frac{j - m\mu}{\sigma\sqrt{m}}\right)$$

$$(5.170)$$

where

$$\Phi(x) = \frac{1}{\sqrt{2\pi}} \int_{-\infty}^{x} \exp(-t^2) dt$$

$$(5.171)$$

is the probability integral [9].

More general results on an unrestricted random walk can be obtained by considering its generating function

$$G(y) = \langle y^{z_i} \rangle = py^1 + (1 - p - q)y^0 + qy^{-1} = py^1 + (1 - p - q) + \frac{q}{y}$$ $$(5.172)$$

Since all increments are independent, one has $\langle y^{\theta_m} \rangle = G^m(y)$. The next step is to introduce a generating function $G(y, s)$ such that $|sG(y)| < 1$ and

$$G(y, s) = \sum_{m=0}^{\infty} s^m [G(y)]^m = \frac{1}{1 - sG(y)} = \frac{y}{-spy^2 + y(1 - s - sp - sq) - sq}$$ $$(5.173)$$

which contains all information about the random walk since the probabilities $P(\theta_m = k)$ are coefficients in the series expansion of $G(y, s)$.

$$G(x, y) = \sum_{m=0}^{\infty} \sum_{k=0}^{\infty} P(\vartheta_m = k) s^m y^k$$ $$(5.174)$$

5.2.2 Markov Processes with Jumps

In this section a few useful Markov processes often used in modelling of various physical, medical and economical phenomena are discussed. A common feature of these processes is that they contain points where the trajectory experiences a finite jump.

5.2.2.1 The Poisson Process

Let $\theta(t)$ be a counting process, i.e. it represents a number of certain event appearances on a given time interval $[t_0, t)$. It is clear that $\theta(t)$ attains only integer values and is a non-decreasing function of time. This implies that the transitional probability $\pi_{ij}(t_0, t)$ vanishes for $j < i$, i.e.

$$\pi_{ij}(t_0, t) = \text{Prob}\{\theta(t) = j \mid \theta(t_0) = i\} = 0 \qquad (5.175)$$

In many applications it is possible to justify that the probability of a change of state of the process $\theta(t)$ on a short interval of time $(t, t + \Delta t)$ is equal to $\lambda \Delta t + o(\Delta t)$ where $(\lambda = \text{const}) > 0$. The probability for this process to remain in the original state during the same time interval is then $1 - \lambda \Delta t + o(t)$. It is also assumed that the probability of more than one change of state during an arbitrary short interval of time is of the order $o(\Delta t)$. Mathematically speaking

$$\text{Prob}\{\theta(t + \Delta t) = j \mid \theta(t) = j - 1\} = \lambda \Delta t + o(t)$$
$$\text{Prob}\{\theta(t + \Delta t) = j \mid \theta(t) = j\} = 1 - \lambda \Delta t - o(t) \qquad (5.176)$$

It is also reasonable to assume that the initial state of the process is zero, i.e.

$$p_j^0 - p_j(t_0) = \text{Prob}\{\theta(t_0) = j\} = \begin{cases} 1, & j = 0 \\ 0, & j = 1, 2, \ldots \end{cases} \qquad (5.177)$$

The first task is to calculate unconditional probabilities of each state, $p_j(t) = \text{Prob}\{\theta(t) = j\}$. This can be accomplished by using eq. (5.106) with $a_{jj} = \lambda$ and $a_{j,j-1} = \lambda$, thus leading to

$$\frac{\partial}{\partial t} p_j(t) = -\lambda p_j(t) + \lambda p_{j-1}(t), \quad j \geq 1$$
$$\frac{\partial}{\partial t} p_0(t) = -\lambda p_0(t), \qquad\qquad j = 0 \qquad (5.178)$$

Once again, a generation function can be used to solve this equation. Indeed, let

$$G(s, t) = \sum_{j=0}^{\infty} s^j p_j(t) \qquad (5.179)$$

be the generating function of the Poisson process. Multiplying both sides of eq. (5.178) by s^j and summing over j one obtains

$$\frac{\partial}{\partial t}p_0(t) + \sum_{j=1}^{\infty} s^j \frac{\partial}{\partial t}p_j(t) = \frac{\partial}{\partial t}G(s,t) = -\lambda p_0(t) - \lambda \sum_{j=1}^{\infty}(s^j p_j(t) - s^j p_{j-1}(t))$$

$$= -\lambda(1+s)G(s,t)$$

(5.180)

The initial condition can be obtained from eq. (5.177) as

$$G(s,0) = \sum_{j=0}^{\infty} s^j p_j(0) = 1$$

(5.181)

The solution of eq. (5.180) with initial condition (5.181) is then

$$G(s,t) = G(s,0)\exp[-\lambda(1+s)t] = \exp[-\lambda(1+s)t]$$

(5.182)

Expanding the latter into the Taylor series and equating coefficients with equal powers of s one obtains

$$G(s,t) = \sum_{n=0}^{\infty} s^n p_n(t) = \exp(-\lambda t)\sum_{n=0}^{\infty}\frac{(\lambda s t)^n}{n!} = \exp(-\lambda t)\sum_{n=0}^{\infty}\frac{(\lambda t)^n}{n!}S^n$$

(5.183)

and thus

$$p_n(t) = \frac{(\lambda t)^n}{n!}e^{-\lambda t}$$

(5.184)

Since the parameter λ does not depend on time or the state, the Poisson process is homogeneous.

Using eqs. (5.101) and (5.109) one obtains the following forward and backward differential equations for transitional probabilities

$$\frac{d}{d\tau}\pi_{ij}(\tau) = -\lambda\pi_{ij}(\tau) + \lambda\pi_{i,j-1}(\tau)$$

(5.185)

$$\frac{d}{d\tau}\pi_{ij}(\tau) = -\lambda\pi_{ij}(\tau) + \lambda\pi_{i+1,j-1}(\tau)$$

(5.186)

It can be shown that the solution of these equations with initial conditions $\pi_{ij}(0) = \delta_{ij}$ is given by the following equation

$$\pi_{ij}(\tau) = \begin{cases} \dfrac{(\lambda\tau)^{j-i}}{(j-i)!}\exp[-\lambda\tau], & j \geq i \\ 0, & j < i \end{cases}$$

(5.187)

It can be pointed out that, for a non-homogeneous Poisson process, parameter $\lambda = \lambda(t)$ changes in time and the expression for probabilities of states becomes

$$p_n(t_0, t) = \frac{1}{n!} \left(\int_{t_0}^{t} \lambda(\tau) d\tau \right)^n \exp \left[-\int_{t_0}^{t} \lambda(\tau) d\tau \right] \tag{5.188}$$

5.2.2.2 A Birth Process

The next step is to consider a process $\theta(t)$ which also attains only integer values $0, 1, 2, \ldots$ but in contrast to the Poisson process, the probability of change of state during a short interval $(t, t + \Delta t)$ depends on the current state. The process considered is known as a birth process. A birth process is defined as follows.

If the system is at the state $\theta(t) = j, j = 0, 1, 2, \ldots$, at a moment of time, then the probability of transition $j \to j + 1$ during a short interval of time $(t, t + \Delta t)$ is equal to $\lambda_j \Delta t + o(\Delta t)$; the probability of remaining in the same state is $1 - \lambda_j \Delta - o(\Delta t)$, where $\lambda_j > 0$. The probability of transition to any other state is of the order $o(\Delta t)$, and thus is negligible. Applying eq. (5.104) to the birth process one obtains a system of linear differential equations

$$\frac{d}{dt} p_j(t) = -\lambda_j p_j(t) + \lambda_{j-1} p_{j-1}(t), \quad j = 1, 2, \ldots$$
$$\frac{d}{dt} p_0(t) = -\lambda_0 p_0(t) \tag{5.189}$$

The last equation must be supplemented by the initial conditions which have the following form

$$p_j^0 = p_j(0) = \delta_{j,1} \tag{5.190}$$

The simplest case of the birth process is the case where the rates λ_j are proportional to the current state j:

$$\lambda_j = \lambda j, \quad \lambda = \text{const} > 0, \quad j > 1 \tag{5.191}$$

which reflects the fact that the probability of transition is proportional to the number of specimens in the total population.[8] Under this assumption eq. (5.189) can be rewritten as

$$\frac{d}{dt} p_j(t) = -\lambda j p_j(t) + \lambda(j-1) p_{j-1}(t), \quad j \geq 1 \tag{5.192}$$

The last system of equations can be solved using sequential integration, thus leading to the following Yule–Farri distribution:

$$p_j(t) = \begin{cases} e^{-\lambda t}(1 - e^{-\lambda t})^{j-1}, & j \geq 1 \\ 0, & j = 0 \end{cases} \tag{5.193}$$

---

[8]This is approximately true for a case of newborn species [10]

The mean value and the variance of this process exponentially increase in time:

$$m_\theta(t) = e^{\lambda t}, \quad \sigma_\theta^2(t) = e^{\lambda t}(e^{\lambda t} - 1) \tag{5.194}$$

Since coefficients λ_j do not explicitly depend on time, the process itself is a homogeneous process, forward and backward Kolmogorov equations (5.109) and (5.110) can be thus written as

$$\frac{d}{d\tau}\pi_{ij}(\tau) = -\lambda_j\pi_{ij}(\tau) + \lambda_{j-1}\pi_{i,j-1}(\tau)$$

$$\frac{d}{d\tau}\pi_{ij}(\tau) = -\lambda_i\pi_{ij}(\tau) + \lambda_j\pi_{i+1,j}(\tau) \tag{5.195}$$

It can be shown [3] that the solution of these equations is given by

$$\pi_{ij}(\tau) = \begin{cases} C_{j-1}^{j-1}e^{-i\lambda\tau}(1 - e^{-\lambda\tau})^{j-1}, & j \geq i \\ 0, & j < i \end{cases} \tag{5.196}$$

5.2.2.3   A Death Process

A death process $\theta(t)$ attains only integer values and is a non-increasing process and is defined by the following conditions:

1. At the initial moment of time a system under consideration is in a state $j_0 \geq 1$, i.e. $\theta(0) = j_0$.

2. If in the moment of time t the system resides at a state j, i.e. $\theta(t) = j$, then the probability of transition $j \to j - 1$ during a short interval of time is equal to $\mu_j\Delta t + o(\Delta t)$ and the probability of remaining at the same state is $1 - \mu_j\Delta t - o(\Delta t)$.

3. The probabilities of other transitions are of order $o(\Delta t)^2$ or less.

Similar to the case of the birth process it is easy to obtain a differential equation for probabilities of each state

$$\frac{d}{dt}p_j(t) = -\mu_j p_j(t) + \mu_{j+1}p_{j+1}(t) \tag{5.197}$$

The solution of the death equation can be obtained in the very same way as the solution of the birth equation.

5.2.2.4   A Death and Birth Process

A birth and death process $\theta(t)$ attains integer values and is allowed to increase or decrease its value according to the following rules:

1. If the system is at the state $\theta(t) = j$ at the moment of time t, then the probability of transition $j \to j + 1$ during a short time interval $(t, t + \Delta t)$ is equal to $\lambda_j\Delta t + o(\Delta t)$.

2. If the system is at the state $\theta(t) = j$ at the moment of time t, then the probability of transition $j \to j - 1$ during a short time interval $(t, t + \Delta t)$ is equal to $\mu_j \Delta t + o(\Delta t)$.

3. The probability of remaining in the current state $\theta(t) = j$ is $1 - (\lambda_j + \mu_j)\Delta t - o(\Delta t)$.

4. The probability of change into an adjacent state is of an order of $o(\Delta t)$.

5. The state $\theta(t) = 0$ is an absorbing state, i.e. if the system reaches this state it remains in this state forever.

Once again resorting to eq. (5.104) one can obtain a system of differential equations describing a death and birth random process:

$$\frac{d}{dt}p_j(t) = \lambda_{j-1}p_{j-1}(t) - (\lambda_j + \mu_j)p_j(t) + \mu_{j+1}p_{j+1}(t) \tag{5.198}$$

In general, for $j = 0$ one can write

$$\frac{d}{dt}p_0(t) = \mu_1 p_1(t) \tag{5.199}$$

Since $j = 0$ is absorbing, only downward trajectories are allowed. It is assumed that at the initial moment of time the system resides in some non-zero initial state $0 < j_0 < \infty$. As a result, the initial conditions are simply

$$p_j(0) = \delta_{jj_0} = \begin{cases} 1, & j = j_0 \\ 0, & j \neq j_0 \end{cases} \tag{5.200}$$

In turn, eq. (5.109) for transitional probabilities becomes

$$\frac{d}{dt}\pi_{ij}(t) = \lambda_{j-1}\pi_{i,i-1}(t) - (\lambda_j + \mu_j)\pi_{ij}(t) + \mu_{j+1}\pi_{j,j+1}(t) \tag{5.201}$$

In general, the solution of eqs. (5.199)–(5.201) is difficult to obtain. However, in a linear case $\lambda_j = \lambda j$ and $\mu_j = \mu j$, $\lambda = \text{const} > 0$, $(\mu = \text{const} > 0)$ and the initial condition $\theta(t) = j_0$ of the process can be treated analytically. It is shown in [3] that the probability of the states can be found as

$$p_0(t) = \alpha(t)$$

$$p_j(t) = [1 - \alpha(t)][1 - \beta(t)]\beta^{j-1}(t), \quad j = 1, 2, \dots \tag{5.202}$$

$$\pi_{ij}(t) = \sum_{n=0}^{i} C_n^i C_{n-1}^{i+i-n-1}[\alpha(t)]^{i-n}[\beta(t)]^{j-n}[1 - \alpha(t) - \beta(t)]^n, \quad i \geq j \tag{5.203}$$

where

$$\alpha(t) = \frac{\mu(e^{(\lambda-\mu)t} - 1)}{\lambda e^{(\lambda-\mu)t} - \mu}, \qquad \beta(t) = \frac{\lambda(e^{(\lambda-\mu)t} - 1)}{\lambda e^{(\lambda-\mu)t} - \mu} \tag{5.204}$$

The mean and variance of the linear death and birth process are given by

$$m_\theta(t) = \exp[(\lambda - \mu)t], \qquad \sigma_\theta^2(t) = \frac{\lambda + \mu}{\lambda - \mu} e^{(\lambda - \mu)}[e^{(\lambda - \mu)t} - 1] \tag{5.205}$$

In particular, if $\lambda = \mu$ and $\theta(0) = j_0 = 1$, the average growth rate is equal to zero, $m_\theta(t) = 1$, and the variance grows linearly, $\sigma_\theta^2(t) = 2\lambda t$. It follows from eqs. (5.202) and (5.204) that the probability that the system resides at the final state $\theta(t) = 0$ at the moment of time t is given by

$$p_0(t) = \mu \frac{e^{(\lambda - \mu)t} - 1}{\lambda e^{(\lambda - \mu)t} - \mu} \tag{5.206}$$

This expression allows one to obtain the condition defining survivability of the system in the long run. Indeed, if $\lambda < \mu$, then

$$\lim_{t \to \infty} p_0(t) = 1, \quad \lambda < \mu \tag{5.207}$$

and the system evolution stops in the absorbing state $\theta(t) = 0$ with certainty. On the other hand, if $\lambda > \mu$ then

$$\lim_{t \to \infty} p_0(t) = \frac{\mu}{\lambda}, \quad \lambda > \mu \tag{5.208}$$

and the system would have a non-zero probability $p = (\lambda - \mu)/\lambda$ to never reach the absorbing state. This result can be easily visualized if λ and μ represent the birth and death rate of the population θ. If the birth rate is less than the death rate, then, eventually, the population becomes extinct. If the birth rate exceeds the death rate then it is possible that the population would never be extinct.

It is possible to show [3] that eqs. (5.202) and (5.203) remain valid for a more general case where coefficients λ and μ depend on time. In this case one has to use the following expressions for $\alpha(t)$ and $\beta(t)$

$$\alpha(t) = 1 - \frac{1}{\varepsilon(t)} \exp[-\gamma(t)], \qquad \beta(t) = 1 - \frac{1}{\varepsilon(t)} \tag{5.209}$$

where

$$\gamma(t) = \int_0^t [\mu(\tau) - \lambda(\tau)]d\tau \text{ and } \varepsilon(t) = \exp[-\gamma(t)]\left[1 + \int_0^t \mu(\tau)\exp[\gamma(\tau)]d\tau\right] \tag{5.210}$$

For such non-homogeneous processes, the mean and variance can be expressed as [3]

$$m_\theta(t) = \exp[-\gamma(t)], \sigma_\theta^2(t) = \exp[-2\gamma(t)]\int_0^t [\mu(\tau) + \lambda(\tau)]\exp[\gamma(\tau)]d\tau \tag{5.211}$$

The probability $p_0(t)$ that a system reaches its absorbing state $\theta(t) = 0$ is given by

$$p_0(t) = \frac{\int_0^t \mu(\tau) \exp[\gamma(\tau)] d\tau}{1 + \int_0^t \mu(\tau) \exp[\gamma(\tau)] d\tau} \tag{5.212}$$

It approaches unity every time when the integral $\int_0^t \mu(\tau) \exp[\gamma(\tau)] d\tau$ diverges for $t \to \infty$.

All the results shown are valid for $j_0 = 1$. If $j_0 > 1$ all equations for the mean value and the variance can be obtained by multiplying the right-hand side by j_0. The probability of reaching the absorbing state at $\theta(t) = 0$ is still one if $\lambda < \mu$ and equals $(\mu/\lambda)^{j_0}$ if $\lambda > \mu$. If λ and μ depend on time, then the probability $p_0(t)$ is obtained by raising expression (5.212) to power j_0.

5.3   THE FOKKER–PLANCK EQUATION

5.3.1   Preliminary Remarks

The transitional probability density $\pi(x, t \mid x_0, t_0), t > t_0$ of a continuous diffusion Markov process satisfies the following partial differential equations (PDE)

$$-\frac{\partial}{\partial t} \pi(x, t \mid x_0, t_0) = -\frac{\partial}{\partial x}[K_1(x, t)\pi(x, t \mid x_0, t_0)] + \frac{1}{2}\frac{\partial^2}{\partial x^2}[K_2(x, t)\pi(x, t \mid x_0, t_0)] \tag{5.213}$$

$$-\frac{\partial}{\partial t_0} \pi(x, t \mid x_0, t_0) = K_1(x_0, t_0)\frac{(\partial)}{\partial x_0}\pi(x, t \mid x_0, t_0) + \frac{1}{2}K_2(x_0, t_0)\frac{\partial^2}{\partial x_0^2}\pi(x, t \mid x_0, t_0) \tag{5.214}$$

The first of these two equations is known as the Fokker–Planck (FPE), or direct, equation. The second equation is called the Kolmogorov equation or the backward equation. It is important to mention that the direct equation involves a time derivative with respect to t (forward time), while the second one involves a time derivative with respect to t_0 (backward time). It will be shown later that these two equations form a conjugate pair.

5.3.2   Derivation of the Fokker–Planck Equation

The direct Fokker–Planck equation can be obtained directly from the differential Kolmogorov–Chapman equation (5.160) by setting $w(x \mid y, t) = 0$. However, in this section an alternative proof is provided. It is assumed for a moment that the transitional probability density $\pi(x, t \mid x_0, t_0)$ has all the necessary derivatives needed to formally write eqs. (5.213) and (5.214). Both equations can be derived from the Smoluchowski equation (5.139)

$$\pi(x, t \mid x_0, t_0) = \int_{-\infty}^{\infty} \pi(x, t \mid x', t')\pi(x', t' \mid x_0, t_0) dx' \tag{5.215}$$

The direct equation can be obtained by choosing t' close to t, while the reversed equation can be obtained by choosing t' close to t_0.

In order to obtain the Fokker–Planck equation, the Smoluchowski equation (5.139) can be rewritten in the following form

$$\pi(x, t + \Delta t \mid x_0, t_0) = \int_{-\infty}^{\infty} \pi(x, t + \Delta t \mid x', t)\pi(x', t \mid x_0, t_0)dx', \quad t + \Delta t > t > t_0 \quad (5.216)$$

where the time interval Δt is very short. Let $\Theta(\Omega, t + \Delta \mid x', t)$ represent a conditional characteristic function of the random increment $\Delta x = x - x'$ over a short interval Δt given x'. Using the definition of the characteristic function one can write

$$\Theta(\Omega, t + \Delta \mid x', t) = \langle \exp[j\Omega(x - x')] \mid x', t \rangle = \int_{-\infty}^{\infty} e^{j\Omega(x-x')}\pi(x, t + \Delta t \mid x', t)dx \quad (5.217)$$

Using the inverse Fourier transform one obtains

$$\pi(x, t + \Delta t \mid x', t) = \frac{1}{2\pi} \int_{-\infty}^{\infty} e^{-j\Omega(x-x')}\Theta(\Omega, t + \Delta \mid x', t)d\Omega \quad (5.218)$$

The conditional characteristic function can be expanded into the Taylor series as

$$\Theta(\Omega, t + \Delta \mid x', t) = \sum_{n=0}^{\infty} \frac{m_n(x', t)}{n!}(j\Omega)^n \quad (5.219)$$

where

$$m_n(x', t) = \langle \{\xi(t + \Delta t) - \xi(t)\}^n \mid \xi(t) \rangle = \int_{-\infty}^{\infty} (x - x')^n \pi(x, t + \Delta t \mid x', t)dx \quad (5.220)$$

are conditional moments of the increment $\Delta x = x - x'$. Using eq. (5.220) in expression (5.219) one can derive that

$$\pi(x, t + \Delta t \mid x', t)$$

$$= \frac{1}{2\pi} \int_{-\infty}^{\infty} e^{-j\Omega(x-x')} \left(\sum_{n=0}^{\infty} \frac{m_n(x', t)}{n!}(j\Omega)^n \right) d\Omega = \sum_{n=0}^{\infty} \frac{m_n(x', t)}{n!} \frac{1}{2\pi} \int_{-\infty}^{\infty} e^{-j\Omega(x-x')}(j\Omega)^n d\Omega$$

$$= \sum_{n=0}^{\infty} \frac{m_n(x', t)}{n!} \frac{(-1)^n}{2\pi} \frac{\partial^n}{\partial \lambda^n} \int_{-\infty}^{\infty} e^{-j\Omega(x-x')}d\Omega = \sum_{n=0}^{\infty} (-1)^n \frac{m_n(x', t)}{n!} \frac{\partial}{\partial x^n} \delta(x - x')$$

$$(5.221)$$

Finally, substituting series (5.221) into the Smoluchowski equation (5.216) one obtains

$$\pi(x, t + \Delta t \mid x_0, t_0) = \sum_{n=0}^{\infty} \frac{(-1)^n}{n!} \frac{\partial^n}{\partial x^n} [m_n(x, t)\pi(x, t \mid x_0, t_0)] \quad (5.222)$$

or, equivalently,

$$\pi(x, t + \Delta t \mid x_0, t_0) - \pi(x, t \mid x_0, t_0) = \sum_{n=1}^{\infty} \frac{(-1)^n}{n!} \frac{\partial^n}{\partial x^n} [m_n(x, t)\pi(x, t \mid x_0, t_0)] \qquad (5.223)$$

Dividing both parts by Δt and taking the limit when $\Delta t \to 0$ the following partial differential equation is obtained

$$\frac{\partial}{\partial t} \pi(x, t \mid x_0, t_0) = \sum_{n=1}^{\infty} \frac{(-1)^n}{n!} \frac{\partial^n}{\partial x^n} [K_n(x, t)\pi(x, t \mid x_0, t_0)] \qquad (5.224)$$

Here

$$K_n(\lambda, t) = \lim_{\Delta t \to 0} \frac{1}{\Delta t} \langle \{\xi(t + \Delta t) - \xi(t)\}^n \mid \xi(t) \rangle = \lim_{\Delta t \to 0} \frac{1}{\Delta t} \int_{-\infty}^{\infty} (x - x')^n \pi(x, t + \Delta t \mid x', t) dx \qquad (5.225)$$

It is worth noting here that since only the Chapman equation is used to derive eq. (5.224), it is valid for any Markov process as long as quantities $K_n(x, t)$ are finite.

Most attention in this book is paid to Markov processes with finite coefficients $K_1(\lambda, t)$ and $K_2(\lambda, t)$, and zero higher order coefficients $K_n(x, t)$:

$$K_n(x, t) \neq 0, \quad n = 1, 2, \qquad K_n(x, t) = 0, \quad n > 2 \qquad (5.226)$$

It follows from eq. (5.225) that condition (5.226) describes the rate of decay in the probability of large deviations with decreasing Δt. Large deviations of the process $\xi(t)$ are permitted, but they must appear in opposite directions with approximately the same probability. Thus, the average increment of the random process over a short time interval Δt is of the order of $\sqrt{\Delta t}$ [4]. Finite jumps of this process appear with zero probability; all the trajectories are continuous in a regular sense with probability one.

It can be proven that if, for a Markov process with finite coefficients $K_n(x, t) < \infty$ for all n, some even coefficient is zero, i.e. $K_{2n}(x, t) = 0$, then all the higher order coefficients $n \geq 3$ are equal to zero. This statement is known as the Pawula theorem [2,11]. Thus one may just require $K_4(x, t) = 0$. Thus, for a Markov process with continuous range to be a continuous Markov process it is necessary and sufficient that $K_n(x, t) = 0$ for all $n \geq 3$.

For continuous Markov processes, the Fokker–Planck equation becomes

$$\frac{\partial}{\partial t} \pi(x, t \mid x_0, t_0) = -\frac{\partial}{\partial x} [K_1(x, t)\pi(x, t \mid x_0, t_0)] + \frac{1}{2} \frac{\partial^2}{\partial x^2} [K_2(x, t)\pi(x, t \mid x_0, t_0)] \qquad (5.227)$$

Historically, the Fokker–Planck equation in the form of eq. (5.227) was obtained to study Brownian motion and is sometimes called a diffusion equation [1,2]. The coefficient $K_1(x, t)$ describes the average value of the particle displacement and is also known as the drift. Likewise, $K_2(x, t)$ can be interpreted as a variance of a particle's displacement from the average value and is called the diffusion coefficient.

Partial differential equation (5.227) is a linear partial differential equation of a parabolic type and a variety of methods can be used to obtain its solution. A few of these methods will be reviewed later. For a detailed analysis the interested reader can refer to [1,2]. Any solution $\pi(x, t \mid x_0, t_0)$ of the Fokker–Planck equation must satisfy the following conditions: a transitional probability density should be positive for all values of the arguments and be normalized to one after integration over the destination state, i.e.

$$\pi(x, t \mid x_0, t_0) \geq 0, \qquad \int_{-\infty}^{\infty} \pi(x, t \mid x_0, t_0)\mathrm{d}x = 1 \tag{5.228}$$

In addition, this density must be reduced to a δ function if $t = t_0$, i.e.

$$\pi(x, t_0 \mid x_0, t_0) = \delta(x - x_0) \tag{5.229}$$

A solution of the Fokker–Planck equation for unbounded space $x \in (-\infty, \infty)$ and the initial condition (5.229) is called the fundamental solution of the Fokker–Planck equation. If the initial value of the process x_0 is not fixed but distributed according to probability density function

$$p(x, t_0) = p_0(x) \tag{5.230}$$

then this probability density must be considered as a part of the initial conditions. An unconditional marginal probability density function can be calculated in two ways.

1. Using the property of a joint probability function to normalize to a marginal probability density one readily obtains

$$p(x, t) = \int_{-\infty}^{\infty} p_2(x, t, x_0, t_0)\mathrm{d}x_0 = \int_{-\infty}^{\infty} p(x_0, t_0)\pi(x, t \mid x_0, t_0)\mathrm{d}x_0 \tag{5.231}$$

This implies that in order to find the marginal probability density it is necessary to know the initial condition (5.229) and the fundamental solution $\pi(x, t \mid x_0, t_0)$ of the Fokker–Planck.

2. It is possible to calculate $p(x, t)$ directly from the Fokker–Planck equation. Indeed, multiplying both sides of eq. (5.227) by $p(x_0, t_0)$ and integrating over x_0 one obtains

$$\frac{\partial}{\partial t}p(x, t) = -\frac{\partial}{\partial x}[K_1(x, t)p(x, t)] + \frac{1}{2}\frac{\partial^2}{\partial x^2}[K_2(x, t)p(x, t)] \tag{5.232}$$

This indicates that the marginal probability density also satisfies the same Fokker–Planck equation as the transitional probability density. If the initial condition is expressed by eq. (5.229) then the marginal probability density coincides with the fundamental solution of the Fokker–Planck equation. This statement follows directly from eq. (5.231). Taking the latter property into account, as well as the fact that it is the marginal probability density which is usually of interest for applications, it is possible to concentrate on studying the Fokker–Planck equation in the form (5.232). This equation must be solved with initial conditions

(5.230). The solution must also satisfy the positivity and normalization conditions given by

$$p(x,t) \geq 0, \qquad \int_{-\infty}^{\infty} p(x,t)\mathrm{d}x = 1 \tag{5.233}$$

Furthermore, if the covariance function $C_{\xi\xi}(t,t')$, $t > t'$ of the Markov random process $\xi(t)$ is to be found, one has to find both the transitional probability density $\pi(x,t \mid x',t')$ and the marginal probability density $p(x',t')$. Once this is accomplished, the desired covariance function can be found as

$$C_{\xi\xi}(t,t') = \langle [\xi(t) - \langle \xi(t) \rangle][\xi(t') - \langle \xi(t') \rangle] \rangle$$
$$= \int\int [x - m_\xi(t)][x' - m_\xi(t')]\pi(x,t \mid x',t')p(x',t')\mathrm{d}x\mathrm{d}x' \tag{5.234}$$

5.3.3 Boundary Conditions

In order to obtain the solution of the Fokker–Planck equation (5.232) with the initial conditions (5.230) one must also consider appropriate boundary conditions, which depend on a particular problem and can vary from case to case.

In order to choose proper boundary conditions the following interpretation of the Fokker–Planck equation can be helpful. In particular, one can associate the value of probability density with some quantity proportional to a number of particles in a flow of an abstract substance [2]. In this case the probability density $p(x,t)$ reflects concentration of this substance at the point x at the moment of time t. The flow (or current) of the probability $G(x,t)$ (flow of substance) along the axis x consists of two parts: a local average flow $K_1(x,t)p(x,t)$ of particles moving in the field of a potential and a random (diffusion) flow $-0.5\frac{\partial}{\partial t}[K_2(x,t)p(x,t)]$, i.e.

$$G(x,t) = K_1(x,t)p(x,t) - \frac{1}{2}\frac{\partial}{\partial t}[K_2(x,t)p(x,t)] \tag{5.235}$$

Using the notion of probability current $G(x,t)$, the Fokker–Planck equation can be rewritten as

$$\frac{\partial}{\partial t}p(x,t) + \frac{\partial}{\partial x}G(x,t) = 0 \tag{5.236}$$

which can be now interpreted as a law of conservation of the total probability. The conservation equation (5.236) can be rewritten using small increments Δx and Δt instead of derivatives as

$$\frac{p(x,t+\Delta t) - p(x,t)}{\Delta t} + \frac{G(x+\Delta x,t) - G(x,t)}{\Delta x} \approx 0 \tag{5.237}$$

or

$$[p(x, t + \Delta t) - p(x, t)]\Delta x \approx [G(x, t) - G(x + \Delta x, t)]\Delta t \tag{5.238}$$

Thus, it can be said that the increment of the probability (substance) during a small interval of time Δt at the location x is equal to a difference between the incoming flow $G(x, t)$, entering on the left of the interval $[x, x + \Delta x]$ and the outcoming flow $G(x + \Delta x, t)$, leaving it, as shown in Fig. 5.7. The expression (5.238) can be interpreted as a law of conservation of probability, or continuity of probability. This also allows one to write similar expressions for the case of discontinuous Markov processes, both in continuous and discrete time.

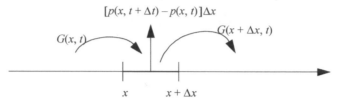

Figure 5.7   Illustration of the concept of probability current.

If a random process $\xi(t)$ attains all values on the real axis, i.e. $x \in (-\infty, \infty)$, then eq. (5.236) is valid on the whole real axis. As the boundary condition one must specify the values of the process at $x = \pm\infty$. Integrating eq. (5.236) over x from $-\infty$ to ∞ and using the normalization condition (5.233) one deduces that the following conditions must be satisfied

$$G(-\infty, t) = G(\infty, t) \tag{5.239}$$

However, in many practical problems, the latter condition is substituted by somewhat stronger conditions

$$G(-\infty, t) = G(\infty, t) = 0, p(-\infty, t) = p(\infty, t) = 0 \tag{5.240}$$

which are known as zero boundary conditions or natural conditions.

In a case when the random process $\xi(t)$ attains values from a finite interval $[c, d]$, eq. (5.236) should be considered only on this interval, with the solution vanishing outside this interval. However, in this case a number of different boundary conditions can be considered, similar to those used in section 5.2.1. In particular, one can distinguish between the absorbing and reflecting wall. For some problems periodic boundary conditions can be imposed [2]. Detailed discussions of the boundary conditions can be found in [1,2]. An absorbing boundary condition at $x = c$ or $x = d$ requires that the value of the probability density vanishes on this boundary at all times, i.e.

$$p(c, t) = 0 \quad \text{or} \quad p(d, t) = 0 \tag{5.241}$$

while a reflecting boundary condition requires that the probability current vanishes on the boundary

$$G(c, t) = 0 \quad \text{or} \quad G(d, t) = 0 \tag{5.242}$$

Many problems can be described by an absorbing boundary on one end and a reflecting boundary on the other end. If one of the boundaries is at infinity, natural boundary conditions must be applied on that boundary. The periodic boundary conditions

$$p(x, t) = p(x + L, t), G(x, t) = G(x + L, t) \tag{5.243}$$

can be found in such applications as the theory of matter, crystals, phase lock loops, etc. [1,2,12].

For a sub-class of stationary processes, the transitional probability density $\pi(x, t \mid x_0, t_0)$ depends only on the difference of the time arguments $\tau = t - t_0$, i.e.

$$\pi(x, t \mid x_0, t_0) = \pi(x, x_0, \tau) \tag{5.244}$$

while the drift and the diffusion do not explicitly depend on time, i.e.

$$K_1(x, t) = K_1(x), \qquad K_2(x, t) = K_2(x) \tag{5.245}$$

The marginal stationary probability density function $p_{st}(x)$, if it exists, depends neither on time nor on the initial distribution $p_0(x) = p(x, t_0)$. Thus, for a stationary distribution

$$\frac{\partial}{\partial t} p_{st}(x) = 0 \tag{5.246}$$

and, due to the conservation property of the Fokker–Planck equation

$$\frac{\partial}{\partial x} G(x) = 0, \qquad G(x) = G = \text{const.} \tag{5.247}$$

As a result, the Fokker–Planck equation is reduced to the following ordinary differential equation

$$\frac{d}{dx} [K_2(x) p_{st}(x)] - 2K_1(x) p_{st}(x) = -2G \tag{5.248}$$

The general solution of this equation can be found elsewhere (see [13] for example)

$$p_{st}(x) = \frac{C}{K_2(x)} \exp\left[2 \int_{x'}^{x} \frac{K_1(s)}{K_2(s)} ds\right] - \frac{2G}{K_2(x)} \int_{x'}^{x} \exp\left[2 \int_{y'}^{y} \frac{K_1(s)}{K_2(s)} ds\right] dy \tag{5.249}$$

Here, a constant C can be defined from the normalization property of the probability density function. The lower limit of integration is an arbitrary point on the real axis where the process $\xi(t)$ is defined.

For zero boundary conditions on the probability flow $(G = 0)$ eq (5.248) becomes

$$\frac{d}{dx} [K_2(x) p_{st}(x)] - 2K_1(x) p_{st}(x) = 0 \tag{5.250}$$

The general solution of this equation is

$$p_{st}(x) = \frac{C}{b(x)} \exp\left[2 \int_{x'}^{x} \frac{K_1(s)}{K_2(s)} ds\right] \tag{5.251}$$

and the constant C can be obtained to normalize the probability density function $p_{st}(x)$ to unity.

The latter results illustrate the importance of the Fokker–Planck equation for applications. Indeed, once the drift $K_1(x)$ and diffusion $K_2(x)$ coefficients are known, it is possible to write an expression for the stationary probability density. In many cases integral (5.251) can be evaluated in quadratures. Unfortunately, investigation of the non-stationary solution is much more complicated. However, as will be shown later, a number of useful results can be obtained.

5.3.4 Discrete Model of a Continuous Homogeneous Markov Process

In order to visualize the nature of time variation of a continuous homogeneous Markov process $\xi(t)$ for small time increments Δt, it is useful to consider an approach of a discrete Markov process to its continuous limit. In order to do so, a discrete Markov chain with transitions only to one of the neighbouring states is considered. In other words, if the chain is in the state j at the moment of time t, then at the moment of time $t + \Delta t$ the chain can be found in states $j - 1, j$, or $j + 1$. Let the probabilities of these transitions be $\alpha_j, 1 - \alpha_j - \beta_j$ and β_j, respectively. In addition, let $\pi_{ij}(m)$ be the probability of transition from the state i to the state j in m steps. Then

$$\pi_{ij}(m) = \alpha_{j-1}\pi_{i,j-1}(m-1) + (1 - \alpha_j - \beta_j)\pi_{i,j}(m-1) + \beta_j\pi_{i,j+1}(m-1) \tag{5.252}$$

In a similar way the continuous process can be approximated as a discrete process with increments $-\Delta x$, 0, and Δx with probabilities $\alpha(x), 1 - \alpha(x) - \beta(x)$ and $\beta(x)$, respectively. If $\pi(x, t \mid x_0, t_0)$ stands for the conditional probability to find the trajectory of the process inside an interval $(x, x + \Delta x)$ at the moment of time t, given that this trajectory was at x_0 at the moment of time t_0, then the equation for evolution of $\pi(x, t \mid x_0, t_0)$ in time can be written as

$$\pi(x, t \mid x_0, t_0)\Delta x$$
$$= \pi(x - \Delta x, t - \Delta t \mid x_0, t_0)\Delta x \alpha(x - \Delta x) + \pi(x, t - \Delta t \mid x_0, t_0)\Delta x[1 - \alpha(x) - \beta(x)]$$
$$+ \pi(x + \Delta x, t - \Delta t \mid x_0, t_0)\Delta x \beta(x + \Delta x) \tag{5.253}$$

or

$$\pi(x, t \mid x_0, t_0) = \pi(x - \Delta x, t - \Delta t \mid x_0, t_0)\alpha(x - \Delta x) + \pi(x, t - \Delta t \mid x_0, t_0)$$
$$\times [1 - \alpha(x) - \beta(x)] + \pi(x + \Delta x, t - \Delta t \mid x_0, t_0)\beta(x + \Delta x) \tag{5.254}$$

In order for the approximation to approach a continuous process, the time increment Δt has to approach zero, $\Delta t \to 0$. It will be shown below that certain restrictions must be applied on probabilities $\alpha(x)$ and $\beta(x)$.

Since the particle located at x at the moment of time t is allowed only to remain in the current position or move to the closest location separated by Δx, average displacement and its variance during a small time interval Δt are thus

$$m(\Delta t) = [\alpha(x) - \beta(x)]\Delta x \tag{5.255}$$

$$\sigma^2(\Delta t) = [\alpha(x) + \beta(x) - [\alpha(x) - \beta(x)]^2](\Delta x)^2 \tag{5.256}$$

The drift and diffusion coefficients now can be obtained as the limit of $m(\Delta t)$ and $\sigma^2(\Delta t)$

$$K_1(x) = \lim_{\Delta x, \Delta t \to 0} [\alpha(x) - \beta(x)]\frac{\Delta x}{\Delta t}, \; K_2(x) = \lim_{\Delta x, \Delta t \to 0} [\alpha(x) + \beta(x) - [\alpha(x) - \beta(x)]^2]\frac{(\Delta x)^2}{\Delta t}$$

$$\tag{5.257}$$

In order to obtain a finite drift and diffusion, certain restrictions must be imposed on how $\alpha(x)$ and $\beta(x)$ behave for small Δt and Δx.

Assuming that the diffusion coefficient $K_2(x)$ is limited

$$K_2(x) < C = \text{const.} \tag{5.258}$$

and choosing Δx such that

$$(\Delta x)^2 = C\Delta t \tag{5.259}$$

one can easily verify that

$$\alpha(x) - \frac{1}{2C}[K_2(x) + K_1(x)\Delta x], \qquad \beta(x) - \frac{1}{2C}[K_2(x) - K_1(x)\Delta x] \tag{5.260}$$

satisfies eq. (5.257). Thus, in a discrete model of a continuous Markov process, spacial increments Δx must be of the order of $O(\sqrt{\Delta t})$ and the probabilities $\alpha(x)$ and $\beta(x)$ differ only by a quantity of the same order

$$\alpha(x) - \beta(x) \sim \frac{K_1(x)}{\sqrt{K_2(x)}}\sqrt{\Delta t} \sim O(\sqrt{\Delta t}) \tag{5.261}$$

5.3.5   On the Forward and Backward Kolmogorov Equations

Since the forward and the backward Kolmogorov equations (5.213) and (5.214) define the same transitional probability density $\pi(x, t \mid x_0, t_0)$, these two equations are not independent. In particular, it can be shown that these two equations define two adjoint operations.

According to [14], two differential functionals

$$L[u(x)] = \frac{d}{dx}\left[k(x)\frac{d}{dx}u(x)\right] - l(x)\frac{d}{dx}u(x) + \frac{d}{dx}[\mu(x)u(x)] - \gamma(x)u(x) \tag{5.262}$$

and

$$M[v(x)] = \frac{d}{dx}\left[k(x)\frac{d}{dx}v(x)\right] + l(x)\frac{d}{dx}v(x) - \mu(x)\frac{d}{dx}v(x) - \gamma(x)v \tag{5.263}$$

where $k(x)$, $l(x)$, $\mu(x)$ and $\nu(x)$ are given functions of x; $u(x)$ and $v(x)$ are two functions which satisfy the following condition

$$\int (\nu(x)L[u(x)] - u(x)M[\nu(x)])\mathrm{d}x = \left[k(x) \left(\nu(x) \frac{\mathrm{d}}{\mathrm{d}x} u(x) - u(x) \frac{\mathrm{d}}{\mathrm{d}x} \nu(x) \right) \right. $$
$$\left. + (\mu(x) - l(x))u(x)\nu(x) \right] \tag{5.264}$$

and are called conjugate. In this case a differential operator $M[\nu]$ is uniquely defined by the differential operator $L[\nu]$. It is easy to show that if $u(x) = \pi(x, t \mid x_0, t_0)$ in eq. (5.213) and $v(x) = \pi(x, t \mid x_0, t_0)$ in eq. (5.214) then direct and inverse Fokker–Planck operators are conjugate with $k(x) = 0.5K_2(x, t)$, $l(x) = 0$, $\mu(x) = 0.5K_2(x, t) - K_1(x, t)$ and $v(x) = 0$.

Depending on a particular problem it is wise to use either the direct equation or the reversed equation. For example, if the initial state $x_0 = x(t_0)$ is given, it is more appropriate to use the direct equation (5.213). However, if it is required to calculate the probability of the first level crossing it is better to use the reversed equation (5.214). Furthermore, since the final value of the process is known, it can be found that if certain limitations are imposed on the solution $x(t)$, the direct equation cannot be used, while the reversed equation is still valid. For example, one can consider movement of a particle on a limited interval of its range. When the particle reaches one of the two boundaries of this interval it remains on the boundary for a certain random time τ, distributed exponentially. At the end of the time interval τ the particle instantaneously returns to a point x inside the specified interval described by a known probability density. Further motion of the particle is described by a regular Fokker–Planck equation. The described process is a Markov process. However, the direct equation cannot be immediately applied since the transitions are not local.[9] Meanwhile the reversed equation would not change and can be applied [1,2].

5.3.6 Methods of Solution of the Fokker–Planck Equation

Since the Fokker–Planck equation is an equation of a parabolic type, one can apply all the known methods [14]. Only four methods are considered below: (a) method of separation of variables, (b) the Laplace transform, (c) method of characteristic function, and (d) changing the independent variables.

5.3.6.1 Method of Separation of Variables

This method is reasonable to apply when both the drift and the diffusion coefficients do not depend on time (homogeneous Markov process). In this case eq. (5.232) becomes

$$\frac{\partial}{\partial t} p(x, t) = -\frac{\partial}{\partial x} [K_1(x)p(x, t)] + \frac{1}{2} \frac{\partial^2}{\partial x^2} [K_2(x)p(x, t)] \tag{5.265}$$

---

[9]Since jumps are allowed.

The solution of this equation can be thought of in the form of an anzatz [2,6]

$$p(x, t) = \Phi(x)T(t) \tag{5.266}$$

where $\Phi(x)$ and $T(t)$ are functions of x and t only. Substituting eq. (5.266) into eq. (5.265) and dividing both sides by $p(x, t) = \Phi(x)T(t)$ one obtains

$$\frac{1}{T(t)}\frac{\partial}{\partial t}T(t) = \frac{-\dfrac{\partial}{\partial x}[K_1(x)\Phi(x)] + \dfrac{1}{2}\dfrac{\partial^2}{\partial x^2}[K_2(x)\Phi(x)]}{\Phi(x)} \tag{5.267}$$

The left-hand side of this equation depends only on t, while the right-hand side depends only on x. This implies that both sides of this equation are equal to some constant, say $-\gamma^2$. Thus, the partial differential equation can be substituted by an equivalent system of two ordinary differential equations

$$\frac{1}{T(t)}\frac{d}{dt}T(t) = -\gamma^2 = -\lambda, \quad \lambda \geq 0 \tag{5.268}$$

$$\frac{1}{2}\frac{d^2}{dx^2}[K_2(x)\Phi(x)] - \frac{d}{dx}[K_1(x)\Phi(x)] + \gamma^2\Phi(x) = 0 \tag{5.269}$$

Differential equation (5.268) has an exponential solution[10]

$$T(t) = \exp(-\gamma^2 t) \tag{5.270}$$

and the second order differential equation (5.269) can be solved using traditional methods [13,14]. This solution $\Phi(x, A, B, \gamma)$ depends on two arbitrary constants A and B as well as on the separation parameter γ. Since the Fokker–Planck equation is a linear equation, its general solution is a linear combination of particular solutions, i.e.

$$p(x, t) = \sum_{n=0}^{\infty} \Phi(x, A_n, B_n, \gamma_n) \exp(-\gamma_n^2 t) \tag{5.271}$$

where constants A_n, B_n and γ_n are defined by the corresponding boundary and initial conditions, imposed by a particular physical problem. It will be seen that it is often possible to satisfy the boundary conditions only for particular values of γ, called the eigenvalues, while the corresponding function $\Phi_n(x, A, B) = \Phi(x, A, B, \gamma_n)$ is called the eigenfunction corresponding to the eigenvalue γ_n. A set of all possible eigenvalues is also known as the spectrum of the Fokker–Planck equation. If there is a countable number of isolated eigenvalues, the spectrum is called a discrete spectrum; if there is a continuum of eigenvalues then the spectrum is called a continuous spectrum. Finally, if there are both isolated eigenvalues and a region of continuous eigenvalues, the spectrum is called a mixed

---

[10] The weight of the exponent is arbitrarily chosen to be unity with a proper constant absorbed by the function Φ.

spectrum. A comprehensive treatment of the so-called eigenvalue problem can be found in [15].

It can be shown that if the flow $G(x, t)$ across the boundaries is zero and the spectrum of the Fokker–Planck operator is discrete, the solution can be further simplified to

$$p(x, t) = P_0 p_{st}(x) + \sum_{n=1}^{\infty} \Phi_n(x) T_n \exp[-\gamma_n^2(t - t_0)], \qquad \Phi_0(x) = p_{st}(x) \qquad (5.272)$$

where $\Phi_n(x)$ are eigenfunctions of eq. (5.269) corresponding to the eigenvalues γ_n^2. P_0 and T_n are some constant coefficients to be defined from the initial conditions. The eigenfunction which corresponds to the smallest eigenvalue $\lambda_0 = 0$ is the stationary solution $p_{st}(x)$ of the stationary Fokker–Planck equation as given by eqs. (5.250) and (5.251). Functions $\Phi_n(x)$ form an orthonormal set with weight $1/p_{st}(x)$. Mathematically speaking

$$\int \frac{\Phi_n(x) \Phi_n(x)}{p_{st}(x)} dx = \delta_{mn} = \begin{cases} 1, & n = m \\ 0, & n \neq m \end{cases} \qquad (5.273)$$

In particular

$$\int \frac{\Phi_0(x) \Phi_n(x)}{p_{st}(x)} dx = \int \Phi_n(x) dx = 0 \quad \text{for } n > 0 \qquad (5.274)$$

If the initial probability density is specified, $p(x, t_0) = p_0(x)$ then the coefficients T_n and P_0 can be obtained simply by setting $t = t_0$ in eq. (5.272) and by multiplying both parts of the latter by $\Phi_m(x)$ followed by integration over x. Due to the orthogonality property (5.272) one obtains that the coefficients P_0 and T_n are uniquely defined by

$$P_0 = 1 \text{ and } T_n = \int \frac{\Phi_n(x) p_0(x)}{p_{st}(x)} dx, \quad n > 0 \qquad (5.275)$$

If the initial value of the process is fixed at x_0, i.e. $p_0(x) = \delta(x - x_0)$ then the transitional density $\pi(x, t \mid x_0, t_0)$ is recovered

$$p(x, t) = \pi(x, t \mid x_0, t_0) = \sum_{n=0}^{\infty} \frac{\Phi_n(x) \Phi_n(x_0)}{p_{st}(x_0)} \exp[-\gamma_n^2(t - t_0)] \qquad (5.276)$$

It is sometimes beneficial to consider normalized eigenfunctions $\psi_n(x)$, defined by the following rule[11]

$$\psi_n(x) = \frac{\Phi(x)}{p_{st}(x)} \qquad (5.277)$$

---

[11] Another useful normalization is $\psi_n(x) = \Phi_n(x)/\sqrt{p_{st}(x)}$.

In this case eqs. (5.272)–(5.276) can be rewritten as

$$p(x,t) = p_{st}(x)\left[1 + \sum_{n=1}^{\infty} \psi_n(x)T_n \exp[-\gamma_n^2(t-t_0)]\right], \qquad \psi_0(x) = 1 \quad (5.278)$$

$$\int \psi_n(x)\psi_n(x)p_{st}(x)dx = \delta_{mn} = \begin{Bmatrix} 1, & n = m \\ 0, & n \neq m \end{Bmatrix} \qquad (5.279)$$

$$T_n = \int \psi_n(x)p_0(x)dx, \qquad n > 0 \qquad (5.280)$$

and

$$p(x,t) = \pi(x,t \mid x_0,t_0) = p_{st}(x) \sum_{n=0}^{\infty} \psi_n(x)\psi_n(x_0) \exp[-\gamma_2^n(t-t_0)], \quad t \geq t_0 \qquad (5.281)$$

Once the transitional density is known, one is able to calculate all parameters of the corresponding Markov process. For example, the joint PDF $p_2(x,x_0;t,t_0)$ is thus

$$p_2(x,x_0;t,t_0) = \pi(x,t \mid x_0,t_0)p(x_0,t_0) = p_{st}(x)p(x_0,t_0) \sum_{n=0}^{\infty} \psi_n(x)\psi_n(x_0) \exp[-\gamma_n^2(t-t_0)]$$

$$(5.282)$$

and the correlation function is given by

$$R_{\xi\xi}(t,t_0) = \int\int xx_0 p_2(x,x_0;t,t_0)dxdx_0$$

$$= \sum_{n=0}^{\infty} \exp[-\gamma_n^2(t-t_0)]\left[\int x\psi_n(x)p_{st}(x)dx \int x_0\psi_n(x_0)p(x_0,t_0)\right] \qquad (5.283)$$

In the stationary case, $p(x_0,t_0) = p_{st}(x_0)$ and the correlation function depends only on the time difference $\tau = t - t_0$

$$R_{\xi\xi}(\tau) = \sum_{n=0}^{\infty} h_n^2 \exp[-\gamma_n^2|\tau|] = h_0^2 + \sum_{n-1}^{\infty} h_n^2 \exp[-\gamma_n^2|\tau|] \qquad (5.284)$$

where

$$h_n = \int x\psi_n(x)p_{st}(x)dx \qquad (5.285)$$

Since $\psi_0(x) = 1$ one can obtain from eq. (5.285) that h_0 is just the mean value m of the process and, thus, the expression for the covariance function $C_{\xi\xi}(\tau)$ becomes simply

$$C_{\xi\xi}(\tau) = R_{\xi\xi}(\tau) - m^2 = \sum_{n=1}^{\infty} h_n^2 \exp[-\gamma_n^2|\tau|] \qquad (5.286)$$

Finally, taking the Fourier transform of both sides of eq. (5.286) one obtains the expression for the power spectral density of the Markov process $\xi(t)$

$$S(\omega) = \int_{-\infty}^{\infty} C_{\xi\xi}(\tau) \exp(-j\omega\tau) d\tau = \sum_{n=1}^{\infty} \frac{4\gamma_n^2 h_n^2}{\omega^2 + \gamma_n^4} \tag{5.287}$$

Higher order statistics are derived in section 5.6.

Example: *Gaussian process [6].* Let the following Fokker–Planck equation be considered

$$\frac{\partial}{\partial t} p(x,t) = \frac{1}{\tau_c} \frac{\partial}{\partial x} ([xp(x,t)]) + \frac{\sigma^2}{\tau_c} \frac{\partial^2}{\partial x^2} p(x,t) \tag{5.288}$$

with natural boundary conditions at $x = \pm\infty$, i.e. $G(\pm\infty, t) = 0$. In this case eq. (5.269) becomes

$$\sigma^2 \frac{\partial^2}{\partial x^2} \Phi(x) + \frac{\partial}{\partial x} [x\Phi(x)] + \gamma^2 \tau_c \Phi(x) = 0 \tag{5.289}$$

This problem has a discrete spectrum given by [15,16]

$$\gamma_n^2 = \frac{n}{\tau_c}, \qquad n = 0, 1, 2, \ldots \tag{5.290}$$

and the eigenfunctions

$$\Phi_n(x) = \frac{1}{\sqrt{n!}} \frac{1}{\sigma} F^{(n+1)} \left(\frac{x}{\sigma}\right) \tag{5.291}$$

where function $F^{(n+1)}(z)$ can be easily expressed in terms of the Hermitian polynomials using the Rodrigues formula [9]

$$F^{(n+1)}(z) = \frac{1}{\sqrt{2\pi}} \frac{d^n}{dz^n} e^{-\frac{z^2}{2}} = \frac{1}{\sqrt{2\pi}} \tag{5.292}$$

The stationary solution corresponds to $\Phi_0(x)$ and is thus

$$p_{st}(x) = \Phi_0(x) = \frac{1}{\sigma} F^{(1)} \left(\frac{x}{\sigma}\right) = \frac{1}{\sqrt{2\pi\sigma^2}} \exp\left(-\frac{x^2}{2\sigma^2}\right) \tag{5.293}$$

i.e. the stationary distribution is a Gaussian one with zero mean and variance σ^2. Having this in mind, the normalized eigenfunctions $\psi_n(x)$ are given by

$$\psi_n(x) = \frac{\Phi_n(x)}{p_{st}(x)} = \frac{(-1)^n 2^{-n/2}}{\sqrt{n!}} H_n \left(\frac{x}{\sqrt{2}\sigma}\right) \tag{5.294}$$

This, in turn, provides the following expression for the transitional density

$$\pi(x, t \mid x_0, t_0) = \frac{1}{\sqrt{2\pi\sigma^2}} \exp\left(-\frac{x^2}{2\sigma^2}\right) \sum_{n=0}^{\infty} \frac{1}{2^n n!} H_n\left(\frac{x}{\sqrt{x}\sigma}\right) H_n\left(\frac{x_0}{\sqrt{2}\sigma}\right) \exp\left[-\frac{\pi\tau}{\tau_c}\right] \quad (5.295)$$

Taking into account that $H_1(x) = \sqrt{2}x$ one obtains that the expression for the covariance function is reduced to a single exponent

$$C_{\xi\xi}(\tau) = \sigma^2 \exp\left[-\frac{\tau}{\tau_c}\right] \quad (5.296)$$

Example: one-sided Gaussian distribution [6]. The Fokker–Planck equation (5.288), considered with different boundary conditions, leads to a different expression for the eigenvalues and eigenfunctions. Let there be a reflecting wall at $x = 0$ and the natural boundary conditions at $x = \infty$, i.e.

$$G(0, t) = G(\infty, t) = 0 \quad (5.297)$$

It is easy to verify that in this case one can keep only eigenvalues (5.290) and eigenfunctions (5.291) with even indices, i.e.

$$\gamma_n^2 = \frac{2n}{\tau_c} \quad (5.298)$$

thus resulting in the following expression for the stationary PDF

$$p_{st}(x) = \sqrt{\frac{2}{\pi\sigma^2}} \exp\left(-\frac{x^2}{2\sigma^2}\right), \quad x \geq 0 \quad (5.299)$$

and the transitional density

$$\pi(x, t \mid x_0, t_0) = \sqrt{\frac{2}{\pi\sigma^2}} \exp\left(-\frac{x^2}{2\sigma^2}\right) \sum_{n=0}^{\infty} \frac{1}{2^{2n}(2n)!} H_{2n}\left(\frac{x}{\sqrt{2}\sigma}\right) H_{2n}\left(\frac{x_0}{\sqrt{2}\sigma}\right) \exp\left[-\frac{2\pi\tau}{\tau_c}\right] \quad (5.300)$$

Example: Nakagami distributed process [6]. Let the Fokker–Planck equation have the following form

$$\frac{\partial}{\partial t} p(x, t) = \frac{1}{2\tau_c} \frac{\partial}{\partial x}\left[\left(x - \frac{\Omega(2m - 1)}{2mx}\right) p(x, t)\right] + \frac{\Omega}{4\tau_c m} \frac{\partial^2}{\partial x^2} p(x, t) \quad (5.301)$$

with the boundary conditions

$$G(0, t) = G(\infty, t) = 0 \quad (5.302)$$

Separation of variables produces the following eigenvalue problem

$$\frac{\Omega}{2m}\frac{d^2}{dx^2}\Phi(x) + \frac{\partial}{\partial x}\left[\left(x - \frac{\Omega(2m-1)}{2mx}\right)\Phi(x)\right] + 2\tau_c\gamma^2\Phi(x) = 0 \tag{5.303}$$

Using the following substitution

$$z = \frac{2mx^2}{\Omega} \quad \text{and} \quad u(z) = z^{-(m-1)/2}\Phi\left(\sqrt{\frac{\Omega z}{2m}}\right) \tag{5.304}$$

Eq. (5.303) can be transformed to a more standard form [6]

$$\frac{d^2}{dz^2}u(z) + \frac{d}{dz}u(z) + \left[\frac{\frac{m}{2} + \gamma^2\tau_c}{z} + \frac{\frac{1}{4} - \left(\frac{m-1}{2}\right)^2}{z^2}\right]u(z) = 0 \tag{5.305}$$

Once again, using [9], one can obtain that the eigenvalues of eq. (5.305) are given by

$$\gamma_n^{2'} = \frac{n}{\tau_c} \tag{5.306}$$

with corresponding eigenfunctions expressed in terms of the generalized Laguerre polynomials [9]

$$u_n(z) = z^{m/2}\exp(-z)L_n^{m-1}(z) \tag{5.307}$$

or, returning to the original variables and normalizing the eigenfunctions, one obtains [6]

$$\Phi_n(x) = \frac{1}{\sqrt{n!\Gamma(n+m)\Gamma(m)}}x^{2m-1}\exp\left(-\frac{mx^2}{\Omega}\right)L_n^{m-1}\left(\frac{mx^2}{\Omega}\right) \tag{5.308}$$

and the joint stationary PDF is given by

$$p_2(x, x_0; \tau) = \frac{4}{\Omega}\left(\frac{mxx_0}{\Omega}\right)^{2m-1}\exp\left[-\frac{m(x^2 + x_0^2)}{\Omega}\right]\sum_{n=0}^{\infty}\frac{L_n^{m-1}\left(\frac{mx^2}{\Omega}\right)L_n^{m-1}\left(\frac{mx_0^2}{\Omega}\right)}{n!\Gamma(n+m)\Gamma(m)}\exp\left(-\frac{n|\tau|}{\tau_c}\right) \tag{5.309}$$

Thus, the stationary distribution described by the Fokker–Planck equation (5.301) is the Nakagami distribution

$$p_{st}(x) = \frac{2}{\Gamma(m)}\left(\frac{m}{\Omega}\right)^m x^{2m-1}\exp\left(-\frac{mx^2}{\Omega}\right) \tag{5.310}$$

5.3.6.2 The Laplace Transform Method

The effectiveness and popularity of the Laplace transform is explained by the fact that it eliminates the dependence of the solution on time t and the partial differential equation becomes an ordinary differential equation. This approach can be directly applied to the Fokker–Planck equation. In order to do so, it can be written as

$$\frac{\partial}{\partial t} p(x, t) = \frac{1}{2} K_2(x) \frac{\partial^2}{\partial x^2} p(x, t) + \left[\frac{\partial}{\partial x} K_2(x) - K_1(x)\right] \frac{\partial}{\partial x} p(x, t)$$
$$+ \frac{1}{2} \left[\frac{\partial^2}{\partial x^2} K_2(x) - \frac{\partial}{\partial x} K_1(x)\right] p(x, t)$$

(5.311)

Next, let $P(x, s)$ stand for the Laplace transform of the probability density $p(x, t)$, i.e.

$$P(x, s) = \int_0^\infty p(x, t) e^{-st} dt$$

(5.312)

Taking the Laplace transformation of both sides of the Fokker–Planck equation one obtains the following ordinary differential equation

$$\frac{1}{2} K_2(x) \frac{d^2}{dx^2} P(x, s) + [K_2'(x) - K_1(x)] \frac{d}{dx} P(x, s) + \left[\frac{1}{2} K_2''(x) - K_1'(x) - s\right] P(x, s) + p_0(x) = 0$$

(5.313)

where $p_0(x)$ is the initial PDF of the process under consideration, $p(x, t) = p_0(x)$. The corresponding boundary conditions are also obtained by taking the Laplace transform of these conditions.

Example: *thermal driven motion in a well with an absorbing wall.* If a particle is moving in a limited region with a reflecting wall at $x = 0$ and an absorbing wall under the influence of the thermal force, the corresponding Fokker–Planck equation can be written as

$$\frac{\partial}{\partial t} p(x, t) = \frac{\sigma^2}{T_c} \frac{\partial^2}{\partial x^2} p(x, t)$$

(5.314)

with the boundary conditions

$$\frac{\partial}{\partial x} p(x, t) \bigg|_{x=0} = p(L, t) = 0$$

(5.315)

Assuming that the initial position of the particle is uniformly distributed over the interval $[0, L]$ the Laplace transform of eq. (5.314) becomes

$$sP(x, s) - \frac{1}{L} = \frac{\sigma^2}{T_c} \frac{\partial^2}{\partial x^2} P(x, s), \frac{\partial}{\partial x} P(x, s) \bigg|_{x=0} = P(L, s) = 0$$

(5.316)

After simple manipulation the solution of the problem (5.316) is given by

$$P(x, s) = \frac{1}{sL} - \frac{\cosh\left(\frac{\sqrt{s\tau}}{\sigma x}\right)}{sL\cosh\left(\frac{\sqrt{s\tau}}{\sigma}L\right)} \tag{5.317}$$

Since there is an absorbing wall, the particle eventually will be removed from the system. This conclusion is confirmed by the fact that[12]

$$\lim_{\tau \to \infty} p(x, t) = \lim_{s \to 0} sP(x, s) = 0 \tag{5.318}$$

Furthermore, one can define a residency (survival) distribution

$$P_r(t) = \frac{d}{dt}\int_0^L p(x, t)dx \tag{5.319}$$

Since it is difficult to find the exact inverse Laplace transform of eq. (5.317), instead of finding the exact distribution $p_r(t)$ one can settle for finding the average residency time T_c which can be expressed in terms of $P(x, s)$ as

$$T_c = \int_0^\infty t p_r(t)dt = \int_0^\infty \int_0^L p(x, t)dxdt = \lim_{s \to 0} \frac{\int_0^L P(x, s)dx}{s} = \frac{T_c L^2}{2\sigma^2} \tag{5.320}$$

The latter equation is derived using the finite limit theorem for the Laplace transform. As expected, the average residency time depends on the width of the well and the intensity of noise. Higher order moments of the distribution $p_r(t)$ can also be calculated in a similar manner.

5.3.6.3   Transformation to the Schrödinger Equations

Let the process considered have a constant diffusion coefficient $K_2(x, t) = K_2 = $ const. In this case the eigenvalue problem for the Fokker–Planck equation becomes

$$-\frac{d}{dx}[K_1(x)\psi(x)] + \frac{K_2}{2}\frac{d^2}{dx^2}\psi(x) = -\lambda\psi(x) \tag{5.321}$$

or

$$\frac{K_2}{2}\frac{d^2}{dx^2}\psi(x) - K_1(x)\frac{d}{dx}\psi(x) + \left(\lambda - \frac{d}{dx}K_1(x)\right)\psi(x) = 0 \tag{5.322}$$

---

[12]Here the initial and the final value theorem for the Laplace transform is used [17].

It is easy to see that substitution [18]

$$\psi(x) = u(x) \exp\left(\int \frac{K_1(x)}{K_2} dx\right) \tag{5.323}$$

reduces eq. (5.322) to the canonical form

$$\frac{d^2}{dx^2} u(x) + \left[2\frac{\lambda - \frac{d}{dx} K_1(x)}{K_2} - \frac{K_1^2(x)}{K_2^2} \right] u(x) = \frac{d^2}{dx^2} u(x) + [E - V(x)]u(x) = 0 \tag{5.324}$$

where

$$E = \frac{2\lambda}{K_2} \tag{5.325}$$

and

$$V(x) = 2\frac{\frac{d}{dx} K_1(x)}{K_2} + \frac{K_1^2(x)}{K_2^2} \tag{5.326}$$

The importance of such a representation is obvious when approximate methods of solution of the Fokker–Planck are to be found: there is a wealth of approximate methods developed for the Schrödinger equation in the form (5.324).

Example: Gaussian process and Hermitian polynomials. Let the drift of $K_1(x)$ be a linear function of its argument

$$K_1(x) = -\alpha x \tag{5.327}$$

and let the natural boundary conditions be used for the corresponding Fokker–Planck equation. In this case eq. (5.326) is reduced to

$$\frac{d^2}{dx^2} u(x) + \left[2\frac{\lambda + \alpha}{K_2} - \frac{\alpha^2 x}{K_2^2} \right] u(x) = 0 \tag{5.328}$$

which is well studied [15].

5.4 STOCHASTIC DIFFERENTIAL EQUATIONS

Let a deterministic system be described by a differential equation known as the Liouville equation

$$\frac{d}{dt} x(t) = f(x, t) + g(x, t)u(t), x(t_0) = x_0 \tag{5.329}$$

where $x(t)$ is the state of the system, x_0 is the initial state of the system, $f(x, t)$ and $g(x, t)$ are some deterministic functions, and $u(t)$ is external force, for a moment considered deterministic. In this case the state evolution is purely deterministic and can be found through a closed form solution of eq. (5.329). Of course, such a solution is deterministic as well.[13] Equation (5.329) can be rewritten in equivalent differential and integral forms as

$$dx(t) = f(x(t), t)dt + g(x(t), t)d\nu(t), x(t_0) = x_0 \qquad (5.330)$$

and

$$x(t) = x(t_0) + \int_{t_0}^{t} f(x(\tau), \tau)d\tau + \int_{t_0}^{t} g(x(\tau), \tau)d\nu(\tau) \qquad (5.331)$$

where

$$d\nu(\tau) = u(\tau)d\tau \qquad (5.332)$$

Randomness can be brought in by a number of ways: one can consider initial conditions $x(t_0) = \xi$, with functions $f(x, t)$ and $g(x, t)$ being random. However, one of the most common situations is the case where the external excitation $u(t)$ is itself a random process. In particular, if $u(t)$ is a white Gaussian noise $\xi(t)$ or a Poisson flow $\eta(t)$, the corresponding equation (5.329) is called a *Stochastic Differential Equation (SDE)*, expression (5.331) is called a *Stochastic Differential*, and eq. (5.331) is called a *Stochastic Integral*. Treatment of stochastic integrals is somewhat different from regular integrals and is considered in some detail below. Comprehensive discussions can be found in [6,8].

5.4.1 Stochastic Integrals

Let $f(x, t)$ and $g(x, t)$ be smooth functions of their arguments. In the case when both t and x are deterministic variables the Riemann integral and the Stielties integrals are defined through the limits of the following sums

$$\int_{t_0}^{t} f(x(\tau), \tau)d\tau = \lim_{\Delta \to 0} \sum_{i=0}^{m-1} f(x(\tau_i), \tau_i)(t_{i+1} - t_i) \qquad (5.333)$$

$$\int_{t_0}^{t} g(x(\tau), \tau)d\nu(\tau) = \lim_{\Delta \to 0} \sum_{i=0}^{m-1} g(x(\tau_i), \tau_i)[\nu(t_{i+1}) - \nu(t_i)] \qquad (5.334)$$

Here $t_0 < t_1 < \cdots < t_m = t, \Delta = \max(t_{i+1} - t_i)$, and τ_i are some points belonging to intervals $[t_i, t_{i+1}]$, i.e. $t_i \le \tau_i \le t_{i+1}$. For any piece-wise continuous function $f(x(t), t)$ and continuous function $g(x(t), t)$ limits in eqs. (5.333) and (5.334) do not depend on the choice of points τ_i inside each interval.

---

[13]In chaotic systems such a statement is not valid if one allows for small errors in knowledge of the initial conditions [19].

It can be shown [8] that integrals involving random functions can be defined using eqs. (5.333) and (5.334) if the following conditions are met.

1. Random functions $f(x, t)$ and $g(x, t)$ are uniformly continuous in the mean square sense on the time interval $[t_0, t]$, i.e. $E\{[f(x(\tau + \Delta), \tau + \Delta) - f(x(\tau), \tau)]^2\} \to 0$ as $\Delta \to 0$ uniformly with respect to $\tau \in [t_0, t]$. A similar equality must hold for $g(x, t)$.

2. The following integrals exist and are finite

$$E\left\{\int_{t_0}^{t} f^2(x(t), t)dt\right\} < \infty, \qquad E\left\{\int_{t_0}^{t} g^2(x(t), t)dt\right\} < \infty \tag{5.335}$$

3. Limits in eqs. (5.333) and (5.334) are understood in the mean square sense.

The definition of an integral of a random function does not differ much from the definition of the integrals of deterministic functions. However, in a striking contrast, the value of the integral (5.334) depends on the way in which points τ_i are chosen.

Definition (5.334) implies that a continuous function $g(x(t), t)$ is substituted on the interval $[t_0, t]$ by a piece-wise constant function $g_\Delta(t)$ such that

$$g_\Delta(\tau_i) = g(x(\tau_i), \tau_i), \qquad \tau_i \in [t_i, t_{i+1}] \tag{5.336}$$

and τ_i are chosen according to the following rule

$$\tau_i = \tau_{i\nu} = (1 - \nu)t_i + \nu t_{j+1}, \quad 0 \le \nu \le 1 \tag{5.337}$$

Since it is assumed that $x(t)$ is a solution of an SDE in the form (5.331) its trajectory can be approximated by a straight line on a small time interval

$$x(\tau) = x(t_i) + \frac{x(t_{i+1}) - x(t_i)}{t_{i+1} - t_i}(\tau - t_i) + o(\Delta) \tag{5.338}$$

Using expression (5.337) in eq. (5.338) one obtains

$$x(\tau) = (1 - \nu)x(t_i) + \nu x(t_{i+1}) + o(\Delta) \tag{5.339}$$

If one further assumes that the function $g(x(t), t)$ is continuously differentiable, then the following approximation can be easily obtained from eqs. (5.337) and (5.339)

$$g(x(\tau_{i\nu}), \tau_{i\nu}) = g((1 - \nu)x(t_i) + \nu x(t_{i+1}), t_i)x(t_i) + \nu x(t_{i+1}) + o(\Delta) \tag{5.340}$$

Finally, eq. (5.340) allows one to approximate the stochastic integral (5.334) as

$$J_\nu = \int_{t_0}^{t} g(x(\tau), \tau)d_\nu\nu(\tau) = \lim_{\Delta \to 0} \sum_{i=0}^{m-1} g(x(\tau_i), \tau_i)[\nu(t_{i+1}) - \nu(t_i)]$$

$$= \lim_{\Delta \to 0} \sum_{i=0}^{m-1} g((1 - \nu)x(t_i) + \nu x(t_{i+1}), (1 - \nu)t_i + \nu t_{i+1})[\nu(t_{i+1}) - \nu(t_i)] \tag{5.341}$$

Here the sub-index ν emphasizes the choice of the time $\tau_{i\nu}$ inside the interval $[t_i, t_{i+1}]$. If $\nu = 0$, eq. (5.341) defines the so-called Ito stochastic integral [1,2]

$$J_0 = \lim_{\Delta \to 0} \sum_{i=0}^{m-1} g(x(t_i), t_i)[\nu(t_{i+1}) - \nu(t_i)] \tag{5.342}$$

which is widely used in the theory of Markov processes.

The next step is to evaluate the difference $J_\nu - J_0$ for an arbitrary value of ν from the interval $[0, 1]$. The assumption of differentiability of function $g(x, t)$ allows one to represent it by the first two terms of the corresponding Taylor expansion, i.e.

$$g((1 - \nu)x(t_i) + \nu x(t_{i+1}), t_i)$$

$$\approx g(x(t_i), t_i) + \nu \left[\frac{\partial}{\partial x} g(x, t_i) \bigg|_{x = x(t_i)} \right] [x(t_{i+1}) - x(t_i)] + o[x(t_{i+1}) - x(t_i)] \tag{5.343}$$

Since the trajectory $x(t)$ is the solution of a differential equation (5.329), one can substitute the corresponding differential (5.330) by a finite difference for a small time interval $t_{i+1} - t_i$

$$x(t_{i+1}) - x(t_i) = f(x(\tau_{i\nu}), \tau_{i\nu})(t_{i+1} - t_i) + g(x(\tau_{i\nu}), \tau_{i\nu})[\nu(t_{i+1}) - \nu(t_i)] + o(\Delta) \tag{5.344}$$

The latter can be used to transform eq. (5.341) to

$$J_\nu - J_0 = \lim_{\Delta \to 0} \left\{ \nu \lim_{\Delta \to 0} \sum_{i=0}^{m-1} g(x(t_i), t_i) \frac{\partial}{\partial x} g(x(t_i), t_i)[\nu(t_{i+1}) - \nu(t_i)]^2 + o(\Delta) \right\} \tag{5.345}$$

If the external force $\nu(t)$ is white Gaussian noise, this implies that (see section 3.15 or [2])

$$E\{[\nu(t_{i+1}) - \nu(t_i)]^2\} = \frac{N_0}{2}(t_{i+1} - t_i) \tag{5.346}$$

and eq. (5.346) approaches in the limit to

$$J_\nu - J_0 = \nu \frac{N_0}{2} \int_{t_0}^{t} g(x(\tau), \tau) \frac{\partial}{\partial x} (g(x(\tau), \tau)) d\tau \tag{5.347}$$

Thus, the relationship between J_ν and the Ito stochastic integral J_0 is given by

$$J_\nu = \int_{t_0}^{t} g(x(\tau), \tau) d_\nu \nu(\tau) = \int_{t_0}^{t} g(x(\tau), \tau) d_0 \nu(\tau) + \nu \frac{N_0}{2} \int_{t_0}^{t} g(x(\tau), \tau) \frac{\partial}{\partial x} g(x(\tau), \tau) d\tau \tag{5.348}$$

In more general settings it can be proven that if $x(t)$ is solution of a stochastic differential equation (5.330), $\phi(x(t), t)$ is [8]

$$\int_{t_0}^t \phi(x(\tau), \tau) d_\nu \nu(\tau) = \int_{t_0}^t \phi(x(\tau), \tau) d_0 \nu(\tau) + \nu \frac{N_0}{2} \int_{t_0}^t g(x(\tau), \tau) \frac{\partial}{\partial x} \phi(x(\tau), \tau) d\tau \quad (5.349)$$

It is important to note that if the function $g(x, t)$ does not explicitly depend on the current value of the random process $x(t)$ then $J_\nu = J_0$ for all ν, i.e. all definitions of stochastic integrals are identical.

While the Ito integrals are mostly used in theoretical analysis due to the fact that they are martingales [4,8], other definitions (with $\nu \neq 0$) are very important. In particular, the case of $\nu = 0.5$ is known as the Stratonovich form of stochastic integrals [6]

$$J_{0.5} = \int_{t_0}^t g(x(\tau), \tau) d_{0.5} \nu(\tau) = \lim_{\Delta \to 0} \sum_{i=0}^{m-1} g\left(\frac{x(t_i) + x(t_{i+1})}{2}, \frac{t_i + t_{i+1}}{2} \right) [\nu(t_{i+1}) - \nu(t_i)]$$

$$(5.350)$$

The formal difference between Ito and Stratonovich integrals lies in the fact that in order to evaluate values of the function $g(x, t)$ in the pre-limit sum (5.334) one picks points at the beginning of the time intervals (Ito form) or in the middle of the time intervals (Stratonovich form). Ito and Stratonovich integrals lead to different results for integration of random functions. For example, integrals of a Wiener process can be obtained using a straightforward equality

$$w(t_i)[w(t_{i+1}) - w(t_i)] = \frac{1}{2} \{ w^2(t_{i+1}) - w^2(t_i) - [w(t_{i+1}) - w(t_i)]^2 \} \quad (5.351)$$

Thus

$$J_0 = \int_{t_0}^t w(\tau) d_0 w(\tau) = \frac{1}{2} \left[w^2(t) - w^2(t_0) - \frac{N}{2}(t - t_0) \right] \quad (5.352)$$

while the Stratonovich definition will produce

$$J_{0.5} = J_0 + \frac{N}{4}(t - t_0) = \frac{1}{2} [w^2(t) - w^2(t_0)] \quad (5.353)$$

It is possible to prove that for any "good" function $\phi(x(t), t)$ [8]

$$\int_{t_0}^t \phi(x(\tau), \tau) d_{0.5} x(\tau) = \int_{t_0}^t \phi(x(\tau), \tau) d_{0.5} x(\tau) + \frac{1}{2} \int_{t_0}^t g^2(x(\tau), \tau) \frac{\partial}{\partial x} \phi(x(\tau), \tau) d\tau \quad (5.354)$$

It follows from the discussion above that different definitions of the stochastic integrals would lead to a different solution of the stochastic differential equation, eqs. (5.329)–(5.331). That is why it is always necessary to indicate which form of stochastic integrals are used. Sometimes, different notations are used for Stratonovich and Ito forms, as discussed in Appendix C on the web.

While the Ito form is convenient in theoretical studies, it can be seen from the example below that the usual rules of calculus of integrals do not apply to the Ito form of stochastic integrals. The rules of differentiation also become more complicated than those of classical calculus. In particular, it can be proven that

$$d\phi(x(t), t) = \left[\frac{\partial}{\partial t}\phi(x(t), t) + f(x(t), t)\frac{\partial}{\partial x}\phi(x(t), t) + \frac{1}{2}g^2(x(t), t)\frac{\partial^2}{\partial x^2}\phi(x(t), t)\right]dt$$
$$+ \frac{1}{2}g(x(t), t)\frac{\partial}{\partial x}g(x(t), t)d_0\nu(t) \tag{5.355}$$

while the usual rules of calculus result in

$$d\phi(x(t), t) = \left[\frac{\partial}{\partial t}\phi(x(t), t) + f(x(t), t)\frac{\partial}{\partial x}\phi(x(t), t)\right]dt + \frac{1}{2}g(x(t), t)\frac{\partial}{\partial x}\phi(x(t), t)d\nu(t) \tag{5.356}$$

This fact makes Ito integrals unattractive for engineering calculations. Interestingly enough, the Stratonovich form of stochastic calculus preserves the rules of classical calculus for derivatives and substitution of variables. Further discussion will be provided later in this section.

The next step is to realize that a random process described by a stochastic differential equation (5.329) defines a diffusion Markov process if the excitation is white Gaussian noise. This fact is known as Doob's first theorem [8]. Indeed, let $t_3 > t_2 > t_1 \geq t_0$ be three distinct moments of time. Using the stochastic integral (5.331) one can write

$$x(t_3) = x(t_2) + \int_{t_2}^{t_3} f(x(\tau), \tau)d\tau + \int_{t_2}^{t_3} g(x(\tau), \tau)d_\nu w(t) \tag{5.357}$$

Since increments of the Wiener process on non-intersecting intervals are independent, the value $x(t_3)$ [given value of $x(t_2)$] is independent of the value $x(t_1)$, i.e. the process $x(t)$ is indeed a Markov process.

Since both the SDE and the Fokker–Planck describe a diffusion Markov process it is possible to relate the coefficients of SDE (5.329) with those of the Fokker–Planck equation. It will be assumed that the Ito form of stochastic differentials is used, i.e.

$$dx(t) = f(x(t), t)dt + g(x(t), t)d_0w(t) \tag{5.358}$$

Integrating both sides of eq. (5.358) over a short interval $[t, t + \Delta t]$ one obtains

$$x(t + \Delta t) - x(t) = \int_t^{t+\Delta t} f(x(\tau), \tau)d\tau + \int_t^{t+\Delta t} g(x(t), t)d_0w(t) \tag{5.359}$$

For a short interval Δt, the value of the first integral in eq. (5.359) is just

$$E\left\{\int_t^{t+\Delta t} f(x(\tau),\tau)d\tau \,\bigg|\, x(t)=x\right\}=f(x,t)\Delta t \tag{5.360}$$

while from the definition of the Ito integral it follows that

$$E\left\{\int_t^{t+\Delta t} f(x(\tau),\tau)d_0w(t)\,\bigg|\, x(t)=x\right\}=0 \tag{5.361}$$

The latter is the consequence of the fact that the increment of the Wiener process over the time interval $[t,t+\Delta t]$ does not depend on the value of the function $x(t)=x$ at the moment of time t. This property also allows one to write the expression for the mean root square value of the integral of $g(x,t)$

$$E\left\{\left[\int_t^{t+\Delta t} f(x(\tau),\tau)d_0w(t)\right]^2\,\bigg|\, x(t)=x\right\}=\frac{N_0}{2}g^2(x,t)\Delta t \tag{5.362}$$

Thus, the drift and diffusion of the solution of SDE (5.358) are given by

$$K_{10}(x,t)=\lim_{\Delta t\to 0} E\left\{\frac{x(t+\Delta t)-x(t)}{\Delta t}\right\}=f(x,t) \tag{5.363}$$

$$K_{20}(x,t)=\lim_{\Delta t\to 0} E\left\{\frac{[x(t+\Delta t)-x(t)]^2}{\Delta t}\right\}=\frac{N_0}{2}g^2(x,t) \tag{5.364}$$

In case of general ν the relation between J_ν and J_0 allows one to write the drift and diffusion as

$$K_{1\nu}(x,t)=f(x,t)+\frac{\nu}{2}\frac{\partial}{\partial x}b_\nu(x,t)=K_{10}(x,t)+\frac{\nu}{2}\frac{\partial}{\partial x}b_\nu(x,t) \tag{5.365}$$

$$K_{2\nu}(x,t)=\lim_{\Delta t\to 0} E\left\{\frac{[x(t+\Delta t)-x(t)]^2}{\Delta t}\right\}=\frac{N_0}{2}g^2(x,t)=K_{20}(x,t) \tag{5.366}$$

One-to-one correspondence between the drift and the diffusion and functions $f(x,t)$ and $g(x,t)$ allows one to rewrite SDE (5.358) directly in terms of parameters of a Markov process:

$$dx(t)-\left[K_1(x,t)-\frac{\nu}{2}\frac{\partial}{\partial x}K_2(x,t)\right]dt+\sqrt{K_2(x,t)}d_\nu w(t) \tag{5.367}$$

In particular, the Ito and the Stratonovich forms of the SDE thus become

$$dx(t)=K_1(x,t)dt+\sqrt{K_2(x,t)}d_0w(t) \tag{5.368}$$

$$dx(t)=\left[K_1(x,t)-\frac{1}{4}\frac{\partial}{\partial x}K_2(x,t)\right]dt+\sqrt{K_2(x,t)}d_{0.5}w(t) \tag{5.369}$$

Formula (5.367) also shows that equations with different ν can be transformed into each other simply by changing the value of function $f(x,t)$. However, it also shows that the same Markov process can be described by various SDEs and such ambiguity creates some uncertainty in using SDEs as models for realistic physical processes. As has been pointed out, the Ito form is attractive in theoretical investigations since it defines the solutions of SDEs as martingales. However, rules of differentiation and changes of variables are more complicated than those of deterministic calculus. The latter explains the popularity of the Stratonovich form among engineers and physicists: the rules of calculus are the same as in the deterministic case.

It is often important to evaluate the adequacy of the model (5.329) which uses white Gaussian noise as an excitation, since none of the realistic waveforms can be WGN. In principle it is possible to avoid such a problem by considering excitation to be a correlated Gaussian noise which is produced by a linear SDE. In this case the Langevin equation will become a system of two SDEs

$$\frac{\mathrm{d}}{\mathrm{d}t}x(t) = f(x,t) = g(x,t)u(t) \tag{5.370a}$$

$$\frac{\mathrm{d}}{\mathrm{d}t}u(t) = -\frac{1}{\tau_c}u(t) + \sqrt{\frac{2\sigma^2}{\tau_c}}\xi(t) \tag{5.370b}$$

It can be shown that the random vector $[x(t), u(t)]^{\mathrm{T}}$ is a Markov vector and can be described in a similar way as a scalar Markov process (see section 5.7 for more details). As a result, it has been shown that if $\tau_c \to 0$ the solution of SDE (5.370a) approaches the Stratonovich form of idealized SDE (5.329). Thus, the Stratonovich form of integrals is probably closer to a physical model of the system excited by a very wide band noise than the Ito form. This fact is an additional reason why the Stratonovich form is often used in applied research—one can use deterministic steady-state equations of systems under investigation and then simply assume that the excitation is WGN. No additional changes of non-linear function $f(x,t)$ are then necessary.

Numerical simulation of SDEs is considered in more detail in Appendix C on the web and in much greater detail in a number of monographs [20,21]. It is important to mention here that a different numerical scheme would converge to a different form of the solution of SDE: thus the second and the fourth order Runge–Kutta scheme converges to the Stratonovich form of the solution, while the Euler scheme converges to the Ito form of the solution. In general, on each step of simulation $[t_i, t_{i+1}]$ WGN is substituted by a Gaussian random variable n_i

$$n_i = \frac{1}{\Delta}\int_{t_i}^{t_{i+1}} \xi(t)\mathrm{d}t, \qquad \Delta = t_{i+1} - t_i \tag{5.371}$$

It is easy to see from the definition that this random variable has zero mean and variance $N_0/2\Delta$. Samples of the Gaussian random variable are independent of each other. For uniform sampling WGN is substituted by an independent identically distributed (iid) Gaussian sequence with variance $N_0/2\Delta t$.

Example: *log-normal process.* Let the following equation be considered

$$\mathrm{d}x(t) = x(t)\mathrm{d}u(t) \tag{5.372}$$

Its solution satisfying the initial condition $x(t_0) = x_0$ is given by [13,18]

$$x(t) = x_0 \exp[u(t) - u(t_0)] \tag{5.373}$$

For simplicity it is also assumed that $x_0 > 0$. If $u(t) = \xi(t)$ is white Gaussian noise, the process $w(t) = u(t) - u(t_0)$ is the Wiener process, and its PDF is then

$$p_w(x) = \frac{1}{\sqrt{\pi N_0(t - t_0)}} \exp\left[-\frac{x^2}{N_0(t - t_0)}\right] \tag{5.374}$$

Using rules of zero memory non-linear transformation of densities one obtains the following expression for the PDF of the solution of SDE (5.373)

$$p(x,t) = \frac{1}{x} \frac{1}{\sqrt{\pi N_0(t - t_0)}} \exp\left[-\frac{\ln^2(x/x_0)}{N_0(t - t_0)}\right] \tag{5.375}$$

Thus, the resulting process is distributed according to a log-normal law. It is interesting to find out which form of SDE produces the same solution. In order to do that one can obtain solutions in both Ito and Stratonovich forms and compare the results with eq. (5.375).

Stratonovich form of the solution. Let time interval $[t_0, t]$ be partitioned by points $t_0 < t_1 < \cdots < t_m < t$. For the *i-th* sub-interval one can write

$$x(t_i) = x(t_{i-1}) + [x(t_i) + x(t_{i-1})]\frac{[w(t_i) - w(t_{i-1})]}{2} \tag{5.376}$$

or, equivalently

$$x(t_i) = x(t_{i-1})\frac{1 + \dfrac{w(t_i) - w(t_{i-1})}{2}}{1 - \dfrac{w(t_i) - w(t_{i-1})}{2}} \tag{5.377}$$

The last equality, recursively applied i times, leads to

$$x(t_i) = x(t_0)\prod_{i=1}^{m}\frac{1 + \dfrac{w(t_i) - w(t_{i-1})}{2}}{1 - \dfrac{w(t_i) - w(t_{i-1})}{2}} \tag{5.378}$$

The limit in eq. (5.378) can be calculated by noting that if $\max|t_i - t_{i-1}| \to 0$ and $m \to \infty$ then $w(t_i) - w(t_{i-1}) \to 0$. Thus, taking the logarithm of both sides of eq. (5.378) and using approximation $\ln(1 + x) \approx x - x^2/2$ for small x one obtains

$$\ln x(t) = \ln x_0 + \sum_{i=1}^{m}\left[\ln\left(1 + \frac{w(t_i) - w(t_{i-1})}{2}\right) - \ln\left(1 - \frac{w(t_i) - w(t_{i-1})}{2}\right)\right]$$

$$\approx \ln x_0 + \sum_{i=1}^{m}[w(t_i) - w(t_{i-1})] = \ln x_0 + w(t) - w(t_0) \tag{5.379}$$

and thus

$$x(t) = x(t_0) \exp[w(t) - w(t_0)] \tag{5.380}$$

The latter coincides with eq. (5.374). Using expressions for univariate and bivariate characteristic functions of the Wiener process one can obtain expressions for the mean value and the covariance function of non-stationary process $x(t)$:

$$m_x(t) = x_0 E\{\exp[w(t) - w(t_0)]\} = x_0 \exp\left[\frac{N_0}{4}(t - t_0)\right] \tag{5.381}$$

$$C_{xx}(t_1, t_2) = x_0^2 E\{\exp[(w(t_1) - w(t_0) + (w(t_2) - w(t_0)))]\}$$
$$= x_0^2 \exp\left\{\frac{N_0}{4}[t_1 + t_2 - 2t_0 + 2\min(t_1 - t_0, t_2 - t_0)]\right\} \tag{5.382}$$

Ito solution: Since in this case the point is taken on the left of each sub-interval, one obtains that

$$x(t_i) = x(t_{i-1}) + x(t_{i-1})[w(t_i) - w(t_{i-1})] = x(t_{i-1})[1 + w(t_i) - w(t_{i-1})] \tag{5.383}$$

(Euler scheme) and, after recursive application of the latter

$$x(t) = x(t_0) \prod_{i=1}^{m}(1 + w(t_i) - w(t_{i-1})) \tag{5.384}$$

Taking the logarithm of both sides and using approximation $\ln(1 + x) \approx x - x^2/2$ once again one easily concludes that

$$\ln x(t) = \ln x_0 + \sum_{i=1}^{m}\ln(1 + w(t_i) - w(t_{i-1})) = \ln x_0 + w(t) - w(t_0) + \sum_{i=1}^{m}\frac{[w(t_i) - w(t_{i-1})]^2}{2} \tag{5.385}$$

If $m \to \infty$ and $\max|t_i - t_{i-1}| \to 0$, then the last term approaches $N_0(t - t_0)/2$ thus leading to

$$x(t) = x_0 \exp\left[w(t) - w(t_0) - \frac{N_0}{4}(t - t_0)\right] \tag{5.386}$$

It is easy to confirm that this solution is described by the following PDF, average and correlation function

$$p(x, t) = \frac{1}{x}\frac{1}{\sqrt{\pi N_0(t - t_0)}}\exp\left\{-\frac{1}{N_0(t - t_0)}\left[\ln\left(\frac{x}{x_0}\right) + \frac{N_0}{4}(t - t_0)\right]^2\right\} \tag{5.387}$$

$$m_x(t) = x_0 \tag{5.388}$$

$$C_{xx}(t_1, t_2) = x_0^2 \exp\left[\frac{N_0}{4}\min(t_1 - t_0, t_2 - t_0)\right] \tag{5.389}$$

•

The same result can be obtained by changing variables in the original eq. (5.372). Since conventional rules of calculus do not apply for the Ito form of SDE one has to use a special rule for substitution: if one considers the SDE (Ito form)

$$\frac{d}{dt}x(t) = f(x,t) + g(x,t)\xi(t) \tag{5.390}$$

and the following substitution $y(t) = \phi(x(t))$ is required, then the Ito form SDE for $y(t)$ can be written as

$$\frac{d}{dt}y(t) = f(x,t)\phi'(x(t)) + g(x,t)\phi'(x(t))\xi(t) + \frac{N_0}{4}g^2[\phi(x(t))]\phi''(x(t)) \tag{5.391}$$

Applying this rule to eq. (5.373) and choosing $y(t) = \ln[x(t)]$ one obtains

$$\frac{d}{dt}y(t) = -\frac{N_0}{4} + \xi(t) \tag{5.392}$$

This, in turn, has an exact solution

$$y(t) = y(t_0) - \frac{N_0}{4}(t - t_0) + \int_{t_0}^{t}\xi(\tau)d\tau \tag{5.393}$$

which immediately produces the result obtained earlier (5.386)

$$x(t) = x(t_0)\exp\left[w(t) - w(t_0) - \frac{N_0}{4}(t - t_0)\right] \tag{5.394}$$

In a general case, one can write

$$x(t_i) = x(t_{i-1}) + [(1 - \nu)x(t_{i-1}) + \nu x(t_i)][w(t_i) - w(t_{i-1})] \tag{5.395}$$

which leads to a solution in the form

$$x(t) = x(t_0)\exp\left[w(t) - w(t_0) + \left(\nu - \frac{1}{2}\right)\frac{N_0}{2}(t - t_0)\right] \tag{5.396}$$

Thus, solutions of the SDE (5.372) obtained by methods applied to regular differential equations correspond to the Stratonovich form of SDE. Physically, this can be explained by the fact that the white Gaussian noise term $\xi(t)$ in the Stratonovich form of stochastic integrals can be treated as a zero mean Gaussian process $\xi(t)$ with covariance function $C(\tau)$ with a very small correlation interval τ_c (much smaller than the characteristic time of the system under consideration, $\tau_s \gg \tau_c$). The impact of such wide band noise has approximately the same effect as pure WGN if one substitutes $C(\tau) \to (N_0/2)\delta(\tau)$, where

$$\frac{N_0}{2} = \int_{-\infty}^{\infty} C(\tau)d\tau \tag{5.397}$$

However, substitution of WGN by a correlated process allows one to treat an SDE as a regular differential equation, thus explaining why the Stratonovich solution is obtained as a result of application of the regular rules of calculus.

It is beneficial to illustrate this fact on the example considered above. Let the covariance function of $\xi(t)$ be represented in the form $\mu C(\mu\tau)$ where μ is a time scale factor. For example, the exponential covariance function can be written as

$$C(\tau) = \mu N_0 \Delta f_1 \exp(-\mu\Delta f_1|\tau|) \tag{5.398}$$

where $\mu = \Delta f/\Delta f_1$ and Δf_1 is the "bandwidth" corresponding to $\mu = 1$. On a fixed subinterval $[t_{i-1}, t_i]$ eq. (5.373) now can be rewritten as

$$\frac{d}{dt}\tilde{x}(t) = \tilde{x}(t)\tilde{\xi}(t) \tag{5.399}$$

Its solution is given by a simple formula

$$\tilde{x}(t_i) = \tilde{x}(t_{i-1}) \int_{t_{i-1}}^{t_i} \xi(t)dt = \tilde{x}(t_{i-1})[\tilde{w}(t_i) - \tilde{w}(t_{i-1})] \tag{5.400}$$

If μ approaches infinity, the process $\tilde{w}(t)$ approaches the Wiener process and the solution (5.400) approaches that of the Stratonovich solution (5.380). Indeed, the variance of the increment of the process $\tilde{w}(t)$ over time interval T is given by

$$\begin{aligned}
E\{[\tilde{w}(t+T) - \tilde{w}(t)]^2\} &= \int_t^{t+T}\int_t^{t+T} E\{\tilde{\xi}(\tau_1)\tilde{\xi}(\tau_2)\}d\tau_1 d\tau_2 \\
&= \int_t^{t+T}\int_t^{t+T} \mu R(\mu(\tau_1 - \tau_2))d\tau_1 d\tau_2 \\
&= T\int_{-T}^{T}\left(1 - \frac{|\tau|}{T}\right)R(\mu|\tau|)d(\mu\tau) \\
&= T\int_{-\mu T}^{\mu T}\left(1 - \frac{|s|}{\mu T}\right)R(s)d(s) \rightarrow T\int_{-\infty}^{\infty} R(s)d(s) = \frac{N_0}{2}T \quad (5.401)
\end{aligned}$$

if $\mu \rightarrow \infty$. Thus the variance of the process $\tilde{\xi}(t)$ approaches that of the Wiener process implying that $\tilde{\xi}(t)$ itself approaches the Wiener process. The same conclusion can be reached for a general case of SDE: if WGN is substituted by wide band noise, the solution of the corresponding SDE approaches the Stratonovich form of the solution as bandwidth of the noise increases.

Finally, it is possible to summarize the feasibility of modelling of real physical processes using stochastic differential equations into the following three key points.

1. Such modelling is possible if the bandwidth of the driving noise is much wider than the bandwidth of the system, or equivalently $\tau_c \ll \tau_s$, where τ_s is the characteristic time constant of the system and τ_c is the correlation interval of the noise. If a continuous time system without delays is described by a system of differential state equations under deterministic input, the same equations can be used if the input is substituted by WGN. In this case one has to treat the obtained SDE as the Stratonovich form of SDE.

2. If the discrete system is described by a difference equation under deterministic input, such as

$$x_{i+1} = F_i(x_i) + g_i(x_i)\xi_i \tag{5.402}$$

where ξ_i is an iid Gaussian process with variance $D = \sigma^2$ and discretization step Δt, such a system can be described by a continuous SDE

$$\frac{dx}{dt} \approx \frac{x_{i+1} - x_i}{\Delta t} = \frac{F_i(x_i) - x_i}{\Delta t} + g_i(x_i)\xi_i = f(x,t) + g(x,t)\xi(t) \tag{5.403}$$

In this case SDE (5.403) must be treated as the Ito form of SDE since the Euler numerical scheme is used (see Appendix C on the web). However, if the input sequence ξ_i has correlation interval τ_c exceeding Δt, $\tau_c > \Delta t$ one has to treat the approximate continuous SDE as the Stratonovich form of SDE.

3. If a system under consideration is driven by wide band noise (i.e. again $\tau_c \ll \tau_s$) but contains delay τ_d one has to carefully examine the amount of delay. If such delays are small compared to the time constant of the system $\tau_d \ll \tau_s$ they can be omitted from the differential equations describing the model. However, the delay must also be compared to the correlation interval of the noise. If $\tau_d \ll \tau_c$ then the system of SDE inferred from the state model must be treated as the Stratonovich SDE, while in the opposite case the Ito form must be used.

5.5 TEMPORAL SYMMETRY OF THE DIFFUSION MARKOV PROCESS

In this section the so-called temporally symmetric property of diffusion Markov processes is considered. It has been found [22–26] that not all actual world processes have this property. Fortunately, stationary processes in communication channels can be thought of as temporally symmetric. This enables one to apply diffusion Markov processes as models of the real ones.

Temporal symmetry in a strict sense stationary stochastic process is defined in terms of its joint distributions.

A stationary process $\xi(t)$, $t \in T$ is temporally symmetric if, for every n and every $t_1, \ldots, t_n$ in T, the random vectors $\{\xi(t_1), \ldots, \xi(t_n)\}$ and $\{\xi(-t_1), \ldots, \xi(-t_n)\}$ have the same probability distributions [23].

The set of time indices, T, is assumed to be the entire real line for continuous-time processes, and the set of all integers for discrete-time processes. Thus, whenever t_0 is in T, so is $-t_0$. Although this facilitates the definition of temporal symmetry given above, it should be noted that if T does not have this property, the definition can be suitably modified, essentially retaining the same ideas.

It is said that a stationary process is second order temporally symmetric if the joint distributions of $\{\xi(t_1), \xi(t_2)\}$ and $\{\xi(-t_1), \xi(-t_2)\}$ are identical for every t_1 and t_2. The relationship of second order temporal symmetry to the more general symmetry is similar to that of a wide sense stationary process to a strict sense stationary process.

By analogy with [24] the following proposition is proven

Proposition 1. *A stationary process is second order temporally symmetric if and only if its bivariate distributions are symmetric.*

Proof: If $\xi(t)$ is second order temporally symmetric, by definition

$$P(\xi(t_2), \xi(t_1)) = P(\xi(-t_2), \xi(-t_1)) \tag{5.404}$$

for every t_1 and t_2, where $P(\xi(t_2), \xi(t_1))$ denotes a bivariate distribution function. Shifting the time origin by $t_1 + t_2$ and using the stationary condition, one can write

$$P(\xi(-t_1), \xi(-t_2)) = P(\xi(t_2), \xi(t_1)) \tag{5.405}$$

Combining eqs. (5.404) and (5.405) one obtains

$$P(\xi(t_2), \xi(t_1)) = P(\xi(t_1), \xi(t_2)) \tag{5.406}$$

The converse can be proved by retracing the above steps in reverse order.

All stationary Gaussian processes are temporally symmetric [24]. This follows from the fact that all joint distributions of these processes depend only on the covariance matrices and that due to stationarity, those matrices are Hermitian. It is possible to prove the following.

Proposition 2. *A stationary Markov process $\xi(t)$, generated by the Fokker–Planck equation is temporally symmetric.*

Proof: We prove this proposition for the case of a discrete spectrum for a one-dimensional process. The proof in the case of a continuous spectrum, as well as a vector case, repeats the same steps. Taking into account the expression for bivariate PDF (5.282)

$$p_\tau(x, x_0) = \sum_{m=0}^{\infty} \Phi_m(x) \Phi_m(x_0) \exp[-\lambda_m \tau] \tag{5.407}$$

one can conclude that this bivariate PDF is symmetric. Then, following Proposition 1, it is possible to conclude that the Markov process generated by the corresponding SDE is temporally symmetric. This means that this approach does not allow one to model temporally asymmetric random processes. One possible approach to modelling asymmetric-in-time processes, staying within the SDE framework, is the use of the Poisson flow of delta pulses as the standard excitation.

5.6 HIGH ORDER SPECTRA OF MARKOV DIFFUSION PROCESSES

The high order spectra of random processes became a topic of great interest in recent decades [25]. Higher order statistical analysis involves high order distributions which, in the case of Markov diffusion processes generated by proper SDEs, are defined by transient density

$$\pi_\tau(x, x_0) = \sum_{m=0}^{\infty} \frac{\Phi_m(x)\Phi_m(x_0)}{p_{\xi \, \mathrm{st}}(x_0)} \exp[-\lambda_m|\tau|] \tag{5.408}$$

Let t_0, $t_0 + \tau_1$, and $t_0 + \tau_1 + \tau_2$ be three sequential moments of time and the corresponding values of the Markov diffusion process $\xi(t)$ are $x_0 = \xi(t_0)$, $x_1 = \xi(t_0 + \tau_1)$, and $x_2 = \xi(t_0 + \tau_1 + \tau_2)$. In this case, expression (5.408), written for two intervals τ_1 and τ_2, has the following form

$$\pi_{\tau_1}(x_1, x_0) = \sum_{m=0}^{\infty} \frac{\Phi_m(x_1)\Phi_m(x_0)}{p_{\xi \, \mathrm{st}}(x_0)} \exp[-\lambda_m|\tau_1|] \tag{5.409}$$

and

$$\pi_{\tau_2}(x_2, x_1) = \sum_{n=0}^{\infty} \frac{\Phi_n(x_2)\Phi_n(x_1)}{p_{\xi \, \mathrm{st}}(x_1)} \exp[-\lambda_n|\tau_2|] \tag{5.410}$$

respectively. The trivariate density can be then obtained by using the Markov property of the process $\xi(t)$, i.e.

$$\begin{aligned}
p_{\tau_2\tau_1}(x_2, x_1, x_0) &= \pi_{\tau_2}(x_2, x_1)\pi_{\tau_1}(x_1, x_0)p_{\xi \, \mathrm{st}}(x_0) \\
&= \sum_{m=0}^{\infty}\sum_{n=0}^{\infty} \frac{\Phi_n(x_2)\Phi_n(x_1)\Phi_m(x_1)\Phi_m(x_0)}{p_{\xi \, \mathrm{st}}(x_1)} \exp[-\lambda_n|\tau_2| - \lambda_m|\tau_1|]
\end{aligned} \tag{5.411}$$

The knowledge of the trivariate density allows one to calculate moment $m_{k_2k_1k_0}(\tau_2, \tau_1)$ of order k_2, k_1, k_0

$$\begin{aligned}
m_{k_2k_1k_0}(\tau_2, \tau_1) &= \int_{\Omega^3} x_2^{k_2} x_1^{k_1} x_0^{k_0} p_{\tau_2\tau_1}(x_2, x_1, x_0) dx_2 dx_1 dx_0 \\
&= \int_{\Omega^3} \sum_{m=0}^{\infty}\sum_{n=0}^{\infty} x_2^{k_2} x_1^{k_1} x_0^{k_0} \frac{\Phi_n(x_2)\Phi_n(x_1)\Phi_m(x_1)\Phi_m(x_0)}{p_{\xi \, \mathrm{st}}(x_1)} \\
&\quad \times \exp[-\lambda_n|\tau_2| - \lambda_m|\tau_1|] dx_2 dx_1 dx_0 \\
&= \sum_{m=0}^{\infty}\sum_{n=0}^{\infty} x_2^{k_2} x_1^{k_1} x_0^{k_0} \exp[-\lambda_n|\tau_2| - \lambda_m|\tau_1|] \\
&\quad \times \int_{\Omega}\int_{\Omega}\int_{\Omega} \frac{\Phi_n(x_2)\Phi_n(x_1)\Phi_m(x_1)\Phi_m(x_0)}{p_{\xi \, \mathrm{st}}(x_1)} dx_2 dx_1 dx_0 \\
&= \sum_{m=0}^{\infty}\sum_{n=0}^{\infty} \exp[-\lambda_n|\tau_2| - \lambda_m|\tau_1|] \int_{\Omega} x_2^{k_2}\Phi_n(x_2)dx_2 \int_{\Omega} x_1^{k_1} \frac{\Phi_n(x_1)\Phi_m(x_1)}{p_{x \, \mathrm{st}}(x_1)} dx_1 \\
&\quad \times \int_{\Omega} x_2^{k_2}\Phi_n(x_2)dx_2 = \sum_{m=0}^{\infty}\sum_{n=0}^{\infty} h_{n;k_2} h_{m;k_0} h_{n,m;k_1} \exp[-\lambda_n|\tau_2| - \lambda_m|\tau_1|]
\end{aligned} \tag{5.412}$$

where Ω is the support of the marginal PDF $p_{\xi\,\mathrm{st}}(x)$, and the following notation is used

$$h_{n;k} = \int_\Omega x^k \Phi_n(x)\,dx \tag{5.413}$$

$$h_{n,m;k} = \int_\Omega x^k \frac{\Phi_n(x)\Phi_m(x)}{p_{\xi\,\mathrm{st}}(x)}\,dx \tag{5.414}$$

Thus the following proposition is proven.

Proposition 3. *The bivariate moment function of order* k_2, k_1, k_0 *of the Markov diffusion process has the form*

$$m_{k_2 k_1 k_0}(\tau_2, \tau_1) = \sum_{m=0}^{\infty}\sum_{n=0}^{\infty} \alpha_{m,n;k_2 k_1 k_0} \exp[-\lambda_n|\tau_2| - \lambda_m|\tau_1|] \tag{5.415}$$

Corollary. *The third order moment of the diffusion Markov process is given by*

$$m_3(\tau_1, \tau_2) = m_{111}(\tau_1, \tau_2) = \sum_{m=0}^{\infty}\sum_{n=0}^{\infty} h_n h_m h_{nm} \exp[-\lambda_n|\tau_2| - \lambda_m|\tau_1|] \tag{5.416}$$

where

$$h_n = \int_\Omega x\Phi_n(x)\,dx \quad\text{and}\quad h_{nm} = \int_\Omega x\frac{\Phi_n(x)\Phi_m(x)}{p_{\xi\,\mathrm{st}}(x)}\,dx \tag{5.417}$$

Proposition 4. $h_{n;k} = h_{n,0;k}$
Proof: Since $\Phi_0(x) = p_{\xi\,\mathrm{st}}(x)$, one obtains

$$h_{n,0;k} = \int_\Omega x^k \frac{\Phi_n(x)\Phi_0(x)}{p_{\xi\,\mathrm{st}}(x)}\,dx = \int_\Omega x^k \Phi_n(x)\,dx = h_{n;k} \tag{5.418}$$

The two-dimensional moment spectra $M_{k_2 k_1 k_0}(\omega_2, \omega_1)$ can be found as a two-dimensional Fourier transform of $m_{k_2 k_1 k_0}(\tau_2, \tau_1)$

$$\begin{aligned}
M_{k_2 k_1 k_0}(\omega_2, \omega_1) &= \frac{1}{(2\pi)^2}\int_{-\infty}^{\infty}\int_{-\infty}^{\infty} m_{k_2 k_1 k_0}(\tau_2, \tau_1) \exp[-j\omega_2\tau_2 - j\omega_1\tau_1] \\
&= \sum_{m=0}^{\infty}\sum_{n=0}^{\infty} \frac{16\lambda_n\lambda_m h_{n;k_2} h_{m;k_0} h_{n,m;k_1}}{(\omega_2^2 + \lambda_n^2)(\omega_1^2 + \lambda_m^2)}
\end{aligned} \tag{5.419}$$

thus proving the following proposition.

Proposition 5.

$$M_{k_2 k_1 k_0}(\omega_2, \omega_1) = \sum_{m=0}^{\infty}\sum_{n=0}^{\infty} \frac{16\lambda_n\lambda_m h_{n;k_2} h_{m;k_0} h_{n,m;k_1}}{(\omega_2^2 + \lambda_n^2)(\omega_1^2 + \lambda_m^2)} \tag{5.420}$$

Corollary. *For the third order moment, two-dimensional spectra have the form*

$$M_3(\omega_2, \omega_1) = \sum_{m=0}^{\infty} \sum_{n=0}^{\infty} \frac{16\lambda_n \lambda_m h_n h_m h_{nm}}{(\omega_2^2 + \lambda_n^2)(\omega_1^2 + \lambda_n^2)} \qquad (5.421)$$

A third order cumulant can be expressed in terms of moments of order up to 3 as [25,26]

$$c_3(\tau_2, \tau_1) = m_3(\tau_2, \tau_1) - m_1[m_2(\tau_2) + m_2(\tau_1) + m_2(\tau_2 - \tau_1)] + 2m_1^3 \qquad (5.422)$$

or, after substitution of eq. (5.416) into eq. (5.422),

$$c_3(\tau_2, \tau_1) = \sum_{m=1}^{\infty} \sum_{n=1}^{\infty} h_m h_n h_{mn} \exp[-\lambda_n|\tau_2| - \lambda_m|\tau_1|] - m_1 \sum_{m=1}^{\infty} h_m^2 \exp[-\lambda_m(|\tau_2| - |\tau_1|)]$$

$$\qquad (5.423)$$

and consequently the following proposition is proved.

Proposition 6. *Bispectrum $C_3(\omega_2, \omega_1)$ [25] of the diffusion Markov process is given by the following formula*

$$C_3(\omega_2, \omega_1) = \sum_{m=1}^{\infty} \sum_{n=1}^{\infty} \frac{16\lambda_n \lambda_m h_m h_n h_{mn}}{(\omega_2^2 + \lambda_n^2)(\omega_1^2 + \lambda_m^2)} - m_1 \sum_{m=1}^{\infty} \frac{16\lambda_n^2 h_m^2}{(\omega_2^2 + \lambda_m^2)(\omega_1^2 + \lambda_m^2)} \qquad (5.424)$$

The coefficients $h_{n,m;k}$ and eigenvalues λ_m play a significant role in third order statistics of the diffusion Markov process. We now show that all spectral properties of any order are defined by the same coefficients.

Proposition 7. *All spectral properties of a diffusion Markov process are defined by spectra of eigenvalues λ_m and the coefficients h_{nm}.*
Proof: Consider n samples of the process $\xi(t)$ at the moments $t_0, t_0 + \tau_1, \ldots,$ $t_0 + \tau_1 + \ldots + \tau_{n-1}$. Using Markov properties of the process, the n-variate density is given by

$$p_{\tau_1, \ldots, \tau_{n-1}}(x_{n-1}, x_{n-2}, \ldots, x_0)$$

$$= p_{\xi \, st}(x_0) \prod_{i=1}^{n-1} \pi_{\tau_i}(x_i, x_{i-1})$$

$$= \sum_{k_{n-1}} \sum_{k_{n-2}} \cdots \sum_{k_0 = 0}^{\infty} \frac{\Phi_{k_{n-1}}(x_{n-1}) \Phi_{k_{n-1}}(x_{n-2}) \Phi_{k_{n-2}}(x_{n-2}) \Phi_{k_{n-2}}(x_{n-3}) \ldots \Phi_{k_1}(x_0)}{p_{\xi \, st}(x_{n-2}) p_{\xi \, st}(x_{n-3}) \ldots p_{\xi \, st}(x_1)}$$

$$\times \exp[-\lambda_{k_{n-1}}|\tau_{n-1}| - \lambda_{k_{n-3}}|\tau_{n-3}| - \cdots - \lambda_{k_1}|\tau_1|] \qquad (5.425)$$

To obtain the expression for the moment of order n we should integrate eq. (5.425) n times over real axis R

$$m_n(\tau_{n-1}, \tau_{n-2}, \ldots, \tau_1)$$

$$= \int_{R^n} x_{n-1} x_{n-2} \ldots x_0 p_{\tau_1,\ldots,\tau_{n-1}}(x_{n-1}, \ldots, x_0) dx_{n-1} \ldots dx_0$$

$$= \sum_{k_{n-1}}^{\infty} \sum_{k_{n-2}}^{\infty} \ldots \sum_{k_0=0}^{\infty} \exp[-\lambda_{k_{n-1}} |\tau_{n-1}| - \lambda_{k_{n-3}} |\tau_{n-3}| - \cdots - \lambda_{k_1} |\tau_1|]$$

$$\times \int_R x_{n-1} \Phi_{k_{n-1}}(x_{n-1}) dx_{n-1} \int_R x_{n-2} \frac{\Phi_{k_{n-2}}(x_{n-2}) \Phi_{k_{n-3}}(x_{n-2})}{p_{x\,\mathrm{st}}(x_{n-2})} dx_{n-2} \cdots$$

$$\times \int_R x_j \frac{\Phi_{k_j}(x_j) \Phi_{k_{j-1}}(x_j)}{p_{x\,\mathrm{st}}(x_j)} dx_j \times \cdots \times \int_R x_1 \frac{\Phi_{k_1}(x_1) \Phi_{k_0}(x_1)}{p_{x\,\mathrm{st}(x_1)}} dx_1 \int_R x_0 \Phi_{k_0}(x_0) dx_0 \quad (5.426)$$

thus

$$m_n(\tau_{n-1}, \tau_{n-2}, \ldots, \tau_1)$$

$$= \sum_{k_{n-1}}^{\infty} \sum_{k_{n-2}}^{\infty} \ldots \sum_{k_0=0}^{\infty} h_{k_{n-1}} h_{k_{n-2}} k_{n-3} \ldots h_{k_2 k_1} h_{k_0} \times \exp[-\lambda_{k_{n-1}} |\tau_{n-1}| - \lambda_{k_{n-3}} |\tau_{n-3}| - \cdots - \lambda_{k_1} |\tau_1|]$$

$$(5.427)$$

Since

$$m_{k_{n-1}\ldots k_1 k_0}(\tau_{n-1}, \ldots, \tau_2, \tau_1) = m_{k_{n-1}} + \cdots + k_1 + k_0(\tau_{n-1}, 0, \ldots, 0, \tau_{n-2}, \ldots, 0, \tau_1, \ldots, 0)$$

$$(5.428)$$

an n-variate moment can be expressed in terms of $k_{n-1} + \cdots + k_1 + k_0$ variate moment with some values of time equal to zero. The last statement together with eq. (5.427) proves the proposition.

Example: The random process with Rayleigh distribution is generated by the following SDE

$$\dot{x} = \frac{K^2}{2} \left(\frac{1}{x} - \frac{x}{\sigma^2} \right) + K\xi(t) \quad (5.429)$$

In this case [6]

$$\lambda_m = \frac{2Km}{\sigma^2} \quad (5.430)$$

and

$$\Phi_m(x) = \frac{x}{n!\sigma^2} \exp\left[-\frac{x^2}{2\sigma^2} \right] L_m\left(\frac{x^2}{2\sigma^2} \right) \quad (5.431)$$

Table 5.1 Coefficients h_{mn}/σ for Rayleigh process (5.429)

m/n	0	1	2	3	4
0	1.2533	−0.6267	−0.0783	−0.0131	−0.0020
1	−0.6267	2.1933	−0.4308	−0.0326	−0.0039
2	−0.0783	−0.4308	0.7099	−0.0873	−0.0048
3	−0.0131	−0.0326	−0.0873	0.0934	−0.0084
4	−0.0020	−0.0039	−0.0048	−0.0084	0.0066

Then, using eq. (5.417) one obtains

$$
\begin{aligned}
h_{mn} &= \frac{1}{\sigma^2 n!m!} \int_0^\infty x^2 \exp\left[-\frac{x^2}{2\sigma^2}\right] L_n\left(\frac{x^2}{2\sigma^2}\right) L_m\left(\frac{x^2}{2\sigma^2}\right) dx \\
&= \frac{\sqrt{2}\sigma}{n!m!} \int_0^\infty x^{1/2} \exp[-x] L_n(x) L_m(x) dx
\end{aligned}
\tag{5.432}
$$

Values of coefficients h_{mn} in eq. (5.432) for some integers n and m are presented in Table 5.1.

It can be seen from this table that the most important terms h_{mn} are those whose index does not exceed 3. Thus, even for higher order statistics, it is possible to take into account only a few first terms in expansion (5.427).[14]

5.7 VECTOR MARKOV PROCESSES

5.7.1 Definitions

The notion of a scalar Markov process can be extended to vector processes. In particular, let $\xi(t) = \{\xi_1(t), \xi_2(t), \dots, \xi_M(t)\}$ be a random vector of dimension M which can be described by a series of probability density functions $p_{n+1}(x_0, \dots, x_n; t_0, \dots, t_n)$ of order n, as defined in Chapter 2, or in terms of a series of conditional probability densities

$$
\pi_n(x_n, t_n \mid x_{n-1}, t_{n-1}, \dots, x_0, t_0) = \frac{p_{n+1}(x_0, \dots, x_n; t_0, \dots, t_n)}{p_n(x_0, \dots, x_{n-1}; t_0, \dots, t_{n-1})}
\tag{5.433}
$$

The Markov property is defined by the fact that the transitional density, defined by eq. (5.429), depends only on the value of the process at the moment of time t_{n-1}, i.e.

$$
\pi_n(x_n, t_n \mid x_{n-1}, t_{n-1}, \dots, x_{n-1}, t_{n-1}) = \pi(x_n, t_n \mid x_{n-1}, t_{n-1})
\tag{5.434}
$$

All the properties of a scalar Markov process discussed in sections 5.1–5.5 can be readily extended to the vector case. In particular, one can show that the transitional density

---

[14]For the case of the correlation function this result was considered in [26,29].

$\pi(\boldsymbol{x}, t \mid \boldsymbol{x}_0, t_0)$ must obey the Smoluchowski–Chapman equation

$$\pi(\boldsymbol{x}, t \mid \boldsymbol{x}_0, t_0) = \int_{-\infty}^{\infty} \pi(\boldsymbol{x}, t \mid \boldsymbol{x}_1, t_1) \pi(\boldsymbol{x}_1, t_1 \mid \boldsymbol{x}_0, t_0) d\boldsymbol{x}_1, \quad t_0 < t_1 < t \tag{5.435}$$

Here the integration is taken over R^M. It is also possible to show that if $\xi(t)$ is a Markov vector process with M components and continuous trajectories, i.e. only the first two conditional moments of increments are non-zero

$$K_{1i}(\boldsymbol{x}, t) = \lim_{\Delta t \to 0} \frac{1}{\Delta t} E\{[\xi_i(t + \Delta t) - \xi_i(t)] \mid \xi(t) = \boldsymbol{x}\} \tag{5.436}$$

$$K_{2ij}(\boldsymbol{x}, t) = \lim_{\Delta t \to 0} \frac{1}{\Delta t} E\{[\xi_i(t + \Delta t) - \xi_i(t)][\xi_j(t + \Delta t) - \xi_j(t)] \mid \xi(t) = \boldsymbol{x}\} \tag{5.437}$$

then the marginal PDF $p(\boldsymbol{x}, t)$ and the transitional density $\pi(\boldsymbol{x}, t \mid \boldsymbol{x}_0, t_0)$ must obey the Fokker–Planck equation

$$\frac{\partial}{\partial t} p(\boldsymbol{x}, t) = \mathcal{L}\{p(\boldsymbol{x}, t)\} \tag{5.438}$$

where the Fokker–Planck operator is defined as

$$\mathcal{L}\{p(\boldsymbol{x}, t)\} = -\sum_{m=1}^{M} \frac{\partial}{\partial x_m} [K_{1m}(\boldsymbol{x}, t) p(\boldsymbol{x}, t)] + \frac{1}{2} \sum_{m=1}^{M} \sum_{k=1}^{M} \frac{\partial^2}{\partial x_m \partial x_k} [K_{2mk}(\boldsymbol{x}, t) p(\boldsymbol{x}, t)] \tag{5.439}$$

In addition to continuity of $K_{1m}(\boldsymbol{x}, t)$ and $K_{2mk}(\boldsymbol{x}, t)$ and their derivatives, it is required that the quadratic form

$$H(\boldsymbol{x}) = \sum_{m=1}^{M} \sum_{k=1}^{M} x_m x_k K_{2mk}(\boldsymbol{x}, t) \tag{5.440}$$

is positively defined, i.e. for all x_i and x_j $H(\boldsymbol{x}) \geq 0$. Markov processes which satisfy the corresponding Fokker–Planck equations are called diffusion Markov processes.

Partial differential equations (5.438) and (5.439) must be supplemented by a proper set of initial and boundary conditions and their solution, even in a stationary case, is much more complicated than the case of a scalar diffusion Markov process.

In contrast to the one-dimensional case, the exact solution of the Fokker–Planck equation, even in a stationary case, is impossible in a general case [1,2]. In this section, we consider the solution for one particular case, called the *potential isotropic case* [6].

In this case the Fokker–Planck equation can be rewritten in the form [6]

$$\frac{\partial}{\partial t} p_x(\boldsymbol{x}, t) = -\sum_{i=1}^{M} \frac{\partial}{\partial x_i} [K_{1i}(\boldsymbol{x}) p_x(\boldsymbol{x}, t)] + \frac{1}{2} \sum_{i,j=1}^{M} \frac{\partial^2}{\partial x_j^2} [K_{2ij}(\boldsymbol{x}) p_x(\boldsymbol{x}, t)] = -\sum_{i=1}^{M} \frac{\partial}{\partial x_i} G_i(\boldsymbol{x})$$

$$\tag{5.441}$$

where the following notation is used

$$G_i(\boldsymbol{x}) = K_{1i}(\boldsymbol{x})p_x(\boldsymbol{x},t) - \frac{1}{2}\sum_{j=1}^{M}\frac{\partial}{\partial x_j}[K_{2ij}(\boldsymbol{x})p_x(\boldsymbol{x},t)] \qquad (5.442)$$

It can be seen from eq. (5.441) that the stationary PDF, if it exists, satisfies the equation

$$\sum_{i=1}^{M}\frac{\partial}{\partial x_i}G_i(\boldsymbol{x}) = 0 \qquad (5.443)$$

In the vector case the probability current $\boldsymbol{G} = \{G_1, G_2, \ldots, G_M\}$ does not have to vanish inside the region R under consideration, even if $\boldsymbol{G}$ satisfies zero boundary conditions [6]

$$G_i(\boldsymbol{x}) = 0, \quad i = 1, \ldots, M \qquad (5.444)$$

since rotational probability flows can occur. In fact, the current $\boldsymbol{G}$ vanishes in the whole region R, i.e.

$$G_i(\boldsymbol{x}) = K_{1i}(\boldsymbol{x})p_x(\boldsymbol{x},t) - \frac{1}{2}\sum_{j=1}^{M}\frac{\partial}{\partial x_j}[K_{2ij}(\boldsymbol{x})p_x(\boldsymbol{x},t)] \qquad (5.445)$$

only in the special case which is called the *potential* one [6] when

$$\frac{\partial}{\partial x_i}\left(\frac{K_{1j}(\boldsymbol{x})}{K_2(\boldsymbol{x})}\right) = \frac{\partial}{\partial x_j}\left(\frac{K_{1i}(\boldsymbol{x})}{K_2(\boldsymbol{x})}\right) \qquad (5.446)$$

If the conditions (5.446) are met, the stationary PDF can be represented in the form [6]

$$p_{x\,\mathrm{st}}(\boldsymbol{x}) = \frac{C}{K_2(\boldsymbol{x})}\exp\left[-2\int_{x_1}^{x_2}\frac{\sum_{i=1}^{M}K_{1i}(\boldsymbol{x})\mathrm{d}x_i}{K_2(\boldsymbol{x})}\right] \qquad (5.447)$$

The non-stationary PDF $p_x(\boldsymbol{x},t)$ can sometimes be obtained using Fourier methods in the same manner as in a one-dimensional case.[15] By using the representation $p(\boldsymbol{x},t) = T(t)\Phi(\boldsymbol{x})$ one can separate eq. (5.441) into the system

$$\dot{T}(t) = -\lambda T(t) - \sum_{i=1}^{M}\frac{\partial}{\partial x_i}K_{1i}(\boldsymbol{x})\Phi(\boldsymbol{x}) + \frac{1}{2}\sum_{i,j=1}^{M}\frac{\partial^2}{\partial x_j^2}K_{2ij}(\boldsymbol{x})\Phi(\boldsymbol{x}) + \lambda\Phi(\boldsymbol{x}) = 0 \qquad (5.448)$$

---

[15]In the case of a potential isotropic system the differential operator of the generating SDE is a symmetric one and this means that application of the Fourier method is possible [14].

The transition density, bivariate density, and cross-covariance function between two deterministic functions $F_1(x)$ and $F_2(x)$ of the process $x(t)$ can be given in terms of the eigenvalues and eigenfunction of the boundary problem as [27,29][16]

$$\pi_\tau(x, x_0) = \sum_{m=0}^{\infty} \frac{\Phi_m(x)\Phi_m(x_0)}{p_{xst}(x_0)} \exp[-\lambda_m|\tau|] \qquad (5.449)$$

$$p_\tau(x, x_0) = \sum_{m=0}^{\infty} \Phi_m(x)\Phi_m(x_0) \exp[-\lambda_m|\tau|] \qquad (5.450)$$

$$C_{F_1 F_2}(\tau) = \sum_{m=0}^{\infty} h_{m1} h_{m2} \exp[-\lambda_m \tau] \qquad (5.451)$$

where

$$h_{mj} = \int_{x_1}^{x_2} F_j(x)\Phi_m(x)dx, \quad j = 1, 2 \qquad (5.452)$$

Having the transient and stationary, or initial PDF, one can construct the PDF of any dimension using the Markov properties of the process $x(t)$.

In addition to the Fourier method of solution of FPE (5.441), some other methods, such as the Laplace transform, transition to the Volterra integral equation, and different numerical schemes can be considered [2]. However, these methods do not give sufficient information for synthesis of the SDE and therefore they are not considered here.

As was shown earlier, any diffusion Markov process can be considered as a solution of a properly defined stochastic differential equation as discussed in Section 5.4. A similar statement can be made about vector diffusion Markov processes. Indeed, Doob proved [8] that the following system of SDEs

$$dx_i(t) = f_i(x, t)dt + \sum_{m=1}^{M} g_{im}(x, t)d_\nu w_k(t), x(t_0) = x_0, \quad 0 \le \nu \le 1 \qquad (5.453)$$

defines a diffusion Markov process with the drift and diffusion defined as

$$K_{1i}(x, t) = f_i(x, t) + \nu \sum_{k=1}^{M} \sum_{j=1}^{M} \frac{N_k}{2} g_{jk}(x, t) \frac{\partial}{\partial x_j} g_{jk}(x, t)$$

$$K_{2ij}(x, t) = \frac{1}{2} \sum_{k=1}^{M} N_k g_{ik}(x, t) g_{jk}(x, t) \qquad (5.454)$$

Here $w_k(t)$ is the Wiener process with zero mean and linear cross-covariance

$$E\{w_k(t)\} = 0,$$

$$E\{[w_k(t_2) - w_k(t_1)][w_l(t_2) - w_l(t_1)]\} = \delta_{kl}\sqrt{N_k N_l}|t_2 - t_1| = g_{kl}N_k|t_2 - t_1|, \ 1 \le k \le M$$

$$(5.455)$$

---

[16]The continuous spectrum case can also be easily obtained as a generalization of eqs. (5.449) and (5.450).

Equivalently, $w_k(t)$ can be thought of as a time integral of a zero mean white Gaussian noise

$$w_k(t) = \int_{-\infty}^{t} \xi_k(t)dt, \quad E\{\xi_k(t_1)\xi_k(t_2)\} = N_n\delta(t_2 - t_1), \quad E\{\xi_k(t_1)\xi_l(t_2)\} = 0, \quad k \neq l \quad (5.456)$$

Having this in mind, one can formally rewrite the system of SDE (5.453) as

$$\frac{d}{dt}x_i(t) = f_i(\boldsymbol{x}, t) + \sum_{m=1}^{M} g_{im}(\boldsymbol{x}, t)\xi_m(t), \boldsymbol{x}(t_0) = \boldsymbol{x}_0, \quad 0 \leq \nu \leq 1 \quad (5.457)$$

with the additional specification of the form of stochastic integrals used. The latter form of SDE is particularly useful if a deterministic system excited by a narrow band random input is studied. In this case one can reuse the deterministic state equations to write SDE (5.457), of course keeping in mind that such a procedure would lead to the solution of SDE (5.457) in the Stratonovich form.[17] In particular, two types of the state equation are often found in applications. In the first case the state equation is resolved with respect to the higher order derivative, i.e. it is written in the following form

$$\frac{d^m}{dt^m}x(t) = F(x, x', \dots, x^{(m-1)}, t; \xi(t)) \quad (5.458)$$

Using change of variables $x_0(t) = x(t)$, $x_k(t) = (d^k/dt^k)x(t)$, one can rewrite eq. (5.458) as a system of m first order SDEs, thus converting it to the form (5.457). In the second case, the system is linear and thus it can be described by a system of linear SDEs

$$\frac{d}{dt}x_j(t) = -\sum_{k=1}^{N} a_{jk}x_k(t) + \xi_j(t), \quad 1 < k, \quad j < N \quad (5.459)$$

where a_{jk} are some constant coefficients and $\xi_j(t)$ are elements of a zero mean vector Gaussian process with covariance matrix

$$C_{ij}(t_2, t_1) = \frac{1}{2}R_{ij}\sqrt{N_iN_j}\delta(t_2 - t_1), \quad R_{kk} = 1 \quad (5.460)$$

It is assumed that coefficients R_{ij} and N_i do not depend on time. Since the system of equations describes linear transformation of a Gaussian process, the resulting process is also a Gaussian one. This significantly simplifies the solution of the problem since one has to find only the mean vector $\boldsymbol{m}(t)$ and the covariance matrix $\boldsymbol{K}(t)$. In order to do so one can rewrite the system of SDEs (5.459) in the matrix form

$$\frac{d}{dt}X(t) = -AX(t) + \Xi(t) \quad (5.461)$$

---

[17]If the excitation is just an additive function, i.e. $g_{ij}(\mathbf{x}, t) = g_{ij}(t)$, both the Stratonovich and the Ito form coincide and one does not need to worry about proper interpretation of SDE (5.457).

where $X(t) = [x_1(t), \ldots, x_M(t)]^T$, $A = \{a_{ij}\}$ and $\Xi(t) = [\xi_1(t), \ldots, \xi_M(t)]$. The initial state of the system is described by deterministic initial conditions $X(t_0) = X_0$.

Averaging both parts of SDE (5.461) over all possible realizations, and taking into account that linear operations (such as differentiation and multiplication by a number) and averaging can be interchanged, one readily obtains equations for the mean vector

$$\frac{d}{dt} m(t) = -Am(t) + E\{\Xi(t)\} = -Am(t) \tag{5.462}$$

Thus

$$m(t) = m_0 \exp(-At) \tag{5.463}$$

and

$$\lim_{t \to \infty} m(t) = 0 \tag{5.464}$$

if the matrix A represents a stable system, i.e. all its eigenvalues have a negative real part. In other words, the mean value of the process $X(t)$ in the stationary regime is equal to zero.

In a similar manner one can derive a differential equation for the variance matrix $C(t, t + \tau)$. Indeed

$$
\begin{aligned}
\frac{d}{dt} C(t, t + \tau) &= \frac{d}{dt} E\{(X(t) - m(t))(X(t + \tau) - m(t + \tau))^T\} \\
&= E\left\{\frac{d}{dt}(X(t) - m(t))(X(t + \tau) - m(t + \tau))^T\right\} \\
&\quad + E\left\{(X(t) - m(t))\frac{d}{dt}(X(t + \tau) - m(t + \tau))^T\right\} \\
&\quad E\{(-A(X(t) - m(t))) + \Xi(t) \cdot (X(t + \tau) - m(t + \tau))^T\} \\
&\quad + E\{(X(t) - m(t))(-A(X(t + \tau) - m(t + \tau)) + \Xi(t + \tau))^T\} \\
&= -AC(t, t + \tau) - C(t, t + \tau)A^T + B
\end{aligned}
\tag{5.465}
$$

where

$$B = \left\{\frac{1}{2} R_{ij} \sqrt{N_i N_j}\right\} \tag{5.466}$$

A solution of the vector equation (5.465) can be obtained by known methods [28]. In particular, by setting $\tau = 0$ one can obtain matrix of variances $D(t) = C(t, t)$ as a solution of a system of linear ordinary differential equations

$$\frac{d}{dt} D(t) = -AD(t) - D(t)A^T + B, \quad D(t_0) = 0 \tag{5.467}$$

$$D(t) = \int_{t_0}^{t} e^{-As} B e^{-A^T s} ds \tag{5.468}$$

while the time dependent covariance matrix $K(t, \tau)$ is then given by

$$C(t, \tau) = \begin{cases} \Phi(t, \tau)D(\tau) & \text{if } t \geq \tau \\ D(t)\Phi^{\mathrm{T}}(t, \tau) & \text{if } t < \tau \end{cases} \tag{5.469}$$

where $\Phi(t, \tau)$ is the matrix satisfying the vector homogeneous equation

$$\frac{\mathrm{d}}{\mathrm{d}t}\Phi(t, \tau) = -A\Phi(t, \tau) \tag{5.470}$$

with initial conditions $\Phi(t_0, t_0) = I$. Here I is an identity matrix.

Some simplification can be achieved if the matrix A can be represented in the diagonal form as

$$A = T^{-1}\Lambda T, \Lambda = \begin{bmatrix} \lambda_1 & 0 & \dots & 0 \\ 0 & \lambda_2 & & 0 \\ \dots & \dots & \dots & 0 \\ 0 & 0 & \dots & \lambda_M \end{bmatrix} \tag{5.471}$$

where T is the transformation matrix. In this case

$$\exp(-At) = T^{-1}\exp(-\Lambda t)T = T^{-1}\begin{bmatrix} \exp(-\lambda_1 t) & 0 & \dots & 0 \\ 0 & \exp(-\lambda_2 t) & \dots & 0 \\ \dots & \dots & \dots & 0 \\ 0 & 0 & \dots & \exp(-\lambda_M t) \end{bmatrix}T \tag{5.472}$$

Thus

$$m(t) = T^{-1}\begin{bmatrix} \exp(-\lambda_1 t) & 0 & \dots & 0 \\ 0 & \exp(-\lambda_2 t) & \dots & 0 \\ \dots & \dots & \dots & 0 \\ 0 & 0 & \dots & \exp(-\lambda_M t) \end{bmatrix}TX_0 \tag{5.473}$$

and

$$C(t) = T^{-1}R(T^{-1})^{\mathrm{T}}, R = \{r_{ij}\} \tag{5.474}$$

where

$$r_{ij} = \frac{g_{ij}}{\lambda_i + \lambda_j}[1 - \exp(-\lambda_i - \lambda_j)t], G = \{g_{ij}\} = CBC^{\mathrm{T}} \tag{5.475}$$

Thus, calculation of parameters of the PDF $p(x, t)$ is reduced to matrix calculus.

5.7.1.1   A Gaussian Process with a Rational Spectrum

A linear system with a rational transfer function

$$H(j\omega) = \frac{P_m(j\omega)}{Q_n(j\omega)} = \frac{\beta_0 x^m + \beta_1 x^{m-1} + \cdots + \beta_m}{\alpha_0 x^n + \alpha_1 x^{n-1} + \cdots + \alpha_n}, \quad m < n \tag{5.476}$$

driven by a WGN with spectral density $N_0/2$ produces a Gaussian random process $\eta(t)$ with rational power spectral density given, according to eq. (4.141), by

$$S_\eta(\omega) = \frac{N_0}{2} \left| \frac{P_m(j\omega)}{Q_n(j\omega)} \right|^2 \tag{5.477}$$

It is easy to see that such a system can be described by the following SDE

$$\frac{d^n}{dt^n}\eta(t) + \alpha_1 \frac{d^{n-1}}{dt^{n-1}}\eta(t) + \cdots + \alpha_n\eta(t) = \beta_0 \frac{d^m}{dt^m}\xi(t) + \beta_1 \frac{d^{m-1}}{dt^{m-1}}\xi(t) + \cdots + \beta_m\xi(t)$$

$$\tag{5.478}$$

It is possible to provide an equivalent system of n first order differential equations instead of a single differential equation (5.477) of n-th order. This can be accomplished by introducing $(n-1)$ auxiliary processes

$$\dot{\eta}(t) = \eta_2(t), \quad \dot{\eta}_{n-m}(t) = \eta_{n-m+1}(t) + \gamma_{n-m}\xi(t), \quad m = 1, \ldots, n-1 \tag{5.479}$$

Constants γ_k could be chosen in such a way that no derivatives of WGN $\xi(t)$ would appear in the equivalent system. Substituting new processes defined by eq. (5.479) into the original eq. (5.477) and equating the coefficients in the form of the equal order derivatives of $\xi(t)$ one obtains a recursive equation for calculating coefficients γ_k

$$\gamma_k = \beta_{k+m-n} - \sum_{i=1}^{k+m-n} \alpha_i\gamma_{k-i}, k = n - m, \ldots, n \tag{5.480}$$

Given that the solution $\eta(t)$ of the original eq. (5.477) can be represented as a component of the vector solution of the following system (5.477) augmented with an additional equation

$$\dot{\eta}_n(t) = -\sum_{i=1}^{n} \alpha_{n+1-i}\eta_i(t) + \gamma_n\xi(t) \tag{5.481}$$

Since the linear system of SDEs defines a Markov vector, a Gaussian random process with a rational spectrum can always be considered as a component of a vector Markov process.

Furthermore, if a deterministic differential system is driven by a colored Gaussian noise $\xi_c(t)$ with a rational spectrum of order n

$$\dot{x}(t) = f(x, t) + g(x, t)\xi_c(t) \tag{5.482}$$

it is possible to represent the solution $x(t)$ as a component of a Markov vector $\{x(t), \xi_1(t), \ldots, \xi_n(t)\}^T$ where $\xi_k(t)$ are obtained as components of a Markov vector representing the colored noise.

In addition, it is important to mention that if $\eta(t)$ is a Markov diffusion process, its zero memory one-to-one non-linear transformation $\zeta(t) = F[\eta(t)]$ is also a diffusion Markov process. Its characteristics can be found by standard rules, as described in this section.

5.8   ON PROPERTIES OF CORRELATION FUNCTIONS OF ONE-DIMENSIONAL MARKOV PROCESSES

It can be seen from eq. (5.286) that for a random process generated by SDE (5.329) the correlation function is the infinite series of exponents.[18] Nevertheless, the asymptotic property of eigenvalues indicates that this series rapidly converges [29]. In fact, if one takes into account the asymptotic relationship

$$\frac{\lambda_q}{\lambda_1} = q^m, \qquad m = 1, 2 \tag{5.483}$$

between eigenvalues of the Fokker–Planck operator [2], it follows that in the range $\tau > \tau_{\mathrm{corr}}$, where τ_{corr} is the correlation interval of the process $x(t)$, one can restrict considerations to the first three terms in eq. (5.286) [29]. This fact raises the problem of the approximate exponential representation of the correlation function of a continuous Markov process being a solution of the non-linear first order SDE (5.329).

In order to achieve this goal, one can introduce the following function

$$V_k(t - t_0) = \int_{-\infty}^{\infty} \int_{-\infty}^{\infty} x(t_0) x^{k-1}(t) p_x(x(t_0), x(t), t_0) d[x(t_0)] d[x(t)] \tag{5.484}$$

which becomes the covariance function of the process $x(t)$

$$V_2(t - t_0) = C_{xx}(\tau) \tag{5.485}$$

where $\tau = t - t_0$, if $k = 2$ and $\langle x(t) \rangle = 0$. Taking into account eqs. (5.329) and (5.485) one obtains the following equation [6]

$$\frac{\partial}{\partial t} V_2(t - t_0) = \left\langle x(t_0) \frac{d}{dx} g(x) \right\rangle \tag{5.486}$$

It was shown in [6] that

$$\langle x(t_0) g(x) \xi(t) \rangle = \frac{1}{2} \left\langle x(t_0) g(x) \frac{d}{dx} g(x) \right\rangle \tag{5.487}$$

---

[18] This fact proves that the random process generated by SDE (5.329) has a monotone spectrum. Consequently, it is impossible to generate a narrow band process using SDE (5.329).

for the Stratonovich definition of the stochastic integral. Thus, taking into account that for the Stratonovich form of SDE its drift is given by

$$K_1(x) = f(x) + g(x)\frac{\mathrm{d}}{\mathrm{d}x}g(x) \tag{5.488}$$

one obtains that

$$\frac{\partial}{\partial t}V_2(t - t_0) = \langle x(t_0)K_1(x)\rangle \tag{5.489}$$

Expanding $f(x)$ and

$$g_1(x) = g(x)\frac{\mathrm{d}}{\mathrm{d}x}g(x) \tag{5.490}$$

into the Taylor series, one can rewrite eq. (5.489) as

$$\frac{\partial}{\partial t}V_2(t - t_0) = \sum_{k=2}^{\infty}a_{k-1}V_k(t - t_0) \tag{5.491}$$

where [6]

$$a_{k-1} = \frac{1}{(k-1)!}\left[\frac{\mathrm{d}^{k-1}}{\mathrm{d}x^{k-1}}f(x) + \frac{1}{2}\frac{\mathrm{d}^{k-1}}{\mathrm{d}x^{k-1}}g(x)\right]\Bigg|_{x=0} \tag{5.492}$$

It can be shown that for the symmetrical PDF, function $V_3(t - t_0)$ is close to zero [6] and $V_4(t - t_0) \approx n_0 V_2(t - t_0)$ [6], where

$$n_0 = \frac{\int_{-\infty}^{\infty} x^4 p_{x\,\mathrm{st}}(x)\mathrm{d}x}{\int_{-\infty}^{\infty} x^2 p_{x\,\mathrm{st}}(x)\mathrm{d}x} \tag{5.493}$$

Leaving only the first two even terms in eq. (5.491), one obtains the following approximate representation of the correlation function of the solution of SDE (5.329) [29]

$$\hat{C}_{xx}(\tau) \approx \sigma^2 \exp[(a_1 + a_3 n_0)|\tau|] \tag{5.494}$$

The last formula has only a limited application and should not be considered as a proof of the fact that the correlation function of the Markov process generated by SDE (5.329) can be closely approximated by a single exponent. At the same time, if the drift of the Markov process can be accurately approximated by a linear function as

$$K_1(x) = -\lambda x + \varepsilon(x) \tag{5.495}$$

where $|\lambda x| \gg \varepsilon(x)$, then $C_{xx}(\tau)$ in fact can be approximated by a simple exponent. In this case, according to eq. (5.489)

$$\frac{\mathrm{d}}{\mathrm{d}\tau} C_{xx}(\tau) = -\lambda C_{xx}(\tau) + \langle x\varepsilon(x_\tau)\rangle \qquad (5.496)$$

The last term in eq. (5.496) is relatively small, so the solution of eq. (5.496) is relatively close to solution of the equation

$$\frac{\mathrm{d}}{\mathrm{d}\tau} C_{xx}(\tau) = -\lambda C_{xx}(\tau) \qquad (5.497)$$

which is an exact exponent. More detailed discussion of this problem, as well as some examples, can be found in Chapter 7.

REFERENCES

1. C. W. Gardiner, *Handbook of Stochastic Methods for Physics, Chemistry and the Natural Sciences*, Berlin: Springer, 1994.
2. R. Risken, *The Fokker–Planck Equation: Methods of Solution and Applications*, Berlin: Springer, 1996.
3. A. T. Bharucha-Reid, *Elements of The Theory of Markov Processes and Their Applications*, New York: McGraw-Hill, 1960.
4. W. Feller, *An Introduction to Probability Theory and Its Applications*, New York: Wiley, 1966.
5. A. Papoulis and S. Pillai, *Probability, Random Variables, and Stochastic Processes*, Boston: McGraw-Hill, 2002.
6. R. L. Stratonovich, *Topics in the Theory of Random Noise*, New York: Gordon and Breach, 1967.
7. J. Kemeny and L. Snell, *Finite Markov Chains*, Princeton: Van Nostrand, 1960.
8. J. Doob, *Stochastic Processes*, New York: Wiley, 1953.
9. A. Tikhonov and A. Samarski, *Partial Differential Equations of Mathematical Physics*, San Francisco, Holden-Day, 1964.
10. A. Abramowitz and I. Stegun, *Handbook of Mathematical Functions with Formulas, Graphs, and Mathematical Tables*, New York: Dover, 1972.
11. J. Murray, *Mathematical Biology*, New York: Springer, 1993.
12. E. Kamke, *Differentialgleichungen, Lösungsmethoden und Lösungen*, Leipzig: Akademische Verlagsgesellschaft Geest and Portig, 1959.
13. A. Polianin and V. Zaitsev, *Handbook of Exact Solutions for Ordinary Differential Equations*, Boca Raton, FL: Chapman & Hall/CRC, 2003.
14. P. E. Kloeden and E. Platen, *The Numerical Solution of Stochastic Differential Equations*, Springer-Verlag, 1992.
15. P. E. Kloeden, E. Platen and H. Schurz, *The Numerical Solution of Stochastic Differential Equations through Computer Experiments*, Springer-Verlag, 1994.
16. V. Tikhonov and N. Kulman, *Nonlinear Filtration and Quasicoherent Signal Reception*, Moscow: Sovradio, 1975 (in Russian).
17. R. Pawula, Approximation of the linear Boltzmann equation by the Fokker–Planck equation, *Physical Review* **162**(1–5), 186–188, 1967.
18. W. Lyndsey, *Synchronization Systems in Communication and Control*, Englewood Cliffs: Prentice-Hall, 1972.

19. E. Titchmarsh, *Eigenfunction Expansions Associated with Second-order Differential Equations*, Oxford: Clarendon Press, 1962.
20. I. S. Gradsteyn and I. M. Ryzhik, *Table of Integrals, Series, and Products*, New York: Academic Press, 1980.
21. A. Berger, *Chaos and Chance: An Introduction to Stochastic Aspects of Dynamics*, New York: Walter de Gruyter, 2001.
22. J. Schiff, *The Laplace Transform: Theory and Applications*, New York: Springer, 1999.
23. G. Golub and C. Van Loan, *Matrix Computations*, Baltimore: Johns Hopkins University Press, 1996.
24. P. Rao, D. Johnson and D. Becker, Generation and analysis of non-Gaussian Markov time series, *IEEE Transactions on Signal Processing*, **40**(4), 845–855, 1992.
25. G. Weiss, Time-reversibility of linear stochastic processes, *Journal of Applied Probability*, **12**(2), 831–836, 1975.
26. R. Rao, *Non-Gaussian Time Series*, M. Sc. Thesis, Houston: Rice University, 1988.
27. C. Nikias and A. Petropulu, *Higher-Order Spectra Analysis,* New Jersey: Prentice-Hall, 1993.
28. A. N. Malakhov, *Cumulant Analysis of Non-Gaussian Random Processes and Their Transformations*, Moscow: Sovetskoe Radio, 1978 (in Russian).
29. D. Klovsky V. Kontorovich, and S. Shirokov, *Models of Continuous Communication Channels Based on Stochastic Differential Equations*, Moscow: Radio i Sviaz, 1984 (in Russian).

6

Markov Processes with Random Structures

6.1 INTRODUCTION

Markov processes with random structure form a sub-class of mixed Markov processes. They can be characterized by significant changes of their "local"[1] probability density considered at different time instants. Changes of the local PDF appear at random. It is assumed here that the number of possible states (i.e. number of different PDFs) is finite and is described by a discrete component $\vartheta(t)$ with M possible states $\{\theta_i\}, i = 1, \ldots, M$. Changes in the properties of a Markov process with random structure can be attributed to changes in the state of this discrete component. In each state, the process can be either continuous or discrete, "locally" stationary in a strict or a wide sense, etc. An example of a realization of the process with random structure is shown in Fig. 6.1.

Essentially, if no change appears on a given time interval $[t_i, t_{i+1}]$, the process described above can be considered as a piece-wise continuous or a piece-wise discrete random process. Thus for its description it is possible to utilize a well developed mathematical apparatus, used to describe continuous and discrete Markov processes. This description should be augmented by assigning particular indices (the state index) to the trajectories of a particular continuous or discrete Markov process in each state, running from $i = 1$ to $i = M$, and providing moments of time t_j at which switching from one state to another takes place, as shown in Fig. 6.1. Referring to Fig. 6.2, the process $\xi^{(1)}(t)$ exists on the time intervals $[t_0, t_1]$ and $[t_3, t_4]$, the process $\xi^{(2)}(t)$ exists on the time interval $[t_2, t_3]$, and the process $\xi^{(3)}(t)$ exists on the time interval $[t_1, t_2]$, etc. It is also assumed that the value of the process $\xi^{(l)}(t)$ is zero if $\vartheta(t) \neq \theta_l$. It can be easily seen that the probability density $p_{\xi;l}(x)$ of such an augmented process is related to the probability density $p^{(l)}(x)$ of the realization considered only in a given state by the following rule

$$p_{\xi;l}(x) = P_l p^{(l)}(x) + (1 - P_l)\delta(x) \tag{6.1}$$

[1]The "local" properties here refer to the statistical properties (such as mean, variance, PDF, etc.) estimated through the time averaging of a segment of the realization of the process $\xi(t)$ for a fixed value of the switching process $\vartheta(t) = \theta_l$. These properties are assumed to be constant as long as the state of the switching process remains the same, but are allowed to change once the switching process moves to a new state.

Stochastic Methods and Their Applications to Communications.
S. Primak, V. Kontorovich, V. Lyandres
© 2004 John Wiley & Sons, Ltd   ISBN: 0-470-84741-7

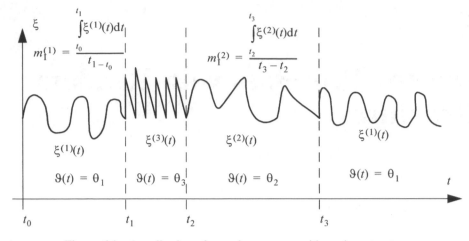

Figure 6.1   A realization of a random process with random structure.

where P_l is the probability of the state θ_l. It is assumed that $\xi^{(l)}(t)$ and the switching process are both stationary.

The order in change of indices is random and it is described by the state of the discrete component $\vartheta(t)$ with M possible states $\{\theta_i\}, i = 1, \ldots, M$. In future it is assumed that the switching process $\vartheta(t)$ is also a Markov process. As a result, an observer would see pieces of trajectories of different random processes, commutated at random. The resulting process is called a Markov process with random structure.

It follows from the definition of a process with random structure, that it reflects an approach to the description of a non-stationary random process by using locally stationary pieces of random duration [1]. The transition from one stationary piece to another happens according to the statistical properties of a non-stationary process.

The following examples emphasize reasons to consider processes with random structure as a useful tool in modeling. A number of physical situations can be found to illustrate the validity of such an approach.

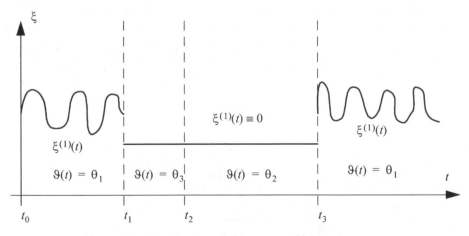

Figure 6.2   A realization of the process $\xi(t)$ in a given state.

1. A receiver operating in an environment contaminated by interference and noise (unwanted radiation) from different sources can be considered as the first example. It is assumed that the receiver has bandwidth B_R (or, equivalently, the response time T_R). Furthermore, it is assumed that the interference has energy spectrum with a spread $\Delta \omega$ much less than the bandwidth of the receiver $\Delta \omega_I \ll B_R$. Interference from a particular source goes on and off at random. Multiple interference at the input is treated as a sequence of individual interferences as shown in Fig. 6.3. This problem can be treated as a narrow band process driving a wide band system, thus implying that the response of the receiver to different

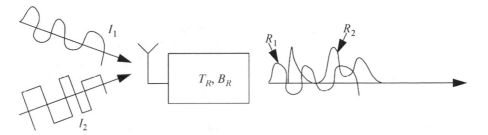

Figure 6.3 Illustration of the impact of interference on a receiver. R_i is the receiver's response to the interferences I_i.

interferers does not overlap, or such overlap is not a significant one, since the transient time $\Delta T_R \sim 1/B_R$ introduced by the receiver itself is much smaller than the duration of the interference. Statistical characteristics of the interferers are different and since they appear at the output of the receiver at separate moments of time, the output process can be treated as non-stationary. According to classification suggested by D. Middleton, interference is called Class A interference if the switching between different sources appears according to the Poisson law [2]. In practice, such situations can be encountered at the base station (wide band receiver) with cross-channel interference from individual mobiles.

2. The second example arises when the conditions on a communication channel vary greatly during one communication session (or frame, block or packet) duration. As an example, the Lutz model [3,4] of a satellite communication link takes into account the fact that during one telephone call the position of a satellite may change significantly to change fading from Rician to Rayleigh or vice versa. A proper model may include a number of states, each describing a particular severity of the channel conditions. These conditions may not only affect marginal distributions but also differ in second order statistics, for example in the level crossing rates [see Section 4.6].

3. It is often important to model speech signals on a computer. Given the specifics of the speech formation one can suggest a model where a set of states reflect particular characteristics of the speech modality. A slow speech detector, i.e. a device which reacts to significant changes in speech behaviour, is modeled as two (talk–silence) or three (talk–minitalk–silence) state continuous time Markov chains, with random transition from one state to another [5]. More detailed analysis may track variations in articulation, pitch, instantaneous power, etc. [6,7], resulting in models with a higher number of states.

4. It is possible to approximate a continuous non-stationary fading channel by a discrete process with random structure. In this case, states of the channel are chosen based on the required quality of the reception, such as probability of errors P_{err}, error burst duration, etc. A large class of models found in the literature (see Chapter 8) use the so-called Gilbert–Elliot model of a channel, which can be considered as a particular class of systems with random structure [8,9]. In this case, the channel under study is considered to reside only in one of two possible states: the "Good" state with a low probability of error and the "Bad" state with a higher probability of error. More details of this model are considered in Section 8.2.

Of course, it is possible to provide even a greater number of examples where the suggested approach may be useful. In future discussions, it is assumed that a proper modeling of the process in each state is available. In other words, it is assumed that if proper state variables are chosen, dynamic equations are known and the parameters of the models are either known or properly estimated, etc. [8,10,11]. It is important to mention here that the concept of the Markov process with random structure can be completely considered in the framework of so-called Hidden Markov Models (HMM) [6], characterized by a set of the following characteristics

$$\{p^{(l)}(\mathbf{0}), \mathbf{T} = [p_{lj}], p^{(l)}(\mathbf{x})\}, \quad l = 1, \ldots, M \tag{6.2}$$

Here $p^{(l)}(\mathbf{0})$ is a vector column of initial distributions of the states of HMM, p_{lj} are elements of the transitional matrix $\mathbf{T}$, describing the probability of transitions between states (sub-structures) of the HMM. Finally, $p^{(l)}(\mathbf{x})$ is a vector row of joint probability densities (of order more than one in general) in each state. It can be seen from eq. (6.2) that the HMM approach is effective when a particular class of distributions $p^{(l)}(\mathbf{x})$ allows for an effective analysis of the system in each state separately. For example, if each state is described by a Gaussian distribution, this fact significantly broadens the class of problems which can be analytically analyzed based on the knowledge of second order moments. However, if the distributions in each state are non-Gaussian, the multivariate probability densities $p^{(l)}(\mathbf{x})$ are known only for a limited number of cases [12].

This problem can be somewhat simplified if Markov processes (or their components) are adopted as model processes in each of the M states (sub-structures). It is assumed in the following that such models can be adequately represented by continuous Markov processes, and thus, can be considered as solutions of certain generating SDEs. As has been discussed earlier, the solution of an SDE can be treated as the response of a dynamic system to random excitation. Extending this view to the case of systems with random structure, it is possible to visualize it as a set of M separate systems, whose outputs are randomly connected (commutated) to an observer. Since such a view refers to a number of separate dynamic sub-systems, such a process is sometimes called a "polystructure process", which degenerates to a "monostructure" if there is only one state.

In addition to the representation of isolated systems, some description of the commutation process must be given. This is often accomplished by means of the matrix of transition probabilities $\{P_{lj}\}$. If the commutation processes are point processes with different intensities ν_{lj} then the matrix $\{P_{lj}\}$ can be expressed in terms of ν_{lj} and is often referred as the $\mathbf{Q}$ matrix [6,7]. The formation of the output signal is shown in Fig. 6.1.

6.2 MARKOV PROCESSES WITH RANDOM STRUCTURE AND THEIR STATISTICAL DESCRIPTION

6.2.1 Processes with Random Structure and Their Classification

This section is devoted to the mathematical description of Markov processes with random structure, similar to the one shown in Fig. 6.4. A sudden transition from the l-th state to the

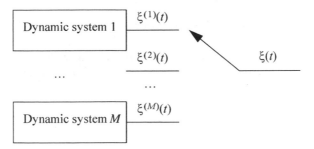

Figure 6.4   Formation of a process with random structure.

k-th state can be interpreted as absorption (termination) of the l-th state trajectory $\xi^{(l)}(t)$, and regeneration (re-creation) of the k-th state trajectory $\xi^{(k)}(t)$. Since, by the definition, these trajectories do not overlap in time (the system is only in one state at any given moment of time), one can formally write

$$\xi(t) = \sum_{l=1}^{M} \xi^{(l)}(t) \tag{6.3}$$

That is, it is assumed that $\xi^{(l)}(t) = 0$ if the system is at a state different from the l-th state. In the instant of time when the state of the system changes from l to k, certain initial conditions for regenerating process $\xi^{(k)}(t)$ must be specified. These conditions, random in general, can be described by certain probability densities

$$q_{lk}(x^{(k)}, t|x^{(l)}, t) \equiv q_{lk}(x, t|x', t) \tag{6.4}$$

conditional on the value of the absorbed process at the switching time, if the trajectory of the l-th process is normalized to unity[2]

$$\int_{-\infty}^{\infty} q_{lk}(x, t|x', t)\mathrm{d}x = 1 \tag{6.5}$$

The switching process $\vartheta(t)$ is an ordinary discrete process without memory, i.e. during a short interval of time, Δt, the probability that $\vartheta(t)$ would change its state more than once is of the order of $(\Delta t)^2$. This assumption implies that the switching process $\vartheta(t)$ is a Markov process.

---

[2]All integrals hereafter are to be treated as integrals over R^n with $n = \max\{n^{(1)}, \ldots, n^{(M)}\}$

Classification of Markov processes with random structure can be carried out based on the values of the process $\xi(t)$ at which the state change can appear. In particular, the process $\xi(t)$ is called a process with distributed transitions if the transitions from the l-th state to the k-th state may appear for any value of the process $\xi^{(l)}(t)$. If the state change may occur only if the process $\xi^{(l)}(t)$ reaches certain threshold levels, forming a discrete set of values, such a process is called a process with lumped transitions. In the following, mainly processes with distributed transitions are considered.

6.2.2   Statistical Description of Markov Processes with Random Structure

A Markov model can be adopted to describe the process with random structure if the absorption and the regeneration of the trajectories are described as above. In this case, an extended state vector

$$\zeta = [\xi, \vartheta(t)] \tag{6.6}$$

is a Markov vector and can be described by means of marginal probability densities

$$p_\zeta(x, l, t) = p^{*(l)}(x, t) \tag{6.7}$$

and the transitional PDF

$$\pi_\zeta(x_2, l, t_2 | x_1, l, t_1) = \pi^{*(l)}(x_2, t | x_1, t_1) \tag{6.8}$$

The PDF (eq. 6.8) describes behaviour of the process under the assumption that there is no change of state (no absorption) on the interval $[t_1, t_2]$. Since probability densities, defined by eq. (6.7), are in fact particular cases of the joint PDF of the extended process $\zeta(t)$, the integration just over x variable would not produce unity:

$$\int_{-\infty}^{\infty} p^{*(l)}(x, t)dx \leq 1, \quad \text{and} \quad \int_{-\infty}^{\infty} \pi^{*(l)}(x_2, t_2 | x_1, t_1)dx_2 \leq 1 \tag{6.9}$$

This reflects the fact that a change of the current state θ_l may appear on the time interval $[t_1, t_2]$ and a part of the realizations could be absorbed. However, the normalization to unity is obtained by summation over all possible states l, i.e.

$$\sum_{l=1}^{M} \int_{-\infty}^{\infty} p^{*(l)}(x, t)dx = 1 \tag{6.10}$$

and

$$\sum_{l=1}^{M} \int_{-\infty}^{\infty} \pi^{*(l)}(x_2, t_2 | x_1, t_1)dx_2 = 1 \tag{6.11}$$

It can be seen that integration just over x of the PDF $P^{*(l)}(x,t)$ produces the probability of the l-th state of the switching process $\vartheta(t)$:

$$\int_{-\infty}^{\infty} p^{*(l)}(x,t)\mathrm{d}x = P_l(t), \quad \text{and} \quad \sum_{l=1}^{M} P_l(t) = 1 \tag{6.12}$$

The quantity $P_l(t)$ describes the probability that the trajectory of the process belongs to the l-th state and it is not absorbed at the moment of time t, i.e. $P_l(t) = \mathrm{Prob}\{\xi^l(t), \theta = \vartheta_l\}$.

The next step is to consider conditions which can be applied on the transitional density $q_{lk}(x_2, t_2 | x_1, t_1)$, describing the process of regeneration of the trajectories. In many cases, it is reasonable to assume that the new trajectory corresponding to the k-th state picks up from the level achieved by the absorbed l-th trajectory[3], i.e.

$$q_{lk}(x_2, t_2 | x_1, t_1) = \delta(x_2 - x_1) \tag{6.13}$$

This condition, called continuous regeneration, is illustrated in Fig. 6.5. Of course there are other ways to define this density; some examples can be found in [1].

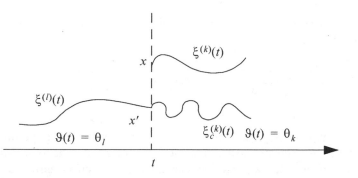

Figure 6.5 Jump conditions for a random process with random structure. Regenerated trajectory $\xi^{(k)}(t)$ corresponds to the process with continuous regeneration of trajectories, while $\xi^{(k)}(t)$ corresponds to a general case. In the general case, $\mathrm{Prob}\{\xi^{(l)}(t) \in [x', x' + \mathrm{d}x'], \xi^{(k)}(t) \in [x, x + \mathrm{d}x]\} = q_{kl}(x, t | x', t) \, \mathrm{d}x \, \mathrm{d}x'$.

6.2.3 Generalized Fokker–Planck Equation for Random Processes with Random Structure and Distributed Transitions

It is often required to obtain an equation which describes the evolution of the probability density $p^{*(l)}(x,t)$ as a function of time. The discussions in this section are limited to a case of a scalar random process $\xi(t)$ with random structure and with distributed transitions.

---

[3]In general, this assumption is not needed for the derivation of the generalized Fokker–Planck or Kolmogorov–Feller equations. However, here only this case is treated in some detail due to its importance in practical applications. Another important case often considered due to its relative simplicity is the case when the initial value of the regenerated trajectory does not depend on the final value of the absorbed trajectory. In this case $q_{kl}(x, t | x', t') = p_k(x, t)$.

For an ordinary switching process, which governs the absorbtion and regeneration of the trajectories, one can write

$$P_{lk}(t) = \nu_{lk}(t)\Delta t + o(\Delta t^2) \tag{6.14}$$

Here $\nu_{lk}(t)$ is the average number per unit time of transitions (intensity of transitions) from the l-th state into the k-th state at the moment of time t and $P_{lk}(t)$ is the probability of change of the state from $\theta = \vartheta_l$ to $\theta = \vartheta_k$. It is also assumed here that these intensities are independent of the process $\xi(t)$. In addition, the trajectories in each isolated state[4] are assumed for now to be diffusion Markov processes, described by a set of Fokker–Planck equations

$$\frac{\partial}{\partial t}\pi^{(l)}(x,t|x',\tau)$$
$$= -\frac{\partial}{\partial x}[K_1^{(l)}(x,t)\pi^{(l)}(x,t|x',\tau)] + \frac{1}{2}\frac{\partial^2}{\partial x^2}[K_2^{(l)}(x,t)\pi^{(l)}(x,t|x',\tau)], \quad l = 1,\ldots,M \tag{6.15}$$

Here $\tau < t$, and $x' = \xi(\tau)$ is the value of the process ξ at a given moment of time τ. The drift $K_1^{(l)}(x,t)$ and diffusion $K_2^{(l)}(x,t)$ have the same meaning as for an ordinary Markov diffusion process as described in Section 5.3.2.

For a short interval of time $\Delta t = t - \tau$, the Fokker–Planck equation (6.15) can be rewritten as [13]

$$\pi^{(l)}(x,t|x',\tau) = \delta(x-x') - \Delta t\frac{\partial}{\partial x}[K_1^{(l)}(x,t)\delta(x-x')] + \frac{1}{2}\Delta t\frac{\partial^2}{\partial x^2}[K_2^{(l)}(x,t)\delta(x-x')] + o(\Delta t^2) \tag{6.16}$$

Keeping this in mind, it is possible to obtain an equation describing the time evolution of $p^{*(l)}(x,t)$, taking into account possible transitions from the k-th state to the l-th state. During a short interval of time Δt, a particular realization can be either absorbed, if a switching event occurs, or remain unabsorbed. Given the fact that the switching process is ordinary[5], the probability of no switching is then

$$P_N = 1 - P_S = 1 - \Delta t\sum_{k=1}^{M}\nu_{lk}(\tau) = 1 - \Delta t\nu_l(\tau) \tag{6.17}$$

where

$$\nu_l(\tau) = \sum_{k=1}^{M}\nu_{lk}(\tau) \tag{6.18}$$

is the net intensity of transitions (switching) from the state l and $\nu_{ll} = 0$. As has been discussed earlier, the process of regeneration proceeds according to a certain transitional

---

[4] That is assuming that there is no switching.
[5] That is, the probability of more than two switchings in a short time interval Δt is of the order of $(\Delta t)^2$ or less.

PDF $q_{kl}(x, t|x', t')$ which is now to be used to relate the probability densities $p^{*(l)}(x, t)$ and $p^{*(l)}(x', t')$ as

$$p^{*(l)}(x, t) \approx [1 - \Delta t \nu_l(\tau)] \int_{-\infty}^{\infty} p^{*(l)}(x', \tau) \pi^{(l)}(x, t|x', \tau) \mathrm{d}x'$$
$$+ \Delta t \sum_{k=1}^{M} \nu_{kl}(\tau) \int_{-\infty}^{\infty} p^{*(k)}(x', \tau) q_{kl}(x, t|x', \tau) \mathrm{d}x' \qquad (6.19)$$

The first term in this expansion represents evolution of the trajectory if no switching appears, while each term under the sum sign represents evolution of the remaining part of the trajectory after absorption, switching and regeneration. Furthermore, taking into account approximation (eq. 6.16) this equation can further be simplified to the following form

$$p^{*(l)}(x, t)$$
$$= p^{*(l)}(x, \tau) - \Delta t \frac{\partial}{\partial x} [K_1^{(l)}(x, t) p^{*(l)}(x, t)] + \frac{1}{2} \Delta t \frac{\partial^2}{\partial x^2} [K_2^{(l)}(x, t) p^{*(l)}(x, t)] - \Delta t \nu_l(t)$$
$$+ \Delta t \sum_{k=1}^{M} \nu_{kl}(\tau) \int_{-\infty}^{\infty} q_{kl}(x, t|x', \tau) p^{*(k)}(x', \tau) \mathrm{d}x' \qquad (6.20)$$

Finally, subtracting $p^{*(l)}(x, \tau)$ from both sides, dividing both sides of eq. (6.20) by Δt and taking a limit when $\Delta t \to \infty$ one obtains the so-called generalized Fokker–Planck equation for a process with random structure and distributed transitions.

$$\frac{\partial}{\partial t} p^{*(l)}(x, t) = \lim_{t \to \infty} \frac{p^{*(l)}(x, t) - p^{*(l)}(x, \tau)}{\Delta t}$$
$$= -\frac{\partial}{\partial x} [K_1^{(l)}(x, t) p^{*(l)}(x, t)] + \frac{1}{2} \frac{\partial^2}{\partial x^2} [K_2^{(l)}(x, t) p^{*(l)}(x, t)]$$
$$- \nu_l(t) p^{*(l)}(x, t) + \sum_{k=1}^{M} \nu_{kl}(t) \int_{\infty} q_{kl}(x, t|x', t) p^{*(k)}(x', t) \mathrm{d}x' \qquad (6.21)$$

The same equation can be written for every state θ_l of the switching process, thus eq. (6.21) should be considered as a system of equations. This is in striking contrast to the Fokker–Planck equation for a vector Markov process with continuous components when the joint PDF satisfies a scalar partial differential equation. Unfortunately, this property also complicates solution of the generalized Fokker–Planck equation. It should be mentioned here that by varying the form of the transitional density $q(x, t|x', t)$ it is possible to model a great variety of random processes with random structure. Detailed investigation of these possibilities is beyond the scope of this book. More detail can be found in [1].

The generalized eq. (6.21) can also be interpreted in terms of the probability currents, as for an ordinary Fokker–Planck equation. Indeed, defining absorption functions as

$$U_l^*(x, t) = \sum_{k=1}^{M} U_{lk}^*(x, t) = \sum_{k=1}^{M} \nu_{lk}(t) p^{*(l)}(x, t) \qquad (6.22)$$

and regenerating functions as

$$V_{kl}^*(x,t) = \int_{-\infty}^{\infty} \nu_{kl}(t)p^{*(l)}(x',t)q_{kl}(x,t|x',t)\mathrm{d}x', \qquad V_l^*(x,t) = \sum_{k=1}^{M} V_{kl}^*(x,t) \qquad (6.23)$$

the generalized Fokker–Planck equation can be rewritten in the form of conservation of probability currents

$$\frac{\partial}{\partial t}p^{*(l)}(x,t) = -\frac{\partial}{\partial x}G^{(l)}(x,t) - U_l^*(x,t) + V_l^*(x,t) \qquad (6.24)$$

That is, the changes in probability are due to changes in the diffusion current $G^{(l)}(x,t)$ minus the flow of the absorption of the trajectory plus the flow of the regenerated trajectories. This fact can be explained by the absence of dependence between values of $\xi(t)$ and the moment of switching t_k. This also means that the statistical properties of the switching process do not depend on the prehistory of the process $\xi(t)$. In other words, behaviour of the process before the switching and at the moment of switching are statistically independent. This fact allows one to assume that the structure of the equation for $p^{*(l)}(x,t)$ would remain similar to the non-diffusion type of Markov process, such as a Markov process with jumps: the right-hand side of the generalized Fokker–Planck equation will contain a sum of an operator corresponding to each process without switching and an additional term describing switching of trajectories. The same can be said about generalization of the FPE to the vector case. In this case, the FPE becomes

$$\frac{\partial}{\partial t}p^{*(l)}(\boldsymbol{x},t) = -\sum_{i=1}^{n} \frac{\partial}{\partial x_i}[K_1^{(l)}(\boldsymbol{x},t)p^{*(l)}(\boldsymbol{x},t)]$$

$$+ \frac{1}{2}\sum_{i=1}^{n}\sum_{j=1}^{n} \frac{\partial^2}{\partial x_i \, \partial x_j}[K_{2ij}^{(l)}(\boldsymbol{x},t)p^{*(l)}(\boldsymbol{x},t)] - U_l^*(\boldsymbol{x},t) + V_l^*(\boldsymbol{x},t) \qquad (6.25)$$

Here $n = \max\{n_1, \ldots, n_M\}$ and

$$U_l^*(\boldsymbol{x},t) = \sum_{k=1}^{M} U_{lk}^*(\boldsymbol{x},t) = \sum_{k=1}^{M} \nu_{lk}(\boldsymbol{x},t)p^{*(l)}(\boldsymbol{x},t) \qquad (6.26)$$

$$V_l^*(\boldsymbol{x},t) = \sum_{k=1}^{M} V_{kl}^*(\boldsymbol{x},t) = \sum_{k=1}^{\infty}\int_{-\infty}^{\infty} \nu_{kl}(\boldsymbol{x}',t)q_{kl}(\boldsymbol{x},t|\boldsymbol{x}',t)p^{*(l)}(\boldsymbol{x}',t)\mathrm{d}\boldsymbol{x}' \qquad (6.27)$$

For a specific form of the transitional density $q_{kl}(\boldsymbol{x},t|\boldsymbol{x}',t)$, given by eq. (6.13), the term representing the regeneration flow in eq. (6.27) can be further simplified to produce

$$V_l^*(\boldsymbol{x},t) = \sum_{k=1}^{M} \nu_{kl}(t)p^{*(l)}(\boldsymbol{x},t) \qquad (6.28)$$

If, on the other hand, $q_{kl}(x, t | x', t) = q_{kl}(x, t)$, the expression for the same term becomes

$$V_l^*(x, t) = \sum_{k=1}^{M} V_{kl}^*(x, t) = \sum_{k=1}^{\infty} q_{kl}(x, t) \int_{-\infty}^{\infty} \nu_{kl}(x', t) p^{*(l)}(x', t) dx' \qquad (6.29)$$

Furthermore, if the switching rates do not depend on the values of the process $\xi(t)$, i.e. if $\nu_{kl}(x', t) = \nu_{kl}(t)$, then eq. (6.29) is simplified even more to produce

$$V_l^*(x, t) = \sum_{k=1}^{\infty} q_{kl}(x, t) \nu_{kl}(t) \int_{-\infty}^{\infty} p^{*(l)}(x', t) dx' = \sum_{k=1}^{\infty} q_{kl}(x, t) \nu_{kl}(t) \qquad (6.30)$$

Finally, if the jump conditions are such that the amount of change of the coordinate is distributed according to some PDF $p_A(a, t)$ which does not depend on the level of the process $\xi(t)$ then

$$q_{kl}(x, t | x', t) = p_A(x' - a, t) \qquad (6.31)$$

and the regeneration flow is described by

$$V_l^*(x, t) = \sum_{k=1}^{M} V_{kl}^*(x, t) = \sum_{k=1}^{\infty} \int_{-\infty}^{\infty} \nu_{kl}(x', t) p^{*(l)}(x', t) p_A(x' - a, t) dx' \qquad (6.32)$$

The next step is to generalize the Fokker–Planck equation to a case of the Markov process with jumps. It is assumed that between two jumps the process is a diffusion Markov process and that the appearance of the jumps is described by a Poisson process with intensity $\nu(t)$. This process can be treated as a process with random structure with two states and distributed transitions. In each state, statistical properties of the trajectories are the same and coincide with those of the process with jumps. The discontinuity is taken into account by the initial conditions immediately after the jump, which do not coincide with the final conditions of the absorbed trajectory. In other words, the form of the PDF $q_{kl}(x, t | x', t) = q_{lk}(x, t | x', t) = q(x, t | x', t)$ is different from the delta function.

For each of two states the following Fokker–Planck equations can be written

$$\frac{\partial}{\partial t} p^{*(l)}(x, t) = -\sum_{i=1}^{n} \frac{\partial}{\partial x_i} [K_{1i}^{(l)}(x, t) p^{*(l)}(x, t)] + \frac{1}{2} \sum_{i=1}^{n} \sum_{j=1}^{n} \frac{\partial^2}{\partial x_i \partial x_j} [K_{2ij}^{(l)}(x, t) p^{*(l)}(x, t)]$$

$$- \nu(t) p^{*(l)}(x, t) + \nu(t) \int_{-\infty}^{\infty} q(x, t | x', t) p^{*(k)}(x', t) dx' \qquad (6.33)$$

where $k, l = 1, 2, k \neq l$. Summing both sides of the eq. (6.33) over the index l and taking into account relation (6.18), the well known Kolmogorov–Feller equation can be obtained

$$\frac{\partial}{\partial t} p(x, t) = -\sum_{i=1}^{n} \frac{\partial}{\partial x_i} [a_i(x, t) p(x, t)] + \frac{1}{2} \sum_{i=1}^{n} \sum_{j=1}^{n} \frac{\partial^2}{\partial x_i \partial x_j} [b_{ij}(x, t) p(x, t)] - \nu(t) p(x, t)$$

$$+ \nu(t) \int_{-\infty}^{\infty} q(x, t | x', t) p(x, t) dx' \qquad (6.34)$$

since

$$p(x,t) = p^{*(1)}(x,t) + p^{*(2)}(x,t) \tag{6.35}$$

is the PDF of the process $\xi(t)$. If the diffusion (continuous) component is absent then the Kolmogorov–Feller equation is reduced to

$$\frac{\partial}{\partial t}p(x,t) = -\sum_{i=1}^{n}\frac{\partial}{\partial x_i}[K_{1i}(x,t)p(x,t)] - \nu(t)p(x,t) + \nu(t)\int_{-\infty}^{\infty} q(x,t|x',t)p(x,t)dx' \tag{6.36}$$

If jumps are induced by external δ-pulses with magnitude distribution $p_A(A)$ then $q(x,t|x',t) = p_A(x - x')$. All of the above can be generalized on the vector case when the switching is performed between processes with jumps and no diffusion component

$$\frac{\partial}{\partial t}p^{*(l)}(x,t) = -\sum_{i=1}^{n}\frac{\partial}{\partial x_i}[K_{1i}^{(l)}(x)p^{*(l)}(x,t)] - \nu^{(l)}(t)p^{*(l)}(x,t)$$

$$+ \nu^{(l)}(t)\int_{-\infty}^{\infty} p^{*(l)}(x - A,t)p_A(A)dA + V_l^*(x,t) - U_l^*(x,t) \tag{6.37}$$

Elements $u_{lk}^*(x,t)$ and $v_{lk}^*(x,t)$ can be organized into the following matrices

$$U^*(x,t) = \begin{bmatrix} 0 & u_{12}^*(x,t) & \cdots & u_{1M}^*(x,t) \\ \cdots & & & \cdots \\ \cdots & & & \cdots \\ u_{M1}^*(x,t) & \cdots & \cdots & 0 \end{bmatrix} \tag{6.38}$$

$$V^*(x,t) = \begin{bmatrix} 0 & v_{12}^*(x,t) & \cdots & v_{1M}^*(x,t) \\ \cdots & 0 & & \cdots \\ \cdots & & & \cdots \\ v_{M1}^*(x,t) & & & 0 \end{bmatrix} \tag{6.39}$$

known as absorption and regeneration matrices. It can be easily seen that if the transitional density $q(x,t|x',t)$ is a delta function as in eq. (6.13) then the absorption and regeneration functions are related

$$V^*(x,t) = U^{*T}(x,t) \tag{6.40}$$

The next step is to describe the evolution of probabilities of the states $P_l(t)$ as defined by eq. (6.12). It is assumed that the intensities of the transitions do not depend on time t and the current state x of the process, i.e. $\nu_{kl}(t) = \nu_{kl} = \text{const}$. This assumption is not very restrictive from an application point of view and it significantly simplifies the analysis of the problem. Having this in mind and integrating both sides of the Kolmogorov–Chapman eq. (6.25) over x, one obtains an equation for the time evolution of probabilities of the states as

$$\dot{P}_l(t) = \nu_l P_l(t) + \sum_{k=1}^{M}\nu_{kl}P_k(t) \tag{6.41}$$

This is consistent with the fact that $\vartheta(t)$ is a continuous time Markov chain. If $P_l(t)$ is known then the conditional probability density $p^{(l)}(x,t)$ of the process at a given state can be expressed as

$$p^{(l)}(x,t) = \frac{p^{*(l)}(x,t)}{P_l(t)} \tag{6.42}$$

Substituting expression (6.42) into the generalized Fokker–Planck eqs. (6.25) and (6.27), it is possible to obtain the Fokker–Planck equation for the conditional probabilities $p^{(l)}(x,t)$ of the process in a given state

$$\frac{\partial}{\partial t}p^{(l)}(x,t) = -\sum_{i=1}^{n}\frac{\partial}{\partial x_i}[K^{(l)}_{1i}(x)(x,t)p^{(l)}(x,t)] + \frac{1}{2}\sum_{i=1}^{n}\sum_{j=1}^{n}\frac{\partial^2}{\partial x_i\,\partial x_j}[EK^{(l)}_{2ij}(x)(x,t)p^{(l)}(x,t)]$$
$$+ \sum_{k=1}^{M}\left[\frac{P_k(t)}{P_l(t)}\nu_{kl}(x,t) - u_{kl}(x,t)\right] - \frac{\dot{P}_l(t)}{P_l(t)}p^{(l)}(x,t) \tag{6.43}$$

where

$$\nu_{kl}(x,t) = \frac{\nu^*_{kl}(x,t)}{P_l(t)},\ u_{kl}(x,t) = \frac{u^*_{kl}(x,t)}{P_l(t)} \tag{6.44}$$

are normalized absorption and regeneration flows.

A similar generalization can be made for the Kolmogorov–Feller eq. (6.37), which becomes

$$\frac{\partial}{\partial t}p^{(l)}(x,t) = -\sum_{i=1}^{n}\frac{\partial}{\partial x_i}[a^{(l)}_i(x)p^{(l)}(x,t)] - \nu^{(l)}(t)p^{(l)}(x,t) + \nu^{(l)}(t)\int_{-\infty}^{\infty}p^{(l)}(x-A,t)p^{(l)}_A(A)dA$$
$$+ \sum_{k=1}^{M}\left[\frac{P_k(t)}{P_l(t)}\nu_{kl}(x,t) - u_{kl}(x,t)\right] - \frac{\dot{P}_l(t)}{P_l(t)}p^{(l)}(x,t) \tag{6.45}$$

The transitional density $\pi^{(l)}(x,t+\tau|x',t)$ also satisfies eqs. (6.43)–(6.45) (as a function of x and t) with the initial conditions

$$\pi^{(l)}(x,t|x',t) = \delta(x-x') \tag{6.46}$$

Assuming that the resulting process is stationary and multiplying both sides of eq. (6.43) by $p^{(l)}(x_0,t_0)$, the joint conditional PDF $p^{(l)}_2(x,x_0;\tau)$ considered at two moments of time must satisfy the following equation

$$\frac{\partial}{\partial t}p^{(l)}_2(x,x_0;\tau) = -\sum_{i=1}^{n}\frac{\partial}{\partial x_i}[K^{(l)}_{1i}(x)p^{(l)}_2(x,x_0;\tau)] + \frac{1}{2}\sum_{i=1}^{n}\sum_{j=1}^{n}\frac{\partial^2}{\partial x_i\,\partial x_j}[K^{(l)}_{2ij}(x)p^{(l)}_2(x,x_0;\tau)]$$
$$- \nu_l p^{(l)}_2(x,x_0;\tau) + \sum_{k=1}^{M}\nu_{kl}\frac{P_k}{P_l}p^{(l)}(x)pK^{(l)}(x) - \frac{d}{dt}\ln P_l(t)p^{(l)}_2(x,x_0;\tau) \tag{6.47}$$

To shorten the notation, the equations can be rewritten in the operator form

$$\frac{\partial}{\partial t} p_2^{(l)}(x, x_0; \tau) = \mathscr{L}[p_2^{(l)}(x, x_0; \tau)] \tag{6.48}$$

where the notation $\mathscr{L}[\cdot]$ is used to denote the right-hand side of eq. (6.45). An operator, conjugate[6] to $\mathscr{L}[\cdot]$ is denoted as $\mathscr{M}[\cdot]$.

$$\mathscr{M}[p_2^{(l)}(x, x_0; \tau)] = \sum_{i=1}^{n} K_{1i}^{(l)}(x) \frac{\partial}{\partial x_i} p_2^{(l)}(x, x_0; \tau) + \frac{1}{2} \sum_{i=1}^{n} \sum_{j=1}^{n} K_{2ij}^{(l)}(x) \frac{\partial^2}{\partial x_i \partial x_j} p_2^{(l)}(x, x_0; \tau)$$

$$- \nu_l p_2^{(l)}(x, x_0; \tau) + \sum_{k=1}^{M} \nu_{kl} \frac{P_k}{P_l} p^{(l)}(x) pK^{(l)}(x) - \frac{d}{dt} \ln P_l(t) p_2^{(l)}(x, x_0; \tau) \tag{6.49}$$

In order to solve system of equation (6.45), proper initial and boundary conditions must be specified. All the equations must be solved simultaneously, including the equations for evolution of the switching process. Given that all the equations are interdependent, obtaining an analytical solution remains an open question. However, there are two limiting cases when a qualitative and quantitative analysis can be performed. In particular, this includes a case of a very slow and a very fast switching, treated in [14].

The distinction between a fast and a slow switching can be formally expressed in terms of the product of intensity of switching ν_l and the characteristic time $\tau_s^{(l)}$ of an isolated system in the l-th state. A switching process is considered slow if

$$\max\{\nu_l \tau_s^{(l)}\} \ll 1 \tag{6.50}$$

Similarly, the switching is considered to be fast if

$$\max\{\nu \tau_s^{(l)}\} \gg 1 \tag{6.51}$$

It can be seen from eqs. (6.20)–(6.21) that the terms containing the intensities of the transition are small compared to the diffusion terms if the intensity of switching is low. Interdependence between the states will also become negligible. In other words, it is reasonable to expect that the conditional density describing the evolution of the process in a given state differs little from the isolated process alone in the same state. In the fast switching case, the particular contribution to the internal dynamics of the individual states would become negligible.

6.2.4 Moment and Cumulant Equations of a Markov Process with Random Structure

Without loss of generality, equations for the evolution of moments of a process with random structure can be derived for a scalar case. Using the same technique as in Section 4.4 and the

---

[6] This implies that $\int_{-\infty}^{\infty} f_1(x) \mathscr{L}[p_2^{(l)}(x, x_0; \tau)] dx = \int_{-\infty}^{\infty} p_2^{(l)}(x, x_0; \tau) \mathscr{M}[f_1(x)] dx.$

following notation

$$\langle f(x,t)\rangle_l = \int_{-\infty}^{\infty} f(x,t)p^{(l)}(x,t)dx \tag{6.52}$$

the time derivative of the average of function $f(x,t)$ in the l-th state can be written as

$$\frac{\partial}{\partial t}\langle f(x,t)\rangle_l = \int_{-\infty}^{\infty} p^{(l)}(x,t)\mathcal{M}^{(l)}[f(x,t)]dx - \nu_l(t)\langle f(x,t)\rangle l$$
$$+ \sum_{k=1}^{M} \nu_{kl}(t)\frac{P_k(t)}{P_l(t)}\langle f(x,t)\rangle_k - \frac{d}{dt}\ln P_l(t)\langle f(x,t)\rangle_l \tag{6.53}$$

or, equivalently

$$\frac{\partial}{\partial t}\langle f(\lambda,t)\rangle_l - \sum_{n=1}^{\infty} \frac{1}{n!}\langle K_n^{(l)}(\lambda,t)\cdot\frac{\partial^n}{\partial x^n}f(\lambda,t)\rangle_l \quad \nu_l\langle f(\lambda)\rangle_l \; | \; \sum_{k=1}^{M}\nu_{kl}\frac{P_k(t)}{P_l(t)}\langle f(x,t)\rangle_k \quad \frac{\dot{P}_l(t)}{P_l(t)}\langle f(x,t)\rangle_l \tag{6.54}$$

Setting $f(x) = x^n$ in eq. (6.54) allows one to obtain evolution equations for the moments of the distribution in each state

$$\dot{m}_s^{(l)}(t) = s\langle x^{s-1}\cdot K_1^{(l)}(x)\rangle + \frac{s(s-1)}{2}\langle x^{s-2}\cdot K_2^{(l)}(x)\rangle$$
$$- \left(\nu_l(t) + \frac{d}{dt}\ln P_l(t)\right)m_s^{(l)}(t) + \sum_{k=1}^{M}\nu_{kl}(t)\frac{P_k(t)}{P_l(t)}m_s^{(k)(t)} \tag{6.55}$$

It can be seen that the structure of the moment equation (6.55) coincides with the structure of the corresponding Fokker–Planck and Kolmogorov–Feller equations for the conditional probability densities of the processes in each state. This is not surprising since the functions $f(x) = x^k$ can be directly brought inside the Fokker–Planck or Kolmogorov–Feller operators[7]. Equation (6.55) is to be solved using the initial conditions

$$m_s^{(l)}(t)|_{t=t_0} = m_{s0}^{(l)} \tag{6.56}$$

All equations must be solved simultaneously and jointly with the corresponding Chapman eq. (6.41) governing probabilities in each state. Thus, this procedure is similar to that of the ordinary Fokker–Planck or Kolmogorov–Feller equations. It is important to mention that the structure of the equations, with terms belonging to the partial process in the l-th state added to the terms reflecting absorption and regeneration, is to a certain degree the simplest possible structure to account for the internal dynamics of each isolated system and interaction between different states.

[7]Assuming that $\lim_{x \to a} xp(x) = 0$ for $a = 0$ or $a = \pm\infty$.

It can be seen that unless

$$K_1^{(l)}(x) = a_0^{(l)} + a_1^{(l)}x \quad \text{and} \quad K_2^{(l)}(x) = b_0^{(l)} + b_1^{(l)}x + b_2^{(l)}x^2 \tag{6.57}$$

the system of moment equations is not closed and its solution, therefore, is impossible. The Pearson system of distributions[8] in the isolated states, defined by equality (6.57), results in the following set of moment equations

$$\dot{m}_s^{(l)}(t) = s[a_1^{(l)}m_s^{(l)}(t) + a_0^{(l)}m_{s-1}^{(l)}(t)] + \frac{s(s-1)}{2}[b_2^{(l)}m_s^{(l)}(t) + b_1^{(l)}m_{s-1}^{(l)}(t) + b_0^{(l)}m_{s-2}^{(l)}(t)]$$
$$- \left(\nu_l(t) + \frac{\mathrm{d}}{\mathrm{d}t}\ln P_l(t)\right)m_s^{(l)}(t) + \sum_{k=1}^{M}\nu_{kl}(t)\frac{P_k(t)}{P_l(t)}m_s^{(k)}(t) \tag{6.58}$$

It should be pointed out that the equation for $m_s^{(l)}$ contains only moments of equal or lesser order, thus the system of linear eqs. (6.58) can be solved in a sequential manner, i.e. first one can obtain expressions for the mean values $m_1^{(l)}$, then for the second order moments, etc.

Example: *evolution of the mean value of an exponentially correlated random process.* If all random processes in the isolated state are exponentially correlated random processes, then their drifts are a linear function[9]

$$K_1^{(l)}(x) = -\frac{x - m_l}{\tau_{s;l}} \tag{6.59}$$

where m_l and $\tau_{s;l}$ are the mean value and the correlation interval of the process in the l-th isolated state. If the switching process achieves a stationary state, probabilities $P_l(t)$ do not depend on time and, thus, the evolution equations for the mean values become

$$\dot{m}_1^{(l)} = -\left(\frac{1}{\tau_{s;l}} + \nu_l\right)m_1^{(l)}(t) + \sum_{k=1}^{M}\nu_{kl}\frac{P_k}{P_l}m_1^{(k)}(t) + \frac{m_l}{\tau_{s;l}} \tag{6.60}$$

or, in a matrix form

$$\dot{m}_1(t, \nu) = -\Gamma m_1(t, \nu) + Tm_0 \tag{6.61}$$

where

$$\Gamma = \begin{bmatrix} \nu_1 + \frac{1}{\tau_{s;1}} & -\nu_{21}\frac{P_2}{P_1} & -\nu_{31}\frac{P_3}{P_1} & \cdots & -\nu_{M1}\frac{P_M}{P_1} \\ -\nu_{12}\frac{P_1}{P_2} & \nu_2 + \frac{1}{\tau_{s;2}} & -\nu_{32}\frac{P_3}{P_2} & \cdots & -\nu_{M2}\frac{P_M}{P_2} \\ \cdots & \cdots & \cdots & \cdots & \cdots \\ -\nu_{1M}\frac{P_1}{P_M} & -\nu_{2M}\frac{P_2}{P_M} & \cdots & \cdots & \nu_M + \frac{1}{\tau_{s;M}} \end{bmatrix} \tag{6.62}$$

---

[8]See section 2.1.1.3.
[9]See section 7.2.4.1 for details.

and

$$m_0 = [m_1 \, m_2 \cdots m_M]^\mathrm{T}, \quad T = \mathrm{diag}\{1/\tau_{s;1}, \ldots, 1/\tau_{s;M}\} \tag{6.63}$$

Assuming that the initial conditions for the system of eqs. (6.62) are those given by the stationary moments in the isolated states, i.e.

$$m_1(0) = m_0 \tag{6.64}$$

and the matrix Γ is invertible, one can easily obtain the solution of the system of equations as

$$m_1(t, \nu) = (I - e^{-\Gamma t})\Gamma^{-1} T m_0 + e^{-\Gamma t} m_1(0) \tag{6.65}$$

The stationary vector of mean values is thus given by

$$m_1(\nu) = \lim_{t \to \infty} m_1(t) = \Gamma^{-1} T m_0 \tag{6.66}$$

if the eigenvalues of Γ are all negative. Since the time needed for the mean value to reach its steady state is the largest of all moments, and it is of the order of the correlation interval, the smallest non-zero eigenvalue λ_1 of the matrix Γ defines the characteristic time of a system with random structure.

Some additional details about behaviour of moments can be extracted by considering a case of only two equations $M = 2$ and when $P_2 + P_1 = 1, \nu_{21} = P_1\nu, \nu_{12} = P_2\nu$. In this case

$$\Gamma = \begin{bmatrix} \nu P_2 + \frac{1}{\tau_{s1}} & -\nu P_2 \\ -\nu P_1 & \nu P_1 + \frac{1}{\tau_{s2}} \end{bmatrix} \quad \text{and} \quad \Gamma^{-1} = \frac{\tau_{s1}\tau_{s2}}{1 + \nu(P_1\tau_{s2} + P_2\tau_{s1})} \begin{bmatrix} \frac{1}{\tau_{s2}} + \nu P_1 & \nu P_2 \\ \nu P_1 & \frac{1}{\tau_{s1}} + \nu P_2 \end{bmatrix} \tag{6.67}$$

i.e. the matrix Γ is always invertible and the inverse matrix has only positive entries. As a result, the steady-state values of the moments always exist and are given by the following expression

$$m_1(\nu) = \Gamma^{-1} T m_0 = \frac{1}{1 + \nu(P_1\tau_{s2} + P_2\tau_{s1})} \begin{bmatrix} (1 + \nu\tau_2 P_1)m_1 + \nu\tau_1 P_2 m_2 \\ \nu\tau_2 P_1 m_1 + (1 + \nu\tau_1 P_2)m_2 \end{bmatrix} \tag{6.68}$$

If no switching appears, i.e. $\nu = 0$, then $m_1(0) = m_0$ which is consistent with the fact that m_0 represents a steady-state solution of isolated systems. If the switching intensity is so large that $\min\{\nu\tau_2 P_1, \nu\tau_1 P_2\} \gg 1$ then

$$m_1(\infty) = m_{1\infty} = \frac{1}{P_1\tau_{s2} + P_2\tau_{s1}} \begin{bmatrix} \tau_2 P_1 m_1 + \tau_1 P_2 m_2 \\ \tau_2 P_1 m_1 + \tau_1 P_2 m_2 \end{bmatrix} = \begin{bmatrix} m_{1\infty} \\ m_{1\infty} \end{bmatrix} \tag{6.69}$$

i.e. the mean values are equal in both states[10]. In order to understand how the mean values behave for a finite value of ν, one can rewrite eq. (6.68) as

$$m_1(\nu) = \frac{1}{1 + \nu(P_1 \tau_{s2} + P_2 \tau_{s1})} \begin{bmatrix} m_1 \\ m_2 \end{bmatrix} + \frac{\nu}{1 + \nu(P_1 \tau_{s2} + P_2 \tau_{s1})} \begin{bmatrix} \tau_2 P_1 m_1 + \tau_1 P_2 m_2 \\ \tau_2 P_1 m_1 + \tau_1 P_2 m_2 \end{bmatrix}$$

$$\frac{1}{1 + \nu(P_1 \tau_{s2} + P_2 \tau_{s1})} m_0 + \frac{\nu(P_1 \tau_{s2} + P_2 \tau_{s1})}{1 + \nu(P_1 \tau_{s2} + P_2 \tau_{s1})} m_\infty = w_1(\nu)m_0 + [1 - w_1(\nu)]m_\infty$$

$$(6.70)$$

Thus, for a finite value of ν the first moments are weighted sums of moments in the isolated states and the values of the same moments in the case of infinite intensity of switching. The weighting function is given by

$$0 \le w_1(\nu) = \frac{1}{1 + \nu(P_1 \tau_{s2} + P_2 \tau_{s1})} \le 1 \qquad (6.71)$$

and is the same for both sub-systems. Expression (6.71) also provides a true measure of the intensity of switching: it is slow if

$$\nu(P_1 \tau_{s2} + P_2 \tau_{s1}) \ll 1 \qquad (6.72)$$

and fast if

$$\nu(P_1 \tau_{s2} + P_2 \tau_{s1}) \gg 1 \qquad (6.73)$$

Condition (6.72) is equivalent to that used earlier, while condition (6.73) is much more conservative and depends on the probability of each state.

Example: Evolution of second moments in the case of two linear SDEs. If in isolated states both processes are Gaussian, they can be described by the following drift and diffusion

$$K_1^{(1)}(x) = -\frac{x - m_1}{\tau_{s1}}, \qquad K_1^{(2)}(x) = -\frac{x - m_2}{\tau_{s2}}, \qquad K_2^{(1)} = \frac{2\sigma_1^2}{\tau_{s1}}, \qquad K_2^{(2)} = \frac{2\sigma_2^2}{\tau_{s2}} \quad (6.74)$$

The evolution of first moments has already been discussed in the previous examples and is given by eq. (6.65). A differential equation for the second moment can then be written as

$$\dot{m}_2(t, \nu) = -\Gamma_2 m_2(t, \nu) + 2 T M m_1(t) = 2 T \Sigma \qquad (6.75)$$

where

$$m_2(t) = \begin{bmatrix} m_2^{(l)}(t) \\ m_2^{(2)}(t) \end{bmatrix}, \qquad \Gamma_2 = \begin{bmatrix} \frac{2}{\tau_{s1}} + \nu P_2 & -\nu P_2 \\ -\nu P_1 & \frac{2}{\tau_{s2}} + \nu P_1 \end{bmatrix},$$

$$M = \begin{bmatrix} m_1 & 0 \\ 0 & m_2 \end{bmatrix} \quad \text{and} \quad \Sigma = \begin{bmatrix} \sigma_1^2 & 0 \\ 0 & \sigma_2^2 \end{bmatrix} \qquad (6.76)$$

---

[10]This fact will be discussed in detail in Section 6.3.

with other quantities having the same meaning as in the previous example. The stationary solution $m_2(\nu)$ of this equation can be found by setting the right-hand side to zero and using stationary values of the first moments, i.e. $m_2(\nu)$ satisfies the following equation

$$-\Gamma_2 \, m_2(\nu) + 2\, T M \, m_1(\nu) + 2\, T \Sigma = 0 \tag{6.77}$$

which can be solved to produce

$$m_2(\nu) = 2\, \Gamma_2^{-1} T M \, m_1(\nu) + 2\, \Gamma_2^{-1} T \Sigma \tag{6.78}$$

It can be easily shown that

$$
2\,\Gamma_2^{-1} = \frac{\tau_{s1}\,\tau_{s2}}{2 + \nu(P_1\,\tau_{s2} + P_2\tau_{s1})}
\begin{bmatrix} \frac{2}{\tau_{s2}} + \nu P_1 & \nu P_2 \\ \nu P_1 & \frac{2}{\tau_{s1}} + \nu P_2 \end{bmatrix}
= w_2(\nu)\Gamma_{20}^{-1} + [1 - w_2(\nu)]\Gamma_{2\infty}^{-1}
$$

$$
\times \frac{1}{1 + \frac{\nu}{2}(P_1\,\tau_{s2} + P_2\tau_{s1})}
\begin{bmatrix} \tau_{s1} & 0 \\ 0 & \tau_{s2} \end{bmatrix}
\; \Bigg|\; \frac{1}{2}\frac{\tau_{s1}\,\tau_{s2}}{(P_1\,\tau_{s2} + P_2\,\tau_{s1})}
\begin{bmatrix} P_1 & P_2 \\ P_1 & P_2 \end{bmatrix}
\tag{6.79}
$$

where

$$w_2(\nu) = \frac{1}{1 + \frac{\nu}{2}(P_1\,\tau_{s2} + P_2\,\tau_{s1})} \tag{6.80}$$

and

$$
\Gamma_{20}^{-1} = \begin{bmatrix} \tau_{s1} & 0 \\ 0 & \tau_{s2} \end{bmatrix}, \qquad
\Gamma_{2\infty}^{-1} = \frac{1}{2}\frac{\tau_{s1}\,\tau_{s2}}{(P_1\,\tau_{s2} + P_2\,\tau_{s1})}
\begin{bmatrix} P_1 & P_2 \\ P_1 & P_2 \end{bmatrix} \tag{6.81}
$$

It can be noted that the matrix Γ_2 is always invertible. Finally, the stationary solution can be represented as

$$
\begin{aligned}
m_2(\nu) ={}& (w_2(\nu)\Gamma_{20}^{-1} + [1 - w_2(\nu)]\Gamma_{2\infty}^{-1})TM[w_1(\nu)m_0 + [1 - w_1(\nu)]m_\infty] + (w_2(\nu)\Gamma_{20}^{-1} \\
&+ [1 - w_2(\nu)]\Gamma_{2\infty}^{-1})T\,\Sigma\, w_2(\nu)w_1(\nu)\Gamma_{20}^{-1}TMm_0 + w_2(\nu)[1 - w_1(\nu)]\Gamma_{20}^{-1}TMm_\infty \\
&+ w_1(\nu)[1 - w_2(\nu)]\Gamma_{2\infty}^{-1}TMm_0 + [1 - w_1(\nu)][1 - w_2(\nu)]\Gamma_{2\infty}^{-1}TMm_\infty \\
&+ w_2(\nu)\Gamma_{20}^{-1}T\,\Sigma + [1 - w_2(\nu)]\Gamma_{2\infty}^{-1}T\,\Sigma
\end{aligned}
$$

Without further calculations one can deduce the nature of the dependence of the second moment on the parameter ν. Since $w_1(0) = w_2(0) = 1$ and $w_1(\infty) = w_2(\infty) = 0$ one can obtain

$$m_2(0) = \Gamma_{20}^{-1}TMm_0 + \Gamma_{20}^{-1}T\,\Sigma = \begin{bmatrix} m_1^2 + \sigma_1^2 \\ m_2^2 + \sigma_2^2 \end{bmatrix} \tag{6.82}$$

$$m_2(\infty) = \Gamma_{2\infty}^{-1}TMm_\infty + \Gamma_{2\infty}^{-1}T\,\Sigma = \begin{bmatrix} \dfrac{P_1\,m_1^2\,\tau_{s2} + P_2\,m_2^2\,\tau_{s1}}{P_1\,\tau_{s2} + P_2\,\tau_{s1}} + \dfrac{P_1\,\sigma_1^2\,\tau_{s2} + P_2\,\sigma_2^2\,\tau_{s1}}{P_1\,\tau_{s2} + P_2\,\tau_{s1}} \\[3mm] \dfrac{P_1\,m_1^2\,\tau_{s2} + P_2\,m_2^2\,\tau_{s1}}{P_1\,\tau_{s2} + P_2\,\tau_{s1}} + \dfrac{P_1\,\sigma_1^2\,\tau_{s2} + P_2\,\sigma_2^2\,\tau_{s1}}{P_1\,\tau_{s2} + P_2\,\tau_{s1}} \end{bmatrix}$$

$$\tag{6.83}$$

It can be noted that, for intermediate values of ν, the second moment depends on a set of two weights $w_1(\nu)$ and $w_2(\nu)$ as well as additional terms which cannot be attributed either to isolated dynamics or to the properties of the system at the infinite rate of switching. This explains why it is relatively difficult to analyze a case of intermediate values of ν.

Equations for the cumulants of the marginal PDF can be obtained in a very similar way to that shown in Section 4.4. However, taking into account that the moments and the cumulants are related through a non-linear transformation, it would be impossible to retain the minimum structure of such equations [15]. This conjecture can be illustrated by considering the following examples. Since $\kappa_1^{(l)} = m_1^{(l)}$ the structure of the cumulant equations for the first order cumulant is preserved. However, for the second cumulant one can write

$$\kappa_2^{(l)} = m_2^{(l)} - [m_1^{(l)}]^2, \qquad \dot{\kappa}_2^{(l)}(t) = \dot{m}_2^{(l)}(t) - 2\dot{m}_1^{(l)}(t)m_1^{(l)}(t) \qquad (6.84)$$

Thus, the evolution equation for the second cumulant becomes

$$\dot{\kappa}_2^{(l)} = 2\langle x \cdot K_1^{(l)}(x)\rangle - 2\langle x^{(l)}\rangle\langle K_1^{(l)}(x)\rangle + \langle K_2^{(l)}(x)\rangle - \left(\nu_l + \frac{d}{dt}\ln P_l(t)\right)(m_2^{(l)} - 2m_1^{2(l)})$$

$$+ \sum_{k=1}^{M} \nu_{kl}\frac{P_k(t)}{P_l(t)}(m_2^{(k)} - 2m_1^{(k)}m_1^{(l)}) \qquad (6.85)$$

Formally, eq. (6.85) can be rewritten as

$$\dot{\kappa}_2^{(l)} = 2\langle x, K_1^{(l)}(x)\rangle + \langle K_2^{(l)}(x)\rangle - \left(\nu_l + \frac{d}{dt}\ln P_l(t)\right)[\kappa_2^{(l)} - (\kappa_1^{(l)})^2]$$

$$+ \sum_{k=1}^{M} \nu_{kl}\frac{P_k(t)}{P_l(t)}\left[\kappa_2^{(k)} + \kappa_1^{(k)}(\kappa_1^{(k)} - 2\kappa_1^{(l)})\right] \qquad (6.86)$$

However, its structure is different from a typical one.

The situation becomes even more complicated for the third cumulant, since stronger non-linearities appear in the expression relating moments and cumulants of the order three.

$$\kappa_3 = m_3 - 3m_1m_2 + 2m_1^3, \qquad \dot{\kappa}_3 = \dot{m}_3 - 3\dot{m}_1m_2 - 3m_1\dot{m}_2 + 6m_1^2\dot{m}_1 \qquad (6.87)$$

and the corresponding equation for $\kappa_3^{(l)}(t)$ becomes intractable. Thus, it can be concluded that investigation of systems with random structure is better conducted either by using Gaussian approximation or by obtaining the first four moments through eq. (6.55). The former requires a simultaneous solution of non-linear equations for $\kappa_1^{(l)}, \kappa_2^{(l)}, m_1^{(l)}$ and $m_2^{(l)}$ with corresponding initial conditions. The latter is well known to produce a significant error, since reducing the number of equations by assuming higher order moments equal to zero is not consistent with the fact that the magnitude of even higher order moments increases[11]. Thus it seems difficult to obtain productive methods of investigating processes with random

---

[11]This statement is not true if a closed system of equations can be obtained. One such case is the Pearson system discussed above.

structure through cumulant analysis. This fact clearly distinguishes processes with random structure from regular Markov processes, when the description using lower order cumulants is a simple and effective means of analysis of non-linear systems.

6.3 APPROXIMATE SOLUTION OF THE GENERALIZED FOKKER–PLANCK EQUATIONS

In this section, the question of approximate solution of a system of generalized Fokker–Planck equations

$$
\frac{\partial}{\partial t} p^{*(l)}(x,t) = -\frac{\partial}{\partial x} [K_1^{(l)}(x) p^{*(l)}(x,t)] + \frac{1}{2} \frac{\partial^2}{\partial x^2} [K_2^{(l)}(x) p^{*(l)}(x,t)] - \nu_l p^{*(l)}(x,t)
$$
$$
+ \sum_{k=1}^{M} \nu_{kl} p^{*(k)}(x,t) \tag{6.88}
$$

is considered. In particular, the dependence of the solution on the intensity parameters ν_{kl} is of interest here. Each isolated system ($\nu = 0$ and no switching) produces a stationary solution in the form

$$
p_{0\,\mathrm{st}}^{(l)}(x) = \frac{C_l}{K_2^{(l)}(x)} \exp\left[2 \int_{-\infty}^{\infty} \frac{K_1^{(l)}(x)}{K_2^{(l)}(x)} \, dx \right] \tag{6.89}
$$

and the transient solution of the form[12]

$$
p_0^{(l)}(x,t) = p_{0\,\mathrm{st}}^{(l)}(x) \left[1 + \sum_{n=1}^{\infty} \alpha_n e^{-\lambda_n t} \varphi_n(x) \right] \tag{6.90}
$$

Assuming that this switching process $\theta(t)$ of zero intensity is a stationary process, the stationary solution of eq. (6.88) in isolated states is given by

$$
p_{0\,\mathrm{st}}^{*(l)}(x) = P_l p_{0\,\mathrm{st}}^{(l)}(x) \tag{6.91}
$$

$$
p_0^{*(l)}(x,t) = P_l p_{0\,\mathrm{st}}^{l}(x) \left[1 + \sum_{n=1}^{\infty} \alpha_n e^{-\lambda_n t} \varphi_n(x) \right] = p_{0\,\mathrm{st}}^{*(l)}(x) \left[1 + \sum_{n=1}^{\infty} \alpha_n e^{-\lambda_n t} \varphi_n(x) \right] \tag{6.92}
$$

The material in the next section contains various approximations of the solution for low and high intensity of switching between states.

[12]Assuming that the spectrum is discrete.

6.3.1 Gram–Charlier Series Expansion

6.3.1.1 Eigenfunction Expansion

In this section, an approximate method of solution based on the Gram–Charlier series is presented. While not being the most efficient, this method allows one to understand certain aspects of dependence of the solution on the switching intensity.

Let $p_{0\,\mathrm{st}}^{(l)}(x)$ be a stationary solution of the isolated FPE and let us assume that the spectrum of the FPE is discrete and the corresponding eigenfunctions $\psi_{n,l}(x) = p_{0\,\mathrm{st}}^{(l)}(x)Q_{k,l}(x)$ form a basis[13]. In other words, any PDF $p(x)$ can be expanded into the following series

$$p(x) = \sum_{k=0}^{\infty} \alpha_{k,l}\,\psi_{k,l}(x) = p_{0\,\mathrm{st}}^{(l)}(x)\left[1 + \sum_{k=0}^{\infty}\alpha_{k,l}\,Q_{k,l}(x)\right] \qquad (6.93)$$

where $\alpha_{k,l}$ are some coefficients. Since the eigenvalues of the FPE operator are distinct, non-negative, real and satisfy the following normalization condition

$$\int_{-\infty}^{\infty}\frac{\psi_{n,l}(x)\psi_{m,l}(x)}{p_{0\,\mathrm{st}}^{(l)}(x)}\,\mathrm{d}x = \int_{-\infty}^{\infty}p_{0\,\mathrm{st}}^{(l)}(x)Q_{n,l}(x)Q_{m,\,l}(x)\mathrm{d}x = \delta_{mn} \qquad (6.94)$$

multiplication of both sides of eq. (6.93) by $\psi_{k,l}(x)/p_{0\,\mathrm{st}}^{(l)}(x)$ and integration over the real axis produces the following expression for the coefficients $\alpha_{k,l}$

$$\alpha_{k,l} = \int_{-\infty}^{\infty}p(x)\frac{\psi_{k,l}(x)}{p_{0\,\mathrm{st}}^{(l)}(x)}\,\mathrm{d}x = \int_{-\infty}^{\infty}p(x)Q_{k,\,l}(x)\mathrm{d}x \qquad (6.95)$$

The next step is to represent a stationary solution of the generalized Fokker–Plank equation in each state as a series

$$p_{\mathrm{st}}^{*(m)}(x) = P_m\sum_{k=1}^{\infty}\alpha_{k,l}^{(m)}\psi_{k,l}(x) \qquad (6.96)$$

and substitute approximation (6.95) in the l-th equation of the system of the generalized FPE. This produces

$$L_{l\,\mathrm{FP}}[p_{\mathrm{st}}^{(l)}(x)] - \nu_l p_{\mathrm{st}}^{*(l)}(x) + \sum_{n\neq 1}^{M}\nu_{n\,l}p_{\mathrm{st}}^{*(n)}(x)$$

$$= L_{l\,\mathrm{FP}}\left[P_l\sum_{k=1}^{\infty}\alpha_{k,l}^{(l)}\psi_{k,l}(x)\right] - \nu_l P_l\sum_{k=1}^{\infty}\alpha_{k,l}^{(l)}\psi_{k,l}(x) + \sum_{n\neq 1}^{M}\nu_{n\,l}P_n\sum_{k=1}^{\infty}\alpha_{k,l}^{(n)}\psi_{k,l}(x)$$

$$= -P_l\sum_{k=1}^{\infty}\lambda_{k,l}\alpha_{k,l}^{(l)}\psi_{k,l}(x) - \nu_l P_l\sum_{k=1}^{\infty}\alpha_{k,l}^{(l)}\psi_{k,l}(x) + \sum_{n\neq 1}^{M}\nu_{n,l}P_n\sum_{k=1}^{\infty}\alpha_{k,l}^{(n)}\psi_{k,l}(x) = 0 \qquad (6.97)$$

---

[13]In the case of Pearson's PDF, $Q_{m,l}(x)$ are classical orthogonal polynomials [16,17].

where $\lambda_{k,l}$ is the eigenvalue of L_{lFP}, i.e.

$$L_{lFP}[\psi_{k,l}(x)] = -\lambda_{k,l}\psi_{k,l}(x), 0 = \lambda_{0,l} < \lambda_{1,l} < \cdots < \lambda_{k,l} < \cdots \tag{6.98}$$

Collecting terms with equal $\psi_{k,l}(x)$, one obtains

$$-P_l(\lambda_{k,l} + \nu_l)\alpha_{k,l}^{(l)} + \sum_{n \neq l}^{M} \nu_{n,l}P_n \alpha_{k,l}^{(n)} = 0 \tag{6.99}$$

or, equivalently, if $\lambda_{k,l}\nu_l \neq 0$

$$\alpha_{k,l}^{(l)} = \frac{1}{\lambda_{k,l} + \nu_l} \sum_{n \neq l}^{M} \nu_{n,l} \frac{P_n}{P_l} \alpha_{k,l}^{(n)} \tag{6.100}$$

While, in general, eq. (6.100) expresses the unknown coefficient $\alpha_{k,l}^{(l)}$ in terms of still unknown coefficients $\alpha_{k,l}^{(n)}$, it allows one to understand the effect of the switching intensity on the changes in the PDF of the solution due to the imbalance between distributions in each state. Since $\lambda_{0,l} = 0$ and $\psi_{0,l}(x) = p_{0\,st}^{(l)}(x)$, equality (6.95) produces

$$\alpha_{0,l}^{(m)} = 1 \tag{6.101}$$

for any m and l. As a result eq. (6.99) is reduced to the identity defining stationary probability of states as can be seen from eq. (6.41)

$$\sum_{n \neq l}^{M} \nu_{n,l}P_n = P_l\nu_l \tag{6.102}$$

6.3.1.2 Small Intensity Approximation

Since for a very small intensity of switching, perturbation of the solution is relatively small, it is possible to assume that

$$\alpha_{k,l}^{(m)} = \bar{\alpha}_{k,l}^{(m)} \tag{6.103}$$

where

$$\bar{\alpha}_{k,l}^{(m)} = \int_{-\infty}^{\infty} p_{0\,st}^{(m)}(x) \frac{\psi_{k,l}(x)}{p_{0\,st}^{(l)}(x)} dx = \int_{-\infty}^{\infty} p_{0\,st}^{(m)} Q_{k,l}(x) dx \tag{6.104}$$

are coefficients of the expansion of the conditional PDF $p_{0\,\text{st}}^{(m)}(x)$ in the m-th isolated state. Thus the first order approximation of the PDF in the l-th state is

$$p^{*(l)}(x) \approx P_l p_{0\,\text{st}}^{(l)}(x)\left[1 + \sum_{k=1}^{\infty} \frac{Q_{k,l}(x)}{\lambda_{k,l} + \nu_l} \sum_{n \neq 1}^{M} \nu_{n,l} \frac{P_n}{P_l} \bar{\alpha}_{k,l}^{(n)}\right]$$

$$\approx P_l p_{0\,\text{st}}^{(l)}(x)\left[1 + \sum_{k=1}^{\infty} \frac{Q_{k,l}(x)}{\lambda_{k,l}} \sum_{n \neq 1}^{M} \nu_{n,l} \frac{P_n}{P_l} \bar{\alpha}_{k,l}^{(n)}\right] \qquad (6.105)$$

This equation provides a good approximation if $\nu_l \ll \lambda_{1,l} < \lambda_{k,l}$, since in general the value $\bar{\alpha}_{k,l}^{(n)}$ decreases with k.

It is important to mention here that the expansion (6.105) suggested in this section is not an expansion in powers of a small parameter $\nu = \max\{\nu_{k,l}\}$ and thus cannot be refined to include second order (in ν) correction terms. The perturbation type solutions are considered below in Section 6.3.2.1.

Example: two Gaussian processes. Let two Gaussian processes described by the following SDEs

$$\dot{x} = -\frac{x - m_1}{\tau_{s1}} + \frac{2\sigma_1^2}{\tau_{s1}}\xi_1(t)$$

$$\dot{x} = -\frac{x - m_2}{\tau_{s2}} + \frac{2\sigma_2^2}{\tau_{s2}}\xi_2(t) \qquad (6.106)$$

in isolated states be commutated by a switching process described by $\nu_{21} = \nu P_1 = \nu_2, \nu_{12} = \nu P_2 = \nu_1, P_1 + P_2 = 1$. In this case, the solution for an isolated system is given by

$$p_{0\,\text{st}}^{(1)}(x) = \frac{1}{\sqrt{2\pi\sigma_1^2}}\exp\left[-\frac{(x - m_1)^2}{2\sigma_1^2}\right]$$

$$p_{0\,\text{st}}^{(2)}(x) = \frac{1}{\sqrt{2\pi\sigma_1^2}}\exp\left[-\frac{(x - m_2)^2}{2\sigma_2^2}\right] \qquad (6.107)$$

The corresponding stationary generalized Fokker–Planck equations are then

$$0 = \frac{d}{dx}\left[\frac{x - m_1}{\tau_{s1}}p^{*(1)}(x)\right] + \frac{d^2}{dx^2}\left[\frac{\sigma_1^2}{\tau_{s1}}p^{*(1)}(x)\right] - \nu P_2 p^{*(1)}(x) + \nu P_1 p^{*(2)}(x)$$

$$0 = \frac{d}{dx}\left[\frac{x - m_2}{\tau_{s2}}p^{*(2)}(x)\right] + \frac{d^2}{dx^2}\left[\frac{\sigma_2^2}{\tau_{s2}}p^{*(2)}(x)\right] - \nu P_1 p^{*(2)}(x) + \nu P_2 p^{*(1)}(x) \qquad (6.108)$$

Probability density $p_{0\,\text{st}}^{(1)}(x)$ generates a system of Hermitian polynomials $H_n\left(\frac{x-m_1}{\sigma_1}\right)$, i.e. the eigenfunctions $\psi_{k,1}(x)$ are given by

$$\psi_{k,1}(x) = \frac{1}{\sqrt{2\pi\sigma_1^2}}\exp\left[-\frac{(x - m_1)^2}{2\sigma_1^2}\right]\frac{1}{n!}H_n\left(\frac{x - m_1}{\sqrt{2}\sigma_1}\right) \qquad (6.109)$$

corresponding to the eigenvalues [17]

$$\lambda_{k,l} = \frac{k}{\tau_{s1}} \tag{6.110}$$

As the result of application of eq. (6.95), the following expression for the coefficients $\bar{\alpha}_{k,l}^{(2)}$ of expansion of $p_{0\,\mathrm{st}}^{(2)}(x)$ can be obtained

$$\bar{\alpha}_{k,l}^{(2)} = \frac{1}{n!} \int_{-\infty}^{\infty} \frac{1}{\sqrt{2\pi\sigma_2^2}} \exp\left[\frac{-(x-m_2)^2}{2\sigma_2^2}\right] H_n\left(\frac{x-m_1}{\sqrt{2}\sigma_1}\right) dx$$

$$= \frac{1}{n!\sqrt{x}} \int_{-\infty}^{\infty} \exp\left[-\left(z - \frac{m_2 - m_1}{\sqrt{2}\sigma_2}\right)^2\right] H_n\left(\frac{\sigma_2}{\sigma_1}z\right) dz \tag{6.111}$$

Here, the following substitution has been used

$$x = \sqrt{2}\sigma_2 z - m_1 \tag{6.112}$$

If $\sigma_1 \neq \sigma_2$, expression (6.111) can be evaluated using a standard integral 7.374–8 in [18]

$$\int_{-\infty}^{\infty} \exp[-(x-y)^2] H_n(\alpha x) dx = \sqrt{\pi}(1-\alpha^2)^{n/2} H_n\left(\frac{\alpha y}{\sqrt{1-\alpha^2}}\right) \tag{6.113}$$

with

$$y = \frac{m_2 - m_1}{\sqrt{2}\sigma_2}, \alpha = \frac{\sigma_2}{\sigma_1} \tag{6.114}$$

This results in the following expression for the coefficients $\bar{\alpha}_{k,l}^{(2)}$

$$\bar{\alpha}_{k,l}^{(2)} = \frac{1}{n!}\left(1 - \frac{\sigma_2^2}{\sigma_1^2}\right)^{n/2} H_n\left(\frac{m_2 - m_1}{\sqrt{2}\sqrt{\sigma_1^2 - \sigma_2^2}}\right)$$

$$= \begin{cases} \dfrac{1}{n!}\left(1 - \dfrac{\sigma_2^2}{\sigma_1^2}\right)^{n/2} H_n\left(\dfrac{m_2 - m_1}{\sqrt{2}\sqrt{\sigma_1^2 - \sigma_2^2}}\right) & \text{if } \sigma_2 < \sigma_1 \\[3mm] \dfrac{1}{n!}\left(\dfrac{\sigma_2^2}{\sigma_1^2} - 1\right)^{n/2} j^n H_n\left(j\dfrac{m_2 - m_1}{\sqrt{2}\sqrt{\sigma_2^2 - \sigma_1^2}}\right) & \text{if } \sigma_2 > \sigma_1 \end{cases} \tag{6.115}$$

If $\sigma_2 = \sigma_1 = \sigma$, one can use the following standard integral 7.374–6 in [18]

$$\int_{-\infty}^{\infty} \exp[-(x-y)^2] H_n(x) dx = \sqrt{\pi} y^n 2^n \tag{6.116}$$

to obtain

$$\bar{\alpha}_{k,l}^{(2)} = \frac{1}{n!\sqrt{\pi}} \int_{-\infty}^{\infty} \exp\left[-\left(z - \frac{m_2 - m_1}{\sqrt{2}\sigma}\right)^2\right] H_n(z) dz = \frac{2^{n/2}}{n!}\left(\frac{m_2 - m_1}{\sigma}\right)^n \tag{6.117}$$

Finally, eq. (6.105) produces a series for the solution of the generalized Fokker–Planck equation

$$p^{*(1)}(x) \approx P_1 \frac{\exp\left[-\frac{(x-m_1)^2}{2\sigma_1^2}\right]}{\sqrt{2\pi\sigma_1^2}}\left[1 + P_2\nu\tau_{s1}\sum_{k=1}^{\infty}\frac{H_k\left(\frac{x-m_1}{\sqrt{2}\sigma_1}\right)}{k}\bar{\alpha}_{k,1}^{(2)}\right]$$

$$p^{*(2)}(x) \approx P_2 \frac{\exp\left[-\frac{(x-m_2)^2}{2\sigma_2^2}\right]}{\sqrt{2\pi\sigma_2^2}}\left[1 + P_1\nu\tau_{s2}\sum_{k=1}^{\infty}\frac{H_k\left(\frac{x-m_2}{\sqrt{2}\sigma_2}\right)}{k}\bar{\alpha}_{k,2}^{(1)}\right] \tag{6.118}$$

where an expression for $\bar{\alpha}_{k,1}^{(1)}$ can be easily obtained from the expression for $\bar{\alpha}_{k,2}^{(1)}$ by permutation of indices. It is worth noting that the correction terms depend not only on the marginal distributions in the isolated states but also on correlation intervals of the partial processes. Characteristically in the case $\sigma_2 = \sigma_1$ and $m_2 = m_1$, coefficients $\bar{\alpha}_{k,2}^{(1)}$ and $\bar{\alpha}_{k,2}^{(2)}$ become zero as follows from eq. (6.117), thus confirming an intuitive conclusion that switching between two processes with identical PDFs must preserve the marginal PDF and the difference will appear only in the second order statistics. Results of numerical simulations are shown in Fig. 6.6.

Example: Switching between two Gamma distributed processes. Let each of two isolated processes be described by the Gamma PDF

$$p_{0\,st}^{(1)}(x) = \frac{\beta_l^{\alpha_l}}{\Gamma(\alpha_1)} x^{\alpha_l} \exp(-\beta_l x), \quad l = 1, 2 \tag{6.119}$$

In this case, a system of SDEs and corresponding generalized Fokker–Planck equations can be written as

$$\dot{x} = -\frac{\beta_1 x - \alpha_1}{\beta_1 \tau_{s1}} + \frac{2}{\beta_1 \tau_{s1}} x \xi_1(t)$$

$$\dot{x} = -\frac{\beta_2 x - \alpha_2}{\beta_2 \tau_{s2}} + \frac{2}{\beta_2 \tau_{s2}} x \xi_2(t) \tag{6.120}$$

and

$$0 = \frac{d}{dx}\left[\frac{\beta_1 x - \alpha_1}{\beta_1 \tau_{s1}} p^{*(1)}(x)\right] + \frac{d^2}{dx^2}\left[\frac{1}{\beta_1 \tau_{s1}} x p^{*(1)}(x)\right] - \nu P_2 p^{*(1)}(x) + \nu P_1 p^{*(2)}(x)$$

$$0 = \frac{d}{dx}\left[\frac{\beta_2 x - \alpha_2}{\beta_2 \tau_{s2}} p^{*(2)}(x)\right] + \frac{d^2}{dx^2}\left[\frac{1}{\beta_2 \tau_{s2}} p^{*(2)}(x)\right] - \nu P_1 p^{*(2)}(x) + \nu P_2 p^{*(1)}(x)$$

$$\tag{6.121}$$

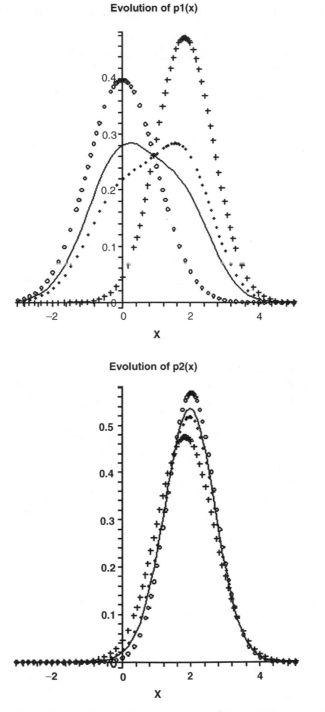

Figure 6.6 Numerical simulation of two switched Gaussian processes. Parameters of the simulation: $p_1 = 0.3$, $p_2 = 0.7$, $m_1 = 0$, $m_2 = 2$, $\sigma_1 = 1$, $\sigma_2 = 0.7$, $\tau_1 = 0.2$, $\tau_2 = 0.3$. $\nu = 0$ ○, $\nu = 0.5$, ◆, $\nu = 1$ ..., $\nu = \infty$ + +.

respectively. Each isolated FPE defines a set of eigenfunctions [17]

$$\psi_{k,l}(x) = \sqrt{\frac{\Gamma(\alpha+k)}{k!}} \exp(-\beta_l x) x^{\alpha_l-1} L_k^{(\alpha_l-1)}(\beta_l x) \tag{6.122}$$

Application of eq. (6.104) produces

$$\bar{\alpha}_{k,l}^{(2)} = \frac{\beta_2^{\alpha_2}}{\Gamma(\alpha_2)} \int_{-\infty}^{\infty} \exp(-\beta_2 x) x^{\alpha_2-1} L_k^{(\alpha_1-1)}(\beta_1 x) dx = \frac{\Gamma(k+\alpha_1)}{k!\Gamma(\alpha_1)} {}_2F_1\left(\alpha_2, -k; \alpha_1; \frac{\beta_1}{\beta_2}\right) \tag{6.123}$$

Here the following standard integral, 7.414–7 in [18], has been used

$$\int_0^\infty e^{-st} t^\beta L_n^{(\alpha)}(t) dt = \frac{\Gamma(\beta+1)\Gamma(n+\alpha+1)}{n!\Gamma(\alpha+1)} s^{-\beta-1} F\left(-n, \beta+1; \alpha+1; \frac{1}{s}\right) \tag{6.124}$$

where $F(a, b; c; z) = {}_2F_1(a, b; c; z)$ is the hypergeometric function [17]. Results of a numerical simulation are shown in Fig. 6.7.

6.3.1.3   Form of the Solution for Large Intensity

The case of large intensities of switching also provokes great interest from the point of view of applications. While the perturbation analysis of this case is conducted in some detail in section 6.3.3, here some conclusions are drawn based on the analysis of eq. (6.100):

$$\alpha_{k,l}^{(l)} = \frac{1}{\lambda_{k,l} + \nu_l} \sum_{n \neq 1}^M \nu_{n,l} \frac{P_n}{P_l} \alpha_{k,l}^{(n)}$$

Let ν_l be a large but finite number such that $\lambda_1 \ll \nu_l$. Then, it is possible to find an index $N(\nu) \gg 1$ such that $\lambda_{N(\nu)} \geq \nu$. If $1 < k \ll N(\nu)$ then $\lambda_k \ll \nu_l$ and, thus

$$\nu_l P_l \alpha_{k,l}^{(l)} \approx \sum_{n \neq l}^M \nu_{n,l} P_n \alpha_{k,l}^{(n)} \tag{6.125}$$

Comparing eq. (6.125) with eq. (6.102), one concludes that, for small values of k, the coefficients of expansion (6.96) are equal for every state of the process with random structure, i.e.

$$\alpha_{k,l}^{(n)} = \alpha_{k,l}^{(l)} \tag{6.126}$$

Once ν_l approaches infinity, $N(\nu)$ also must approach infinity and thus the number of equal terms in expansion (6.96) approaches infinity as well. As a result, it can be contended that for a very large intensity of switching, the resulting distributions in each state are proportional to the same function.

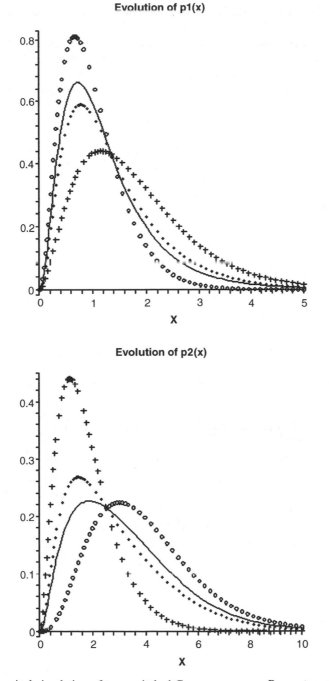

Figure 6.7   Numerical simulation of two switched Gamma processes. Parameters of the simulation: $p_1 = 0.3$, $p_2 = 0.7$, $\alpha_1 = 0$, $\alpha_2 = 3$, $\beta_1 = 1$, $\beta_2 = 0.5$, $\tau_1 = 0.2$, $\tau_2 = 0.3$. $\nu = 0$ ○, $\nu = 0.5$ ◆, $\nu = 1$ ..., $\nu = \infty$ ++.

6.3.2 Solution by the Perturbation Method for the Case of Low Intensities of Switching

In this section, a small perturbation solution of the time-variant system of two generalized Fokker–Planck equations is considered [19,20]. In order to obtain such an approximation, one has to find correction terms for the eigenvalues and eigenvectors in the isolated states. At this stage, it is assumed that the isolated Fokker–Planck operators are described by a discrete spectrum, i.e. there are two discrete sets of eigenvalues $\{\lambda_n\}$, $\{\mu_n\}$ and eigenfunctions $\{\psi_n(x)\}$ and $\{\phi_n(x)\}$ such that

$$
L_{1\,\mathrm{FP}}[\psi_n(x)] = -\frac{d}{dx}[K_1^{(1)}(x)\psi_n(x)] + \frac{1}{2}\frac{d^2}{dx^2}[K_2^{(1)}(x)\psi_n(x)] = -\lambda_n\psi_n(x)
$$

$$
L_{2\,\mathrm{FP}}[\phi_n(x)] = -\frac{d}{dx}[K_1^{(2)}(x)\phi_n(x)] + \frac{1}{2}\frac{d^2}{dx^2}[K_2^{(2)}(x)\phi_n(x)] = -\mu_n\phi_n(x)
$$

(6.127)

and

$$
\int_{-\infty}^{\infty} \psi_n(x)\frac{\psi_m(x)}{p_{\mathrm{st}}^{*(1)}(x)}\,dx = \int_{-\infty}^{\infty} \phi_n(x)\frac{\phi_m(x)}{p_{\mathrm{st}}^{*(2)}(x)}\,dx = \delta_{mn} = \begin{cases} 1 & \text{if } m = n \\ 0 & \text{if } m \neq n \end{cases}
$$

(6.128)

The stationary solutions $\psi_0(x) = p_{\mathrm{st}}^{*(1)}(x)$ and $\phi_0(x) = p_{\mathrm{st}}^{*(2)}(x)$ correspond to the zero eigenvalue for each isolated equation, i.e. $\lambda_0 = \mu_0 = 0$.

6.3.2.1 General Small Parameter Expansion of Eigenvalues and Eigenfunctions

The next step is to consider the eigenvalues and the eigenfunctions of the perturbed Fokker–Planck operators

$$
L_1[\psi(x)] = L_{1\,\mathrm{FP}}[\psi(x)] - \nu P_2\,\psi(x) + \nu P_1 p^{*(2)}(x)
$$

$$
L_2[\phi(x)] = L_{2\,\mathrm{FP}}[\phi(x)] - \nu P_1\phi(x) + \nu P_2 p^{*(1)}(x)
$$

(6.129)

While both equations must be solved simultaneously, the assumption of a small ν allows one to treat these equations separately. Due to a similar structure of both operators, the discussion is limited to the operator L_1 only. At this point, the exact value of the function $p_2(x)$ is not specified; however, it is assumed to be a known function.

A further assumption is that the eigenvalues Λ_n and the eigenvectors $\Psi_n(x)$ of the perturbed operator are holomorphic functions of the small parameter

$$
\nu\tau_s = \frac{\nu P_1 P_2 \tau_{s1} \tau_{s2}}{P_2 \tau_{s1} + P_1 \tau_{s2}}
$$

(6.130)

where τ_{s1} and τ_{s2} are the characteristic time constants of the isolated systems. The holomorphic property allows the following expansions for small values of $\nu\tau_s$

$$
\Lambda_n = \lambda_n + (\nu\tau_s)\lambda_n^{(1)} + (\nu\tau_s)^2\lambda_n^{(2)} + \cdots = \lambda_n + \sum_{k=1}^{\infty}(\nu\tau_s)^k\lambda_n^{(k)}
$$

$$
\Psi_n(x) = \psi_n(x) + (\nu\tau_s)\psi_n^{(1)}(x) + (\nu\tau_s)^2\psi_n^{(2)}(x) + \cdots = \psi_n(x) + \sum_{k=1}^{\infty}(\nu\tau_s)^k\psi_n^{(k)}(x)
$$

(6.131)

where $\lambda_n^{(k)}$ and $\psi_n^{(k)}(x)$ are to be defined. Since $\Psi_n(x)$ is the eigenfunction corresponding to the eigenvalue Λ_n the following equality must hold

$$
L_1[\Psi_n(x)]
$$

$$
= L_{1\,\mathrm{FP}}[\Psi_n(x)] - \nu\,P_2\,\Psi_n(x) + \nu\,P_1 p^{*(2)}(x)
$$

$$
= \lambda_n\,\psi_n(x) + \sum_{k=1}^{\infty}(\nu\,\tau_s)^k L_{1\,\mathrm{FP}}[\psi_n^{(k)}(x)] - \nu\,P_2\,\psi_n(x) - \nu\,P_2 \sum_{k=1}^{\infty}(\nu\,\tau_s)^k \psi_n^{(k)}(x) + \nu\,P_1\,p^{*(2)}(x)
$$

$$
= -\Lambda_n\,\Psi_n(x) = -\left[\lambda_n\,\psi_n(x) + \nu\,\tau_s\,\lambda_n^{(1)}\,\psi_n(x) + \nu\,\tau_s\,\lambda_n\,\psi_n^{(1)}(x) + \sum_{m+l>1}(\nu\,\tau_s)^{m+l}\lambda_n^{(m)}\psi_n^{(l)}(x)\right]
$$

(6.132)

Equating terms with equal powers of ν on both sides of this expansion, one obtains a system of equations for $\lambda_n^{(k)}$ and $\psi_n^{(k)}(x)$:

$$
L_{1\,\mathrm{FP}}[\psi_n^{(1)}(x)] + \lambda_n\psi_n^{(1)}(x) = \left(\frac{P_2}{\tau_s} - \lambda_n^{(1)}\right)\psi_n(x) - \frac{P_1}{\tau_s}p^{*(2)}(x) \tag{6.133}
$$

$$
L_{1\,\mathrm{FP}}[\psi_n^{(k)}(x)] - \frac{P_2}{\tau_s}\psi_n^{(k-1)}(x) = -\sum_{m+l=k}\lambda_n^{(m)}\psi_n^{(l)}(x) \tag{6.134}
$$

6.3.2.2   Perturbation of $\Psi_0(x)$

In the following it is assumed that the solution of the system of the Fokker–Planck eq. (6.129) has a stationary solution. Thus one has to enforce that $\Lambda_0 = \lambda_0 = 0$, while the corresponding eigenfunction $\Psi_n(x)$ still can be expressed as

$$
\Psi_0(x) = p_{\mathrm{st}}^{*(1)}(x) + \sum_{k=1}^{\infty}(\nu\,\tau_s)^k\psi_0^{(k)}(x) \tag{6.135}
$$

A similar expansion must be valid for the perturbed stationary solution of the second Fokker–Planck equation, i.e.

$$
\Phi_0(x) = p_{\mathrm{st}}^{*(2)}(x) + \sum_{k=1}^{\infty}(\nu\,\tau_s)^k\phi_0^{(k)}(x) \tag{6.136}
$$

Taking this into account, one can rewrite the first of eqs. (6.129) as

$$
L_{1\,\mathrm{FP}}[\Psi_0(x)] - \nu\,P_2 p^{*(1)}(x) + \nu_1 P^{*(2)}(x) = L_{1\,\mathrm{FP}}\left[p_{\mathrm{st}}^{*(1)}(x) + \sum_{k=1}^{\infty}(\nu\,\tau_s)^k\psi_0^{(k)}(x)\right]
$$

$$
- \nu P_2\left[p_{\mathrm{st}}^{*(1)}(x) + \sum_{k=1}^{\infty}(\nu\tau_s)^k\psi_0^{(k)}(x)\right] + \nu P_1\left[p_{\mathrm{st}}^{*(2)}(x) + \sum_{k=1}^{\infty}(\nu\,\tau_s)^k\phi_0^{(k)}(x)\right] = 0 \tag{6.137}
$$

Since $L_{1\,\text{FP}}[p_{\text{st}}^{*(1)}(x)] = 0$, one obtains the following system of perturbation equations

$$\sum_{k=1}^{\infty} L_{1\,\text{FP}}[\psi_0^{(1)}(x)] - \frac{P_2}{\tau_s} p_{\text{st}}^{*(1)}(x) + \frac{P_1}{\tau_s} p_{\text{st}}^{*(2)}(x) = 0 \qquad (6.138)$$

$$\sum_{k=1}^{\infty} L_{1\,\text{FP}}[\psi_0^{(k)}(x)] - \frac{P_2}{\tau_s} \psi_0^{(k-1)} + \frac{P_1}{\tau_s} \phi_0^{(k-1)}(x) = 0 \qquad (6.139)$$

Representing $\psi_0^{(1)}(x)$ on the basis of $\{\psi_n(x)\}$ as

$$\psi_0^{(1)}(x) = \sum_{p=0}^{\infty} \alpha_{0,p} \psi_n(x) \qquad (6.140)$$

and plugging it into eq. (6.138) one derives

$$\sum_{k=1}^{\infty} L_{1\,\text{FP}}[\psi_0^{(k)}(x)] = \sum_{p=0}^{\infty} \alpha_{0,p} L_{1\,\text{FP}}[\psi_n(x)] = -\sum_{p=0}^{\infty} \alpha_{0,p} \lambda_p \psi_n(x) = \frac{P_2}{\tau_s} p_{\text{st}}^{*(1)}(x) - \frac{P_1}{\tau_s} p_{\text{st}}^{*(2)}(x)$$

$$(6.141)$$

Furthermore, multiplying both parts of eq. (6.141) by $\psi_m(x)/p_{\text{st}}^{*(1)}(x)$ and integrating over x, the expression for the coefficients $\alpha_{0,m}$ for $m > 0$ can be deduced:

$$\alpha_{0,m} = \frac{P_1}{\tau_s \lambda_m} \int_{-\infty}^{\infty} p_{\text{st}}^{*(2)}(x) \frac{\psi_m(x)}{p_{\text{st}}^{*(1)}(x)} \, dx, \quad m \geq 1 \qquad (6.142)$$

with $\alpha_{0,0}$ still undefined. The latter can be obtained from the normalization condition, since

$$\int_{-\infty}^{\infty} p^{*(1)}(x) dx = \int_{-\infty}^{\infty} p_{\text{st}}^{*(1)}(x) dx = P_1 \qquad (6.143)$$

which is equivalent to $\alpha_{0,0} = 0$. Thus, the first order approximation is

$$p_1(x) \approx p_{\text{st}}^{*(1)}(x) - P_1 \nu \sum_{m=1}^{\infty} \frac{\psi_n(x)}{\lambda_m} \int_{-\infty}^{\infty} p_{\text{st}}^{*(2)}(x) \frac{\psi_m(x)}{p_{\text{st}}^{*(1)}(x)} \, dx \qquad (6.144)$$

Furthermore, approximation of $\lambda_n^{(1)}$ and $\psi_n^{(1)}(x)$ can be obtained directly from eq. (6.133). Since $\{\psi_n(x)\}$ forms a complete orthogonal basis, the function $\psi_n^{(1)}(x)$ can be expanded in terms of this basis:

$$\psi_n^{(1)}(x) = \sum_{p=0}^{\infty} \alpha_{n,p} \psi_p(x) \qquad (6.145)$$

In this case, eq. (6.133) can be rewritten as

$$L_{1\,\text{FP}}[\psi_n^{(1)}(x)] + \lambda_n \psi_n^{(1)}(x)$$

$$= L_{1\,\text{FP}}\left[\sum_{p=0}^{\infty} \alpha_{n,p}\psi_p(x)\right] + \lambda_n \sum_{p=0}^{\infty} \alpha_{n,p}\psi_p(x) = \sum_{p=0}^{\infty} \alpha_{n,p}(\lambda_n - \lambda_p)\psi_p(x)$$

$$= \left(\frac{P_2}{\tau_s} - \lambda_n^{(1)}\right)\psi_n(x) - \frac{P_1}{\tau_s}p^{*(2)}(x) \tag{6.146}$$

Multiplying both sides of eq. (6.146) by $\psi_n(x)/p_{\text{st}}^{*(1)}(x)$, integrating over x and using orthogonality, one obtains

$$0 = \frac{P_2}{\tau_s} - \lambda_n^{(1)} - \int_{-\infty}^{\infty} \frac{P_1}{\tau_s}p^{*(2)}(x)\frac{\psi_n(x)}{p_{\text{st}}^{*(1)}(x)}\,dx \tag{6.147}$$

i.e.

$$\lambda_n^{(1)} = \frac{P_2}{\tau_s} - \int_{-\infty}^{\infty} \frac{P_1}{\tau_s}p^{*(2)}(x)\frac{\psi_n(x)}{p_{\text{st}}^{*(1)}(x)}\,dx \tag{6.148}$$

Furthermore, multiplying both sides of eq. (6.146) by $\psi_m(x)/p_{\text{st}}^{*(1)}(x), m \neq n$, integrating over x and using orthogonality one obtains

$$\alpha_{n,m}(\lambda_m - \lambda_n) = -\frac{P_1}{\tau_s}\int_{-\infty}^{\infty} p^{*(2)}(x)\frac{\psi_m(x)}{p_{\text{st}}^{*(1)}(x)}\,dx \tag{6.149}$$

and, thus

$$\alpha_{n,m} = -\frac{P_1}{\tau_s(\lambda_m - \lambda_n)}\int_{-\infty}^{\infty} p^{*(2)}(x)\frac{\psi_m(x)}{p_{\text{st}}^{*(1)}(x)}\,dx \tag{6.150}$$

Returning to eq. (6.131), one derives

$$\Lambda_n = \lambda_n + \nu P_2 - \nu P_1\int_{-\infty}^{\infty} p^{*(2)}(x)\frac{\psi_n(x)}{p_{\text{st}}^{*(1)}(x)}\,dx$$

$$\Psi_n(x) = (1 - \alpha_{n,n}\nu\,\tau_s)\psi_n - P_1\nu\sum_{\substack{m=0\\m\neq n}}^{\infty}\frac{\psi_m(x)}{\lambda_m - \lambda_n}\int_{-\infty}^{\infty} p^{*(2)}(x)\frac{\psi_m(x)}{p_{\text{st}}^{*(1)}(x)}\,dx \tag{6.151}$$

Integrating the second equation in (6.151) over x and taking into account the normalization conditions, one obtains, for $n = 0$,

$$1 = \int_{-\infty}^{\infty} \Psi_n(x)\,dx = (1 - \alpha_{0,0}\nu\,\tau_s) \tag{6.152}$$

which implies that $\alpha_{0,0} = 0$ and, thus

$$\Psi_0(x) = p_{st}^{*(1)}(x) - P_1 \nu \sum_{\substack{m=0 \\ m \neq n}}^{\infty} \frac{\psi_m(x)}{\lambda_m - \lambda_n} \int_{-\infty}^{\infty} p^{*(2)}(x) \frac{\psi_m(x)}{p_{st}^{*(1)}(x)} dx \tag{6.153}$$

In general $\alpha_{n,n} = 0$ and, therefore

$$\Lambda_n = \lambda_n + \nu P_2 - \nu P_1 \int_{-\infty}^{\infty} p^{*(2)}(x) \frac{\psi_n(x)}{p_{st}^{*(1)}(x)} dx$$

$$\Psi_n(x) = \psi_n - P_1 \nu \sum_{\substack{m=0 \\ m \neq n}}^{\infty} \frac{\psi_m(x)}{\lambda_m - \lambda_n} \int_{-\infty}^{\infty} p^{*(2)}(x) \frac{\psi_m(x)}{p_{st}^{*(1)}(x)} dx \tag{6.154}$$

It is interesting to note how intensity of switching changes the first smallest non-zero eigenvalue Λ_1 which defines the correlation interval. Indeed, if switching is performed between two processes with the same PDF but different correlation properties, i.e. $P_2 p_{st}^{*(1)}(x) = P_1 p_{st}^{*(2)}(x)$ but $\lambda_1 \neq \mu_1$, then it follows from eq. (6.154) that

$$\Lambda_1 = \lambda_1 + \nu P_2 = \lambda_1 + \nu_{12} > \lambda_1 \tag{6.155}$$

thus providing a shorter correlation interval.

The second order approximation can be obtained from eq. (6.134) by setting $k = 2$

$$L_{1\,FP}[\psi_n^{(2)}(x)] - \frac{P_2}{\tau_s} \psi_n^{(1)}(x) = -\sum_{m+l=2} \lambda_n^{(m)} \psi_n^{(l)}(x) = -\lambda_n \psi_n^{(2)}(x) - \lambda_n^{(1)} \psi_n^{(1)}(x) - \lambda_n^{(2)} \psi_n(x)$$

$$\tag{6.156}$$

or, equivalently

$$L_{1\,FP}[\psi_n^{(2)}(x)] + \lambda_n \psi_n^{(2)}(x) = -\left(\lambda_n^{(1)} - \frac{P_2}{\tau_s}\right) \psi_n^{(1)}(x) - \lambda_n^{(2)} \psi_n(x) \tag{6.157}$$

Expanding $\psi_n^{(2)}(x)$ into a series over the basis $\{\psi_n(x)\}$

$$\psi_n^{(2)}(x) = \sum_{l=0}^{\infty} \beta_{n,l} \psi_l(x) \tag{6.158}$$

and plugging it into eq. (6.157), one obtains

$$L_{1\,FP}[\psi_n^{(2)}(x)] + \lambda_n \psi_n^{(2)}(x)$$

$$= -\sum_{l=0}^{\infty} \beta_{n,l} \lambda_l \psi_l(x) + \lambda_n \sum_{l=0}^{\infty} \beta_{n,l} \psi_l(x) = \sum_{l=0}^{\infty} \beta_{n,l} (\lambda_n - \lambda_l) \psi_l(x)$$

$$= -\left(\lambda_n^{(1)} - \frac{P_2}{\tau_s}\right) \psi_n^{(1)}(x) - \lambda_n^{(2)} \psi_n(x) \tag{6.159}$$

Multiplication by $\psi_m(x)/p_{st}^{*(1)}(x)$ and integration over x produces, for $m = n$,

$$\lambda_n^{(2)} = 0 \tag{6.160}$$

and

$$\beta_{n,m}(\lambda_m - \lambda_n) = -\left(\lambda_n^{(1)} - \frac{P_2}{\tau_s}\right)\alpha_{n,m} \tag{6.161}$$

As a result

$$\beta_{n,m} = -\left(\lambda_n^{(1)} - \frac{P_2}{\tau_s}\right)\frac{\alpha_{n,m}}{(\lambda_m - \lambda_n)}, \qquad m \neq n \tag{6.162}$$

Thus, it can be seen that the second order approximation does not produce any significant correction terms. This fact justifies application of the small perturbation analysis.

Example: Two Gaussian processes.   A case of commutation of two Gaussian processes is described by SDE (6.106). In this case

$$\lambda_n = \frac{n}{\tau_{s1}}, \qquad \psi_n(x) = P_1\frac{1}{\sqrt{n!2\pi\sigma_1^2}}\exp\left[-\frac{(x-m_1)^2}{2\sigma_1^2}\right]H_n\left(\frac{x-m_1}{\sqrt{2}\sigma_1}\right) \cdot \tag{6.163}$$

and

$$\mu_n = \frac{n}{\tau_{s2}}, \qquad \psi_n(x) = P_2\frac{1}{\sqrt{n!2\pi\sigma_2^2}}\exp\left[-\frac{(x-m_2)^2}{2\sigma_2^2}\right]H_n\left(\frac{x-m_2}{\sqrt{2}\sigma_2}\right) \tag{6.164}$$

Taking eq. (6.163) into account, the expression for the stationary PDF of the process in the first state becomes

$$\Psi_0(x) = p_{st}^{*(1)}(x)\left[1 - \frac{P_2\nu\tau_{s1}}{\sqrt{n!2\pi\sigma_2^2}}\sum_{\substack{m=0 \\ m\neq n}}^{\infty}\frac{H_m\left(\frac{x-m_1}{\sqrt{2}\sigma_1}\right)}{m-n}\int_{-\infty}^{\infty}\exp\left[-\frac{(x-m_2)^2}{2\sigma_2^2}\right]H_m\left(\frac{x-m_1}{\sqrt{2}\sigma_1}\right)dx\right]$$

$$p_{st}^{*(1)}(x)\left[1 - \frac{P_2\nu\tau_{s1}}{\sqrt{n!}}\sum_{\substack{m=0 \\ m\neq n}}^{\infty}\frac{H_m\left(\frac{x-m_1}{\sqrt{2}\sigma_1}\right)}{m-n}\left(1 - \frac{\sigma_2^2}{\sigma_1^2}\right)^{n/2}H_m\left(\frac{m_2-m_1}{\sqrt{2}\sqrt{\sigma_1^2-\sigma_2^2}}\right)\right] \tag{6.165}$$

which is very similar to that given by eq. (6.118). Here the following standard integral (7.374.8 in [18]) has been used

$$\int_{-\infty}^{\infty}\exp[-(x-y)^2]H_n(\alpha x)dx = \sqrt{\pi}(1-\alpha^2)^{n/2}H_n\left[\frac{\alpha y}{\sqrt{1-\alpha^2}}\right] \tag{6.166}$$

The first approximation of the eigenvalues is given by eq. (6.151) as

$$\Lambda_n = \frac{n}{\tau_{s1}} + \nu P_2 \left[1 - \frac{1}{\sqrt{n!}} \left(1 - \frac{\sigma_2^2}{\sigma_1^2} \right)^{n/2} H_m \left(\frac{m_2 - m_1}{\sqrt{2}\sqrt{\sigma_1^2 - \sigma_2^2}} \right) \right] \tag{6.167}$$

6.3.3 High Intensity Solution

In this section, an approximation of the solution for high intensity of switching is considered. The first step is to obtain conditions for the infinite intensity of switching and then to convert the problem into a small perturbation problem.

6.3.3.1 Zero Average Current Condition

It is easy to see that the Fokker–Planck equations can be written in the form of conservation of probability as

$$P_1 \frac{\partial}{\partial t} p_1(x, t) = -P_1 \frac{\partial}{\partial x} J_1(x) - P_2 \nu p^{*(1)}(x, t) + P_1 \nu p^{*(2)}(x, t)$$

$$P_2 \frac{\partial}{\partial t} p_2(x, t) = -P_2 \frac{\partial}{\partial x} J_2(x) - P_1 \nu p^{*(2)}(x, t) + P_2 \nu p^{*(1)}(x, t) \tag{6.168}$$

where

$$J_1(x, t) = K_1^{(1)}(x) p_1(x, t) - \frac{1}{2} \frac{\partial}{\partial x} [K_2^{(1)}(x) p_1(x, t)]$$

$$J_2(x, t) = K_1^{(2)}(x) p_2(x, t) - \frac{1}{2} \frac{\partial}{\partial x} [K_2^{(2)}(x) p_2(x, t)] \tag{6.169}$$

and

$$p^{*(1)}(x, t) = P_1 p_1(x, t)$$

$$p^{*(2)}(x, t) = P_2 p_2(x, t) \tag{6.170}$$

are the probability current densities, created by the drift and the diffusion of the continuous Markov process. Additional probability flows $\nu P_2 p_1^*(x, t)$ and $\nu P_1 p_2^*(x, t)$ are due to the absorption and regeneration of trajectories. probabilities P_1 and P_2 represent the proportion of time during which a given flow is observed. Adding both eqs. (6.168), one obtains

$$\frac{\partial}{\partial t} p(x, t) = \frac{\partial}{\partial x} J(x, t) \tag{6.171}$$

where the change in the average probability current

$$J(x,t) = P_1 J_1(x,t) + P_2 J_2(x,t) \tag{6.172}$$

is now the only factor responsible for the variation of the probability density

$$p(x,t) = p_1^*(x,t) + p_2^*(x,t) = P_1 p_1(x,t) + P_2 p_2(x,t) \tag{6.173}$$

in time. For a stationary flow, this current must remain constant, i.e.

$$J(x) = P_1 J_1(x) + P_2 J_2(x)$$

$$= P_1 K_1^{(1)}(x) p_1(x) + P_2 K_1^{(2)}(x) p_2(x) - \frac{1}{2} \frac{\partial}{\partial x} [K_2^{(1)}(x) p_1(x)]$$

$$= P_1 K_1^{(1)}(x) p_1(x) + P_2 K_1^{(2)}(x) p_2(x) - \frac{P_1}{2} \frac{\partial}{\partial x} [K_2^{(1)}(x) p_1(x)] - \frac{P_2}{2} \frac{\partial}{\partial x} [K_2^{(2)}(x) p_2(x)]$$

$$= J_0 = \text{const.} \tag{6.174}$$

If $p_1(x)$ and $p_2(x)$ are defined on a semi-infinite or infinite interval, probability densities $p_1(x)$ and $p_2(x)$, as well as their derivatives, must vanish at infinity, together with the probability current density $J(x)$. Thus, the constant J_0 must be equal to zero:

$$J_0 = 0 \tag{6.175}$$

This condition is known as the natural boundary condition in the theory of continuous Markov processes [13 and Section 5.3.3]. However, in the case of processes with random structure it is the average probability current density which vanishes; individual current densities $J_1(x)$ and $J_2(x)$ may remain functions of x in such a way that their sum is zero.

6.3.3.2 Asymptotic Solution $P_\infty(x)$

For a very high intensity ν all the probability densities $p^{*(k)}(x)$ must approach a fixed distribution $p_\infty(x)$

$$p^{*(k)}(x) = a_k p_\infty(x) \tag{6.176}$$

in order to satisfy the balance conditions

$$\sum_{k=1}^{L} \nu_{kl} p^{*(k)}(x) \approx 0 \tag{6.177}$$

for any l and x. The normalizing coefficients are to be chosen to satisfy the normalization condition

$$P_k = \int_{-\infty}^{\infty} p^{*(k)}(x)\,\mathrm{d}x = a_k \int_{-\infty}^{\infty} p_\infty(x)\,\mathrm{d}x = a_k \tag{6.178}$$

As far as a particular form of the asymptotic distribution $p_\infty(x)$ is concerned, nothing can be deduced from the eq. (6.177) itself. In order to obtain the expression for $p_\infty(x)$, one can take advantage of the fact that $p^{*(l)}(x) = P_l p_\infty(x)$. Taking this into account, and adding all M generalized Fokker–Planck equations, one obtains

$$\frac{\partial}{\partial t}\sum_{l=1}^{M} p^{*(l)}(x,t) = \frac{\partial}{\partial t}\sum_{l=1}^{M} p_\infty(x,t)$$

$$= -\frac{d}{dx}\left[p_\infty(x,t)\sum_{l=1}^{M} P_l K_1^{(l)}(x,t)\right] + \frac{1}{2}\frac{d^2}{dx^2}\left[p_\infty(x,t)\sum_{l=1}^{M} P_l K_2^{(l)}(x,t)\right]$$

$$(6.179)$$

or

$$\frac{\partial}{\partial t}\sum_{l=1}^{M} p_\infty(x,t) = -\frac{d}{dx}[K_{1\infty}(x)p_\infty(x,t)] + \frac{1}{2}\frac{d^2}{dx^2}[K_{2\infty}(x)p_\infty(x,t)] \qquad (6.180)$$

Here the average drift $K_{1\infty}(x)$ and the average diffusion are given by

$$K_{1\infty}(x) = \sum_{l=1}^{L} P_l K_1^{(l)}(x), \; K_{2\infty}(x) = \sum_{l=1}^{L} P_l K_2^{(l)}(x) \qquad (6.181)$$

The formal solution of eq. (6.180) is easy to obtain since it is a regular Fokker–Planck equation:

$$p_\infty(x) = C\exp\left[2\int_{-\infty}^{x}\frac{\sum_{l=1}^{L} P_l K_1^{(l)}(s)}{\sum_{l=1}^{L} P_l K_2^{(l)}(s)}ds\right] \qquad (6.182)$$

Example: Two Gaussian processes. Once again a system of two SDEs (6.106) is considered. In this case, $K_1^{(1)} = -\frac{x-m_1}{\tau_{s1}}, K_1^{(2)} = -\frac{x-m_2}{\tau_{s2}}, K_2^{(1)}(x) = \frac{2\sigma_1^2}{\tau_{s1}}$ and $K_2^{(2)}(x) = \frac{2\sigma_2^2}{\tau_{s2}}$. Thus, the averaged drift and diffusion are

$$K_{1\infty}(x) = -\left[P_1\frac{(x-m_1)}{\tau_{s1}} + P_2\frac{(x-m_1)}{\tau_{s1}}\right] = -\frac{x-m_{1\infty}}{\tau_{s\infty}} \qquad (6.183)$$

and

$$K_{2\infty}(x) = P_1\frac{2\sigma_1^2}{\tau_{s1}} + P_2\frac{2\sigma_2^2}{\tau_{s2}} = 2\frac{\sigma_\infty^2}{\tau_\infty} \qquad (6.184)$$

respectively. Here

$$m_{1\infty} = \frac{P_1 m_1 \tau_{s2} + P_2 m_2 \tau_{s1}}{P_1\tau_{s2} + P_2\tau_{s1}}, \quad \sigma_\infty^2 = \frac{P_1\sigma_1^2\tau_{s2} + P_2\sigma_2^2\tau_{s1}}{P_1\tau_{s2} + P_2\tau_{s1}} \quad \text{and} \quad \tau_\infty = \frac{\tau_{s1}\tau_{s2}}{P_1\tau_{s2} + P_2\tau_{s1}}.$$

$$(6.185)$$

are parameters of the limiting PDF. It can be seen that the limiting distribution $p_\infty(x)$ itself is Gaussian with parameters defined by eq. (6.185):

$$p_\infty(x) = \frac{1}{\sqrt{2\pi\sigma_\infty^2}} \exp\left[-\frac{(x - m_\infty)^2}{2\sigma_\infty^2}\right] \tag{6.186}$$

Example: *Two Gamma distributed processes.* In this case, the system of two SDEs (6.120) is analyzed. In this case, $K_1^{(1)} = -\frac{\beta_1 x - \alpha_1}{\beta_1\tau_{s1}}, K_1^{(2)} = -\frac{\beta_2 x - \alpha_2}{\beta_2\tau_{s2}}, K_2^{(1)}(x) = \frac{2}{\beta_1\tau_{s1}}$ and $K_2^{(2)}(x) = \frac{2}{\beta_2\tau_{s2}}$. Thus, the averaged drift and diffusion are

$$K_{1\infty}(x) = -\left[P_1\frac{\beta_1 x - \alpha_1}{\beta_1\tau_{s1}} + P_2\frac{\beta_2 x - \alpha_2}{\beta_2\tau_{s2}}\right] = -\frac{\beta_\infty x - a_\infty}{\beta_\infty\tau_{s\infty}} \tag{6.187}$$

and

$$K_{2\infty}(x) = P_1\frac{2}{\beta_1\tau_{s1}} + P_2\frac{2}{\beta_2\tau_{s2}} = \frac{2}{\beta_\infty\tau_\infty} \tag{6.188}$$

respectively. Here

$$\frac{1}{\beta_\infty} = -\frac{P_1\frac{1}{\beta_1}\tau_{s2} + P_2\frac{1}{\beta_2}\tau_{s1}}{P_1\tau_{s2} + P_2\tau_{s1}}, \qquad \alpha_\infty = \frac{P_1\frac{\alpha_1}{\beta_1}\tau_{s2} + P_2\frac{\alpha_2}{\beta_2}\tau_{s1}}{P_1\tau_{s2} + P_2\tau_{s1}} \qquad \text{and} \qquad \tau_\infty = \frac{\tau_{s1}\tau_{s2}}{P_1\tau_{s2} + P_2\tau_{s1}} \tag{6.189}$$

are the parameters of the limiting PDF. It can be seen that the limiting distribution $p_\infty(x)$ itself is a Gamma PDF with parameters defined by eq. (6.189).

$$p_\infty(x) = \frac{\beta_\infty^{a_\infty}}{\Gamma(\alpha_\infty)}\exp(-\beta_\infty x) \tag{6.190}$$

Taking into account that for the Gamma distribution

$$m = \frac{\alpha}{\beta} \quad \text{and} \quad \sigma = \frac{1}{\beta} \tag{6.191}$$

one can rewrite eq. (6.189) as

$$m_\infty = \frac{P_1 m_1\tau_{s2} + P_2 m_2\tau_{s1}}{P_1\tau_{s2} + P_2\tau_{s1}}, \qquad \sigma_\infty = \frac{P_1\sigma_1\tau_{s2} + P_2\sigma_2\tau_{s1}}{P_1\tau_{s2} + P_2\tau_{s1}} \qquad \text{and} \qquad \tau_\infty = \frac{\tau_{s1}\tau_{s2}}{P_1\tau_{s2} + P_2\tau_{s1}} \tag{6.192}$$

i.e. transformations of the mean value and standard deviation follow the same rules as for the Gaussian process considered above.

Another important conclusion which can be drawn is the importance of correlation properties of the isolated system on the resulting parameters of the distribution. In the case when the distributions in the isolated states are identical, the resulting marginal PDF of course would remain the same as in the isolated states. However, if there is a difference, even in scale (i.e. only in variance), the effect of correlation in each isolated case begins to play a prominent role in the final result.

6.3.3.3 Case of a Finite Intensity ν

Of course, for a finite but large ν, a certain correction term must be added. It is a goal of the considerations of this section to derive an appropriate expression for such a correction term. In order to keep the derivations compact the case of two equations is considered below:

$$0 = -\frac{\partial}{\partial x}[K_1^{(1)}(x)p^{*(1)}(x)] + \frac{1}{2}\frac{\partial^2}{\partial x^2}[K_2^{(1)}(x)p^{*(1)}(x)] - \nu P_2 p^{*(1)}(x) + \nu P_1 p^{*(2)}(x)$$

$$0 = -\frac{\partial}{\partial x}[K_1^{(2)}(x)p^{*(2)}(x)] + \frac{1}{2}\frac{\partial^2}{\partial x^2}[K_2^{(2)}(x)p^{*(2)}(x)] + \nu P_2 p^{*(1)}(x) + \nu P_1 p^{*(2)}(x)$$

(6.193)

This can be rewritten in the vector form as

$$-\frac{\partial}{\partial x}[K_1(x)\vec{P}^*(x)] + \frac{1}{2}\frac{\partial^2}{\partial x^2}[K_2(x)\vec{P}^*(x)] - \nu Q \vec{P}^*(x) = 0$$

(6.194)

or

$$L_{\mathrm{FP}}[\vec{P}^*(x)] = \nu Q \vec{P}^*(x)$$

(6.195)

where

$$K_1(x) = \begin{bmatrix} K_1^{(1)}(x) & 0 \\ 0 & K_1^{(2)}(x) \end{bmatrix}, \quad K_2(x) = \begin{bmatrix} K_2^{(1)}(x) & 0 \\ 0 & K_2^{(2)}(x) \end{bmatrix}, \quad Q = \begin{bmatrix} P_2 & -P_1 \\ -P_2 & P_1 \end{bmatrix}$$

(6.196)

and

$$L_{\mathrm{FP}}[\vec{P}^*(x)] = -\frac{\partial}{\partial x}[K_1(x)\vec{P}^*(x)] + \frac{1}{2}\frac{\partial^2}{\partial x^2}[K_2(x)\vec{P}^*(x)]$$

(6.197)

is the Fokker–Planck operator written in a matrix form. One may look for the solution of eq. (6.194) in the form of an infinite series

$$\vec{P}^*(x) = P_\infty \vec{P}(x) + \sum_{m=1}^{\infty} \frac{P_{-m}}{\nu^m}[\vec{P}_{-m}(x) - \vec{P}_\infty(x)]$$

(6.198)

where

$$P_\infty = \begin{bmatrix} P_1 & 0 \\ 0 & P_2 \end{bmatrix}, \quad \vec{P}_\infty(x) = \begin{bmatrix} p_\infty(x) \\ p_\infty(x) \end{bmatrix}$$

(6.199)

P_{-m} and $\vec{p}_{-m}(x)$ are the matrix and the vector function to be defined. Substituting a solution in the form (6.199) into eq. (6.194), and using the linearity of the Fokker–Planck equation, one obtains

$$L_{FP}[\vec{p}^*(x)]$$

$$= -\frac{\partial}{\partial x}[K_1(x)\vec{p}_\infty(x)] + \frac{1}{2}\frac{\partial^2}{\partial x^2}[K_2(x)\vec{p}_\infty(x)] + \sum_{m=1}^{\infty}-\frac{\partial}{\partial x}\left[K_1(x)\frac{P_{-m}}{\nu^m}[\vec{p}_{-m}(x) - \vec{p}_\infty(x)]\right]$$

$$+ \frac{1}{2}\frac{\partial^2}{\partial x^2}\left[K_2(x)\frac{P_{-m}}{\nu^m}[\vec{p}_{-m}(x) - \vec{p}_\infty(x)]\right] = \nu Q\left[\left(P_\infty - \sum_{m=1}^{\infty}\frac{P_{-m}}{\nu^m}\right)\vec{p}_\infty(x) + \sum_{m=1}^{\infty}\frac{P_{-m}}{\nu^m}\vec{p}_{-m}(x)\right]$$

$$= \sum_{m=1}^{\infty}\frac{QP_{-m}}{\nu^{m-1}}[\vec{p}_{-m}(x) - \vec{p}_\infty(x)] \tag{6.200}$$

since $QP_\infty\vec{p}_\infty = 0$. By equating the terms with equal negative powers of ν, one can obtain a sequence of equations

$$QP_{-1}[\vec{p}_{-1}(x) - \vec{p}_\infty(x)] = -\frac{\partial}{\partial x}[K_1(x)\vec{p}_\infty(x)] + \frac{1}{2}\frac{\partial^2}{\partial x^2}[K_2(x)\vec{p}_\infty(x)], m = 0 \tag{6.201}$$

and

$$QP_{-m}[\vec{p}_{-m}(x) - \vec{p}_\infty(x)] = \sum_{m=1}^{\infty}-\frac{\partial}{\partial x}\{K_1(x)P_{-m+1}[\vec{p}_{-m+1}(x) - \vec{p}_\infty(x)]\}$$

$$+ \frac{1}{2}\frac{\partial^2}{\partial x^2}\{K_2(x)P_{-m+1}[\vec{p}_{-m+1}(x) - \vec{p}_\infty(x)]\}, \quad m > 0 \tag{6.202}$$

It follows from eq. (6.182) that

$$P_1\left[-\frac{\partial}{\partial x}[K_1^{(1)}(x)p_\infty(x)] + \frac{1}{2}\frac{\partial^2}{\partial x^2}[K_2^{(1)}(x)p_\infty(x)]\right]$$

$$= -P_2\left[-\frac{\partial}{\partial x}[K_1^{(2)}(x)p_\infty(x)] + \frac{1}{2}\frac{\partial^2}{\partial x^2}[K_2^{(2)}(x)p_\infty(x)]\right] \tag{6.203}$$

or, equivalently

$$P_1L_{FP}^{(1)}[p_\infty(x)] = -P_2L_{FP}^{(2)}[p_\infty(x)] \tag{6.204}$$

Thus, the set of eqs. (6.201)–(6.202) are just the same equation repeated twice, i.e. a unique solution cannot be obtained just by considering eq.(6.202). This is the consequence of the fact that the matrix Q is degenerate, i.e. $\det(Q) = 0$. As a result, some additional conditions are required to obtain a unique solution of the original problem. This additional condition is a constant probability current condition (6.174), considered in section 6.3.3.1.

To make proper use of this condition, eq. (6.202) can be written as

$$P_1L_{1\,FP}[p_\infty(x)] - P_1P_2[p_{-1}^{(1)}(x) - p_\infty(x)] + P_1P_2[p_{-1}^{(2)}(x) - p_\infty(x)] = 0 \tag{6.205}$$

or, equivalently

$$p_{-1}^{(1)}(x) - p_{-1}^{(2)}(x) = \frac{1}{P_2} L_{1\,\mathrm{FP}}[p_\infty(x)], p_{-1}^{(2)}(x) = p_{-1}^{(1)}(x) - \frac{1}{P_2} L_{1\,\mathrm{FP}}[p_\infty(x)] \qquad (6.206)$$

Using approximation (6.198), the zero current density conditions can now be rewritten as

$$P_1 K_1^{(1)}(x) p_\infty(x) + P_2 K_1^{(2)}(x) p_\infty(x) - \frac{P_1}{2} \frac{\partial}{\partial x} [K_2^{(1)}(x) p_\infty(x)] - \frac{P_2}{2} \frac{\partial}{\partial x} [K_2^{(2)}(x) p_\infty(x)]$$

$$+ \frac{1}{\nu} \left\{ P_1 K_1^{(1)}(x) p_{-1}^{(1)}(x) + P_2 K_1^{(2)}(x) p_{-1}^{(2)}(x) - \frac{P_1}{2} \frac{\partial}{\partial x} [K_2^{(1)}(x) p_{-1}^{(1)}(x)] - \frac{P_2}{2} \frac{\partial}{\partial x} [K_2^{(2)}(x) p_{-1}^{(2)}(x)] \right\} = 0$$

$$(6.207)$$

Furthermore, since $p_\infty(x)$ satisfies the condition

$$[P_1 K_1^{(1)}(x) + P_2 K_1^{(2)}(x)] p_\infty(x) - \frac{1}{2} \frac{\partial}{\partial x} [\{P_1 K_2^{(1)}(x) + P_2 K_2^{(2)}(x)\} p_\infty(x)] = 0 \qquad (6.208)$$

eq. (6.207) can be simplified using expression (6.206) to produce an equation for $p_{-1}^{(1)}(x)$:

$$[P_1 K_1^{(1)}(x) + P_2 K_1^{(2)}(x)] p_{-1}^{(1)}(x) - \frac{1}{2} \frac{\partial}{\partial x} \{[P_1 K_2^{(1)}(x) + P_2 K_2^{(2)}(x)] p_{-1}^{(1)}(x)\}$$

$$= K_1^{(2)}(x) L_{1\,\mathrm{FP}}[p_\infty(x)] - \frac{1}{2} \frac{\partial}{\partial x} \{K_2^{(2)}(x) L_{1\,\mathrm{FP}}[p_\infty(x)]\} \qquad (6.209)$$

Alternatively, using the notation of eq. (6.181), one obtains

$$K_{1\infty}(x) p_{-1}^{(1)}(x) - \frac{1}{2} \frac{\partial}{\partial x} [K_{2\infty}(x) p_{-1}^{(1)}(x)] = K_1^{(2)}(x) L_{1\,\mathrm{FP}}[p_\infty(x)] - \frac{1}{2} \frac{\partial}{\partial x} \{K_2^{(2)}(x) L_{1\,\mathrm{FP}}[p_\infty(x)]\}$$

$$(6.210)$$

which is an ordinary non-homogeneous differential equation of the first order. The solution of the homogeneous equation $\tilde{P}(x)$ is once again proportional to $p_\infty(x)$:

$$\tilde{P}(x) = C p_\infty(x) = \frac{1}{K_{2\infty}(x)} \exp\left[2 \int_{-\infty}^{x} \frac{K_{1\infty}(x)}{K_{2\infty}(x)} dx \right] \qquad (6.211)$$

The next step is to find solution $p_{-1}^{(1)}(x)$ using the method of undefined coefficients, i.e. the solution is thought of in the form

$$p_{-1}^{(1)}(x) = C(x) \frac{1}{K_{1\infty}(x)} \exp\left[2 \int_{-\infty}^{x} \frac{K_{1\infty}(x)}{K_{2\infty}(x)} dx \right] \qquad (6.212)$$

where function $C(x)$ has to be determined. Substituting expression (6.212) into eq. (6.210), one obtains a differential equation for the unknown function $C(x)$

$$-\frac{1}{2} \exp\left[2 \int_{-\infty}^{x} \frac{K_{1\infty}(x)}{K_{2\infty}(x)} dx \right] \frac{d}{dx} C(x) = K_1^{(2)}(x) L_{1\,\mathrm{FP}}[p_\infty(x)] - \frac{1}{2} \frac{\partial}{\partial x} \{K_2^{(2)}(x) L_{1\,\mathrm{FP}}[p_\infty(x)]\}$$

$$(6.213)$$

whose general solution is

$$C(x) = C_0 + \int_{-\infty}^{x} \exp\left[-2 \int_{-\infty}^{s} \frac{K_{1\infty}(y)}{K_{2\infty}(y)} dy\right]$$
$$\times \left\{\frac{\partial}{\partial s}\{K_2^{(2)}(s)L_{1\,\mathrm{FP}}[p_\infty(s)]\} - 2 K_1^{(2)}(s)L_{1\,\mathrm{FP}}[p_\infty(s)]\right\} ds \qquad (6.214)$$

Since $K_1^{(2)}(x)$ and $K_2^{(2)}(x)$ are related through the probability density $p_{0\,\mathrm{st}}^{(2)}(x)$ in the second isolated state as

$$K_1^{(2)}(s) = \frac{1}{2} \frac{\frac{d}{ds}[K_2^{(2)}(s)p_{0\,\mathrm{st}}^{(2)}(s)]}{p_{0\,\mathrm{st}}^{(2)}(s)} \qquad (6.215)$$

then the expression in braces can be further simplified to produce

$$\frac{\partial}{\partial s}\{K_2^{(2)}(s)L_{1\,\mathrm{FP}}[p_\infty(s)]\} - 2 K_1^{(2)}(s)L_{1\,\mathrm{FP}}[p_\infty(s)]$$

$$= \frac{K_2^{(2)}(s)p_{20}(s)\frac{d}{ds}\{K_2^{(2)}(s)L_{1\,\mathrm{FP}}[p_\infty(s)]\} - K_2^{(2)}(s)L_{1\,\mathrm{FP}}[p_\infty(s)]\frac{d}{ds}[K_2^{(2)}(s)p_{20}(s)]}{[K_2^{(2)}(s)p_{20}(s)]^2}[K_2^{(2)}(s)p_{20}(s)]$$

$$= K_2^{(2)}(s)p_{20}(s)\frac{d}{ds}\frac{L_{1\,\mathrm{FP}}[p_\infty(s)]}{p_{20}(s)} \qquad (6.216)$$

Finally, eq. (6.214) can be written as

$$C(x) = C_0 + C_\infty \int_{-\infty}^{x} \frac{K_2^{(2)}(s)p_{20}(s)}{K_{2\infty}(s)p_\infty(s)}\frac{d}{ds}\frac{L_{1\,\mathrm{FP}}[p_\infty(s)]}{p_{20}(s)} ds$$

$$= C_0 + C_\infty \int_{-\infty}^{x} \exp\left[2\int_{-\infty}^{s}\left(\frac{K_1^{(2)}(u)}{K_2^{(2)}(u)} - \frac{K_{1\infty}(u)}{K_{2\infty}(u)}\right)du\right] d\frac{L_{1\,\mathrm{FP}}[p_\infty(s)]}{p_{20}(s)}$$

$$= C_0 + C_\infty \left\{\frac{K_2^{(2)}(s)(L_{1\,\mathrm{FP}}[p_\infty(s)])}{K_{2\infty}(s)p_\infty(x)}\right\}\Bigg|_{-\infty}^{x} - C_\infty \int_{-\infty}^{x}\left(\frac{K_1^{(2)}(s)}{K_2^{(2)}(s)} - \frac{K_{1\infty}(s)}{K_{2\infty}(s)}\right)\frac{K_2^{(2)}(s)L_{1\,\mathrm{FP}}[p_\infty(s)]}{K_{2\infty}(s)p_\infty(s)} ds$$

$$\qquad (6.217)$$

The values of these constants can be determined through the normalization conditions.

6.4 CONCLUDING REMARKS

The following remarks can be added to the results obtained for processes with random structure:

1. The general theory of Markov processes allows the same level of development as the theory of continuous Markov processes. In particular, these results can be extended to the

Pontryagin equations in attainment problems [21], narrow band processes with random structure, etc.

2. Analytical results in a closed form can only be obtained for a case of a very low or a very high intensity of switching. These limiting cases are especially attractive since they can be seen as perturbed solutions of isolated systems.

3. Despite the asymptotic nature of the results discussed, they can be readily applied to model processes with weak non-stationary, mixture models such as Middleton Class A noise, etc. [22].

4 In general, the model of a process with random structure, with a proper choice of the absorption and restoration flows, allows for modelling of a great variety of relatively little studied processes which are very important in applications, such as branching processes, etc.

REFERENCES

1. I. Kazakov and V. Artem'ev, *Optimization of Dynamic Systems with Random Structure*, Moscow: Nauka, 1980 (in Russian).
2. D. Middleton, Non-Gaussian Noise Models in Signal Processing for Telecommunications: New Methods and Results for Class A and Class B Noise Models, *IEEE Trans. Information Theory*, vol. 45, no, 5, May, 1999, pp. 1129–1149.
3. E. Lutz, D. Cugan, M. Dippold, F. Dolaninsky, and W. Papke, The Land Mobile Satellite Communication Channel–Recording, Statistics and Channel Model, *IEEE Trans. Vehicular Technology*, vol. 40, no. 2, 1991, pp. 375–380.
4. A Brusentzov and V. Kontorovich, Radio Channel Modeling Based on Dynamic Systems with Random Structure, *Proc. of the Int. Con. Commsphere-91*, Hertzlia, Israel, December, 1991
5. D. Goodman and S. Wei, Efficiency of Packet Reservation Multiple Access, *IEEE Trans. Vehicular Technology*, vol. 40, no. 1, pp. 170–176, February, 1991.
6. L. R. Rabiner, A Tutorial on Hidden Markov Models and Selected Applications in Speech Recognition, *Proc. of the IEEE*, vol. 77, February, 1989, pp. 257–286.
7. J. Dellec, J. G. Proakis, and H. L. Hausen, *Discrete Time Processing of Speech Signals*, New York: MacMillan, 1993.
8. M. Zorzi, Outage and Error Events in Bursty Channels, *IEEE Trans. Communications*, vol. 46, March, 1998, pp. 349–356.
9. W. Turin and R. van Nobelen, Hidden Markov Modeling of Flat Fading Channels, *IEEE Journal Selected Areas of Communications*, vol. 16, no. 12, December, 1998, pp. 1809–1817.
10. F. Babich and G. Lombardi, A Measurement Based Markov Model for the Indoor Propagation Channel, *Proc. IEEE Vehicular Technology Conferenece Proc*, vol. 1, Phoenix, AZ, May, 1997, pp. 77–81.
11. F. Babich, G. Lombardi, S. Shiavon, and E. Valentinuzzi, Measurements and Analysis of the Digital DECT Propagation Channel, *Proc. of the IEEE Conf. Wireless Personal Communications*, Florence, Italy, October, 1998, pp. 593–597.
12. K. K. Fang, S. Kotz, and K. W. Ng, *Symmetric Multivariate and Related Distributions*, New York: Chapman and Hall, 1990.
13. R. Risken, *The Fokker–Planck Equation: Methods of Solution and Applications*, Berlin: Springer, 1996, 472 p.
14. V. Kontorovich and V. Lyandres, Dynamic Systems with Random Structure: an Approach to the Generation of non-Stationary Stochastic Processes, *Journal of Franklin Institute*, vol. 336, 1999, pp. 939–954.

15. V. Ya. Kontorovich, S. Primak, and K. Almustafa, On Limits Of Applicability of Cumulant Method to Analysis of Dynamic Systems with Random Structure and Deterministic Systems With Chaotic Behaviour, *Proc. DCDIS-2003*, Guelph, Ontario, Canada, May, 2003.

16. J. Ord, *Families of Frequency Distributions*, London: Griffin, 1972.

17. M. Abramovitz and I. Stegun, *Handbook of Mathematical Functions*, Dover, 1965.

18. I. Gradsteyn and I. Ryzhik, *Tables of Integrals, Series, and Products*, New York: Academic Press, 2000.

19. Yu. A. Mitropolskii and N. van Dao, *Applied Asymptotic Methods in Nonlinear Oscillations*, Boston: Kluwer Academic Publishers, 1997

20. T. Kato, *Perturbation Theory for Linear Operators*, New York: Springer-Verlag, 1966.

21. V. Kontorovich, Pontryagin Equations for Non-Linear Dynamic Systems with Random Structure, *Nonlinear Analysis*, vol. 47, 2001, pp. 1501–1512.

22. H. Inoue, K. Sasajima, and M. Tanaka, A simulation of "Class A" Noise by P-CNG, *Proc. IEEE Int Symp. EMC*, 1999, vol. 1, pp. 520–525.

23. O. A. Ladyzhenskaia and N. N. Ural'tseva, *Linear and quasilinear elliptic equation*, Trans. Editor: L. Ehrenpreis, New York: Academic Press, 1968.

24. B. S. Everitt and D. J. Hand, *Finite Mixture Distributions*, New York: Chapman and Hall, 1981.

25. V. Kontorovich and V. Lyandres, Pulse Non-Stationary Process Generated by Dynamic Systems with Random Structure, *Journal of Franklin Institute*, 2003.

26. V. Kontorovich, V. Lyandres, and S. Primak, Non-Linear Methods in Information Processing: Modeling, Numerical Simulation and Applications, *DCDIS, series B: Applications & Algorithms*, vol. 10, October 2003, pp. 417–428.

27. A. N. Malakhov, *Cumulant Analysis of Non-Gaussian Random Processes and Their Transformations*, Moscow: Sovetskoe Radio, 1978 (in Russian).

7

Synthesis of Stochastic Differential Equations

7.1 INTRODUCTION

Modeling of signals and interference in the fields of communications, radar, sonar and speech processing is usually based on an assumption that these processes may be considered as stationary (or at least locally stationary) and that experimental estimations of their simplest statistical characteristics, autocovariance function (ACF) and marginal probability density function (PDF), are available. While the generation of stationary random processes (sequences) with either a specified PDF or a required valid ACF does not present any principal difficulty, the solution of the joint problem requires much more effort. One of the most often used methods was suggested in [1]. This approach utilizes sequential combinations of linear filtering and zero memory non-linear transformations of White Gaussian Noise (WGN). A different technique is based on treatment of a process with the prescribed characteristics as a stationary solution of a proper system of stochastic differential equations (SDE) with the WGN or Poisson process as an external driving force [2,3]. Such an interpretation seems attractive as it takes advantage of Markov process theory and appears to be efficient in the modeling of correlated non-Gaussian processes. Applications of this approach can be found useful in modeling of continuous and impulsive noise in communication systems [4], radar [5,6], biology [7], statistical electromagnetics [8], bursty internet traffic, Middleton Class A noise, intersymbol interference combined with additive noise [9,10–13], speech [14] and stochastic ratchets [15,16], etc. It is important to mention [2,64] that an SDE approach provides a unified framework for simulation of the above mentioned phenomena.

For practical purposes it is important to distinguish between baseband processes with an exponential type covariance function and narrow band processes which often have exponential decaying cosine covariation functions. A class of processes with a certain PDF and exponent ACF was originally considered by E. Wong [18,21], and A. Haddad [22,25] in the sixties and has been extended to narrow band processes in [26].

Stochastic Methods and Their Applications to Communications.
S. Primak, V. Kontorovich, V. Lyandres
© 2004 John Wiley & Sons, Ltd   ISBN: 0-470-84741-7

The following two definitions are important in the further derivations

Definition 7.1. A Markov scalar diffusion random stationary process $\xi(t)$ is called a λ process if its autocovariance function is a pure exponent, i.e.

$$C_{\xi\xi}(\tau) = \sigma^2 \exp(-\lambda|\tau|) \qquad (7.1)$$

Definition 7.2. A stationary random process $\xi(t)$ is called a (λ, ω) process if it is a component of a diffusion Markov process and has the covariance function given by a decaying exponent

$$C_{xx}(\tau) = \sigma^2 \exp[-\lambda|\tau|] \cos\omega\tau \qquad (7.2)$$

Although λ processes are usually referred to as wide sense stationary Markov (WSSM) [27] or separable class [25] processes, the terminology of definitions used is more physically revealing.

7.2 MODELING OF A SCALAR RANDOM PROCESS USING A FIRST ORDER SDE

7.2.1 General Synthesis Procedure for the First Order SDE

The goal of this section is to design a procedure which allows one to synthesize a first order SDE (Ito form)

$$\frac{\mathrm{d}x}{\mathrm{d}t} = f(x) + g(x)\xi(t) \qquad (7.3)$$

such that its stationary solution would have a given PDF $p_{x\,\mathrm{st}}(x)$ and a given correlation interval τ_{corr}.

For a moment one can assume that a non-negative function $g(x) = \sqrt{K_2(x)} \geq 0$ is already chosen and the goal is to find a proper drift $K_1(x) = f(x)$ such that

$$p_{x\,\mathrm{st}}(x) = \frac{C}{K_2(x)} \exp\left[2 \int_{-\infty}^{x} \frac{K_1(s)}{K_2(s)} \mathrm{d}s\right] \qquad (7.4)$$

Multiplying both sides of eq. (7.4) by $K_2(x) = g^2(x)$, taking the natural logarithm and differentiating with respect to x one obtains

$$2\frac{K_1(x)}{K_2(x)} = \frac{\mathrm{d}}{\mathrm{d}x} \ln[K_2(x)p_{x\,\mathrm{st}}(x)] \qquad (7.5)$$

Solving eq. (7.5) with respect to $K_1(x) = f(x)$ the desired synthesis equation becomes [2]

$$f(x) = K_1(x) = \frac{1}{2}K_2(x)\frac{\mathrm{d}}{\mathrm{d}x}\ln[K_2(x)p_{x\,\mathrm{st}}(x)] = \frac{1}{2}\left[K_2(x)\frac{\frac{\mathrm{d}}{\mathrm{d}x}p_{x,\mathrm{st}}(x)}{p_{x\,\mathrm{st}}(x)} + \frac{\mathrm{d}}{\mathrm{d}x}K_2(x)\right] \qquad (7.6)$$

Recalling that $K_2(x) = g^2(x)$, eq. (7.6) can be rewritten as

$$f(x) = K_1(x) = \frac{1}{2}\left[g^2(x)\frac{\frac{d}{dx}p_{x\,st}(x)}{p_{x\,st}(x)} + 2g(x)\frac{d}{dx}g(x)\right] \qquad (7.7)$$

Thus, there are many SDEs which have the same marginal PDF as [28]. It will be seen from future analysis that the choice of function $g(x)$ defines higher order statistics, such as the covariance function, correlation interval, etc.

Further simplification can be achieved if the diffusion is assumed to be constant, i.e. $g(x) = \sqrt{K} = \text{const}$. In this case

$$f(x) = \frac{\frac{K}{2}\frac{d}{dx}p_{x\,st}(x)}{p_{x\,st}(x)} = \frac{K}{2}\frac{d}{dx}\ln p_{x\,st}(x) \qquad (7.8)$$

This was first suggested in [30] and first published in English in [2]. The only parameter yet to be determined is the scale parameter K, defining the correlation interval.

The value of the correlation interval sometimes[1] can be estimated using an approximate formula, derived in [2,31] (see also Section 5.8):

$$T_{\text{corr}} \approx \frac{1}{a_1 + a_3 n_0} \qquad (7.9)$$

where

$$a_k = \frac{1}{k!}\frac{d^{k-1}}{dx^{k-1}}K_1(x)\bigg|_{x=0} = \frac{K}{2k!}\frac{d^k}{dx^k}\ln p_{x\,st}(x)\bigg|_{x=0} \qquad (7.10)$$

and

$$n_0 = \frac{m_4}{m_2} = \frac{\int_{-\infty}^{\infty}x^4 p_{x\,st}(x)dx}{\int_{-\infty}^{\infty}x^2 p_{x\,st}(x)dx} \qquad (7.11)$$

As a result

$$K \approx \frac{1}{T_{\text{corr}}}\frac{2}{\frac{d}{dx}\ln p_{x\,st}(x)\bigg|_{x-0} + \frac{n_0}{6}\frac{d^3}{dx^3}\ln p_{x\,st}(x)\bigg|_{x=0}} \qquad (7.12)$$

[1]This approximation works well for the case when drift is a symmetric non-linear function well approximated by a cubic polynomial, i.e. for $K_1(x) \approx b_1 x + b_3 x^3$. In other cases alternative methods must be used. If the spectrum of the corresponding Fokker–Planck equation is discrete and known, it is appropriate to choose K such that $\lambda_1 = 1/T_{\text{corr}}$.

Example: *Gaussian Markov process.* In this case the stationary marginal PDF $p_{x\,st}(x)$ is given by

$$p_{x\,st}(x) = \frac{1}{\sqrt{2\pi\sigma^2}} \exp\left[-\frac{(x-m)^2}{2\sigma^2}\right] \tag{7.13}$$

and, thus

$$\frac{d}{dx}\ln p_{x\,st}(x) = -\frac{x-m}{\sigma^2} \tag{7.14}$$

In turn, eq. (7.12) provides the value of K equal to

$$K = \frac{2\sigma^2}{\tau_{corr}} \tag{7.15}$$

Finally, plugging expressions (7.14) and (7.15) into eq. (7.8) one obtains the generating SDE for a Markov process with a Gaussian marginal PDF

$$\dot{x} = \frac{x-m}{\tau_{corr}} + \sqrt{\frac{2\sigma^2}{\tau_{corr}}}\xi(t) \tag{7.16}$$

The corresponding Fokker–Planck equation for the transitional density $\pi(x,t \mid x_0,t_0)$ has been solved in section 5.3.6, only the final result is shown below

$$\pi(x,t \mid x_0,t_0) = \frac{1}{\sqrt{2\pi\sigma^2}}\exp\left[-\frac{(s-m^2)}{2\sigma^2}\right]\sum_{k=0}^{\infty}\frac{1}{k!}H_k\left(\frac{x-m}{\sqrt{2}\sigma}\right)H_k\left(\frac{x_0-m}{\sqrt{2}\sigma}\right)\exp\left(-\frac{k|\tau|}{\tau_{corr}}\right) \tag{7.17}$$

Here $H_k(x)$ are Hermitian polynomials [32] and $\tau = t - t_0$.

It is easy to see that the transitional density (7.17) gives rise to an exponential covariance function

$$C_{xx}(\tau) = \int_{-\infty}^{\infty}\int_{-\infty}^{\infty}(x-m)(x_0-m)\pi(x,t \mid x_0,t_0)p_{x\,st}(x_0)dx\,dx_0 = \sigma^2\exp\left(-\frac{|\tau|}{\tau_{corr}}\right) \tag{7.18}$$

i.e. approximation (7.12) produces the exact result.

It is interesting to note that the Markov process defined by SDE (7.16) is also a Gaussian process, i.e. the joint PDF at any number of time instances remains Gaussian. However, choosing $K_2(x) = \alpha x$ in eq. (7.7) one can obtain a non-Gaussian Markov process with a Gaussian marginal PDF. This process would also have a non-exponential covariance function.

Example: a random process with two-sided Laplace marginal PDF. The two-sided Laplace distribution is defined by the following PDF

$$p_{x\,st}(x) = \beta \exp[-2\beta|x|] \tag{7.19}$$

where $\beta > 0$ is a parameter of the distribution. PDF (7.19) is often used to model telephone noise [33] and other phenomena. It is easy to see from eq. (7.7) that the generating SDE has the following form

$$\dot{x} = -K\beta\,\mathrm{sgn}\,x + \sqrt{K}\xi(t) \tag{7.20}$$

with the constant diffusion coefficient K yet to be determined. Since further differentiation of the drift is impossible, the approximation (7.12) cannot be used here. It is also known [38] that the corresponding Fokker–Planck equation has only one discrete eigenvalue $\lambda_0 = 0$ and the rest is a continuous spectrum. Thus it is not possible to use an approximation by a second smallest eigenvalue as well.

A fair estimate can be obtained through the statistical linearization of the non-linear function $f(x)$ in the generating SDE [36]. In other words, one needs to find a slope α such that the mean square error,

$$E\{(f(x) - \alpha x)^2\} = \int_{-\infty}^{\infty} (f(x) - \alpha x)^2 p_{x\,st}(x)\mathrm{d}x \tag{7.21}$$

is minimized. Taking the derivative of both sides with respect to α and equating it to zero one obtains the optimal value of the slope to be

$$\alpha_{\mathrm{opt}} = 2\frac{\int_{-\infty}^{\infty} xf(x)p_{x\,st}(x)\mathrm{d}x}{\int_{-\infty}^{\infty} x^2 p_{x\,st}(x)\mathrm{d}x} = -\frac{1}{\tau_{\mathrm{corr}}} \tag{7.22}$$

In the case of SDE (7.20) this estimate becomes

$$-\alpha_{\mathrm{opt}} = 2K\beta = \frac{1}{\tau_{\mathrm{corr}}} \tag{7.23}$$

Thus, in order to provide a proper correlation interval one has to choose

$$K = \frac{1}{2\beta\tau_{\mathrm{corr}}} \tag{7.24}$$

and the generating SDE becomes

$$\dot{x} = -\frac{1}{2\tau_{\mathrm{corr}}}\,\mathrm{sgn}\,x + \sqrt{\frac{1}{2\beta\tau_{\mathrm{corr}}}}\xi(t) \tag{7.25}$$

A large number of further examples can be constructed and some of them are summarized in Appendix A on the web.

7.2.2 Synthesis of an SDE with PDF Defined on a Part of the Real Axis

It can be seen from eq. (7.8) that the synthesis of SDEs implicitly assumes existence of the logarithmic derivative on the whole real axis, i.e. $p_{xst}(x) \neq 0$. However, many useful PDFs have non-zero values only on a part of the real axis. In this case the direct application of eq. (7.8) is not appropriate. Simply ignoring intervals where $p_{xst}(x) = 0$ leads to controversy. For example, a linear SDE

$$\dot{x} = -\alpha x + \xi(t) \tag{7.26}$$

describes both a Gaussian and a one-sided Gaussian process. Furthermore, numerical simulation of eq. (7.26), using for example the Euler scheme (see Appendix C on the web and [34]), would reproduce only the Gaussian PDF. Interestingly enough, the related Fokker–Planck equation uniquely defines the distributions properly since it is supplemented by a set of boundary conditions, which are different for the Gaussian (natural boundary conditions at $x = \pm\infty$) and one-sided Gaussian (reflecting boundary at $x = 0$ and natural boundary condition at $x = \infty$) [38]. Thus, certain modifications must be made to the synthesis eq. (7.8) to accommodate these boundary conditions. The following procedure has been suggested in [39].

Let $p_{xst}(x)$ be defined only on a finite interval $x \in [a, b]$ and $\zeta(x, \varepsilon)$ be a function which satisfies the following conditions:

- $\zeta(x, \varepsilon) > 0$ on the real axis;
- $\tilde{p}(x, \varepsilon) = [\varepsilon + p_{xst}(x)]\zeta(x, \varepsilon)$ is a PDF for $\varepsilon > 0$;
- the logarithmic derivative of $\tilde{p}(x, \varepsilon)$ is defined on the whole real axis and

$$\lim_{\varepsilon \to 0} \tilde{p}(x, \varepsilon) = p_{xst}(x)$$

One possible candidate for such a function is one defined in [39]

$$\zeta(x, \varepsilon) = \frac{C(\varepsilon)}{1 + \left[\dfrac{2x - a - b}{b - a}\right]^{2/\varepsilon}} \tag{7.27}$$

Here $C(\varepsilon)$ is chosen in such a way that the conditions above are satisfied. It can be easily seen that the logarithmic derivative is defined by

$$\frac{\partial}{\partial x}\tilde{p}(x, \varepsilon) = \frac{\dfrac{\partial}{\partial x}p_{xst}(x)}{p_{xst}(x) + \varepsilon} + \frac{\partial}{\partial x}\zeta(x, \varepsilon) \tag{7.28}$$

The limit of the second term in eq. (7.28) approaches a sum of delta functions as ε approaches zero:

$$\lim_{\varepsilon \to 0} \frac{\partial}{\partial x}\zeta(x, \varepsilon) = 2\delta(x - a) - 2\delta(x + b) \tag{7.29}$$

thus constituting two barriers in the function $f(x)$. Therefore, if the PDF $p_{x\,st}(x)$ is zero on some part of the real axis, one has to augment eq. (7.8) by a correction term (7.29) and, thus, the synthesis equation becomes

$$\dot{x} - \frac{K}{2}\left[\frac{\partial}{\partial x}\ln p_{x\,st}(x) + 2\delta(x - a) - 2\delta(x - b)\right] + \sqrt{K}\xi(t) \tag{7.30}$$

If either a or b are infinite one has to omit the corresponding delta function from the synthesis equation. The suggested modification of SDE would also require modification of corresponding numerical schemes which are considered in Appendix C on the web.

Example: A uniformly distributed correlated process. A uniformly distributed process with the marginal PDF

$$p_{x\,st}(x) = \frac{1}{b - a}, \quad a \le x \le b \tag{7.31}$$

plays a significant role in the modeling of the phase of narrow band signals, as discussed in section 3.12.3. In this case the synthesis eq. (7.30) becomes

$$\dot{x} = \frac{K}{2}[2\delta(x - a) - 2\delta(x - b)] + \sqrt{K}\xi(t) \tag{7.32}$$

It is important to note that direct application of eq. (7.8) would result in the SDE for the Wiener process

$$\dot{x} = \sqrt{K}\xi(t) \tag{7.33}$$

thus, once again, indicating the importance of correction terms.

The Fokker–Planck equation, corresponding to SDE (7.32) is well known to be

$$\frac{\partial}{\partial t}p(x,t) = \frac{K}{2}\frac{\partial^2}{\partial x^2}p(x,t), \quad G(a,t) = G(b,t) = 0 \tag{7.34}$$

and has the following exact solution

$$\pi(x,t \mid x_0, t_0) = \frac{1}{b - a}\sum_{k=0}^{\infty}\cos\left(\pi k\frac{x_0 - a}{b - a}\right)\cos\left(\pi k\frac{x - a}{b - a}\right)\exp\left[-\frac{\pi^2 k^2 K}{2(b - a)^2}\right] \tag{7.35}$$

which implies that the covariance function is given by an infinite series of exponents

$$C_{xx}(\tau) = \sum_{k=1}^{\infty} h_k^2 \exp\left[-\frac{\pi^2 k^2 K}{2(b - a)^2}|\tau|\right] \tag{7.36}$$

where

$$h_k = [(-1)^{k+1} + 1]\frac{(b - a)^2}{k^2\pi^2} \tag{7.37}$$

Using the exponent with the slowest decay $(k = 1)$ one can obtain the following approximation of the coefficient K

$$K \approx \frac{1}{\tau_{\text{corr}}} \frac{2(b-a)^2}{\pi^2} \tag{7.38}$$

Thus, the generating SDE has the following form

$$\dot{x} = \frac{1}{\tau_{\text{corr}}} \frac{2(b-a)^2}{\pi^2} [\delta(x-a) - \delta(x-b)] + \frac{b-a}{\pi} \sqrt{\frac{2}{\tau_{\text{corr}}}} \xi(t) \tag{7.39}$$

Example: *A Nakagami distributed correlated process.* A Nakagami distributed process is one of the most important models of the envelope of fading communication signals. Therefore its modeling is an important issue, especially for proper simulation of communication systems. Since the marginal PDF of the Nakagami process is given by

$$p_{x\text{st}}(x) = \frac{2}{\Gamma(m)} \left(\frac{m}{\Omega}\right)^2 x^{2m-1} \exp\left(-\frac{mx^2}{\Omega}\right) \tag{7.40}$$

the generating SDE becomes

$$\dot{x} = \frac{K}{2} \left[\frac{2m-1}{x} - \frac{2mx}{\Omega} + 2\delta(x)\right] + K\xi(t) \tag{7.41}$$

The solution for the transitional PDF is also known to be [31]

$$\pi(x, t \mid x_0, t_0) = \frac{2}{\Gamma(m)} \left(\frac{m}{\Omega}\right)^m x^{2m-1} \exp\left[-\frac{mx^2}{\Omega}\right]$$

$$\times \sum_{k=0}^{\infty} \sqrt{\frac{k!\Gamma(m)}{\Gamma(m+k)}} L_k^{(m-1)} \left(\frac{x^2}{2\Omega}\right) L_k^{(m-1)} \left(\frac{x_0^2}{2\Omega}\right) \exp\left[-\frac{mkK}{\Omega}|\tau|\right] \tag{7.42}$$

and, thus

$$C_{xx}(\tau) = \frac{\Omega\Gamma^2(m+0.5)}{4m\pi\Gamma(m)} \sum_{k=0}^{\infty} \frac{\Gamma^2(k-0.5)}{\Gamma(m+k)k!} \exp\left[-\frac{2mkK}{\Omega}|\tau|\right] \tag{7.43}$$

Once again estimating the value of the coefficient K through the exponent with the slowest rate of decay one obtains the generating SDE in the following form

$$\dot{x} = \frac{\Omega}{\tau_{\text{corr}}} \left[\frac{2m-1}{x} - \frac{2mx}{\Omega} + 2\delta(x)\right] + \sqrt{\frac{2\Omega}{\tau_{\text{corr}}}} \xi(t) \tag{7.44}$$

If $m = 1$ the Nakagami process coincides with the Rayleigh process whose generating SDE is then

$$\dot{x} = \frac{\Omega}{\tau_{\text{corr}}} \left[\frac{1}{x} - \frac{2x}{\Omega} + 2\delta(x) \right] + \sqrt{\frac{2\Omega}{\tau_{\text{corr}}}} \xi(t) \tag{7.45}$$

7.2.3 Synthesis of λ Processes

It has been seen from the examples above that the solution of the first order SDE with constant diffusion poses an exponentially decaying covariance function. However, approximation of the latter by a single exponent does not always provide good results. In this section a technique is developed to generate a random process with exact exponential correlation and arbitrary marginal PDF (i.e. a λ process as defined on page 322).

The desired λ process is thought to be a solution of the following SDE with further zero memory monotonous non linear transformation $y = \Gamma(x)$.

$$\begin{aligned} \dot{x} &= f(x) + g(x)\xi(t) \\ y &= F(x) \end{aligned} \tag{7.46}$$

At this stage it is assumed that the non-linear monotonous[2] transformation $y = F(x)$ and its inverse $x = F^{-1}(y)$ are given and both PDFs $p_{y\,\text{st}}(y)$ and $p_{x\,\text{st}}(x)$ are also known.[3]

Proposition 1. *The autocovariance function $C_{xx}(\tau)$ of the random process generated by system (7.46) is an exact exponent with damping factor λ, if and only if λ is the eigenvalue of the corresponding Fokker–Planck equation and the corresponding eigenfunction has the following form*

$$\Phi(y) = (F(y) + \eta)p_{y\,\text{st}}(y) \tag{7.47}$$

where

$$\eta = -\int_{y_1}^{y_2} F(y)p_{y\,\text{st}}(y)dy \tag{7.48}$$

Proof: Necessity. Without loss of generality it is possible to assume that λ corresponds to the first non-zero eigenvalue

$$\lambda = \lambda_1 \tag{7.49}$$

---

[2]This requirement is needed to preserve the Markov nature of the transformed process.
[3]This is not a very restrictive assumption since $p_{x\,\text{st}}(x)$ and $p_{y\,\text{st}}(y)$ are related by $p_{x\,\text{st}}(x) = p_{y\,\text{st}}[F(x)]F'(x)$ (see Section 2.3)

and the corresponding eigenfunction is

$$\Phi_1(y) = p_{y\,st}(y)(F(y) + \eta) \tag{7.50}$$

Using expression (7.50) and property (5.273) of eigenfunctions one obtains

$$h_m = \int_{y_1}^{y_2} F(y)\Phi_m(y)dy = \int_{y_1}^{y_2} (F(y) + \eta)\Phi_m(y)dy = \int_{y_1}^{y_2} \frac{\Phi_m(y)\Phi_1(y)}{p_{y\,st}(y)} dy = 0 \tag{7.51}$$

for any constant η and $m \neq 1$. Consequently, the expression for the ACF consists of only one term

$$C_{xx}(\tau) = h_1^2 \exp[-\lambda|\tau|] \tag{7.52}$$

no matter what the other eigenvalues are.

Sufficiency. By virtue of the fact that all the coefficients h_m^2 in expansion (5.276) are non-negative, and the assumption that the function $(F(y) + \eta)p_{y\,st}(y)$ is not an eigenfunction, one obtains that expansion (5.276) contains more than one exponential term. The contradiction proves the sufficiency.

Finally, it follows from eq. (5.273) that

$$\int_{y_1}^{y_2} \Phi_1(y)dy = \int_{y_1}^{y_2} (F(y) + \eta)p_{y\,st}(y)]dy = \int_{y_1}^{y_2} F(y)p_{y\,st}(y)dy \tag{7.53}$$

which, in turn, is equivalent to

$$\eta = -\int_{y_1}^{y_2} p_{y\,st}(y)F(y)dy \tag{7.54}$$

Proposition 2. *If the drift $K_1(y)$ and the diffusion $K_2(y)$ are chosen according to the formulae*

$$K_2(y) = -\frac{2\lambda}{p_{y\,st}(y)\dfrac{d}{dy}F(y)} \int_{y_1}^{y} (F(s) + \eta)p_{y\,st}(s)ds \tag{7.55}$$

$$2K_1(y) = K_2(y)\frac{d}{dy}\ln p_{y\,st}(y) + \frac{d}{dy}K_2(y) \tag{7.56}$$

then the process $x(t)$ generated by the corresponding SDE is a λ process.

Proof: It follows from the synthesis eq. (7.8), that the drift $K_1(y)$ can be expressed in terms of the diffusion $K_2(y)$ and the desired PDF $p_{y\,st}(y)$ as

$$2K_1(y) = K_2(y)\frac{d}{dy}\ln p_{y\,st}(y) + \frac{d}{dy}K_2(y) \tag{7.57}$$

Substituting expression (7.57) into eq. (7.7) and taking equality (7.49) into account, one can derive a differential equation for $K_2(y)$ in the following form

$$\frac{d^2}{dy^2}[K_2(y)(F(y)+\eta)p_{y\,st}(y)] - \frac{d}{dy}\left[\left(K_2(y)\frac{p'_{y\,st}(y)}{p_{y\,st}(y)} + K_2(y)\right)p_{y\,st}(y)\right] = -2\lambda(F(y)+\eta)p_{y\,st}(y)$$

(7.58)

or, equivalently

$$\frac{d}{dy}\left[K_2(y)\frac{d}{dy}F(y)p_{y\,st}(y)\right] = -2\lambda(F(y)+\eta)p_{y\,st}(y)$$

(7.59)

The latter can be easily solved with respect to $K_2(y)$ to produce

$$K_2(y) = -\frac{2\lambda}{p_{y\,st}(y)\dfrac{d}{dy}F(y)}\int_{y_1}^{y}(F(s)+\eta)p_{y\,st}(s)ds$$

(7.60)

Therefore the proposition has been proved.

The next step is to prove that the function $K_2(y)$ defined by eq. (7.54), is non-negative, so there is no contradiction with the definition of the diffusion coefficient.

Proposition 3. $K_2(y) \geq 0$.

Proof: It is sufficient to show that if

$$\Psi(y) = \int_{y_1}^{y}(F(s)+\eta)p_{y\,st}(s)ds$$

(7.61)

then

$$\Psi(y_1) = \Psi(y_2) = 0$$

(7.62)

and the equation

$$\Psi'(y) = F(y) + \eta$$

(7.63)

has a unique solution

$$y_\eta = F^{-1}(\eta)$$

(7.64)

The derivative of the function $\Psi(y)$ increases monotonously, so it is less then zero when $y < \eta$ and greater than zero when $y > \eta$. The last assertion is equal to the following inequalities

$$F(y) < F(y_1), \quad y \in [y_1, y_\eta]$$
$$F(y) < F(y_2), \quad x \in [y_\eta, y_2]$$

(7.65)

and therefore the proposition is proved.

Substituting eq. (7.55) into eq. (7.56) one obtains the expression for the drift $K_1(y)$ and consequently for the coefficients $f(y)$ and $g(y)$ in SDE (7.3). It was shown in [25], that the simple exponential form of the ACF of a λ process is due to its linear drift. This property can be proved also using eqs. (7.55) and (7.56).

Example: *Exponentially correlated processes with the Pearson class PDF.* As has been described in section 2.2, any Pearson PDF $p_{x\,st}(x)$ obeys the following differential equation

$$\frac{d}{dx}\ln p_{x\,st}(x) = \frac{a_1 x + a_0}{b_2 x^2 + b_1 x + b_0} = \frac{A(x)}{B(x)} \tag{7.66}$$

It is easy to see that if $y = x$ then the stationary PDF $p_{x\,st}(x)$ satisfies eq. (7.66) and the desired drift is given by

$$K_2(x) = \alpha(b_2 x^2 + b_1 x + b_0) \tag{7.67}$$

Equation (7.54) is satisfied as well. In fact, expression (7.54) can be rewritten in the form

$$K_2(x)\frac{d}{dx}p_{x\,st}(x) + \frac{d}{dx}K_2(x) = -\lambda(x + \eta) \tag{7.68}$$

if $F(x) = x$, using the definition of the Pearson PDF (7.66) and the fact that

$$\frac{d}{dx}K_2(x) = \alpha(2b_2 x + b_1) \tag{7.69}$$

Furthermore, eq. (7.68) becomes

$$(a_1 + 2b_2)x + a_0 + b_1 = -\frac{\lambda}{\alpha}(x + \eta) \tag{7.70}$$

which is an identity if α and η are defined as

$$\alpha = -\frac{\lambda}{a_1 + 2b_2}$$
$$\eta = \frac{a_0 + b_1}{a_1 + 2b_2} \tag{7.71}$$

It can be seen that for λ processes with a Pearson PDF, $K_2(x)$ is the polynomial of order not higher than 2. All λ processes with a non-Pearson PDF would have a non-polynomial diffusion. A number of such examples can be easily obtained

Example: *Symmetric (two-sided) Laplace distribution.* It is often required to obtain an exponentially correlated process with a specified value of the two-sided Laplace PDF parameter α. It is obvious that this distribution is not a Pearson one. Having $x = y$ in eq. (7.46), one can write out the PDF of $x(t)$ as

$$p_{x\,st}(x) = \frac{\alpha}{2}\exp(-\alpha|x|) \tag{7.72}$$

and consequently

$$K_2(x) = \frac{2\lambda}{\alpha^2}(1 + \alpha|x|) \tag{7.73}$$

$$K_1(x) = -\lambda x \tag{7.74}$$

The corresponding generating SDE is

$$dx = -\lambda x dt + \sqrt{\frac{2\lambda}{\alpha^2}(1 + \alpha|x|)} \, dW(t) \tag{7.75}$$

Example: Generalized gamma distribution. Let the stationary PDF of the λ process be the generalized gamma PDF

$$p_{x \, st}(x) = \frac{\gamma \beta^{\alpha/\gamma} |x|^{\alpha-1}}{2\Gamma(\alpha/\gamma)} \exp[-\beta|x|^\gamma] \tag{7.76}$$

which is used to model radar clutter [40–42], telephone signals [43], and ocean noise [44]. As particular cases, this PDF includes the Rayleigh PDF ($\alpha = 2$, $\gamma = 2$), the Nakagami PDF ($\alpha = 2m$, $\gamma = 2$) and the Weibull one ($\alpha = \gamma$). Using formula (7.55) one obtains the expression for the diffusion in the form

$$K_2(x) = \frac{\lambda(1-\gamma)}{\beta^{\frac{\alpha+1}{\gamma}}\gamma} \Gamma\left[\frac{\alpha+1}{\gamma}, \beta|x|^\gamma\right] |x|^{1-\alpha} \exp[(\beta|x|)^\gamma] \tag{7.77}$$

and the generating SDE is thus

$$dx = -\lambda x dt + \sqrt{\frac{\lambda(1-\gamma)}{\beta^{\frac{\alpha+1}{\gamma}}\gamma} \Gamma\left(\frac{\alpha+1}{\gamma}, \beta|x|^\gamma\right) |x|^{1-\alpha} \exp((\beta|x|)^\gamma)} \, dW(t) \tag{7.78}$$

Example: Bimodal distribution. Let the stationary PDF of the λ process have now the following form

$$p_{x \, st}(x) = C(p, q) \exp(-px^2 - qx^4) \tag{7.79}$$

where constant $C(p, q)$ is to be chosen to normalize $p_{x \, st}(x)$. It can be shown that[4]

$$C(p, q) = 2\sqrt{\frac{q}{p}} \exp\left(-\frac{p^2}{8q}\right) K_{\frac{1}{4}}\left(-\frac{p^2}{8q}\right) \tag{7.80}$$

[4]See integral 3.469-1 in [68].

Such a PDF has two modes when $p < 0$ and $q > 0$. As a particular case, it includes the Gaussian PDF when $p > 0$ and $q = 0$. Using eq. (7.55) and setting $F(x) = x$ it is easy to obtain the expression for the diffusion in the form

$$K_2(x)\frac{\lambda\sqrt{\pi}}{2\sqrt{q}}\exp\left[q\left(x^2 + \frac{p}{2q}\right)^2\right]\text{Erfc}\left[\sqrt{q}\left(x^2 + \frac{p}{2q}\right)\right] \tag{7.81}$$

where

$$\text{Erfc}(z) = \frac{2}{\sqrt{\pi}}\int_z^\infty \exp(-t^2)dt \tag{7.82}$$

is the complementary error function [32]. The corresponding generating SDE is thus

$$dx = -\lambda x dt + \sqrt{\frac{\lambda\sqrt{\pi}}{2\sqrt{q}}\exp\left[q\left(x^2 + \frac{p}{2q}\right)^2\right]\text{Erfc}\left[\sqrt{q}\left(x^2 + \frac{p}{2q}\right)\right]}\,dw(t) \tag{7.83}$$

It is impossible to obtain the exact equation for the transitional probability function, except for a number of cases considered in [47]. However, for a small transitional time τ an approximate formula can be obtained. Indeed, since the solution of the SDE is a diffusion Markov process, it can be approximated with a Gaussian random process with the same local drift and local diffusion as in [47].

$$\pi(x \mid x_0; \tau) = \frac{1}{\sqrt{2\pi K_2(x)\tau}}\exp\left[-\frac{(x - x_0 - K_1(x)\tau)^2}{2K_2(x)}\right] \tag{7.84}$$

or, taking eqs. (7.55) and (7.56) into account

$$\pi(x \mid x_0; \tau) = \frac{\sqrt{p_s(x)}}{\sqrt{4\pi\alpha\left(\int_{-\infty}^x (m_x - x)p_s(x)dx\right)\tau}}\exp\left[-\frac{(x - x_0 + \alpha(x - m_x)\tau)^2 p_s(x)}{4\alpha\left(\int_{-\infty}^x (m_x - x)p_s(x)dx\right)\tau}\right] \tag{7.85}$$

This transitional probability can be used to numerically simulate the random process using the Markov chain method, as described in Appendix C on the web.

7.2.4   Non-Diffusion Markov Models of Non-Gaussian Exponentially Correlated Processes

As has been pointed out by Haddad [25] exponential correlation is due to linear drift of the generating SDE—a fact which can be directly seen from the following equation [49]

$$\frac{d}{d\tau}C_{xx}(\tau) = \langle xK_1(x_\tau)\rangle = -\left\langle x\frac{(x_\tau - m)}{\tau_{\text{corr}}}\right\rangle = \frac{1}{\tau_{\text{corr}}}C_{xx}(\tau) \tag{7.86}$$

thus leading to the unique solution in the form of an exponent

$$C_{xx}(\tau) = \sigma^2 \exp\left(-\frac{\tau}{\tau_{corr}}\right)$$

(7.87)

Thus, in the case of linear drift, the correlation function is always exponential regardless of other kinetic coefficients $K_n(x)$. This fact allows one to extend the synthesis of exponentially correlated processes to non-diffusion processes.

In a general case, a non-linear system can be driven by a mixture of White Gaussian Noise $\xi(t)$ and the Poisson flow of delta pulses $\eta(t)$

$$\frac{dx}{dt} = f(x) + g(x)\xi(t) + \eta(t)$$

(7.88)

Statistical properties of the solution $x(t)$ can be completely described by its transitional probability density $\pi(x, t; x_0, t_0)$, which must obey the differential Chapman–Kolmogorov equation (see section 5.1.5 and [38])

$$\frac{\partial}{\partial t}\pi(x \mid x_0; \tau) = -\frac{\partial}{\partial x}[K_1(x)\pi(x; x_0, t_0)] + \frac{1}{2}\frac{\partial^2}{\partial x^2}[K_2(x)\pi(x, t; x_0, t_0)]$$
$$+ \lambda \int_{-\infty}^{\infty} [W(x \mid z, t)\pi(z, t; x_0, t_0) - W(z \mid x, t)\pi(z, t; x_0, t_0)]dz$$

(7.89)

where drift $K_1(x)$, diffusion $K_2(x)$ and probability of jumps $W(x \mid z, t)$ can be obtained from the corresponding parameters of eq. (7.88) as (Ito form of stochastic integrals)

$$\lim_{t-t_0 \to 0} \pi(z, t; x_0, t_0) = \lambda W(x \mid z, t)$$

(7.90)

$$\int_{-\infty}^{\infty} W(x \mid z, t)dx = 1$$

(7.91)

In the stationary case, if it exists, eq. (7.89) becomes an integro-differential equation of the following form

$$\lambda \int_{-\infty}^{\infty} [W(x \mid z, t)\pi(z \mid x_0; \tau) - W(z \mid x, t)\pi(x \mid x_0; \tau)]dz$$
$$= \frac{\partial}{\partial x}[K_1(x)\pi(x \mid x_0; \tau)] - \frac{1}{2}\frac{\partial^2}{\partial x^2}[K_2(x)\pi(x \mid x_0; \tau)]$$

(7.92)

Different particular cases of SDE (7.88) have been considered in a number of publications [2,3,50,51]. The main goal here is to show how different Markov processes with the same non-Gaussian probability density and exponential covariance function can be obtained.

7.2.4.1 Exponentially Correlated Markov Chain—DAR(1) and Its Continuous Equivalent

In this section the discrete time scheme which generates a non-Gaussian Markov chain with an exponential correlation function and a given arbitrary PDF is considered.

Following [52], it is assumed that the stationary distribution of the Markov chain y_n with N states

$$\gamma_1 < \gamma_2 < \cdots < \gamma_N \tag{7.93}$$

is described by the following probabilities of individual states

$$q_k = \text{Prob}\{y_n = \gamma_k\} \tag{7.94}$$

Any Markov chain can be completely described by its transitional probability matrix $T = [T_{k,l}]$ where

$$T_{k,l} = \text{Prob}\{y_m = \gamma_k \mid y_{m-1} = \gamma_l\}, \quad k, l = 1, 2, \ldots, N \tag{7.95}$$

which is the probability of the event $y_m = \gamma_k$ when $y_{m-1} = \gamma_l$ and satisfies the following conditions

$$T_{k,l} \geq 0, \qquad \sum_{k=1}^{N} T_{k,l} = 1, K = 1, 2, \ldots, N \tag{7.96}$$

The stationary probabilities q_k are obtained as the eigenvectors of the transition probability matrix T corresponding to the eigenvalue $\lambda = 1$ (see section 5.1.1)

$$\sum_{i=1}^{N} T_{k,i} q_i = q_k, \quad k = 1, 2, \ldots, N \tag{7.97}$$

The same matrix defines the power spectrum in the stationary case. However, one may consider the inverse problem, having defined only the stationary probabilities of the states. For a case where the correlation function is an exponential one, the solution has been obtained in [52]. It is possible to follow this procedure to obtain the chain approximation with an infinite number of states as a limit of a finite state Markov chain.

To achieve the first goal, the following matrix Q is defined in terms of the probabilities of the states

$$Q = \begin{bmatrix} q_1 & q_1 & \cdots & q_1 \\ q_2 & q_2 & \cdots & q_2 \\ \cdots & \cdots & \cdots & \cdots \\ q_N & q_N & \cdots & q_N \end{bmatrix} \tag{7.98}$$

$$\underbrace{}_{N \text{ times}}$$

It is easy to check that

$$Q^2 = Q \tag{7.99}$$

and [52]

$$\det[Q - \lambda I] = (1 - \lambda)(-\lambda)^{N-1} \tag{7.100}$$

In terms of the matrix Q, the transition matrix $\{T\}$ can be defined as [52]

$$T = Q + d(I - Q) \tag{7.101}$$

where $0 \leq d < 1$ will define the correlation properties (described below) and I is the identity matrix. At the same time T satisfies the condition (7.97).

For any integer m one can obtain the following expression by making use of eq. (7.101) and eq. (7.99)

$$T^m = Q + d^m(I - Q) \tag{7.102}$$

Since d is a positive number less than 1, one has

$$\lim_{m \to \infty} T^m = Q \tag{7.103}$$

which means that the Markov chain described by eq. (7.101) becomes ergodic [53] and has the stationary probability given by q_k.

The next step is to consider the correlation function R_m of the Markov chain y_m. The average value and the average of the squared value can be obtained in terms of the stationary probability q_k as

$$\langle y_m \rangle = \sum_{i=1}^{N} \gamma_k q_k \tag{7.104}$$

$$\langle y_m^2 \rangle = \sum_{i=1}^{N} \gamma_k^2 q_k \tag{7.105}$$

To calculate the covariance function, one has to consider the two-dimensional probability which may be obtained from eq. (7.102) as

$$Q^{(m)}(k, l) = \text{Prob}\{y_m = \gamma_k, y_0 = \gamma_l\} = \{q_k + d^m(\delta_{k,l} - q_k)\}q_l \tag{7.106}$$

where $m > 0$ and $Q^{(m)}(k, l)$ stands for the m-step transitional probability. It is easy to find that eq. (7.106) fits the consistency relation

$$\sum_{k=1}^{N} Q^{(m)}(k, l) = \sum_{k=1}^{N} Q^{(m)}(l, k) = q_l \tag{7.107}$$

The covariance function C_m is an even function of m defined as

$$C_m = C_{-m} = \langle y_m y_0 \rangle - \langle y_m \rangle^2 \tag{7.108}$$

Substitution of eqs. (7.104)–(7.107) into eq. (7.108) produces

$$
\begin{aligned}
C_m = C_{-m} = \langle y_m y_0 \rangle - \langle y_m \rangle^2 &= \sum_{k,l=1}^{N} \gamma_k \gamma_l \mathrm{Prob}\{y_m = \gamma_k, y_0 = \gamma_l\} - \langle y_m \rangle^2 \\
&= \sum_{k,l=1}^{N} \gamma_k \gamma_l \{q_k + d^m (\delta_{k,l} - q_k)\} q_l - \langle y_m \rangle^2 \\
&= \sum_{k,l=1}^{N} \gamma_k \gamma_l q_k q_l + d^m \sum_{k,l=1}^{N} \gamma_k \gamma_l (\delta_{k,l} - q_k) q_l - \langle y_m \rangle^2 \\
&= \langle y_m \rangle^2 + d^{|m|} (\langle y_m^2 \rangle - \langle y_m \rangle^2) - \langle y_m \rangle^2 \\
&= d^{|m|} (\langle y_m^2 \rangle - \langle y_m \rangle^2)
\end{aligned}
\tag{7.109}
$$

which is an exponential function with the correlation length N_{corr} defined as

$$
N_{\mathrm{corr}} = (-1)/\ln d
\tag{7.110}
$$

Formula (7.102) can be extended to the finite state continuous time Markov chain $y_m(t)$ as in [52] to produce

$$
T(t) = Q + \exp(-\mu t)(I - Q)
\tag{7.111}
$$

The expression for the covariance function can be given in this case as

$$
C_{yy}(\tau) = \exp(-\mu \tau)(\langle y_m^2 \rangle - \langle y_m \rangle^2)
\tag{7.112}
$$

Before turning to the continuous time infinite state Markov chain (a Markov process with a continuum of states) let us point out an important property of the exponentially correlated Markov chain. It follows from the definition of the matrices T and $T(t)$ as in eqs. (7.102) and (7.111) that the transition probability density *does not depend on the current state*. If $N \to \infty$, then the Markov continuous time chain tends to a Markov process, which can be a non-diffusion one. In this case eq. (7.111) can be written as

$$
\pi(x \mid x_0; \tau) = \exp(-\mu \tau)\delta(x - x_0) + (1 - \exp(-\mu \tau))p_s(x)
\tag{7.113}
$$

Here $p_s(x)$ is the stationary distribution of the limit process.

It is important to validate that the latter expression indeed defines a proper PDF of a Markov process. Positivity is obvious, since both summands are positive numbers, thus

$$
\pi(x \mid x_0; \tau) \geq 0
\tag{7.114}
$$

When τ approaches infinity, the transitional PDF must correspond to the stationary PDF of the process since the values of the process far away from the observation points are independent on this observation:

$$
\lim_{\tau \to \infty} \pi(x \mid x_0; \tau) = \lim_{\tau \to \infty} [\exp(-\lambda \tau)\delta(x - x_0) + (1 - \exp(-\lambda \tau))p_s(x)] = p_s(x)
\tag{7.115}
$$

At the same time the limit of the PDF when τ approaches zero must be the delta function $\delta(x - x_0)$, since the process cannot assume two different values at the same moment. Taking the limit of eq. (7.113) one obtains that this condition is indeed satisfied:

$$\lim_{\tau \to 0} \pi(x \mid x_0; \tau) = \lim_{\tau \to 0} [\exp(-\lambda\tau)\delta(x - x_0) + (1 - \exp(-\lambda\tau))p_s(x)] = \delta(x - x_0) \quad (7.116)$$

Finally, in order to represent a Markov process, PDF $\pi(x \mid x_0; \tau)$ must obey the Smoluchowski equation [48]

$$\pi(x \mid x_0; \tau) = \int_{-\infty}^{\infty} \pi(x \mid x_1; \tau_2)\pi(x_1 \mid x_0; \tau_1)dx_1 \quad (7.117)$$

with $\tau = \tau_1 + \tau_2$. It is easy to check that this is the case for PDF given by eq. (7.113). Indeed

$$\int_{-\infty}^{\infty} \pi(x \mid x_1; \tau_2)\pi(x_1 \mid x_0; \tau_1)dx_1 = \int_{-\infty}^{\infty} [\exp(-\lambda\tau_2)\delta(x - x_1) + (1 - \exp(-\lambda\tau_2))p_s(x)]$$

$$\times [\exp(-\lambda\tau_1)\delta(x_1 - x_0) + (1 - \exp(-\lambda\tau_1))p_s(x_1)]dx_1$$

$$= \exp(-\lambda(\tau_1 + \tau_2))\delta(x - x_0)$$

$$+ (1 - \exp(-\lambda(\tau_1 + \tau_2)))p_s(x)$$

$$= \pi(x \mid x_0; \tau_1 + \tau_2) \quad (7.118)$$

Thus, in fact, the PDF (7.113) defines a Markov process. The covariance function of this process is indeed exponential, since

$$C_{xx}(\tau) = \int_{-\infty}^{\infty}\int_{-\infty}^{\infty} xx_1\pi(x \mid x_1; \tau)p_s(x_1)dxdx_1$$

$$= \int_{-\infty}^{\infty}\int_{-\infty}^{\infty} xx_1[\exp(-\lambda\tau)\delta(x_1 - x)]p_s(x_1)dxdx_1$$

$$+ \int_{-\infty}^{\infty}\int_{-\infty}^{\infty} [(1 - \exp(-\lambda\tau))p_s(x)]p_s(x_1)dxdx_1$$

$$= \exp(-\lambda\tau)[\sigma_x^2 + m_x^2] + (1 - \exp(-\lambda\tau))m_x^2$$

$$= \sigma_x^2 \exp(-\lambda\tau) + m_x^2 \quad (7.119)$$

The next step is to understand if the Markov process defined by the PDF (7.113) represents a diffusion Markov process or a process with jumps. In order to do this, one must calculate the following limit, which describes the non-diffusion part of any general Markov process (see Section 5.1.5 and eq. 3.4.1 in [48])

$$W(x \mid x_0; \tau) = \lim_{\substack{\tau \to 0 \\ |x - x_0| > \varepsilon}} \frac{\pi(x \mid x_0; \tau)}{\tau} = \lim_{\substack{\tau \to 0 \\ |x - x_0| > \varepsilon}} \frac{(1 - \exp(-\lambda\tau))p_s(x)}{\tau} = \lambda p_s(x)$$

$$(7.120)$$

The last equation implies that the Markov process defined by the PDF (7.115) is a process with jumps, since $W(x \mid x_0; \tau) \neq 0$. However, it is important to note that the probability of jumps $W(x \mid x_0; \tau)$ does not depend on the current state x_0. This property is inherited from the fact that the pre-limit Markov chain $\{y(t)\}$ has the same property.

It is interesting to find out which SDE, if any, generates such a process. In order to accomplish this, one has to calculate the drift $K_1(x)$ and the diffusion $K_2(x)$ coefficients, which are defined as [48]

$$
\begin{aligned}
K_1(x) + O(\varepsilon) &= \lim_{\tau \to 0} \frac{1}{\tau} \int_{|x-z|<\varepsilon} (z-x)\pi(z \mid x; \tau)dz \\
&= \lim_{\tau \to 0} \frac{1}{\tau} \int_{|x-z|<\varepsilon} (z-x)[\exp(-\lambda\tau)\delta(x-z) + (1 - \exp(-\lambda\tau))p_s(z)]dz \\
&= \lim_{\tau \to 0} \frac{\exp(-\lambda\tau)}{\tau} \int_{|x-z|<\varepsilon} (z-x)\delta(x-z)dz \\
&\quad + \lim_{\tau \to 0} \frac{(1-\exp(-\lambda\tau))}{\tau} \int_{|x-z|<\varepsilon} (z-x)p_s(z)dz = 0 + O(\varepsilon)
\end{aligned}
\tag{7.121}
$$

and

$$
\begin{aligned}
K_2(x) + O(\varepsilon) &= \lim_{\tau \to 0} \frac{1}{\tau} \int_{|x-z|<\varepsilon} (z-x)^2\pi(z \mid x; \tau)dz \\
&= \lim_{\tau \to 0} \frac{1}{\tau} \int_{|x-z|<\varepsilon} (z-x)^2[\exp(-\lambda\tau)\delta(x-z) + (1 - \exp(-\lambda\tau))p_s(z)]dz \\
&= \lim_{\tau \to 0} \frac{\exp(-\lambda\tau)}{\tau} \int_{|x-z|<\varepsilon} (z-x)^2\delta(x-z)dz \\
&\quad + \lim_{\tau \to 0} \frac{(1-\exp(-\lambda\tau))}{\tau} \int_{|x-z|<\varepsilon} (z-x)^2 p_s(z)dz = 0 + O(\varepsilon)
\end{aligned}
\tag{7.122}
$$

The last two equations show that the SDE generating the continuous Markov process with transitional PDF given by eq. (7.113) is

$$
\frac{dx}{dt} = \eta(t)
\tag{7.123}
$$

where $\eta(t)$ is a stream of delta pulses

$$
\eta(t) = \lambda \sum_{t_k} A_k \delta(t - k\Delta t)
\tag{7.124}
$$

with amplitudes A_k distributed according to the stationary PDF $p_s(x)$ and the time between two sequential arrivals is Δt. In order to obtain the continuous time chain, Δt must approach zero.

7.2.4.2   A Mixed Process With Exponential Correlation

Another possible class of non-Gaussian Markov processes with exponential correlation has been considered in [41] and is discussed in this section. In this case, the generating SDE is chosen in the form of a linear system excited by a train of delta functions, similar to eq. (7.124)

$$\frac{dx}{dt} = -\alpha x + \eta(t) \tag{7.125}$$

However, it was found in section 10.4 [51] that a relatively small class of non-Gaussian processes can be represented in this form. The goal here is to identify the class of represented processes. Without loss of generality one may assume that the process has zero mean since a constant value can be easily added to the zero mean random process to account for it. In this case the desired amplitude distribution and its intensity $\eta(t)$ can be adjusted to obtain the desired properties. In the case of eq. (7.84), the differential Chapman–Kolmogorov equation (7.92) becomes

$$\lambda \int_{-\infty}^{\infty} [W(x \mid z, t)\pi(z \mid x_0; \tau) - W(z \mid x, t)\pi(x \mid x_0; \tau)]dz = -\alpha \frac{\partial}{\partial x}[x\pi(x|x_0; \tau)] \tag{7.126}$$

since the diffusion coefficient is zero, and the drift is a linear term. Multiplying both parts by $p_s(x_0)$ and integrating over x_0 one can obtain that

$$\lambda \int_{-\infty}^{\infty} W(x \mid z, t)p_s(z)dz = -\alpha \frac{\partial}{\partial x}([xp_s(x)]) + \lambda p_s(x) = (\lambda - \alpha)p_s(x) - \alpha x \frac{\partial}{\partial x}p_s(x) \tag{7.127}$$

since

$$\int_{-\infty}^{\infty} \pi(x \mid x_0; \tau)p_s(x_0)dx_0 = p_s(x) \tag{7.128}$$

and, according to eq. (7.90)

$$\int_{-\infty}^{\infty} W(z \mid x, t)\pi(x \mid x_0; \tau)dz = \pi(x \mid x_0; \tau) \tag{7.129}$$

Since both $W(x \mid z)$ and $p_s(x)$ must be non-negative functions, the following condition must be satisfied

$$\int_{-\infty}^{\infty} W(x \mid z, t)p_s(z)dz = \left(1 - \frac{\alpha}{\lambda}\right)p_s(x) - \frac{\alpha}{\lambda}x\frac{\partial}{\partial x}p_s(x) \geq 0 \tag{7.130}$$

This is the weakest test of the kind of distribution that can be implemented using this technique. Also it gives a lower bound on the intensity of the jumps λ needed for a stationary distribution to exist (recall that the constant α is defined by the required correlation interval $\tau_{corr} = 1/\alpha$ and cannot be chosen arbitrarily [2,41]):

$$x \frac{\frac{\partial}{\partial x} p_s(x)}{p_s(x)} \leq \frac{\lambda}{\alpha} - 1 \qquad (7.131)$$

It is assumed in the following that the condition (7.131) is indeed satisfied.

Since one has to choose an unknown function $W(x \mid z)$ of two variables having just one eq. (7.130), it is possible that this choice is not unique. Indeed, below, two possibilities are considered. Following the observation in Section 7.2.4.1, one can assume that the jump probability does not depend on the current state, i.e. $W(x \mid z) = W(x)$. As an alternative, a more commonly used kernel, depending on the difference between the current and future states, can be chosen, i.e. $W(x \mid z) = W(x - z)$. Both cases are investigated below.

A. *PDF* $W(x \mid z)$ *does not depend on the current state.* In this case

$$W(x \mid z) = W(x) \qquad (7.132)$$

and the eq. (7.127) is reduced to

$$\lambda \int_{-\infty}^{\infty} W(x \mid z) p_s(z) dz = \lambda W(x) \int_{-\infty}^{\infty} p_s(z) dz = \lambda W(x) = (\lambda - \alpha) p_s(x) - \alpha x \frac{\partial}{\partial x} p_s(x) \qquad (7.133)$$

or, equivalently

$$W(x) = \left(1 - \frac{\alpha}{\lambda}\right) p_s(x) - \frac{\alpha}{\lambda} x \frac{\partial}{\partial x} p_s(x) \qquad (7.134)$$

Since it was assumed that eq. (7.131) is satisfied, the function $W(x)$ is a positive function. The only additional condition would be its normalization to 1, i.e.

$$\int_{-\infty}^{\infty} W(x) dx = \int_{-\infty}^{\infty} \left[\left(1 - \frac{\alpha}{\lambda}\right) p_s(x) - \frac{\alpha}{\lambda} x \frac{\partial}{\partial x} p_s(x)\right] dx = \left(1 - \frac{\alpha}{\lambda}\right) - \frac{\alpha}{\lambda} \int_{-\infty}^{\infty} x \frac{\partial}{\partial x} p_s(x) dx = 1 \qquad (7.135)$$

or, equivalently

$$\int_{-\infty}^{\infty} x \frac{\partial}{\partial x} p_s(x) dx = -1 \qquad (7.136)$$

Using integration by parts, the last integral can be transformed to

$$\int_{-\infty}^{\infty} x \frac{\partial}{\partial x} p_s(x) dx = \int_{-\infty}^{\infty} x dp_s(x) = x p_s(x)|_a^b - \int_{-\infty}^{\infty} p_s(x) dx = x p_s(x)|_a^b - 1 \qquad (7.137)$$

Here a and $b > a$ are the boundaries of the interval $[a, b]$, where $p_s(x)$ differs from zero. Both of them can be infinite. Comparing eq. (7.137) to eq. (7.136) one can conclude that the function $W(x)$ represents a proper PDF if

$$xp_s(x)|_a^b = 0 \qquad (7.138)$$

This condition is satisfied automatically if both boundaries are infinite, since $p_s(x)$ is integrable. If at least one of the boundaries is finite, then eq. (7.138) constitutes yet another restriction on the class of PDF which can be achieved.

B. *Probability depending on the difference between the current and future states.* This case was originally considered in [51]. Equation (7.127) becomes an integral equation of convolution type

$$\int_{-\infty}^{\infty} W(x - z)p_s(z)dz = \left(1 - \frac{\alpha}{\lambda}\right)p_s(x) - \frac{\alpha}{\lambda}x\frac{\partial}{\partial x}p_s(x) \qquad (7.139)$$

and can be solved using the Fourier transform technique. Indeed, let $\Theta_w(j\omega)$ and $\Theta_x(j\omega)$ be characteristic functions, corresponding to the PDF $W(x)$ and $p_s(x)$, respectively

$$\Theta_w(j\omega) = \int_{-\infty}^{\infty} W(x)\exp(j\omega x)dx \qquad (7.140)$$

$$\Theta_x(j\omega) = \int_{-\infty}^{\infty} p_s(x)\exp(j\omega x)dx \qquad (7.141)$$

Taking the Fourier transform of eq. (7.139) and recalling the convolution theorem one could obtain

$$\Theta_w(j\omega)\Theta_x(j\omega) = \Theta_x(j\omega) + \frac{\alpha}{\lambda}\omega\frac{d}{d\omega}\Theta_x(j\omega) \qquad (7.142)$$

and thus

$$\Theta_w(j\omega) = 1 + \frac{\alpha}{\lambda}\omega\frac{\frac{d}{d\omega}\Theta_x(j\omega)}{\Theta_x(j\omega)} = 1 + \frac{\alpha}{\lambda}\omega\frac{d}{d\omega}\Psi_x(j\omega) \qquad (7.143)$$

where $\Psi_x(j\omega)$ is the log-characteristic (cumulant generating) function of the process $x(t)$.

$$\Psi_x(j\omega) = 1 + \sum_{k=1}^{\infty}\frac{\kappa_k}{k!}(j\omega)^k \qquad (7.144)$$

Substitution of eq. (7.144) into eq. (7.143) produces

$$\Theta_w(j\omega) = 1 + \frac{\alpha}{\lambda}\omega\frac{d}{d\omega}\left[1 + \sum_{k=1}^{\infty}\frac{\kappa_k}{k!}(j\omega)^k\right] = 1 + \frac{\alpha}{\lambda}\sum_{k=1}^{\infty}\frac{\kappa_k}{k!k}(j\omega)^k \qquad (7.145)$$

thus implying that

$$m_{kw} = \frac{\alpha}{\lambda k} \kappa_k \qquad (7.146)$$

The last equation allows one to synthesize an SDE if only a few cumulants of the desired output process are known. Since the characteristic function of a proper distribution must obey certain conditions [53], not all PDFs of the solution $x(t)$ can be achieved.

Example: exponentially correlated process with a Gaussian marginal PDF. In this case the stationary PDF is defined by

$$p_s(x) = \frac{1}{\sqrt{2\pi\sigma^2}} \exp\left(-\frac{x^2}{2\sigma^2}\right) \qquad (7.147)$$

In order to synthesize a finite state Markov chain, proper quantization levels γ_i, $1 \le i \le N+1$ are to be chosen, for example, using the entropy method [54]. The corresponding probabilities q_i of the states can be calculated using the error function [32]. Taking an infinite number of states, a continuous time model can be obtained as given by eq. (7.113). It is important to note that, despite the fact that the marginal PDF of this Markov process is a Gaussian one and the correlation function is exponential, the process itself is not Gaussian, i.e. the second order joint probability density is not a Gaussian one.

In contrast, the diffusion process with the same distribution and correlation function is Gaussian [55]. The corresponding SDE of the form (7.55) is a linear one:

$$\dot{x} = -\alpha x + 2\sigma^2 \xi(t) \qquad (7.148)$$

with transitional PDF equal to

$$\pi(x \mid x_0; \tau) = \frac{1}{\sqrt{2\pi\sigma^2}} \frac{1}{\sqrt{1 - \exp(-\alpha\tau)}} \exp\left(-\frac{x^2 - \exp(-\alpha\tau)xx_0}{2\sigma^2(1 - \exp(-2\alpha\tau))}\right) \qquad (7.149)$$

Condition (7.131) becomes

$$-\frac{x^2}{\sigma^2} \le \frac{\lambda}{\alpha} - 1 \qquad (7.150)$$

and is satisfied as long as $\lambda \ge \alpha$. The corresponding density (7.134) of jumps then becomes

$$W(x) = \left(1 - \frac{\alpha}{\lambda}\right)p_s(x) - \frac{\alpha}{\lambda}x\frac{\partial}{\partial x}p_s(x) = \frac{1 - \frac{\alpha}{\lambda} + \frac{\alpha}{\lambda}\frac{x^2}{\sigma^2}}{\sqrt{2\pi\sigma^2}} \exp\left(-\frac{x^2}{2\sigma^2}\right), \quad \lambda > \alpha \qquad (7.151)$$

It was shown in [41] that it is impossible to obtain a Gaussian process using an SDE of the form

$$\dot{x}(t) = \alpha x + \eta(t) \tag{7.152}$$

a Poisson flow of delta pulses. Indeed, in order to achieve a Gaussian process by a linear transformation, one requires the input to be Gaussian as well. However, the Poisson process is not a Gaussian one, unless its intensity is infinite.

Example: Pearson class of PDFs. In this case the stationary PDF $p_{st}(x)$ obeys the following equation

$$\frac{\dfrac{d}{dx} p_{st}(x)}{p_{st}(x)} = \frac{a_1 x + a_0}{b_2 x^2 + b_1 x + b_0} \tag{7.153}$$

It was shown in [32] that the following SDE (Ito form) allows one to generate all possible Pearson processes with exponential correlation function of the type

$$\frac{dx}{dt} = -\lambda \left(x - \frac{a_0 + b_1}{a_1 + 2b_2} \right) + \sqrt{\lambda \frac{(b_2 x^2 + b_1 x + b_0)}{a_1 + 2b_2}} \xi(t) \tag{7.154}$$

In all cases the transition probability density can be expressed in terms of classical orthogonal polynomials [32] or in closed integral form as found in [47]. A few examples are given in Table 7.1.

Table 7.1 Parameters of the continuous Markov process with exponential correlation[a]

$p_{st}(x)$	$\dfrac{d}{dx} \ln p_{st}(x)$	$\pi(x, x_0; \tau)$
$\exp(-x)$	-1	$\dfrac{1}{2\sqrt{\pi\tau}} \exp\left[-\dfrac{x-x_0}{2} - \dfrac{\tau}{4} \right] \times \left\{ \exp\left[-\dfrac{(x-x_0)^2}{4\tau} \right] \right.$ $\left. + \exp\left[-\dfrac{(x+x_0)^2}{4\tau} \right] \right\} + \dfrac{1}{\sqrt{\pi}} e^{-x} \int_{-\frac{x+x_0}{2\sqrt{\tau}}}^{\infty} (x + x_0 - \tau) e^{-z^2} dz$
$\dfrac{x^\beta \exp(-x)}{\Gamma(\beta+1)}$	$\dfrac{\beta - x}{x}$	$\dfrac{1}{1 - e^{-\tau}} \left(\dfrac{x}{x_0 \exp(-\tau)} \right)^{\beta/2}$ $\times \exp\left[-\dfrac{x + x_0 e^{-\tau}}{1 - e^{-\tau}} \right] I_\beta \left(\dfrac{2e^{-\tau/2}\sqrt{xx_0}}{1 - e^{-\tau}} \right)$
$\dfrac{\Gamma(\beta+\gamma+2)}{\Gamma(\beta+1)\Gamma(\gamma+1)}$ $\times \dfrac{(1+x)^\beta (1-x)^\gamma}{2^{\alpha+\gamma+1}}$	$\dfrac{(\beta-\gamma) - (\beta+\gamma)x}{1-x^2}$	$\dfrac{(1+x)^\beta (1-x)^\gamma}{2^{\alpha+\gamma+1}} \sum_{n=0}^{\infty} e^{-n(n+\gamma+\beta+1)}$ $\times A_n P_n^{\beta,\gamma}(x_0) P_n^{\beta,\gamma}(x)$ $A_n = \dfrac{(2n + \alpha + \gamma + 1)\Gamma(n + \alpha + \gamma + 1)}{\Gamma(n + \alpha + 1)\Gamma(n + \gamma + 1)n!}$

[a]Here $I_\nu(x)$ is the modified Bessel function of the first kind and $P_n^{\beta,\gamma}(x)$ are the Jacobi polynomials [32].

Table 7.2 Parameters of Markov processes with exponential correlation and jumps

$p_{st}(x)$	$\dfrac{d}{dx}\ln p_{st}(x)$	$W(x)$
$\exp(-x)$	-1	$\left[\left(1-\dfrac{\alpha}{\lambda}\right)+\dfrac{\alpha}{\lambda}x\right]\exp(-x)$
$\dfrac{x^\beta\exp(-x)}{\Gamma(\beta+1)}$	$\dfrac{\beta-x}{x}$	$\left[\left(1-\dfrac{\alpha}{\lambda}\right)-\dfrac{\alpha}{\lambda}(\beta-x)\right]\dfrac{x^\beta\exp(-x)}{\Gamma(\beta+1)}$
$\dfrac{\Gamma(\beta+\gamma+2)}{\Gamma(\beta+1)\Gamma(\lambda+1)}$ $\times\dfrac{(1+x)^\beta(1-x)^\gamma}{2^{\alpha+\gamma+1}}$	$\dfrac{(\beta-\gamma)-(\beta+\gamma)x}{1-x^2}$	$\left[\left(1-\dfrac{\alpha}{\lambda}\right)-\dfrac{\alpha}{\lambda}\dfrac{(\beta-\gamma)x-(\beta+\gamma)x^2}{1-x^2}\right]$ $\times\dfrac{\Gamma(\beta+\gamma+2)}{\Gamma(\beta+1)\Gamma(\gamma+1)}\dfrac{(1+x)^\beta(1-x)^\gamma}{2^{\alpha+\gamma+1}}$

In order to obtain a convenient expression for the Pearson class of distributions, equation (7.134) can be rewritten as

$$W(x) = \left(1-\frac{\alpha}{\lambda}\right)p_{st}(x) - \frac{\alpha}{\lambda}x\frac{\partial}{\partial x}p_{st}(x) = \left[\left(1-\frac{\alpha}{\lambda}\right) - \frac{\alpha}{\lambda}x\frac{\frac{\partial}{\partial x}p_{st}(x)}{p_{st}(x)}\right]p_{st}(x)$$

$$= \left[\left(1-\frac{\alpha}{\lambda}\right) - \frac{\alpha}{\lambda}\frac{a_1x^2+a_0x}{b_2x^2+b_1x+b_0}\right]p_{st}(x)$$

(7.155)

Table 7.2 contains some of the examples given in Table 7.1.

Example: *Khinchin PDF.* As an example of a non-Pearson distribution one can consider the so-called Khinchin probability density

$$p_{st}(x) = \frac{1-\cos x}{\pi x^2}$$

(7.156)

This distribution is interesting since none of its moments exist. Indeed, the characteristic function, corresponding to this distribution is

$$\Theta(\omega) = \begin{cases} 0 & |\omega| \geq 1 \\ 1 - |\omega| & |\omega| < 1 \end{cases}$$

(7.157)

and it is not differentiable at $\omega = 0$. The last statement is equivalent to the fact that the PDF (7.156) does not have moments which converge. Nevertheless, it was shown in [41] that this distribution can be obtained if

$$W(x\,|\,y) = \frac{1}{\pi\exp(-\gamma)}\left[\alpha^{-\gamma}A^{\gamma-1}C_\gamma\left(\frac{x-y}{\alpha}\right) + \frac{\nu\exp(\gamma) - \nu\cos\left(\frac{x-y}{\alpha}\right) + (x-y)\sin\left(\frac{x-y}{\alpha}\right)}{\nu^2 + (x-y)^2}\right]$$

(7.158)

where

$$C_\gamma(w) = \int_w^\infty \frac{\cos(z)}{z^{-\gamma}} dz \qquad (7.159)$$

and

$$\gamma = \frac{\alpha}{\lambda} \qquad (7.160)$$

7.3   MODELING OF A ONE-DIMENSIONAL RANDOM PROCESS ON THE BASIS OF A VECTOR SDE

7.3.1   Preliminary Comments

It was shown in Section 7.2 that it is possible to synthesize a one-dimensional SDE with the solution being a wide band random process with a given PDF and correlation interval. At the same time many real applications require modeling of random processes with a more complicated correlation function (spectral density). Of a special importance is the problem of modeling narrow band random processes frequently encountered in radio communications and other applications [56–58].

To remain in the SDE framework one should consider the SDE of an order higher than one. In this case a one-dimensional random process can be considered as a component of some vector Markov process.[5] The first part of this section is devoted to the generation of narrow band random processes (processes with an oscillating correlation function) using two simple forms of an SDE of second order, the stochastic analog of the Duffing and Van der Pol equations [31]. The following material shows how the concept of an exponentially correlated process, combined with the concept of compound processes [40,42], can be used to generate one-dimensional processes with a given PDF and rational spectrum.

7.3.2   Synthesis Procedure of a (λ, ω) Process

Since the ACF of the first order SDE solution is always monotonous, a random process with the ACF defined by eq. (7.2) can be generated only by a system of at least two coupling equations. Such a narrow band stationary process, or (λ, ω) process, may be written as

$$\xi(t) = A(t) \cos \varphi(t) \sin(\omega t) + A(t) \sin \varphi(t) \cos(\omega t) \qquad (7.161)$$

with uncorrelated baseband components (considered at the same time instant)

$$\eta_\parallel(t) = A(t) \cos \varphi(t)$$
$$\eta_\perp(t) = A(t) \sin \varphi(t) \qquad (7.162)$$
$$\langle \eta_\parallel(t)\eta_\perp(t) \rangle \equiv 0$$

---

[5]It is important to point out here that in this case the one-dimensional process is not a Markov one.

and

$$\xi_I(t) = \eta_\parallel(t)\sin\omega t$$
$$\xi_R(t) = \eta_\perp(t)\cos\omega t \tag{7.163}$$

The marginal distribution $p_\xi(x)$ defines, according to the Blanc-Lapierre transform[6] [59], the marginal distribution $p_A(A)$ of the envelope

$$A(t) = \sqrt{\eta_\parallel^2(t) + \eta_\perp^2(t)} \tag{7.164}$$

$$p_A(A) = A\int_{-\infty}^{\infty}\int_{-\infty}^{\infty} p_{\xi\,st}(x)\omega J_0(\omega A)\exp(j\omega x)dx\,d\omega \tag{7.165}$$

where $J_0(^\circ)$ stands for a Bessel function of order zero of the first kind [32]. The phase

$$\varphi(t) = \text{atan}\left[\frac{\xi_I(t)}{\xi_R(t)}\right] \tag{7.166}$$

of $\xi(t)$ does not depend on A and is uniformly distributed on $[0, 2\pi]$. Thus

$$p_{A\varphi}(A, \varphi) = \frac{1}{2\pi}p_A(A) \tag{7.167}$$

or, returning back to y and z^7 variables,

$$p_{\xi_R\xi_I}(y, z) = \frac{1}{2\pi}\frac{p_A\left(\sqrt{z^2 + y^2}\right)}{\sqrt{z^2 + y^2}} \tag{7.168}$$

Detailed derivation and discussion of the representation (7.161)–(7.168) is given in Section 4.7. Here it is just emphasized that the components $\xi_R(t)$ and $\xi_I(t)$ have identical distributions and autocovariance functions.

Equation (7.168) allows one to obtain the bivariate PDF $p_{\xi_R\xi_I}(y, z)$ corresponding to the given $p_\xi(x)$. If one denotes

$$\sigma^2 = \int_{-\infty}^{\infty} x^2 p_{\xi_I}(x)dx \tag{7.169}$$

then the variance matrix of the vector process $\zeta = (\xi_R, \xi_I)^T$ has the following form

$$C_{\zeta\zeta} = \begin{bmatrix} \sigma^2 & 0 \\ 0 & \sigma^2 \end{bmatrix} = \sigma^2 I \tag{7.170}$$

The next step is to consider a system of two SDEs

$$d\xi_R = -\lambda\xi_R dt + \omega\xi_I dt + g_{11}(\xi_R, \xi_I)dW_1(t) + g_{12}(\xi_R, \xi_I)dW_2(t)$$
$$d\xi_I = -\omega\xi_R dt - \lambda\xi_I dt + g_{21}(\xi_R, \xi_I)dW_1(t) + g_{22}(\xi_R, \xi_I)dW_2(t) \tag{7.171}$$

---

[6]See also Section 4.7.
[7]It is known that $\eta_\parallel(t)$, $\eta_\perp(t)$, $\xi_R(t)$ and $\xi_I(t)$ have the same PDF (see Section 4.7 for details).

The stationary PDF should satisfy the following Fokker–Planck equation

$$-\frac{\partial}{\partial y}[K_{11}(y,z)p_{\xi_R\xi_I}(y,z)] - \frac{\partial}{\partial z}[K_{12}(y,z)p_{\xi_R\xi_I}(y,z)] + \frac{1}{2}\frac{\partial^2}{\partial y^2}[K_{211}(y,z)p_{\xi_R\xi_I}(y,z)]$$

$$+\frac{1}{2}\frac{\partial^2}{\partial z^2}[K_{222}(y,z)p_{\xi_R\xi_I}(y,z)] + \frac{1}{2}\frac{\partial^2}{\partial z\partial y}[K_{212}(y,z)p_{\xi_R\xi_I}(y,z)] = 0 \tag{7.172}$$

where the drifts $K_{11}(y,z)$ and $K_{12}(y,z)$ are given by

$$K_{11}(y,z) = -\lambda\xi_R + \omega\xi_I \text{ and } K_{12}(y,z) = -\omega\xi_R - \lambda\xi_I \tag{7.173}$$

while the diffusion coefficients can be found according to eqs. (7.176) and (7.178) below.
Since eqs. (7.171) form a system with linear drifts, the solution has the correlation matrix $C(\tau)$ which obeys the following differential equation

$$C'(\tau) = \begin{bmatrix} -\lambda & \omega \\ -\omega & -\lambda \end{bmatrix} C(\tau) \tag{7.174}$$

Taking into account the initial conditions (7.170) one obtains the solution of eq. (7.174) as

$$C(\tau) = \begin{bmatrix} \exp[-\lambda|\tau|]\cos\omega\tau & \exp[-\lambda|\tau|]\sin\omega\tau \\ -\exp[-\lambda|\tau|]\sin\omega\tau & \exp[-\lambda|\tau|]\cos\omega\tau \end{bmatrix} C(0) \tag{7.175}$$

It can be checked by direct substitution that if the drift is chosen according to (7.173) and the diffusions as

$$K_{211}(y,z) = \frac{\lambda}{p_{\xi_R\xi_I}(y,z)} \int_{-\infty}^{y} sp_{\xi_R\xi_I}(s,z)ds \tag{7.176}$$

$$K_{222}(y,z) = \frac{\lambda}{p_{\xi_R\xi_I}(y,z)} \int_{-\infty}^{z} sp_{\xi_R\xi_I}(y,s)ds \tag{7.177}$$

$$K_{212}(y,z) = K_{221}(y,z) = \frac{\omega}{p_{\xi_R\xi_I}(y,z)} \left(\int_{-\infty}^{z} sp_{\xi_R\xi_I}(y,s)ds - \int_{-\infty}^{y} sp_{\xi_R\xi_I}(s,z)ds \right) \tag{7.178}$$

then Fokker–Planck equation (7.172) is satisfied.
Having these parameters defined by eq. (7.173) and eqs. (7.175)–(7.178), one can write the corresponding generating SDE[8] as

$$d\begin{bmatrix} \xi_R \\ \xi_I \end{bmatrix} = \begin{bmatrix} -\lambda & \omega \\ -\omega & -\lambda \end{bmatrix} \begin{bmatrix} \xi_R \\ \xi_I \end{bmatrix} dt + \sqrt{\begin{bmatrix} K_{211}(y,z)K_{212}(y,z) \\ K_{212}(y,z)K_{222}(y,z) \end{bmatrix}} \begin{bmatrix} dw_1(t) \\ dW_2(t) \end{bmatrix} \tag{7.179}$$

[8]In general one has to solve the equation $g(y,z)g^H(y,z) = K_2$. Using the fact that K_2 is a symmetric matrix according to eq. (7.178) and the symmetry in the properties of y and z mentioned on page 348, one can choose the symmetrical solution $g(y,z)$ which is unique in this case. A more general consideration can be the subject of a separate investigation.

Example: *A Gaussian* (λ, ω) *process.* In this case one can write

$$p_x(x) = \frac{1}{\sqrt{2\pi\sigma^2}} \exp\left[-\frac{x^2}{2\sigma^2}\right] \tag{7.180}$$

and

$$p_A(A) = \frac{A}{\sigma^2} \exp\left[-\frac{A^2}{2\sigma^2}\right] \tag{7.181}$$

which results in the joint PDF of the in-phase and quadrature components

$$p_{\xi_R \xi_I}(y, z) = \frac{1}{2\pi\sigma^2} \exp\left[-\frac{y^2 + z^2}{2\sigma^2}\right] \tag{7.182}$$

and the corresponding vector generating SDE is, respectively,

$$\begin{aligned} d\xi_R &= -\lambda\xi_R dt + \omega z dt + \sqrt{2\lambda\sigma^2} dW_1(t) \\ d\xi_I &= -\omega\xi_R dt - \lambda\xi_I dt + \sqrt{2\lambda\sigma^2} dW_2(t) \end{aligned} \tag{7.183}$$

Example: *A* (λ, ω) *process with uniformly distributed envelope.* In this case one has

$$p_A(A) = \frac{1}{A_0}, \quad 0 \leq A \leq a_0 \tag{7.184}$$

$$p_\xi(x) = \frac{1}{\pi A_0} \ln\left(A_0^2 - x^2 + \sqrt{A_0^2 - x^2 + 1}\right) \tag{7.185}$$

$$p_{\xi_R \xi_I}(y, z) = \frac{1}{2\pi A_0 \sqrt{y^2 + z^2}} \tag{7.186}$$

where the corresponding diffusions are

$$K_{211}(y, z) = \lambda\left(\sqrt{\frac{A_0^2 + y^2}{y^2 + z^2}} - 1\right) \tag{7.187}$$

$$K_{222}(y, z) = \lambda\left(\sqrt{\frac{A_0^2 + z^2}{y^2 + z^2}} - 1\right) \tag{7.188}$$

$$K_{211}(y, z) = \lambda\left(\frac{\sqrt{A_0^2 + y^2} - \sqrt{A_0^2 + z^2}}{\sqrt{y^2 + z^2}}\right) \tag{7.189}$$

and the generating SDE is

$$
d\begin{bmatrix} \xi_R \\ \xi_I \end{bmatrix} = \begin{bmatrix} -\lambda & \omega \\ -\omega & -\lambda \end{bmatrix} \begin{bmatrix} \xi_R \\ \xi_I \end{bmatrix} dt + \sqrt{\lambda} \begin{bmatrix} \sqrt{\dfrac{A_0^2 + y^2}{y^2 + z^2} - 1} & \dfrac{\sqrt{A_0^2 + y^2} - \sqrt{A_0^2 + z^2}}{\sqrt{y^2 + z^2}} \\[4mm] \dfrac{\sqrt{A_0^2 + y^2} - \sqrt{A_0^2 + z^2}}{\sqrt{y^2 + z^2}} & \sqrt{\dfrac{A_0^2 + z^2}{y^2 + z^2} - 1} \end{bmatrix} \begin{bmatrix} dW_1(t) \\ dW_2(t) \end{bmatrix}
$$

$$(7.190)$$

7.3.3 Synthesis of a Narrow Band Process Using a Second Order SDE

General properties of a stationary narrow band random process have been discussed in some detail in Section 4.7. It was shown there that a narrow band process $\xi(t)$ can be represented as

$$\xi(t) = A(t)\cos[\omega_0 t + \varphi(t)] \tag{7.191}$$

where $A(t)$ and $\varphi(t)$ are random amplitude and phase, varying slowly [60–62] compared to the "central frequency" ω_0. In this case the spectrum width $\Delta\Omega$ of the process $\xi(t)$ satisfies the following condition

$$\frac{\Delta\Omega}{\omega_0} \ll 1 \tag{7.192}$$

For a narrow band stationary process, the joint distribution of the amplitude and phase has the following form [63]

$$p(A, \varphi) = p_A(A)p_\varphi(\varphi) \tag{7.193}$$

where, $p_\varphi(\varphi) = 1/(2\pi)$, $\varphi \in [-\pi, \pi]$ and the distribution $p_A(A)$ of the envelope $A(t)$ and the distribution $p_x(x)$ of process $x(t)$ are related by the Blanc-Lapierre transform pair [59]

$$p_{\xi\,\mathrm{st}}(x) = \frac{1}{\pi}\int_{|x|}^\infty \frac{p_A(A)}{\sqrt{A^2 - x^2}}\,dA \quad p_A(A) = \frac{A}{2\pi}\int_0^\infty d\omega J_0(\omega A)\omega \int_{-\infty}^\infty p_{x\,\mathrm{st}}(x)e^{-j\omega x}dx \tag{7.194}$$

The covariance function of a narrow band process $\xi(t)$ can be represented in a similar form

$$C_{\xi\xi}(\tau) = B(\tau)\cos(\omega_0\tau) \tag{7.195}$$

where $B(\tau)$ is a slow function compared to $\cos(\omega_0\tau)$.

Assuming that the random process (7.191) is generated by the WGN-excited deterministic dynamic system of minimal order two, the general form of SDE can be written in the form [2]

$$\ddot{x} + f_1(x, \dot{x})\dot{x} + f_2(x, \dot{x}) = f_3(x, \dot{x})\xi(t) \tag{7.196}$$

A full analytical investigation of eq. (7.196) is impossible. To make this problem more manageable only a few particular cases of this equation are considered below.

7.3.3.1　Synthesis of a Narrow Band Random Process Using a Duffing Type SDE

One of the simplest equations to consider is the so-called Duffing SDE [31]

$$\ddot{x} + 2\alpha\omega_0\dot{x} + \omega_0^2 F(x) = \sqrt{K}\omega_0\xi(t) \tag{7.197}$$

Here the power spectral density of the WGN is 1. It can be taken that eq. (7.197) represents a non-linear resonant contour excited by WGN (narrow band form filter). The block diagram of this form filter is given in Fig. 7.1. Equation (7.197) can be written in vector form as

$$\begin{aligned} \dot{x} &= y \\ \dot{y} &= -2\alpha\omega_0 y - \omega_0^2 F(x) + \sqrt{K}\omega_0\xi(t) \end{aligned} \tag{7.198}$$

with the corresponding stationary Fokker–Planck equation

$$0 = -\frac{\partial}{\partial x}[yp(x, y)] + \frac{\partial}{\partial y}[(2\alpha\omega_0 y + F(x)\omega_0^2)p(x, y)] + \frac{K}{2}\frac{\partial^2}{\partial y^2}p(x, y) \tag{7.199}$$

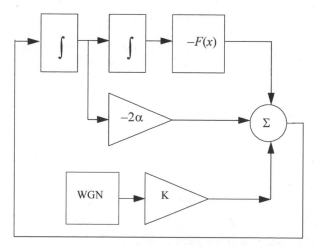

Figure 7.1　A block diagram of a form filter, corresponding to the Duffing equation (7.197).

It can be easily verified [31] that the stationary solution of the Fokker–Planck equation (7.199) has the following one-dimensional joint PDF

$$p_{st}(x, y) = C \exp\left(-\frac{2\alpha}{K\omega_0}y^2 - \frac{4\alpha\omega_0}{K}\int_{x_1}^{x} F(s)ds\right) = p_{x\,st}(x)p_{y\,st}(y) \tag{7.200}$$

From eq. (7.200) one can obtain the expression for the marginal PDF in the form

$$p_{x\,st}(x) = C \exp\left(-\frac{4\alpha\omega_0}{K}\int_{x_1}^{x} F(s)ds\right) \tag{7.201}$$

and, respectively, the unknown non-linearity $F(x)$ can be expressed as

$$F(x) = -\frac{K}{4\alpha\omega_0}\frac{d}{dx}\ln p_{x\,st}(x) \tag{7.202}$$

Once the non-linearity $F(x)$ is chosen, the parameters α and K can be found from the linearized model of the type [17,66]

$$\ddot{x} + 2\alpha\omega_0\dot{x} + \bar{\omega}^2(K, F)\omega_0^2 x = \sqrt{K}\omega_0\xi(t) \tag{7.203}$$

where

$$\alpha = \Delta\omega, \qquad \bar{\omega}^2(K, F) \equiv \frac{1}{\omega_0^2}\left\langle\frac{\partial}{\partial x}F(x)\right\rangle \tag{7.204}$$

It is important to point out here that the parameter $\bar{\omega}$ is defined both by K and by the non-linearity $F(x)$, and this fact complicates the application of the Duffing equation as a model of narrow band processes: a choice of K which satisfies $\bar{\omega}^2(K, F) = 1$ would ensure that the average frequency is equal to the required central frequency ω_0 but it is difficult to define its characteristics in more detail.

Since the PDF $p_x(x)$ must be an even function of its argument[9] x, its log-derivative is an odd function of x. If the function $F(x)$ is well behaved in the vicinity of zero it can be expanded into a Taylor series with only odd powers of x

$$F(x) = \sum_{k=0}^{\infty} \alpha_{2k+1}x^{2k+1} \approx \sum_{k=0}^{K_{max}} \alpha_{2k+1}x^{2k+1} \tag{7.205}$$

Thus, the average derivative of $F(x)$ is then

$$\left\langle\frac{d}{dx}F(x)\right\rangle \approx \sum_{k=0}^{K_{max}}(2k+1)\alpha_{2k+1}\langle x^{2k}\rangle = \sum_{k=0}^{K_{max}}(2k+1)\alpha_{2k+1}m_{2k} \tag{7.206}$$

[9]A narrow band stationary process has a symmetric PDF—see Section 4.7.

If such a derivative does not exist, one can approximate non-linearity in the mean square sense, i.e. minimize the values of the mean square error

$$
\begin{aligned}
\text{MSE}(s) &= \int_{-\infty}^{\infty} [F(x) - sx]^2 p_{x\,\text{st}}(x)dx = \int_{-\infty}^{\infty} [F(x)]^2 p_{x\,\text{st}}(x)dx \\
&\quad - 2s \int_{-\infty}^{\infty} F(x)x p_{x\,\text{st}}(x)dx + s^2 \int_{-\infty}^{\infty} x^2 p_{x\,\text{st}}(x)dx
\end{aligned}
\tag{7.207}
$$

with the minimum achieved at

$$
s_{\text{opt}}(K) = \frac{\int_{-\infty}^{\infty} F(x)x p_{x\,\text{st}}(x)dx}{\int_{-\infty}^{\infty} x^2 p_{x\,\text{st}}(x)dx} = -\frac{K}{4\alpha\omega_0} \frac{\int_{-\infty}^{\infty} x \frac{d}{dx} p_{x\,\text{st}}(x)dx}{\int_{-\infty}^{\infty} x^2 p_{x\,\text{st}}(x)dx} = \frac{K}{4\alpha\omega_0 m_{2x}} = \omega_0^2 \tag{7.208}
$$

Solving for K one obtains

$$
K = 4\alpha\omega_0^3 m_{2x} = 4\alpha\omega_0^3 \sigma_x^2 \tag{7.209}
$$

The models which use the Duffing SDE also have an important property which is often assumed (or indeed, proved) that the derivative of a random process is independent of the value of the process itself at the same moment of time and is a Gaussian random variable. This fact makes models based on the Duffing equation attractive for use in fading modeling.

Example: *A Gaussian narrow band random process.* In this case

$$
p_{x\,\text{st}}(x) = \frac{1}{\sqrt{2\pi\sigma^2}} \exp\left[-\frac{x^2}{2\sigma^2}\right] \tag{7.210}
$$

$$
p_A(A) = \frac{A}{\sigma^2} \exp\left[-\frac{A^2}{2\sigma^2}\right] \tag{7.211}
$$

Using formula (7.201) one obtains the function $F(x)$ as

$$
F(x) = \frac{K}{4\alpha\omega_0\sigma^2} x \tag{7.212}
$$

If the central frequency ω_0 and bandwidth $\Delta\omega$ are desired, then the parameters K and α should be chosen as[10]

$$
K = 4\alpha\sigma^2\omega_0, \quad \alpha = \Delta\omega \tag{7.213}
$$

and finally, the generating SDE is

$$
\ddot{x} + 2\Delta\omega_0\dot{x} + \omega_0^2 x = 2\sigma\omega_0^2 \sqrt{\frac{\Delta\omega}{\omega_0}} \xi(t) \tag{7.214}
$$

[10]The same result is obtained using the statistical linearization, (7.209).

Example: A narrow band process with K_0-distributed envelope. Let the distribution of the envelope of a narrow band random process be given by

$$p_A(A) = \beta^2 A K_0(\beta A), \quad A \geq 0 \tag{7.215}$$

Then, substituting eq. (7.215) into eq. (7.194), one obtains that the distribution of instantaneous values obeys the two-sided Laplace distribution

$$p_x(x) = \frac{\beta}{2} \exp[-\beta |x|] \tag{7.216}$$

and, thus, the corresponding non-linearity is

$$F(x) = \frac{K}{8\alpha\omega_0} \beta \operatorname{sgn} x \tag{7.217}$$

Since the derivative of expression (7.217) does not exist, one has to statistically linearize this expression in the mean square sense. The last function, being statistically linearized using eq. (7.197) [63] produces

$$K = \frac{4\alpha\omega_0^3}{\beta^2} \tag{7.218}$$

Finally, the generating SDE is thus

$$\ddot{x} + 2\Delta\omega\dot{x} + \frac{\omega_0^2}{2\beta} \operatorname{sgn} x = 2\frac{\sqrt{\frac{\Delta\omega}{\omega_0}}}{\beta} \omega_0^2 \xi(t) \tag{7.219}$$

Example: A bimodal distribution. It has been pointed out in Section 7.2.3 that the bimodal distribution

$$p_{x\,\mathrm{st}} = C \exp(-qx^4 - px^2) \tag{7.220}$$

approximates well the distribution of instantaneous values of processes with the Nakagami envelope. In this case the non-linearity $F(x)$ becomes a simple third order polynomial

$$F(x) = -\frac{K}{4\alpha\omega_0} \frac{d}{dx} \ln p_{x\,\mathrm{st}}(x) = \frac{K}{4\alpha\omega_0} (4qx^3 + 2px) \tag{7.221}$$

Estimation of the parameter K can be achieved either by using eq. (7.206)

$$K_1 = \frac{4\alpha\omega_0}{12qm_2 + 2p} \tag{7.222}$$

or by applying eq. (7.209)

$$K_2 = 4\alpha\omega_0 m_2 \tag{7.223}$$

where the second moment m_2 is calculated to be

$$m_2 = \frac{1}{4}\frac{p}{q}\frac{K_{3/4}\left(\frac{p^2}{8q}\right) - K_{1/4}\left(\frac{p^2}{8q}\right)}{K_{1/4}\left(\frac{p^2}{8q}\right)} \tag{7.224}$$

It can be seen from Fig. 7.2 that both methods provide a similar[11] estimate for $|p| > 2q$.

7.3.3.2   An SDE of the Van Der Pol Type

The Duffing SDE is convenient only when the Blanc-Lapierre transformation can be expressed analytically in terms of the known functions. At the same time, as was mentioned before, the central frequency of the power spectral density is strongly dependent on (Fig. 7.3) the function $F(x)$ and is varying when the parameters of the distribution (7.193) are changed.

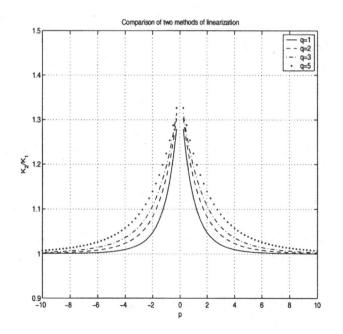

Figure 7.2   Dependence of $K_2/K_1 = m_2(12qm_2 + 2p)$ on p and q.

---

[11]Accuracy of a statistical linearization is often acceptable if it does not exceed 10–15 % [36].

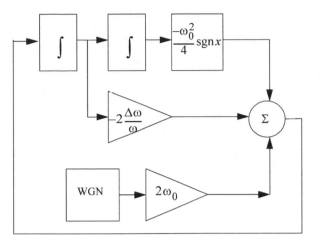

Figure 7.3   Block diagram for a form filter, corresponding to the Duffing eq. (7.219).

The following simple second order equations often produce better results [64,66]

$$\ddot{x} + f_1(x, \dot{x})\dot{x} + \omega_0^2 x = K\xi(t) \tag{7.225}$$

$$\ddot{x} + f_2(x, \dot{x})\dot{x} + \omega_0^2 x = K\dot{x}\xi(t) \tag{7.226}$$

where $f_i(x, \dot{x})$ are some deterministic functions of x and $\dot{x}$. The corresponding form filters are shown in Figs. 7.4 and 7.5.

Once again the narrow band solutions of eqs. (7.225) and (7.226) can be represented in the form (7.191),

$$x(t) = A(t)\cos(\omega_0 t + \varphi(t)) \tag{7.227}$$

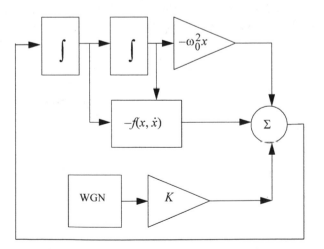

Figure 7.4   Form filter corresponding to the Van der Pol SDE (7.225).

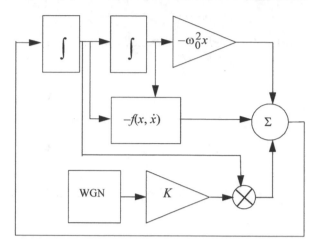

Figure 7.5 Form filter corresponding to the Van der Pol SDE (7.225).

and the adjoined process $\hat{x}(t)$ can be approximated by the Tikhonov formula [67]

$$\hat{x}(t) = \frac{\dot{x}(t)}{\omega_0} \tag{7.228}$$

instead of the exact Hilbert transform (see Section 4.7.1). In this case the envelope can be defined as [67]

$$A(t) = \sqrt{x^2(t) + \hat{x}^2(t)} = \sqrt{x^2(t) + \dot{x}^2(t)} \tag{7.229}$$

Using approximation (7.228) and relationship (7.229) one can rewrite SDEs (7.225) and (7.226) in terms of the envelope $A(t)$ and the process itself as [64,66]

$$\begin{cases} \dot{x} = -\omega_0 \sqrt{A^2 - x^2} \\ \dot{A} = -\dfrac{A^2 - x^2}{A} F_1(x, A) - \dfrac{\sqrt{A^2 - x^2}}{\omega_0 A} \xi(t) \end{cases} \tag{7.230}$$

$$\begin{cases} \dot{x} = -\omega_0 \sqrt{A^2 - x^2} \\ \dot{A} = -\dfrac{A^2 - x^2}{A} F_2(x, A) + \dfrac{A^2 - x^2}{A} \xi(t) \end{cases} \tag{7.231}$$

respectively. Here

$$F_i(x, A) = f_i(x, -\omega_0 \sqrt{A^2 - x^2}), i = 1, 2 \tag{7.232}$$

The corresponding Fokker–Planck equations are then given by

$$\frac{\partial}{\partial t} p_1(A, x, t) = \frac{\partial}{\partial x} \left[\omega_0 \sqrt{A^2 - x^2} p_1(A, x, t) \right]$$
$$- \frac{\partial}{\partial A} \left\{ \left[-\frac{A^2 - x^2}{A} F_1(x, A) + \frac{K^2 x^2}{2\omega_0^2 A^3} \right] p_1(A, x, t) \right\}$$
$$+ \frac{1}{2} \frac{\partial^2}{\partial A^2} \left[K^2 \frac{A^2 - x^2}{\omega_0^2 A^2} p_1(A, x, t) \right] \tag{7.233}$$

and

$$\frac{\partial}{\partial t} p_2(A, x, t) = \frac{\partial}{\partial x} \left[\omega_0 \sqrt{A^2 - x^2} p_2(A, x, t) \right]$$
$$- \frac{\partial}{\partial A} \left\{ \left[-\frac{A^2 - x^2}{A} F_2(x, A) + \frac{K^2 x^2}{2\omega_0^2 A^3} \right] p_2(A, xt) \right\}$$
$$+ \frac{1}{2} \frac{\partial^2}{\partial A^2} \left[K^2 \frac{(A^2 - x^2)^2}{A^2} p_2(A, x, t) \right] \tag{7.234}$$

respectively.

In the stationary case, when $t \to \infty$, the joint PDF can be written as [64,66]

$$p_i(x, A) = \frac{p_i(A)}{\pi \sqrt{A^2 - x^2}}, \quad i = 1, 2 \tag{7.235}$$

taking into account the uniformly distributed phase and eq. (7.193). Here $p_i(A)$ is the stationary distribution of amplitude. After substituting eq. (7.235) into eqs. (7.233)–(7.234) and integrating over interval $(-A, A)$ with respect to x, one obtains the equation for the stationary PDF of amplitude

$$\frac{K^2}{4\omega_0^2} \frac{\partial^2}{\partial A^2} p_1(A) + \frac{\partial}{\partial A} \left\{ \frac{1}{\pi A} \left[\int_{-A}^{A} F_1(x, A) \sqrt{A^2 - x^2} dx - \frac{K^2}{4\omega_0^2} \right] p_1(A) \right\} = 0 \tag{7.236}$$

$$\frac{K^2}{4\omega_0^2} \frac{\partial^2}{\partial A^2} p_2(A) + \frac{\partial}{\partial A} \left\{ \frac{1}{\pi A} \left[\int_{-A}^{A} F_2(x, A) \sqrt{A^2 - x^2} dx - \frac{K^2}{4\omega_0^2} \right] p_2(A) \right\} = 0 \tag{7.237}$$

In turn, eqs. (7.236) and (7.237) can be rewritten to have the unknown functions $F_i(A, x)$ on the left hand side, i.e.

$$\int_{-A}^{A} F_1(x, A) \sqrt{A^2 - x^2} dx = \frac{\pi K^2}{4\omega_0^2} \left[1 - A \frac{d}{dA} \ln p_1(A) \right] \triangleq \mathscr{F}_1(A) \tag{7.238}$$

$$\int_{-A}^{A} F_2(x, A) \sqrt{A^2 - x^2} dx = \frac{\pi K^2}{16} A^2 \left[1 + 3A \frac{d}{dA} \ln p_2(A) \right] \triangleq \mathscr{F}_2(A) \tag{7.239}$$

In order to simplify solution of the integral eqs. (7.238) and (7.239), additional requirements can be imposed on the form of unknown functions $F_i(x,A)$. In the following, considerations are restricted to the case where $F_i(x,A)$ depends only on one variable. In particular, if it is assumed that they depend only on a single variable A, i.e.

$$f_i(x,\dot{x}) = f_i\left(x^2 + \frac{\dot{x}^2}{\omega_0}\right) = F_i(A) \tag{7.240}$$

the left hand side of eqs. (7.238) and (7.239) can be transformed by noting that

$$\int_{-A}^{A} F_i(x,A)\sqrt{A^2 - x^2}\,dx = \int_{-A}^{A} F_i(A)\sqrt{A^2 - x^2}\,dx = \frac{\pi A^2}{2}\mathscr{F}_i(\mathscr{A}) \tag{7.241}$$

and, thus

$$F_i(A) = \frac{2\mathscr{F}_i(A)}{\pi A^2} \tag{7.242}$$

If dependence only on the variable x is assumed in $F_i(x,A)$, i.e.

$$f_i(x,\dot{x}) = f_i(x) = F_i(x) \tag{7.243}$$

eqs. (7.238) and (7.239) are reduced to

$$\int_{-A}^{A} F_1(x)\sqrt{A^2 - x^2}\,dx = \frac{\pi K^2}{4\omega_0^2}\left[1 - A\frac{d}{dA}\ln P_1(A)\right] \overset{\Delta}{=} \mathscr{F}_1(A) \tag{7.244}$$

$$\int_{-A}^{A} F_2(x)\sqrt{A^2 - x^2}\,dx = \frac{\pi K^2}{16}A^2\left[1 + 3A\frac{d}{dA}\ln P_2(A)\right] \overset{\Delta}{=} \mathscr{F}_2(A) \tag{7.245}$$

It can be shown that the solution of eqs. (7.244) and (7.245) has the following form [64,66]

$$f_i(x) = \frac{1}{\pi}\lim_{A\to 0}\frac{\frac{d}{dA}\mathscr{F}_i(A)}{A} + \frac{x}{\pi}\int_0^x \frac{d}{dA}\left[\frac{\frac{d}{dA}\mathscr{F}_i(A)}{A}\right]\frac{dA}{\sqrt{x^2 - A^2}} \tag{7.246}$$

thus providing a convenient tool for modeling narrow band processes with the prescribed PDF of the envelope [64,66].

   Example: *A narrow band process with the Nakagami PDF* [64,66]. For a narrow band process with the envelope distributed according to the Nakagami law with parameters m and Ω, the marginal PDF of the envelope is given by

$$p_A(A) = \frac{2}{\Gamma(m)}\left(\frac{m}{\Omega}\right)^m A^{2m-1}\exp\left(-\frac{mA^2}{\Omega}\right) \tag{7.247}$$

Function $\mathscr{F}_2(A)$ according to eq. (7.239) is then

$$\mathscr{F}_2(A) = \frac{\pi K^2}{16}\left[(6m - 3)A^2 - \frac{6mA^4}{\Omega}\right] \tag{7.248}$$

and eq. (7.242) becomes

$$f_2(A) = \frac{K^2}{8}\left(6m - 3 - \frac{6mA^2}{\Omega}\right) \tag{7.249}$$

Finally, the generating SDE is then

$$\ddot{x} + \frac{K^2}{8}\left[\frac{6m}{\Omega}\left(x^2 + \frac{\dot{x}^2}{\omega_0^2}\right) + 3 - 6m\right]\dot{x} + \omega_0^2 x = K\dot{x}\xi(t) \tag{7.250}$$

Example: A generalized Rayleigh distribution [64,66]. If the PDF of the envelope is given by

$$p_A(A) = CA\exp[-\alpha A^2 - \beta A^4] \tag{7.251}$$

then, according to eq. (7.238), function $F_1(A)$ is defined by

$$\mathscr{F}_1(A) = \frac{\pi K^2}{4\omega_0^2}(2\alpha A^2 + 4\beta A^4) \tag{7.252}$$

The corresponding solution of eq. (7.246) is then

$$f_1(x) = \frac{\pi K^2}{\omega_0^2}(\alpha + 3\beta x^3) \tag{7.253}$$

which, in turn, results in the following generating SDE

$$\ddot{x} + \frac{\pi K^2}{\omega_0^2}(\alpha + 3\beta x^3)\dot{x} + \omega_0^2 x = K\xi(t) \tag{7.254}$$

It is easy to see that, when $\beta = 0$ (Rayleigh distribution), eq. (7.254) becomes a linear equation coinciding with eq. (7.213). These models were used in a hardware simulator of HF communication channels [64,66].

7.4 SYNTHESIS OF A ONE-DIMENSIONAL PROCESS WITH A GAUSSIAN MARGINAL PDF AND NON-EXPONENTIAL CORRELATION

In this section synthesis of a one-dimensional process with Gaussian marginal PDF and a given covariance function is considered. If the covariance function is a pure exponent with decay λ, its mean and variance are m and σ^2 respectively, then the first order SDE

$$\dot{x} = -\lambda(x - m) + \sigma\sqrt{2\lambda}\xi(t) \qquad (7.255)$$

is the generating SDE for this problem.

The problem becomes more difficult if the correlation function differs from the exact exponent. In this case one has to represent this process as a component[12] of the vector Markov Gaussian process which, in turn, is the solution of a vector linear SDE

$$\dot{x} = -Ax + B\xi(t) \qquad (7.256)$$

where the state vector x is composed of components $x(t)$ and its derivatives with an order not higher than $M - 1$

$$x = \begin{bmatrix} x(t) \\ x'(t) \\ \cdots \\ x^{(M-1)}(t) \end{bmatrix} \qquad (7.257)$$

A and B are some constant matrices of dimension $M \times M$ and $M \times 1$, respectively.

It is well known that spectral density $\Phi(\omega)$ of the process $x(t)$ defined by eq. (7.256) is a rational function and it can be written in the form [55]

$$\Phi(\omega) = \frac{N(\omega^2)}{D(\omega^2)} \qquad (7.258)$$

where $N(\omega^2)$ and $D(\omega^2)$ are two polynomials in ω^2.

Any given covariance function $C_{xx}(\tau)$ of the process $x(t)$ can be approximated by a finite series of exponents [55] as[13]

$$C_{xx}(\tau) = \sum_{i=1}^{M} \mu_i \exp(-\lambda_i\tau) \qquad (7.259)$$

The last equality produces a rational spectrum according to the Wiener–Khinchin theorem. If the non-rational power spectral density is needed it can be approximated with some accuracy by the rational spectrum of the form (7.258) [55].

Let h_r, and h_r^* be zeros of the spectral density $\Phi(\omega)$, and λ_r and λ_r^* be its poles. It is assumed that h_r and λ_r are chosen in such a way that their real parts are negative. Then the spectral density can be written as

$$\Phi(\omega) = C\frac{\prod_{r=1}^{m}(\omega^2 + 2\mathrm{Re}(h_r)\omega + |h_r|^2)}{\prod_{s=1}^{n}(\omega^2 + 2\mathrm{Re}(\lambda_r)\omega + |\lambda_r|^2)} \qquad (7.260)$$

---

[12]The process itself is, thus, non-Markov.

[13]Note that here, in contrast to the one-dimensional SDE, both μ_i and λ_i can be complex but are encountered in complex-conjugate pairs.

where C is a constant. The last expression can be factorized to produce

$$\Phi(\omega) = \left|\frac{F(j\omega)}{G(j\omega)}\right|^2 = \frac{F(j\omega)F(-j\omega)}{G(j\omega)G(-j\omega)} \tag{7.261}$$

where $F(z)$ and $G(z)$ are polynomials of degree m and n, respectively, with all their zeros in the left half-plane. In order for the process $\xi(t)$ to have a finite variance, the degree m of the numerator must be less than the degree of the denominator, $m < n$. Without loss of generality, one can fix the leading coefficient of $G(z)$ as equal to 1, so that

$$G(z) = \prod_{r=1}^{n}(z - \lambda_r)$$

$$F(z) = f_0 \prod_{r=1}^{m}(z - h_r) \tag{7.262}$$

where f_0 is the leading coefficient of $F(z)$, $f_0 = c^{1/2}$.[14] Finally the process $x(t)$ can be derived from a WGN $\xi(t)$ of unit spectral density by the equation of n-th degree

$$g\left(\frac{d}{dx}\right)x(t) = F\left(\frac{d}{dx}\right)\xi(t) \tag{7.263}$$

or in the Causchy form

$$\frac{dw_r}{dt} - \lambda_r w_r = M_r\xi(t)$$

$$x(t) = \sum_{r=1}^{n} w_r(t) \tag{7.264}$$

where the coefficients M_r can be found as [55]:

$$M_r = \lim_{p \to \lambda_r} \frac{(p - \lambda_r)F(p)}{G(p)} \tag{7.265}$$

Another, but equivalent, form of eq. (7.264) can be obtained if the coefficients of the polynomials $G(z)$ and $F(z)$ are known. Let

$$G(z) = \sum_{r=0}^{n} g_r z^r \tag{7.266}$$

$$F(z) = \sum_{s=0}^{m} f_s z^s \tag{7.267}$$

[14]It is assumed that all zeros and poles are distinct. If this is not so, the appropriate formulas can be derived by allowing certain zeros or poles to coalesce [55].

Then the generating SDE has a form

$$
\dot{w} = \frac{1}{f_n}
\begin{bmatrix}
-f_{n-1} & f_n & 0 & \cdots & 0 \\
-f_{n-2} & 0 & f_n & \cdots & 0 \\
\cdots & \cdots & \cdots & \cdots & \cdots \\
-f_2 & 0 & 0 & \cdots & f_n \\
-f_1 & 0 & 0 & \cdots & 0
\end{bmatrix}
w + \frac{1}{f_n}
\begin{bmatrix}
f_n g_m \\
f_{n-1} g_{m-1} \\
\cdots \\
f_{n-m} g_0
\end{bmatrix}
\xi(t)
\tag{7.268}
$$

$$
x = \frac{w_1}{f_n}
\tag{7.269}
$$

where w is a vector of hidden states.

If one wants to avoid the calculations, then the form (7.264) should be used if the correlation function is given (or approximated) as the sum of exponents, and the forms (7.268) and (7.269) should be used if the energetic spectrum is given (or approximated) as a rational function of the frequency.

Example: a Gaussian process with a rational spectrum. Let the following power spectral density of a Gaussian process to be generated [36]:

$$
S_x(\omega) = \frac{N_0}{\pi} \frac{(a + \gamma \omega_0) b^2 + (a - \gamma \omega_0)\omega^2}{b^4 + 2(a^2 - \omega_0^2)\omega^2 + \omega^4}
\tag{7.270}
$$

where $w_0^2 = b^2 - a^2$. In this case the numerator has two roots $w = \pm i b_1$, $b_1^2 = b^2(a + \gamma \omega_0)(a - \gamma \omega_0)^{-1}$, and the denominator has four roots $\pm(\omega_0 \pm ia)$. Selecting the roots disposed in the upper half-plane one obtains

$$
\Phi(z) = \frac{g(z)}{F(z)} = \frac{\sqrt{2K^2(a - \gamma \omega_0)}(z + b_1)}{z^2 + 2az + b^2}
\tag{7.271}
$$

The corresponding generating SDE is then

$$
\dot{w} =
\begin{bmatrix}
0 & 1 \\
-b^2 & -2a
\end{bmatrix}
w +
\begin{bmatrix}
q_1 \\
q_2
\end{bmatrix}
\xi(t)
\tag{7.272}
$$

$$
x = w_1
$$

where $q_1 = \sqrt{2K^2(a - \gamma \omega_0)}$ and $q_2 = \sqrt{2K^2(a - \gamma \omega_0)}(b_1 - 2a)$

7.5   SYNTHESIS OF COMPOUND PROCESSES

This section builds on the theory of compound processes presented in Section 4.8. The results shown here are based on the synthesis of a process with exponential correlation and a Gaussian process with a rational spectrum, which have been discussed in Sections 7.2.3 and 7.4, respectively.

7.5.1 A Compound Λ Process

In this section we show how to obtain the exponentially correlated zero mean random process to be a compound one, and, consequently, to be a component of the vector random process:

$$x(t) = x_1(t)x_2(t) \tag{7.273}$$

If processes $x_1(t)$ and $x_2(t)$ in eq. (7.273) have exponential covariance function

$$C_{x_1}(\tau) = \langle x_1(t)x_1(t+\tau)\rangle = \sigma_{x_1}^2 \exp(-\lambda_1|\tau|) \tag{7.274}$$
$$C_{x_2}(\tau) = \langle x_2(t)x_2(t+\tau)\rangle = \sigma_{x_2}^2 \exp(-\lambda_2|\tau|) \tag{7.275}$$

then the correlation function of the process $x(t)$ is

$$\begin{aligned}
C_x(\tau) &= \langle x(t)x(t+\tau)\rangle = \langle x_1(t)x_1(t+\tau)\rangle\langle x_2(t)x_2(t+\tau)\rangle) \\
&= \sigma_{x_1}^2 \exp(-\lambda_1|\tau|)\sigma_{x_2}^2 \exp(-\lambda_2|\tau|) = \sigma_x^2 \exp(-\lambda|\tau|)
\end{aligned} \tag{7.276}$$

where $\lambda = \lambda_1 + \lambda_2$, and, consequently, the random process $x(t)$ is again a λ process.

Being exponentially correlated, processes $x_1(t)$ and $x_2(t)$ are solutions of the following SDEs

$$\dot{x}_1 = -\lambda_1 x_1 + g_1(x_1)\xi_1(t) \tag{7.277}$$

and

$$\dot{x}_2 = -\lambda_2 x_2 + g_2(x_2)\xi_2(t) \tag{7.278}$$

respectively. Here $\xi_1(t)$ and $\xi_2(t)$ are independent WGN and $g_1(x)$ and $g_2(x)$ are to be defined to provide proper distribution densities for $x_1(t)$ and $x_2(t)$. Using the equality

$$\dot{x}(t) = \dot{x}_1(t)x_2(t) + x_1(t)\dot{x}_2(t) \tag{7.279}$$

one can obtain the following vector SDE

$$\begin{cases}
\dot{x} = -\lambda_1 x_1 + g_1(x_1)\xi_1(t) \\
\dot{x}_2 = -\lambda_2 x_2 + g_2(x_2)\xi_2(t) \\
\dot{x} = -\lambda_2 x_2 x_1 - \lambda_1 x_1 x_2 + x_1 g_2(x_2)\xi_2(t) + x_2 g_1(x_1)\xi_1(t)
\end{cases} \tag{7.280}$$

One of the SDEs (7.280) is dependent on two others, so it can be rewritten as a two-dimensional SDE

$$\begin{cases}
\dot{x}_1 = -\lambda_1 x_1 + g_1(x_1)\xi_1(t), \\
\dot{x} = -\lambda_2 x - \lambda_1 x + x_1 g_2\left(\dfrac{x}{x_1}\right)\xi_2(t) + \dfrac{x}{x_1} g_1(x_1)\xi_1(t)
\end{cases} \tag{7.281}$$

or

$$
\begin{cases}
\dot{x}_2 = -\lambda_2 x_2 + g_2(x_2)\xi_2(t), \\
\dot{x} = -\lambda_2 x - \lambda_1 x + x_2 g_1\left(\dfrac{x}{x_2}\right)\xi_1(t) + \dfrac{x}{x_2} g_2(x_2)\xi_2(t)
\end{cases}
\tag{7.282}
$$

When representation (7.273) is found, the synthesis of the SDE (7.280)–(7.282) can be made by taking the following steps:

- choose the correlation time $\tau_1 = \frac{1}{\lambda_1} > \tau_{x\,\mathrm{corr}}$, where $\tau_{x\,\mathrm{corr}}$ is the desired correlation time of the process $\xi(t)$;

- find the generating SDE corresponding to $p_1(x_1)$ and λ_1 using eqs. (7.54)–(7.56);

- find the correlation time of the process $x_2(t)$ according to

$$
\tau_2 = \frac{1}{\lambda - \lambda_1} = \frac{1}{\lambda_2}
$$

- find the generating SDE corresponding to $p_2(x_2)$ and λ_2 using eqs. (7.54)–(7.56);

- write the generating SDE in the form of eq. (7.281) or eq. (7.282).

Example: K_0-distributed λ process. The marginal PDF of K_0-distributed random processes is given by

$$
p_x(x) = \frac{a^2|x|}{16} K_0(\sqrt{a|x|}), \quad x > 0
\tag{7.283}
$$

The direct application of the methodology of Section 7.2.3 in synthesizing the corresponding Λ process with the correlation time $\tau_{\mathrm{corr}} = 1/\lambda$ leads to complicated calculations employing higher transcendental functions. At the same time, the product of two symmetric Gamma distributions

$$
p_{x_1}(x_1) = \frac{\alpha^2}{2}|x_1| \exp(-\alpha|x_1|)
\tag{7.284}
$$

$$
p_{x_2}(x) = \frac{\beta^2}{2}|x_2| \exp(-\beta|x_2|)
\tag{7.285}
$$

gives the K_0-distributed random variable with PDF

$$
p_x(x) = \alpha^2\beta^2|x|K_0(2\sqrt{\alpha\beta|x|})
\tag{7.286}
$$

The SDEs generating a Λ process with PDFs (7.284) and (7.285) are the following

$$
\dot{x} = -\lambda_1 x_1 + \sqrt{\lambda_1 F_1(x_1)}\,\xi_1(t)
\tag{7.287}
$$

$$
\dot{x}_2 = -\lambda_2 x_2 + \sqrt{\lambda_2 F_2(x_2)}\,\xi_2(t)
\tag{7.288}
$$

Here

$$
F_1(x) = -\frac{1}{P_{x_1}(x)} \int_{-\infty}^{x} x p_{x_1}(x) dx =
\begin{cases}
-\dfrac{2 + \alpha^2 x^2 - 2\alpha x}{\alpha^2 x}, & \text{if } x < 0 \\[3mm]
\dfrac{2 + \alpha^2 x^2 + \alpha x}{\alpha^3 x}, & \text{if } x > 0
\end{cases}
\tag{7.289}
$$

and

$$
F_2(x) = \frac{1}{P_{x_2}(x)} \int_{-\infty}^{x} x p_{x_2}(x) dx =
\begin{cases}
-\dfrac{2 + \beta^2 x^2 - 2\beta x}{\alpha^3 x}, & \text{if } x < 0 \\[3mm]
\dfrac{2 + \beta^2 x^2 + \beta x}{\beta^3 x}, & \text{if } x > 0
\end{cases}
\tag{7.290}
$$

By setting $\lambda_1 = \lambda_2 = \lambda/2$ and $\alpha = \beta = \sqrt{b}$ one obtains the desired system of SDEs.

7.5.2 Synthesis of a Compound Process With a Symmetrical PDF

In this section we show how the compound model can be used to obtain the symmetrical random process with a prescribed correlation function and marginal distribution. The following statement lays the foundation for such an approach:

Proposition 4. Let $\xi(t)$ be a zero mean stationary compound random process with a symmetrical PDF and the correlation (covariance) function $C_{\xi\xi}(\tau)$ which is bounded by some exponent

$$
|C_{\xi\xi}(\tau)| \leq \sigma^2 \exp(-\lambda_m |\tau|)
\tag{7.291}
$$

Then the process $\xi(t)$ can be represented as a component of the random vector Markov process being the solution of the proper SDE if it can be represented as a product of a Gaussian process and some modulating process.

Proof: By assumption it is possible to find a representation of the random process $\xi(t)$ in the form of products of a normal distributed random process $\xi_n(t)$, and a properly distributed and independent of $\xi_n(t)$ zero mean symmetric random process $s(t)$

$$
\xi(t) = \xi_n(t)s(t)
\tag{7.292}
$$

If $p_s(s)$ is the PDF of the random process $s(t)$ then the PDF of the process $\xi(t)$ is [42]

$$
p_\xi(x) = \frac{1}{\sqrt{2\pi D_n}} \int_{\infty}^{\infty} |s|^{-1} \exp\left[-\frac{x^2}{2D_n s^2}\right] p_s(|s|) ds = \frac{2}{\sqrt{2\pi D_n}} \int_{0}^{\infty} s^{-1} \exp\left[-\frac{x^2}{2D_n s^2}\right] p_s(s) ds
\tag{7.293}
$$

The corresponding relation for the characteristic function is [42]

$$
\Theta_\xi(\varphi) = 2\varphi^{1/4} \int_{0}^{\infty} \Theta_s(\omega) \omega^{1/2} J_{-1/2}(\omega\sqrt{\varphi}) d\omega
\tag{7.294}
$$

where $\Theta_s(\omega)$ and $\Theta_\xi(\omega)$ are the characteristic functions of $s(t)$ and $\xi(t)$, respectively. As has been shown above, the covariance function of the compound process $\xi(t)$ is simply the product of the correlation functions of its components $\xi_n(t)$ and $s(t)$

$$C_\xi(\tau) = C_n(\tau)C_s(\tau) \tag{7.295}$$

If $s(t)$ is the λ process with decay $\lambda_s < \lambda_m$ then $x_n(t)$ is the normal process with correlation function

$$C_n(\tau) = C_\xi(\tau)\exp(-\lambda_s|\tau|) \tag{7.296}$$

It is easy to show that $C_n(\tau) \to 0$ when $\tau \to \infty$, so the random process $\xi_n(t)$ can be represented as a component of the random process generated by SDE

$$\dot{w} = Aw + B\xi_n(t) \tag{7.297}$$

where $w_1(t) = \xi_n(t)$, and A and B are the matrices obtained using the algorithm described in the previous section. If the generating SDE for Λ process $s(t)$ is

$$\dot{x} = f(s) + g(s)\xi_s(t) \tag{7.298}$$

then consequently the generating SDE for process $\xi(t)$ is

$$\begin{cases} \dot{w} = Aw + B\xi_1(t) \\ \dot{x} = w_1 f\left(\dfrac{x}{w_1}\right) + \dfrac{x}{w_1}\displaystyle\sum_{i=1}^{n} a_{1i}w_i + w_1 g\left(\dfrac{x}{w_1}\right)\xi_2(t) + b_1\dfrac{x}{w_1}\xi_1(t) \end{cases} \tag{7.299}$$

Example: A narrow band K-distributed random process. Let us to obtain the random process with marginal distribution

$$p_\xi(x) = K_0(ax) \tag{7.300}$$

and the correlation (covariance) function of the following form

$$K_{\xi\xi}(x) = \sigma^2 \exp(-\lambda_0|\tau|)\cos(\omega\tau) \tag{7.301}$$

In this case, the desired process $\xi(t)$ can be represented as a product of a normal distributed random process $\xi_n(t)$ with correlation function

$$C_{\xi_n\xi_n}(x) = \sigma^2 \exp\left(-\frac{\lambda_0}{2}|\tau|\right)\cos(\omega\tau) \tag{7.302}$$

and an m-distributed random λ process $S(t)$ with $m = 0.5$ and $\Omega = 1$.

In this case process $x_n(t)$ is generated by SDE

$$\ddot{x}_n + \frac{\lambda_0}{\omega}\dot{x}_n + \omega^2 x_n = \sigma\xi_n(t) \tag{7.303}$$

and $s(t)$ is generated by[15]

$$\dot{s} = -\frac{\lambda_0}{2}\left(s - \sqrt{\frac{2}{\pi}}\right) + F(s)\xi(t) \tag{7.304}$$

where $F(s)$ is, according to equation (4.60),

$$F(s) = \frac{\lambda_0\sqrt{2}\Gamma\left(\frac{1}{2},0,\frac{s^2}{2}\right)}{\pi} \tag{7.305}$$

where $\Gamma(\circ,\circ,\circ)$ is the generalized Gamma function [32].

7.6 SYNTHESIS OF IMPULSE PROCESSES

Let us now model a stationary impulse process with a given PDF $p_{\eta\,\mathrm{st}}(x)$. Synthesis of the corresponding generating SDE

$$\frac{\mathrm{d}}{\mathrm{d}t}x(t) = f(x) + \eta(t) \tag{7.306}$$

model is based on inversion of the stationary Kolmogorov–Feller equation (see Section 6.2.3)

$$-\frac{\mathrm{d}}{\mathrm{d}x}[K_1(x)p_{\eta\,\mathrm{st}}(x)] + \nu\int_{-\infty}^{\infty} p_A(x - A)p_{\eta\,\mathrm{st}}(A)\mathrm{d}A - \nu p_{\eta\,\mathrm{st}}(x) = 0 \tag{7.307}$$

or the generalized Fokker–Planck equation

$$-\frac{\mathrm{d}}{\mathrm{d}x}[K_1(x)p_{\eta\,\mathrm{st}}(x)] + \frac{1}{2}\frac{\mathrm{d}^2}{\mathrm{d}x^2}[K_2(x)p_{\eta\,\mathrm{st}}(x)] + \nu\int_{-\infty}^{\infty} p_A(x - A)p_{\eta\,\mathrm{st}}(A)\mathrm{d}A - \nu p_{\eta\,\mathrm{st}}(x) = 0 \tag{7.308}$$

if a mixture of Gaussian noise and Poisson impulses are present. For a moment the focus of discussions is on eq. (7.307).

If it is also assumed that a particular form of the distribution of the excitation pulses $p_A(A)$ is chosen, then the synthesis problem is reduced to a simple first order differential

---

[15]This SDE is written in Ito form.

equation (7.307) with respect to $K_1(x)$. Its solution is easy to obtain by simple integration, which produces

$$K_1(x)p_{\eta\,st}(x) = \nu \int_{-\infty}^{x} \left[\int_{-\infty}^{\infty} p_A(s-A)p_{\eta\,st}(A)dA - p_{\eta\,st}(s) \right] ds + C_0 \qquad (7.309)$$

A constant C_0 must be chosen in such a way that the net probability current vanishes, i.e. one has to choose $C_0 = 0$. This implies that

$$K_1(x) = \frac{\nu}{p_{\eta\,st}(x)} \int_{-\infty}^{x} \left[\int_{-\infty}^{\infty} p_A(s-A)p_{\eta\,st}(A)dA - p_{\eta\,st}(s) \right] ds = f(x) \qquad (7.310)$$

7.6.1 Constant Magnitude Excitation

In this case all impulses have the same magnitude, i.e.

$$p_A(A) = \delta(A - A_0) \qquad (7.311)$$

As a result, one can reduce eq. (7.311) using the sifting property of the delta function to a simpler equation

$$f(x) = \frac{\nu}{p_{\eta\,st}(x)} \int_{-\infty}^{x} [p_{\eta\,st}(s - A_0) - p_{\eta\,st}(s)]ds = -\nu \frac{P_{\eta}(x) - P_{\eta}(x - A_0)}{p_{\eta\,st}(x)} \qquad (7.312)$$

Example: An exponentially distributed process. We wish to synthesize an SDE with exponential distribution of the solution:

$$p_{\eta\,st}(x) = \alpha \exp(-\alpha x), \qquad x \geq 0 \qquad (7.313)$$

In this case

$$P_{\eta}(x) = \begin{cases} 0 & \text{if } x < 0 \\ 1 - \exp(-\alpha x) & \text{if } x \geq 0 \end{cases} \qquad (7.314)$$

and, thus, eq. (7.312) produces

$$f(x) = K_1(x) = \begin{cases} -\dfrac{\nu}{\alpha}[\exp(\alpha x) - 1] & \text{if } 0 \leq x \leq A_0 \\ -\dfrac{\nu}{\alpha}[\exp(\alpha A_0) - 1] & x \geq A_0 \end{cases} \qquad (7.315)$$

7.6.2 Exponentially Distributed Excitation

In this case all impulses have the same magnitude, i.e.

$$p_A(A) = \alpha \exp(-\alpha A), \qquad A \geq 0 \tag{7.316}$$

As a result, one can reduce eq. (7.311) using the sifting property of the delta function to a simpler equation

$$f(x) = \frac{\nu}{p_{\eta \,\mathrm{st}}(x)} \int_{-\infty}^{x} \left[\alpha \exp(-\alpha s) \int_{0}^{s} \exp(\alpha A) p_{\eta \,\mathrm{st}}(A) da - p_{\eta \,\mathrm{st}}(s) \right] ds \tag{7.317}$$

Example: A Gamma distributed process [64]. Let one desire to model an impulse process with Gamma PDF

$$p_{\eta \,\mathrm{st}}(x) = \frac{\beta^{\gamma}}{\Gamma(\gamma)} x^{\gamma - 1} \exp(-\beta x) \tag{7.318}$$

Setting the parameter α in eq. (7.316) equal to β in eq. (7.318) one can easily see that eq. (7.317) becomes

$$f(x) = \frac{\nu}{\beta} x \tag{7.319}$$

More examples can be found in Table D.2 in Appendix D on the web.

7.7 SYNTHESIS OF AN SDE WITH RANDOM STRUCTURE

In this section the inverse problem is considered for an SDE with random structure. In addition to the information required for synthesis of an ordinary SDE, such as marginal PDF of the process and the correlation function, one has to obtain *a priori* information about a number of states M (number of sub-structures) and intensity of transitions ν_{kl}. Knowledge of intensities allows one to solve the Kolmogorov–Chapman equation to obtain stationary probabilities of states P_l. As the next step, the joint PDF $p_{\mathrm{st}}^{*(l)}(x)$ can be obtained using the equation

$$p_{\mathrm{st}}^{*(l)}(x) = P_l p_{\mathrm{st}}^{l}(x) \tag{7.320}$$

In the following, only a scalar case is considered. In order to further simplify discussion, systems with a constant drift are considered: $K_2^{(l)}(x) = K_2^{(l)}$.

A stationary solution, if it exists, must satisfy the following system of equations:

$$p_{\mathrm{st}}^{*(l)}(x) \frac{\partial}{\partial x} [K_1^{(l)}(x)] + K_1^{(l)}(x) \frac{\partial}{\partial x} p_{\mathrm{st}}^{*(l)}(x) = Q(x) \tag{7.321}$$

where

$$Q(x) = \frac{K_2^{(l)}}{2} \frac{\partial^2}{\partial x^2} p_{st}^{*(l)}(x) + V_l^*(x) - U_l^*(x)$$

(7.322)

The next step is to rewrite eq. (7.321) as a differential equation with respect to $K_1^{(l)}(x)$, i.e. in the following form

$$\frac{d}{dx} K_1^{(l)}(x) + K_1^{(l)}(x) \frac{d}{dx} \ln p_{st}^{*(l)}(x) = \frac{Q(x)}{p_{st}^{*(l)}(x)}$$

(7.323)

which can be easily solved using the method of undefined coefficients

$$K_1^{(l)}(x) = \frac{K_2^{(l)}}{2} \ln p_{st}^{*(l)}(x) - \left(\nu_l \int_\infty^x p_{st}^{*(l)}(x) dx + \sum_{k=1}^{M} \nu_{kl} \int_{-\infty}^x p_{st}^{*(k)}(x) dx \right) + C$$

(7.324)

The constant of integration in eq. (7.324) can be rewritten for $p^{(l)}(x)$ if the additional requirement

$$K_1^{(l)}(-\infty) = 0$$

(7.325)

is imposed on $K_1^{(l)}(x)$. In this case, the synthesis equation becomes

$$a^{(l)}(x) = \frac{b^{(l)}}{2} \frac{d}{dx} \ln p_{st}^{*(l)}(x) + \nu_l \frac{F_l(x)}{p_{st}^{*(l)}(x)} - \sum_{k=1}^{M} \nu_{kl} \frac{P_k F_k(x)}{P_l p_{st}^{*(k)}(x)}$$

(7.326)

where

$$F_l(x) = \int_{-\infty}^x p_{st}^{*(l)}(x) dx$$

(7.327)

It can be seen that the expression for the drift coefficient $K_1^{(l)}(x)$ contains a term present in the synthesis equation for an ordinary SDE (the first term in eq. (7.326)), while the random structure is represented by additional (the second and the third) terms in eq. (7.326). These additional terms represent contributions of flows of absorption and regeneration of the trajectories of the process with random structure.

The diffusion coefficient $K_2^{(l)}$, as in the case of a monostructure, should be chosen in accordance with the correlation interval of the random process $\xi(t)$. Quantitatively it can only be obtained from the expression for the covariance function, which can be obtained only for a case of high and low intensities of the switching process and under the assumption that such a correlation function is close to exponential.

For the case of a low intensity of switching, this approach produces quite accurate results since the effect of switching can be neglected or accounted for only by first order correction terms. In this case the synthesis of the SDE with random structure is no different from the synthesis of an ordinary SDE.

˙ REFERENCES

1. B. Liu and D. Munson, Generation of a Random Sequence Having Jointly Specified Marginal Distribution and Autocovariance, *IEEE Trans. Acoustics, Speech Signal Processing*, Vol. 30, 1982, pp. 973–983.

2. V. Kontorovich and V. Lyandres, Stochastic Differential Equations: An Approach to the Generation of Continuous Non-Gaussian Processes, *IEEE Trans. Signal Processing*, Vol. 43, 1995, pp. 2372–2385.

3. S. Primak and V. Lyandres, On the Generation of Baseband and Narrowband Non-Gaussian Processes, *IEEE Trans. Signal Processing*, Vol. 46, 1998, pp. 1229–1237.

4. D. Middleton, *An Introduction to Statistical Communication Theory*, New York: McGraw-Hill, 1960.

5. E. Conte, M. Longo, and M. Lops, Modeling and Simulation of Non-Rayleigh Radar Clutter, *IEEE Proc. F* 138, 1991, pp. 121–131.

6. M. Rangaswamy, D. Weiner, and A. Ozturk, Non-Gaussian Random Vector Identification Using Spherically Invariant Random Processes, *IEEE Tran. Aeros. Elec. Systems*, Vol. 29, 1993, pp. 111–124.

7. A. Fulinski, Z. Grzywna, I. Mellor, Z. Siwy, and P.N.R. Usherwood, Non-Markovian Character of Ionic Current Fluctuations in Membrane Channels, *Physical Review E*, Vol. 58, 1998, pp. 919–924.

8. S. Primak and J. LoVetri, *Statistical Modeling of External Coupling of Electromagnetic Waves to printed Circuit Boards Inside a Leaky Cavity, Technical Report*, CRC, Ottawa, Canada.

9. J. Lee, R. Tafazolli, and B. Evans, Erlang Capacity of OC-CDMA with Imperfect Power Control, *IEE Electronics Letters*, Vol. 33, 1997, pp. 259–261.

10. R. J. Polge, E. M. Holliday, and B. K. Bhagavan, Generation of a Pseudorandom Set with Desired Correlation and Probability Distribution, *Simulation*, 1973, pp. 153–158.

11. W. J. Szajnowski, The Generation of Correlated Weibull Clutter for Signal Detection Problems, *IEEE Trans. Aer. Electr. Syst.*, Vol. AES-13, 1977, pp. 536–540.

12. V. G. Gujar and R. J. Kavanagh, Generation of Random Signals with Specified Probability Density Functions and Power Density Spectra, *IEEE Trans. Autom. Contr.*, Vol. AC-13, 1968, pp. 716–719.

13. B. Liu and D. C. Munson, Generation of a Random Sequence Having a Jointly Specified Marginal Distribution and Autocovariation, *IEEE Trans. Acous., Speech, Signal Proc.*, Vol. ASSP-30, 1982, pp. 973–983.

14. R. Rosenfeld, Two Decades of Statistical Language Modeling: Where Do We Go From Here?, *Proc. IEEE*, Vol. 88, 2000, pp. 1270–1278.

15. J. Luczka, T. Czernik, and P. Hanggi, Symmetric White Noise Can Induce Directed Current in Ratchets, *Physical Review E*, Vol. 56, 1997, pp. 3968–3975.

16. M. Popescu, C. Arizmendi, A. Salas-Brito, and F. Family, Disorder Induced Diffusive Transport in Ratchets, *Physical Review Letters*, Vol. 85, 2000, pp. 3321–3324.

17. V. Z. Lyandres and M. Shahaf, Envelope Correlation Function of Narrow-Band Non-Gaussian Process, *Int. J. Nonlinear Mechanics*, Vol. 30, 1995, pp. 359–369.

18. E. Wong and J. B. Thomas, On Polynomial Expansion of Second-Order Distributions, *J. Soc. Industr. Math.*, 1962, Vol. 10, pp. 507–516.

19. T. Ozaki, A Local Linearization Approach to Non-Linear Filtering, *Int. Jour. Contr.*, Vol. 57, 1995, pp. 75–96.

20. S. Primak, V. Lyandres, O. Kaufman, and M. Kliger, On the Generation of Correlated Time Series with a Given PDF, Submitted to *Signal Processing*, February, 1997.

21. E. Wong, "The Construction of a Class of Stationary Markov Processes", *Proc. Symp. Appl. Math.*, Vol. 16, 1964, pp. 264–275.

22. T. T. Soong, *Random Differential Equations in Science and Engineering*. New York: Academic Press, 1971.

23. K. Ito, Stochastic integrals, *Proc. Imp. Acad.*, Tokyo, Vol. 40, 1944, pp. 519–524.
24. I. Stakgold, *Boundary Value Problem of Mathematical Physics*, New York: Macmillan, 1967.
25. A.H. Haddad, Dynamical Representation of Markov Processes of the Separable Class, *IEEE Trans. Information Theory*, Vol. 16, 1970, pp. 529–534.
26. S. Primak and V. Lyandres, On the Generation of the Baseband and Narrowband Non-Gaussian Processes, *IEEE Trans. Signal Processing*, Vol. 46, 1998, pp. 1229–1237.
27. J. L. Doob, *Stochastic Processes*, New York: Wiley, 1953.
28. V. Ya. Kontorovich and V. Z. Lyandres, Fluctuation equations generating a one-dimensional Markovian Process, *Radio Engineering and Electronics Physics*, July 1972, pp. 1093–1096 (in Russian).
29. M.G. Kendall and A. Stuart, *The Advanced Theory of Statistics*, London: Griffin, 1958.
30. R. Gut and V. Kontorovich, On Markov Representation of Random Processes, *Voprosi Radiotechniki*, Vol. 6, 1971, pp. 91–94 (in Russian).
31. R. L. Stratonovich, *Topics in the Theory of Random Noise*, Vol. 1, New York: Gordon and Breach, 1967.
32. M. Abramoviz and I.A. Stegun, *Handbook of Mathematical Functions*, New York: Dover Publications, 1971.
33. B. Stuck and B. Kleiner, A Statistical Analysis of Telephone Noise, *Bell Technical J.*, Vol. 53, pp. 1236–1320.
34. R. Risken, *The Fokker–Planck Equation: Methods of Solution and Applications*, Berlin: Springer, 1996, 472 p.
35. P. E. Kloeden and E. Platen, *Numerical Solution of Stochastic Differential Equations*, New York: Springer-Verlag, 1992.
36. V. Pugachev and I. Sinitsin, *Stochastic Differential Systems: Analysis and Filtering*, New York: Wiley, 1987.
37. P. E. Kloeden and E. Platen, *Numerical Solution of Stochastic Differential Equations*, New York: Springer-Verlag, 1992.
38. C. W. Gardiner, *Handbook of Stochastic Methods for Physics, Chemistry and the Natural Sciences* Berlin: Springer-Verlag, 1983.
39. V. Kontorovich, V. Lyandres, and S. Primak, The Generation of Diffusion Markovian Processes with Probability Density Function Defined on Part of the Real Axis, *IEEE Signal Processing Letters*, Vol. 3, 1996, pp. 19–21.
40. E. Conte and M. Longo, Characterization of Radar Clutter as a Spherically Invariant Random Process, *IEE Proc.*, Part F., Vol. 134, 1987, pp. 191–201.
41. S. Primak, V. Lyandres, and V. Kontorovich, Markov Models of non-Gaussian Exponentially Correlated processes and Their applications, *Physical Review E*, Vol. 63, 2001, 061103.
42. M. Rangaswamy, D. Weiner, and A. Ozturk, Non-Gaussian Random Vector Identification Using Spherically Invariant Random Processes, *IEEE Tran. Aeros. Elec. Syst.*, Vol. AES-29, 1993, pp. 111–201.
43. M. M. Sondhi, Random Processes with Specified Spectral Density and First Order Probability Density, *Bell System Technical J.*, Vol. 62, 1983, pp. 679–701.
44. R. J. Webster, A Random Number Generator for Ocean Noise Statistics, *IEEE J. Ocean Eng.*, Vol. 19, 1994, pp. 134–137.
45. S. Xu, Z. Huang, and Y. Yao, "An Analytically Tractable Model for Videoconferencing," *IEEE Trans. Circuits and Systems for Video Technology*, Vol. 10, 2000, p. 63.
46. S. Xu and Z. Huang, A Gamma Autoregressive Video Model on ATM Networks, *IEEE Trans. Circuits and Systems for Video Technology*, Vol. 8, 1998, pp. 138–142.
47. E. Wong, The Construction of a Class of Stationary Markov Processes, *Proc. Symp. Appl. Math.*, Vol. 16, pp. 264–275.
48. S. Primak, Generation of Compound Non-Gaussian Random Processes with a Given Correlation Function, *Physical Review E*, Vol. 61, 2000, p. 100.

49. A. N. Malakhov, *Cumulantnii Analiz Sluchainich Processov i Ich Preobrazovanii* (*Cumulant Analysis of Random Processes and Their Transformations*), Moscow: Radio, 1978 (in Russian).

50. V. Kontorovich and V. Lyandres, Impulsive Noise: A Nontraditional Approach, *Signal Processing*, Vol. 51, 1996, pp. 121–132.

51. G. Kotler, V. Kontorovich, V. Lyandres, and S. Primak, On the Markov Model of Shot Noise, *Signal Processing*, 1999, pp. 79–88.

52. J. Nakayama, Generation of Stationary Random Signals with Arbitrary Probability Distribution and Exponential Correlation, *IEICE Trans. Fundamentals*, Vol. E77-A, 1994, pp. 917–922.

53. S. Karlin, *A First Course in Stochastic Processes*, New York: Academic Press, 1968.

54. N. S. Jayant and P. Noll, *Digital Coding of Waveforms*, New Jersey: Prentice-Hall, 1984.

55. C. W. Helstrom, *Markov Processes and Application in Communication Theory, in Communication Theory*, New York: McGraw-Hill, 1968, pp. 26–86.

56. T. Rappaport, *Wireless Communications: Principles and Practice*, Upper Saddle River, NJ: Prentice Hall PTR, 2002.

57. G. Stüber, *Principles of Mobile Communication*, Boston: Kluwer Academic Publishers, 2001.

58. S. Primak, V. Lyandres, O. Kaufman, and M. Kliger, On the Generation of Correlated Time Series with a Given Probability Density Function, *Signal Processing*, Vol. 72, 1999, pp. 61–68.

59. A. Blanc-Lapierre, M. Savelli, and A. Tortrat, Etude de Modeles Statistiques Suggeres par la Consideration des Effets des Atmospheriques sur les Amplificateurs, *Ann. Telecomm*, Vol. 9, 1954, pp. 237–248.

60. T. Felbinger, S. Schiller, and J. Mlynek, Oscillation and Generation of Nonclassical States in Three-Photon Down-Conversion, *Physical Review Letters*, Vol. 80, 1998, pp. 492–495.

61. A. C. Marley, M.J. Higgins, and S. Bhattacharya, Flux Flow Noise and Dynamical Transitions in a Flux Line Lattice, *Physical Review Letters*, Vol. 74, 1995, pp. 3029–3032.

62. P. Rao, D. Johnson, and D. Becker, Generation and Analysis of Non-Gaussian Markov Time Series, *IEEE Trans. Signal Processing*, Vol. 40, 1992, pp. 845–855.

63. S. M. Rytov, Yu. A. Kravtsov, and V. I. Tatarskii, *Principles of Statistical Radiophysics, Vol. 2, Correlation Theory of Random Processes*, New York: Springer-Verlag, 1988.

64. D. Klovsky, V. Kontorovich, and S. Shirokov, *Models of Continuous Communications Channels Based on Stochastic Differential Equations*, Moscow: Radio i sviaz, 1984 (in Russian).

65. I. Kazakov and V. Artem'ev, *Optimization of Dynamic Systems with Random Structure*, Moscow: Nauka, 1980 (in Russian).

66. A. Berdnikov, A. Brusentzov, V. Kontorovich, and V. Lyandres, Problems in the theory and design of Instruments for Analog Simulation of Multipath Channels, *Telecomm. Radio. Eng.*, Vol. 33–34, 1978, pp. 60–64.

67. V. Tikhonov and M. Mironov, *Markov Processes*, Moscow: Sov. Radio, 1977 (in Russian).

68. I. S. Gradsteyn and I. M. Ryzhik, *Table of Integrals, Series, and Products*, Academic Press, 1980.

8
Applications

8.1 CONTINUOUS COMMUNICATION CHANNELS

8.1.1 A Mathematical Model of a Mobile Satellite Communication Channel

For some years many different organizations all over the world have been involved in activities of land mobile satellite communication services [1]. The quality of this kind of communication suffers from strong variations of the received signal power due to satellite signal shadowing and multipath fading. The shadowing of the satellite signal by obstacles in the propagation path (buildings, bridges, trees, etc.) results in attenuation over the entire signal bandwidth. This attenuation increases with carrier frequency, i.e. it is more pronounced in the L-band than in the UHF band. The shadowed areas are larger for low satellite elevations than for high elevations. Multipath fading occurs because the satellite signal is received not only via the direct path but also after having been reflected from objects in the surroundings. Due to their different propagation distances, multipath signals can add destructively, resulting in a deep fade.

In analysing the scattered field characteristics the following assumptions are usually considered as valid:

- a large number of partial waves;
- identical partial wave amplitudes;
- no correlation between different partial waves;
- no correlation between the phase and amplitude of one partial wave;
- phase distribution is uniform on $[0, 2\pi]$.

Physically, these assumptions are equivalent to a homogeneous diffuse scattered field resulting from randomly distributed point scatterers. With these assumptions, the central limit theorem leads to a complex stationary Gaussian process $z = x + jy$ with zero mean, equal variance and uncorrelated components. The amplitude distribution thus obeys the Rayleigh PDF [2,3]

$$p_A(A) = \frac{A}{\sigma^2} \exp\left[-\frac{A^2}{2\sigma^2}\right] \tag{8.1}$$

Stochastic Methods and Their Applications to Communications.
S. Primak, V. Kontorovich, V. Lyandres
© 2004 John Wiley & Sons, Ltd   ISBN: 0-470-84741-7

which depends on a single parameter σ^2, equal to the variance of the in-phase and quadrature components.

In the line of sight conditions, i.e. when the direct component is superimposed, $p_A(A)$ is well described by the Rice distribution [2,3]

$$p_A(A) = \frac{A}{\sigma^2} \exp\left[-\frac{A^2 + a^2}{2\sigma^2}\right] I_0\left(\frac{aA}{\sigma^2}\right), \tag{8.2}$$

where $I_0(°)$ is the modified Bessel function of zero order, and a is a parameter describing the strength of the Line-of-Sight (LoS) component (see Section 2.6). The Rice factor K, defined as

$$K = \frac{a^2}{2\sigma^2} \tag{8.3}$$

not only defines the ratio of power of the LoS component to the power of the diffused component, but also can be used to compare different models as shown in Table 8.1.

In general, it is possible that none of the aforementioned assumptions is valid due to correlation between a small number of partial waves. Numerous approximations have been suggested to describe the measurement results [4–17]. It was also found [3] that one of the most convenient and closely fitting distributions of the envelope is the Nakagami PDF

$$p_A(A) = \frac{2}{\Gamma(m)} \left(\frac{m}{\Omega}\right) A^{2m-1} \exp\left[-\frac{mA^2}{2\sigma^2}\right] \tag{8.4}$$

In scattering processes generating merely diffuse wave fields, $m \approx 1$ and the Nakagami distribution is identical to the Rayleigh one (8.1). In the case of a line of sight, the distribution (8.4) with $m > 1$ approximates the Rice distribution.

Another important characteristic of the communication channels is the power spectrum density of the fading signal [2]. Consider an unmodulated carrier being detected by a mobile receiver. Each component arrives at the receiver via different angles to the receiver velocity

Table 8.1 Different models of fading envelope and their comparison

Model	PDF	Rice factor	References
Nakagami	$\frac{2}{\Gamma(m)} \left(\frac{m}{\Omega}\right)^m A^{2m-1} \exp\left(-\frac{mA^2}{\Omega}\right)$	$m^2 - m$	[6]
Suzuki	$\frac{A}{\sqrt{2\pi\sigma^2}\Omega}$ $\times \int_0^\infty \left[\frac{1}{s^3}\exp\left(-\frac{\left(\frac{A}{s}\right)^2 + a^2}{2\Omega}\right) I_0\left(\frac{Aa}{s\Omega}\right) \times e^{-\frac{(\ln s - m)}{2\sigma^2}}\right] ds$	$\frac{a^2}{2\Omega}$	[5]
Loo	$\frac{A}{\sqrt{2\pi\sigma^2}\Omega}$ $\times \int_0^\infty \left[\frac{1}{s}\exp\left(-\frac{A^2 + s^2}{2\Omega}\right) I_0\left(\frac{As}{\Omega}\right) \times e^{-\frac{(\ln s - m)}{2\sigma^2}}\right] ds$	$\frac{a^2}{2\Omega}$	[7]

vector and thus each signal component acquires a certain Doppler shift that is related to the receiver speed ν and arrival angle α_i for the i-th ray according to [2]

$$f_i = \frac{\nu}{\lambda} \cos \alpha_i = f_D \cos \alpha_i \qquad (8.5)$$

Due to the Doppler effect, a single frequency carrier results in multiple signals at the receiver having comparable amplitudes, random phases and a relative frequency shift confined to the Doppler spread about the carrier frequency. The theoretically obtained spectrum $S(f)$ (low pass equivalent) is given by [2]

$$S(f) = \frac{1}{\pi\sqrt{f_D^2 - f^2}} \qquad (8.6)$$

and the covariance function by

$$C_\xi(\tau) = J_0(2\pi f_D |\tau|) \qquad (8.7)$$

A number of alternative models have also been suggested, especially in the COST 207 report [4] which includes a combination of two Gaussian PDFs (see Table 3.2 in Chapter 3) and a combination of the Jakes PSD (8.6) and a delta function. These possibilities are summarized in Table 8.2.

Finally, taking into account the multipath nature of the wave propagation one can represent the low pass equivalent impulse response of the channel as a sum of N rays embedded in white Gaussian noise

$$r(t) = \sum_{k=1}^{N} A_i(t)s(t - \tau_i) + \xi(t) \qquad (8.8)$$

Table 8.2 Typical PSD of a fading signal [4]

Model	PSD $S(f)$	Normalized covariance function $\rho(\tau)$
Jakes	$\dfrac{1}{\pi\sqrt{f_D^2 - f^2}}$	$J_0(2\pi f_D \tau)$
Gauss I	$0.909 \exp\left(-\dfrac{(f + 0.8 f_D)^2}{0.005 f_D^2}\right)$ $+ 0.0909 \exp\left(-\dfrac{(f - 0.4 f_D)^2}{0.02 f_D^2}\right)$	$0.909 \exp\left(-\dfrac{(f_D \tau)^2}{400}\right) \exp(-j 1.6 \pi f_D \tau)$ $+ 0.0909 \exp\left(-\dfrac{(f_D \tau)^2}{20}\right) \exp(j 0.8 \pi f_D \tau)$
Gauss II	$0.968 \exp\left(-\dfrac{(f - 0.7 f_D)^2}{0.02 f_D^2}\right)$ $+ 0.0316 \exp\left(-\dfrac{(f + 0.4 f_D)^2}{0.045 f_D^2}\right)$	$+ 0.968 \exp\left(-\dfrac{(f_D \tau)^2}{200}\right) \exp(j 1.4 \pi f_D \tau)$ $+ 0.0316 \exp\left(-\dfrac{(f_D \tau)^2}{40}\right) \exp(-j 0.8 \pi f_D \tau)$
Rice	$\dfrac{0.681}{\pi\sqrt{f_D^2 - f^2}} + 0.8319\, \delta(f - 0.7 f_D)$	$0.1681\, J_0(2\pi f_D \tau)$

where N is the number of rays considered, $A_i(t)$ is the fading amplitude of the i-th ray and τ_i is the delay of the i-th ray, and $\xi(t)$ is the additive WGN. The number of rays, delay of each ray and their parameters vary greatly for different environments. COST 207 provides tables of corresponding parameters for 4, 6 and 12 path models in rural, bad urban, typical urban and hilly terrain environments. However, regardless of the particular environment, the general structure of the simulator of a mobile communication channel remains the same and is shown in Fig. 8.1.

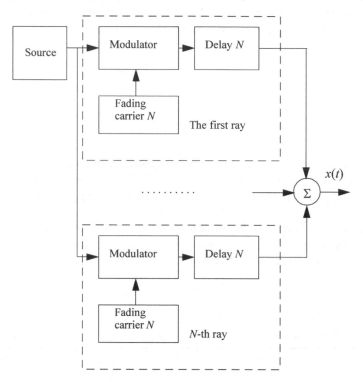

Figure 8.1   General scheme of a mobile communication channel simulator.

8.1.2   Modeling of a Single-Path Propagation

From the diagram shown in Figure 8.1 one can conclude that the main task in the simulation of the multipath channel is to model a single ray propagation with the given characteristics. Two main schemes which allow one to maintain both the PDF of the envelope and the correlation function of the signal are suggested below. Numerous alternative methods can be found elsewhere [18–22].

8.1.2.1   A Process with a Given PDF of the Envelope and Given Correlation Interval

For the first approximation of the power spectrum (8.4) one can use a rational approximation with only one pole. Although this approximation is somewhat poor it is considered here because this scheme facilitates modeling the non-Gaussian random process with any PDF of the envelope.

Let the PDF $p_A(A)$ of the envelope of a narrow band (with bandwidth f_i) signal be given. Then the bandwidth of the narrow band envelope is approximately $2f_i$, and consequently, the correlation interval is

$$\tau_{\text{corr}\,i} = \frac{1}{2f_i} \tag{8.9}$$

If the phase $\varphi(t)$ is uniformly distributed on the interval $[-\pi, \pi]$ and does not depend on $A(t)$ then

$$x(t) = A(t) \cos \varphi(t) \cos 2\pi f_0 + A(t) \sin \varphi(t) \cos 2\pi f_0 \tag{8.10}$$

is a stationary narrow band random process with the envelope $A(t)$ and the power spectrum concentrated around the frequency f_0 (see section 4.7). If $A(t)$ is a λ process with certain PDF and correlation interval $\tau_{\text{corr}\,i}$, the problem is solved. The desired generating SDE can be obtained by applying the results of Chapter 7. The block diagram of the corresponding simulator is shown in Fig. 8.2.

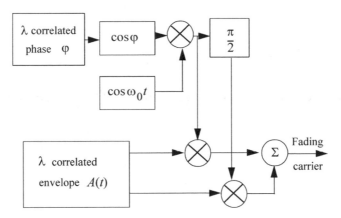

Figure 8.2 Simulator of a narrow band process with given PDF of the envelope and bandwidth.

Example: Nakagami fading. In this case the distribution of the envelope is defined by eq. (8.4) and the corresponding SDE for the envelope is (see Section 7.2.2):

$$\overset{\circ}{A} = \frac{K^2}{2} \left(\frac{2m-1}{A} - \frac{2mA}{\Omega} + 2\delta(x) \right) + K\xi(t) \tag{8.11}$$

The parameter K can be found from the correlation interval to be

$$K = \sqrt{\frac{\Omega}{m\,\tau_{\text{corr}}}} = \sqrt{\frac{\Omega}{m}f_i} = \sqrt{\frac{\Omega\,\nu}{m\,c}f_0}\cos\alpha_i \tag{8.12}$$

where c is the speed of light. Finally, the generating SDE of the envelope has the form

$$\frac{dA}{dt} = \frac{\Omega}{2m}\frac{\nu}{c}f_0 \cos \alpha_i \left(\frac{2m-1}{A} - \frac{2mA}{\Omega} + 2\delta(x)\right) + \sqrt{\frac{\Omega}{m}\frac{\nu}{c}f_0 \cos \alpha_i} \, \xi(t). \qquad (8.13)$$

The results of the numerical simulation are shown in Figs. 8.3 and 8.4.

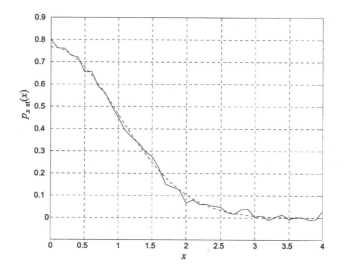

Figure 8.3 Estimated (solid line) and theoretical (dashed line) PDF of the process generated by SDE (8.13); $m = 1/2, \Omega = 1$.

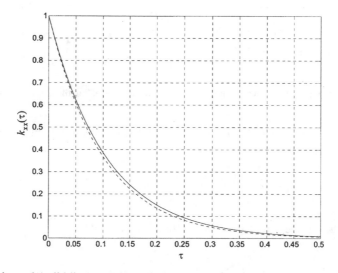

Figure 8.4 Estimated (solid line) and theoretical (dashed line) correlation functions of the envelope generated by SDE (8.13); $\tau = 0.1$ s.

8.1.2.2 A Process with a Given Spectrum and Sub-Rayleigh PDF

A regular procedure to obtain the desired correlation function (power spectrum) is known only for the Gaussian random process. This synthesis procedure was considered earlier in Chapter 7. In the same chapter we showed that the compound representation of the random process could be used to obtain processes with a symmetrical PDF and any desired rational spectrum. In this case the complex (two-dimensional) Gaussian process $x_n(t)$ is modulated by some positive λ process $s(t)$

$$x(t) = x_n(t)s(t) \tag{8.14}$$

Here the Gaussian process $x_n(t)$ defines the spectral (correlation) properties of the process $x(t)$, and the modulating signal $s(t)$ defines the distribution of $x(t)$. Let $p_{Ax}(A_x)$ be a distribution of the envelope of the process $x(t)$. Then the distribution $p_s(s)$ of the modulation is related to $p_{Ax}(A_x)$ as

$$p_{Ax}(A_x) = \frac{A_x}{\sigma^2} \int_0^\infty \exp\left[-\frac{A_x^2}{2\sigma^2\mu^2}\right] s(\mu) \frac{d\mu}{\mu^2} \tag{8.15}$$

where σ^2 is the parameter of the Rayleigh envelope of the process $x_n(t)$.

If the modulation $s(t)$ is Nakagami distributed, then the resulting process has PDF of the envelope in the form

$$p_{Ax}(A_x) = \frac{\left(\dfrac{m}{2\Omega\sigma^2}\right)^{\frac{m+1}{2}} A_x^m K_{m-1}\left(\sqrt{\dfrac{2m}{\Omega\sigma^2}}A_x\right)}{\Gamma(m)} \tag{8.16}$$

where $K_v(°)$ is the modified Bessel function [23]. It is worth mentioning that the mean value of the modulation $s(t)$ should be chosen to be 1, in order to keep the mean energy of the process $x(t)$ the same as the mean energy of the Gaussian process $x_n(t)$. This leads to the following condition

$$\int_0^\infty s p_s(s) ds = 1 \tag{8.17}$$

Which is satisfied if

$$\Omega = m\frac{\Gamma^2(m)}{\Gamma^2(m+0.5)} \tag{8.18}$$

Finally, the PDF of the envelope is

$$p_{Ax}(A_x) = \frac{\left(\dfrac{\Gamma^2(m)}{2\Gamma^2(m+0.5)\sigma^2}\right)^{\frac{m+1}{2}} A_x^m K_{m-1}\left(\sqrt{\dfrac{2\Gamma^2(m)}{\Gamma^2(m+0.5)\sigma^2}}A_x\right)}{\Gamma(m)} \tag{8.19}$$

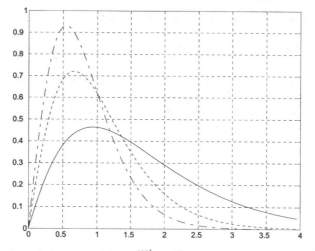

Figure 8.5 Examples of K-PDF $p_x(x) = \frac{b^{\nu+1}}{2^{\nu+1}\Gamma(\nu)} x^{\nu+1} K_\nu(bx)$. (1) $\nu = 1, b = \sqrt{2}$—solid line; (2) $\nu = 2, b = 2\sqrt{2}$—dashed line; (3) $\nu = 3, b = 3\sqrt{2}$—dashed-dotted line; (4) Rayleigh distribution with $\sigma^2 = 1$.

The PDF (8.19) defines the K-distributed envelope, which is a suitable model for sub-Rayleigh fading. When $m = \infty$, PDF (8.19) becomes a Rayleigh one.[1] Some plots of the K distribution are given in Fig. 8.5.

The desired power density spectrum (8.6) can be approximated by the third order FIR filter response

$$y(n) = b_0 x(n) + b_1 x(n-1) + b_2 x(n-2) + b_3 x(n-3) \tag{8.20}$$

where the coefficients $\{b_i\}$ can be calculated as

$$\omega_p = \frac{2\pi F_d}{F_s} \tag{8.21}$$

$$b_0 = 1 \tag{8.22}$$

$$\rho = 0.43 \tag{8.23}$$

$$b_1 = \frac{(\omega_p + 2)(2\omega_p^2 - 8) + (\omega_p - 2)(\omega_p^2 + 4\rho\omega_p + 4)}{(\omega_p + 2)(\omega_p^2 + 4\rho\omega_p + 4)} \tag{8.24}$$

$$b_2 = \frac{(\omega_p - 2)(2\omega_p^2 - 8) + (\omega_p + 2)(\omega_p^2 - 4\rho\omega_p + 4)}{(\omega_p + 2)(\omega_p^2 + 4\rho\omega_p + 4)} \tag{8.25}$$

$$b_3 = \frac{(\omega_p - 2)(\omega_p^2 - 4\rho\omega_p + 4)}{(\omega_p + 2)(\omega_p^2 + 4\rho\omega_p + 4)} \tag{8.26}$$

[1]In this case the modulation $s(t)$ becomes the deterministic function $s(t) = 1, p_s(s) = \delta(s-1)$.

Let z_1, z_2, z_3 be the roots of the polynomial

$$b(z) = b_3 z^3 + b_2 z^2 + b_1 z + 1 \tag{8.27}$$

and s_i and λ be defined as

$$s_i = \alpha_i + j\omega_i = F_s \ln z_i, i = 1, 2, 3 \tag{8.28}$$

$$\lambda = \frac{\max\{\alpha_i\}}{2} \tag{8.29}$$

This allows one to define a new form filter with poles

$$s_{ni} = s_i + \lambda = \alpha_i + \lambda + j\omega_i \tag{8.30}$$

$$z_{ni} = z_i \exp\left[\frac{\lambda}{F_s}\right] \tag{8.31}$$

which is again a stable filter. Then, if process $x_n(t)$ has the spectrum defined by the form filter with poles given by eq. (8.31), and $s(t)$ is a λ process with the decay of the correlation function defined by eq. (8.29), the process $x(t) = x_n(t)s(t)$ has the spectrum defined by eqs. (8.20)–(8.26), i.e. it has the desired spectral properties.

Results of the simulation are shown in Figs. 8.6–8.8.

Example: Nakagami sub-Rayleigh fading.[2] A similar technique can be used for Nakagami sub-Rayleigh fading, i.e. for the case $m < 1$. In this case the modulation process

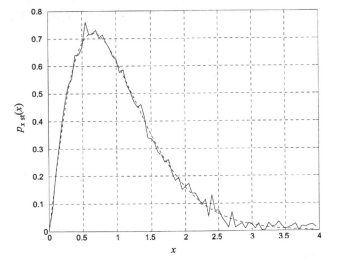

Figure 8.6 Estimated (solid line) and theoretical PDF, corresponding to eq. (8.14).

---

[2]Material for these examples was prepared in collaboration with Ms Vanja Subotic.

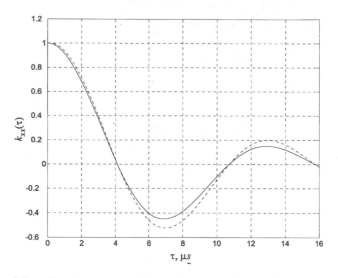

Figure 8.7 Correlation function corresponding to spectrum (8.6) (solid line), and those obtained according to eq. (8.31) (dashed line).

is a square root of a beta distributed random process [24–26]

$$p_{s^2}(y) = \frac{y^{m-1}(1-y)^{-m}}{B(m, 1-m)}, \qquad 0 \le m \le 1 \tag{8.32}$$

Results of the simulation are shown in Figs. 8.6–8.8. Samples of fading have been used to produce an error sequence (see section 8.2) and estimate the so-called $P(n, m)$ characteristic, i.e. the probability that there are exactly n errors in a block of m symbols (Figs. 8.9 and 8.10).

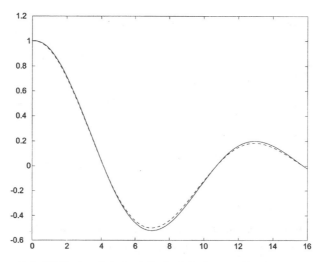

Figure 8.8 Estimated (solid line) and desired (dashed line) normalized correlation function, defined by eq. (8.14).

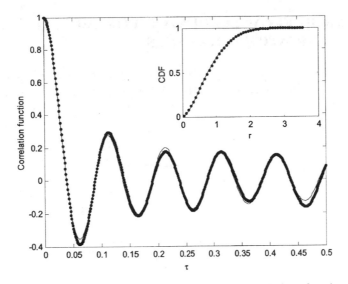

Figure 8.9 Nakagami cumulative distribution (CDF) and correlation functions for $m = 0.7$, $\Omega = 1$, $f_D - 10\,\text{Hz}$ and $\Delta \tau = 1 \times 10^{-3}$ s. Solid line—theory, dotted line—simulation.

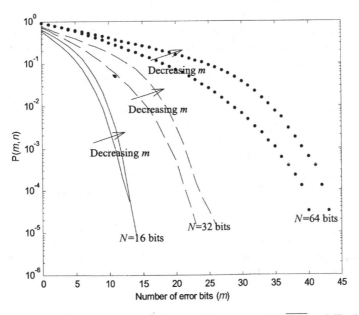

Figure 8.10 Nakagami $P(m, n)$ characteristic for $m = 0.5$ and $m = 0.7$, $\overline{SNR} = 5\,\text{dB}$, $f_D = 10\,\text{Hz}$ and $\Delta \tau = 1 \times 10^{-3}$.

8.2   AN ERROR FLOW SIMULATOR FOR DIGITAL COMMUNICATION CHANNELS

Most modern communications systems use discrete information sources and, consequently, discrete methods of modulation. A general scheme of a communication system of this type is given in Fig. 8.11 [3].

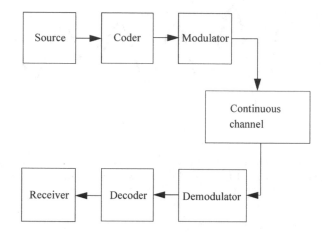

Figure 8.11   Digital communication system.

A large number of encoders–decoders (codecs) have been developed but their practical implementation sometimes leads to the results being far from the theoretical ones. The main reason for this failure is the fact that most scientific work in coding theory was applied to a binary symmetrical channel with independent errors—a model which cannot be considered as adequate for real communication conditions [27–29].

In recent years a number of works were devoted to the experimental investigation of real communication system behaviour and its description by different mathematical models [30–97]. These models allow one to obtain statistical characteristics of the error flow, and then to calculate the efficiency of different correcting codes. It was found that simple mathematical models merely describe channel properties. The complexity of more accurate models arises from the fact that the discrete channel involves the transformation of signals from discrete to continuous, and back again.

It can be seen from Fig. 8.11 that the discrete channel includes a continuous channel,[3] and that the properties of the latter affect the properties of the discrete channel under consideration. The transition from continuous to discrete signals involves a modulator and demodulator and, consequently, a discrete communication channel should be defined by both blocks. Analytical investigation of such models is complicated, but the numerical simulation produces important information for channel designers.

---

[3]This assertion allows the use of all models considered in the previous section as part of the discrete channel.

8.2.1   Error Flow in Digital Communication Systems

In the performance analysis of the modulator–demodulator (modem) where the average error rate is considered as the key parameter, the physical sources of errors are mathematically reasonably tractable and can be found from a mathematical model of the continuous part of the channel. For the coding of a discrete channel involving a modem and the propagation medium, any meaningful analytical description of channel memory in terms of the characteristics of individual physical causes of errors is difficult. It is more convenient to work with sample error sequences obtained from test runs of data transmitted on a given coding channel to obtain statistical descriptions of the error structure, and to use them in the design of error control systems. The required statistical parameters may be obtained from error sequences either by direct processing of the sequences or by forming a parameterized mathematical model that would "generate" similar sequences. The latter involves the following [30]:

- visualizing a model representative of real channel behaviour;

- developing methods to parameterize the model using error sequences;

- deriving the statistics required for code evaluation from the model.

The error process on a digital communication link can be viewed as a binary discrete-time stochastic process, i.e. a family of binary random variables $\{X_t, t \in I\}$ where I is the denumerable set of integers and t denotes time. The sequence

$$x = \ldots, x_{-1}, x_0, x_1, \ldots \tag{8.33}$$

representing a realization of the error process can be related to the channel input and output as follows.

Let $\{a_n\}$ be the digital sequence, and $\{b_n\}$ be the corresponding output sequence. The perturbation due to the channel between the input and output can be considered as an introduction of a noise sequence ξ_n originated from a hypothetical random noise source. This may be represented mathematically as the addition of a noise digit to an input digit, i.e.

$$b_n = a_n + x_n \tag{8.34}$$

where $\{a_n\}$, $\{b_n\}$ and $\{x_n\}$ are elements of the Galois field $GF(q)$ and the addition is over the same field. For binary communications, $q = 2$, and the addition is modulo 2.

An error is said to occur whenever x_n is different from zero. Accordingly, an error sequence $\{x_n\}$ is a stochastic mapping Φ of the noise sequence $\{\xi_n\}$ onto $\{0, 1\}$.

8.2.2   A Model of Error Flow in a Digital Channel with Fading

For the time being a communication channel without fading is considered. In this case the received signal $\hat{s}(t)$ is given by [3]

$$\hat{s}(t) = \mu\, s_k(t - \tau) + K_{ch}\xi(t) \tag{8.35}$$

where $s_k(t)$ is the transmitted signal (logical **0** or **1** in the case of a binary channel), μ is a constant, describing the general attenuation of the signal in the medium of propagation, $\xi(t)$ is WGN of unit intensity and K_{ch}^2 is the intensity of WGN in the channel. After demodulation and decoding, the received signal $\hat{s}(t)$ is transformed to $s_{\hat{k}}(t)$. Error occurs if $s_{\hat{k}}(t)$ does not coincide with $s_k(t)$. The error probability p_{err} depends on the Signal-to-Noise Ratio (SNR), defined as

$$SNR = \frac{\mu^2 E(s_k)}{K_{ch}^2} = h^2 = \gamma \tag{8.36}$$

and on the demodulation scheme. A great number of modulations is treated in [3].

The error probability p_{err} does not depend on time and is equal to the average number of errors per bit transmitted. The error flow in this case can be modeled simply as the Poisson flow with intensity $\lambda = p_{err}$.

If fading is present in the channel, the received signal should be written as

$$\hat{s}(t) = \mu(t)s_k(t - \tau) + K\xi(t) \tag{8.37}$$

Now $\mu(t)$ is a random process and, consequently, SNR and the error probability should be changed in time (i.e. they both are random processes). In the case of slow flat fading, it can be assumed that SNR and p_{err} are slow functions of time, depending only on $\mu(t)$, defined as

$$h(t) = \frac{\mu^2(t)E(s_k)}{K_{ch}^2} \tag{8.38}$$

$$p_{err}(t) = F(h(t)) \tag{8.39}$$

where the structure of the function $F(°)$ is defined by demodulation methods. Corresponding error flow can now be modeled as a Poison flow with slow varying intensity $\lambda(t)$, equal to the instant error probability

$$\lambda(t) = p_{err}(t) \tag{8.40}$$

A corresponding scheme of the error flow simulator is shown in Fig. 8.12.

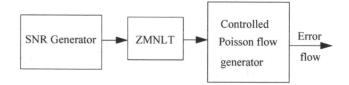

Figure 8.12 Error flow simulator.

It is worth noting that in many cases it is possible to calculate probability of error taking fading into account [3]. In this case the average probability of error would define a constant error rate over the duration of simulation. The model suggested here, while on a large scale leading to the same results, represents a finer structure of error flow since it follows local variations in the signal level. Consequently, it could be more appropriate for simulation of performance of relatively short codes or codes with memory.

Example: Error flow in a channel with Nakagami fading. Let a communication channel experience Nakagami fading, described by PDF

$$p_A(A) = \frac{2}{\Gamma(m)} \left(\frac{m}{\Omega}\right)^m A^{2m-1} \exp\left[-\frac{m A^2}{\Omega}\right] \tag{8.41}$$

Thus, one can consider $\mu(t)$ in eq. (8.37) as a random process generated by the following SDE

$$\dot{\mu} = \frac{\Omega}{2 m \tau_{\text{corr}}} \left(\frac{2m-1}{\mu} - \frac{2m\mu}{\Omega}\right) + \sqrt{\frac{\Omega}{m \tau_{\text{corr}}}} \xi(t) \tag{8.42}$$

Being proportional to the quantity $z(t) = \mu^2(t)$, the SNR $\gamma(t)$ can be thus generated by SDE

$$\dot{\gamma} = \frac{\overline{\Omega}}{m \tau_{\text{corr}}} \left(2m - 1 - \frac{2m\gamma}{\overline{\Omega}}\right) + 2\sqrt{\frac{\overline{\Omega}\gamma}{m \tau_{\text{corr}}}} \xi(t) \tag{8.43}$$

where $\overline{\Omega}/m$ is the instantaneous SNR. The solution of this SDE $\gamma(t)$ can be used to drive the intensity of a Poisson flow of errors.

8.2.3 SDE Model of a Buoyant Antenna–Satellite Link

Small ships often employ a buoyant antenna array that extends into a ship's wake to support satellite communications at sea [98]. Due to the changing surface environment, the instantaneous gain and phase of each array element is unknown because of multiple paths, changing interelement geometry and wash-over effects. The complex nature of communication systems based on this type of antenna system requires efficient numerical simulation to realistically assess performance during the communication system design process. The aim of this section is to present a mathematical model of the fading in the buoyant antenna array–satellite link which is useful both for numerical simulation and analytical calculations.[4]

8.2.3.1 *Physical Model*

A buoyant antenna array is often used for ship–satellite communications, especially by small boats and submerged submarines. Under the ideal conditions, such an array is represented by N equally spaced antenna elements, all transmitting at full power. In practice, some fraction of the elements are washed over by the sea due to wave motion on the sea surface. This results in random fluctuations in the antenna pattern and the signal received at the satellite. For a relatively slow bit rate it is possible to adjust directivity of the antenna in such a way that there is always a direct line-of-sight (LoS) link; however, the energy level of this link varies. Additionally, there are also some diffusion components present due to reflection from the sea surface and random scattering in the atmosphere. Under such conditions it is possible

---

[4]This section is written in collaboration with Mr J. Weaver.

to assume that for a short period of time (when there is no change in the number and position of the active elements), the fading is described either by the Rice law [3]

$$p_A(a) = \frac{A}{\sigma^2} \exp\left(-\frac{A^2 + a^2}{2\sigma^2}\right) I_0\left(\frac{aA}{\sigma^2}\right), \qquad A \geq 0 \tag{8.44}$$

or by Nakagami fading [3]

$$p_A(a) = \frac{2}{\Gamma(m)} \left(\frac{m}{\Omega}\right)^m A^{2m-1} \exp\left[-\frac{mA^2}{\Omega}\right], \quad m > 1, \quad A \geq 0 \tag{8.45}$$

with parameters a, σ, Ω and m chosen to properly represent energy levels of the signal and noise. If only short messages are communicated, such as distress signals or very slow rate sensory information, a static model with constant parameters performs satisfactorily. If long-term communication is considered, these parameters cannot be treated as constants due to fluctuations of the geometry of the array which are themselves random processes. It is important to properly represent the effect of such fluctuations on the resulting quality of the communication channel.

The following physical model of the channel is accepted here. The fluctuation of the signal level is described as a continuous random process with its parameters randomly changing over a finite number of possibilities or states. The latter is justified since there is a finite number (however large) of possible combinations of active and passive array elements [98]. Transitions from state to state are random and described by an underlying Markov chain as described in Chapter 6. At this stage it is assumed that the parameters of fading in each state are known, either from measurements or from detailed simulation, and the main engineering focus is the effect of state changes on the overall performance of communication systems.

One of the most interesting questions that needs to be addressed is the effect of the intensity of state transitions. These transitions are defined by the roughness of the sea surface and the speed of the craft. While there is no exact formula which relates these parameters to the transitional probabilities of a Markov chain, it is reasonable to assume that faster transitions correspond to a rougher sea and a larger velocity. This chapter provides an analytical expression for the resulting PDF of the fading as a function of transitional intensity in the Markov chain. When experimental data becomes available it will be possible to trace a direct influence of the velocity and sea roughness on the bit error rate of the link.

8.2.3.2   *Phenomenological Model*

The mathematical model of the system is represented as M stochastic differential equations (SDEs), each representing a state of the antenna array

$$\dot{x} = \frac{K_2^{(n)}}{2} \frac{d}{dx} \ln p_n(x) + K_2^{(n)} \xi(t), \quad n = 1, \ldots, M \tag{8.46}$$

These sequences are randomly switched, maintaining a single point continuity between adjacent sequences, according to the state of the underlying Markov chain. Here $\xi(t)$ is the white Gaussian noise with unit spectral density, $p_n(x)$ is the desired marginal PDF of the solution and $K_2^{(n)}$ is a constant defining the correlation interval of the resulting process. The probability density function of the solution can be described by a system of generalized Fokker–Planck equations of the following form (see Section 6.2.3)

$$\frac{\partial}{\partial t} p^{*(l)}(x, t) = -\frac{\partial}{\partial x} [K_1^{(l)} p^{*(l)}(x, t)]$$

$$+ \frac{1}{2} \frac{\partial^2}{\partial x^2} [K_2^{(l)} p^{*(l)}(x, t)] - \nu_l p^{*(l)}(x, t) + \sum_{k \neq l}^{M} \nu_{kl} p^{*(k)}(x, t) \qquad (8.47)$$

Here ν_{kl} is the intensity of transition from the k-th state of the array to the l-th state of the array, and, as the total switching intensity from the l-th state

$$\nu_l = \sum_{k \neq l}^{M} \nu_{kl} \qquad (8.48)$$

in the steady state, the probability densities do not depend on time and so eq. (8.47) is further simplified to become an ordinary linear differential equation of the second order.

Transitional intensities ν_{kl} define the probability of each state of the antenna array. In particular, the vector of such probabilities $P = [P_1, \ldots, P_M]^T$ is the eigenvector of the matrix of transitional intensity $T = \{\nu_{kl}\}$

$$TP = P \qquad (8.49)$$

In general, the solution to eq. (8.47) can only be found numerically. However, for the two limiting cases, it is possible to obtain accurate and simple analytical approximations. Let τ_{sn} be a characteristic (correlation) time of the system in the n-th state. Then the switching is considered to be slow if $\max\{\nu_{kl} \tau_{sk}\} \ll 1$ and fast if $\max\{\nu_{kl} \tau_{sk}\} \gg 1$. In the case of slow fading, the PDF of the process is just a weighted sum of the PDF in isolated states (see Section 6.3.2)

$$p(x) = \sum_{n=1}^{M} P_n p_n(x) \qquad (8.50)$$

such that the switching effects last approximately τ_{sn} seconds and are subsequently dominated by internal dynamics of the system in the same fashion as in an isolated state. On the contrary, if fast switching is predominant, then all states in the distribution approach a single distribution given by the following expression (see Section 6.3.3)

$$p(x) = \frac{C}{\overline{K}_2(x)} \exp \left(2 \int_{x_0}^{x} \frac{\overline{K}_1(s)}{\overline{K}_2(s)} ds \right) \qquad (8.51)$$

where

$$\overline{K}_1(s) = \sum_{n=1}^{M} P_n \frac{K_2^{(n)}}{2} \frac{d}{dx} [\ln p_n(x)] \qquad (8.52)$$

and

$$\overline{K}_1(s) = \sum_{n=1}^{M} P_n K_2^{(n)}$$

(8.53)

To simplify the complexity of the model, a two-state Markov model is used to represent the possible states of the channel. The "Good" state represents the situation when almost all the elements of the array are active, and "Bad" represents situations with very few active arrays. In the former case, the fading could be described with the Nakagami law with a large value of the parameter $m_G > 1$ and the average value of the signal level with respect to noise Ω_G. The latter case is described by a relatively small $m_B \leq 1.5$ and the average level of signal $\Omega_B < \Omega_G$. The probability of the Bad state is P_B and the probability of the Good state is $P_G = 1 - P_B$. The transitional matrix T of the Markov chain is then defined as

$$T = \begin{bmatrix} (1-\lambda)P_G + \lambda & (1-\lambda)P_B \\ (1-\lambda)P_G & (1-\lambda)P_B + \lambda \end{bmatrix}$$

(8.54)

where the parameter $0 \leq \lambda \leq 1$ reflects correlation properties of the chain ($\lambda = 0$ corresponds to uncorrelated transitions). It is assumed that BPSK modulation [3] is used.

In an isolated state (i.e. when no switching is present), it is possible to show that fading can be described by the following stochastic differential equation [29]

$$\dot{x} = \frac{K_2^G}{2} \left(\frac{2m_G - 1}{x} - \frac{m_G x}{\Omega_G} \right) + K_2^G \xi(t)$$

$$\dot{x} = \frac{K_2^B}{2} \left(\frac{2m_B - 1}{x} - \frac{m_B x}{\Omega_B} \right) + K_2^B \xi(t)$$

(8.55)

For a small intensity of fading, the distribution of the magnitude is therefore

$$p(A) = P_G \frac{2}{\Gamma(m_G)} \left(\frac{m_G}{\Omega_G} \right)^{m_G} A^{2m_G - 1} \exp\left[-\frac{m_G A^2}{\Omega_G} \right] + P_B \frac{2}{\Gamma(m_B)} \left(\frac{m_B}{\Omega_B} \right)^{m_B} A^{2m_B - 1} \exp\left[-\frac{m_B A^2}{\Omega_B} \right]$$

(8.56)

For the fast switching it follows from eq. (8.52) that the distribution is again Nakagami-distributed

$$p(A) = \frac{2}{\Gamma(m_\infty)} \left(\frac{m_\infty}{\Omega_\infty} \right)^{m_\infty} A^{2m_\infty - 1} \exp\left[-\frac{m_\infty A^2}{\Omega_\infty} \right]$$

(8.57)

where

$$m_\infty = \frac{P_G m_G K_2^{(G)} + P_B m_B K_2^{(B)}}{P_G K_2^{(G)} + P_B K_2^{(B)}}$$

(8.58)

and

$$\Omega_\infty = \frac{P_G m_G K_2^{(G)} + P_B m_B K_2^{(B)}}{P_G \frac{m_G}{\Omega_G} K_2^{(G)} + P_B \frac{m_B}{\Omega_B} K_2^{(B)}} \qquad (8.59)$$

Finally, since the expression for the bit error rate of BPSK in Nakagami fading is known in terms of the hypergeometric function [2,23]

$$P_{err} = \frac{1}{2} - \sqrt{\frac{\Omega}{\pi m}} \frac{\Gamma(m+0.5)}{\Gamma(m)} {}_2F_1\left(\frac{2m+1}{2}, \frac{1}{2}; \frac{\Omega}{m}\right) \qquad (8.60)$$

one can easily obtain the expression for the BER in fading with changing parameters. For the case of a slow switching, eq. (8.60) produces

$$P_e = \frac{1}{2} - P_G \sqrt{\frac{\Omega_G}{\pi m_G}} \frac{\Gamma(m_G + 0.5)}{\Gamma(m_G)} {}_2F_1\left(\frac{2m_G+1}{2}, \frac{1}{2}; \frac{\Omega_G}{m_G}\right)$$

$$- \sqrt{\frac{\Omega_B}{\pi m_B}} \frac{\Gamma(m_B + 0.5)}{\Gamma(m_B)} {}_2F_1\left(\frac{2m_B+1}{2}, \frac{1}{2}; \frac{\Omega_B}{m_B}\right) \qquad (8.61)$$

For the fast switching, the following is produced

$$P_e = \frac{1}{2} - \sqrt{\frac{\Omega_\infty}{\pi m_\infty}} \frac{\Gamma(m_\infty + 0.5)}{\Gamma(m_\infty)} {}_2F_1\left(\frac{2m_\infty+1}{2}, \frac{1}{2}; \frac{\Omega_\infty}{m_\infty}\right) \qquad (8.62)$$

Since parameters P_1, P_2 and the intensity of transitions directly affect BER and are functions of the roughness of the sea and the velocity of the craft, the developed equations could be a useful tool for system designers.

8.2.3.3   Numerical Simulation

The modified Ozaki scheme (see Section C.3 in Appendix C on the web and [27–29]) could be used to simulate each non-linear SDE

$$f_n = \frac{d}{dx}[p_{st}(x_n)] \qquad (8.63)$$

$$K_t = \frac{1}{\Delta t} \log\left[1 + \frac{f_n}{x_n f_n}[\exp(f_n \Delta t) - 1]\right] \qquad (8.64)$$

$$\nu_t = \sqrt{\frac{\exp(2K_t \Delta t) - 1}{2K_t}} \qquad (8.65)$$

$$x_{n+1} = |\exp(K_t \Delta t)x_n + \nu_t \xi_i| \qquad (8.66)$$

where K_t is a factor inversely proportional to the correlation interval of the fading and ξ_i is a Gaussian iid process. The processes are then commutated by the two-state Markov process. Performance of a few representative signalling techniques is evaluated with the numerical model.

Additionally, the two composite processes are compared on the basis of the Kolmogorov–Smirnov "Goodness-of-Fit" measure [99]. This measure maps the greatest absolute error from the empirical CDF to a reference CDF as shown below

$$D = \sup|\hat{P}_{\text{obs}}(x) - P_{\text{ref}}(x)| \tag{8.67}$$

and

$$P(D > \text{observed}) = Q_{ks}\left[D\left(\sqrt{N} + 0.12 + \frac{0.11}{\sqrt{N}}\right)\right] \tag{8.68}$$

where N is the number of bins in the empirical CDF and Q_{ks} is defined as

$$Q_{ks}(\lambda) = 2\sum_{j=1}^{\infty}(-1)^{j-1}\exp[-2j^2\lambda^2] \tag{8.69}$$

Using this measure, both fast and slow switching processes were highly likely to match their theoretical counterparts. In the slow switching case, the match was a virtual certainty ($\lambda < 0.4$) and in the fast switching case the probability of a theoretical match was 98.5% ($\lambda \approx 0.5$). Below, the results for the slow switching are reviewed first (PDF and K-S performance) followed by similar graphs for the fast switching case. Fig. 8.13 shows the results of numerical simulation of the suggested model.

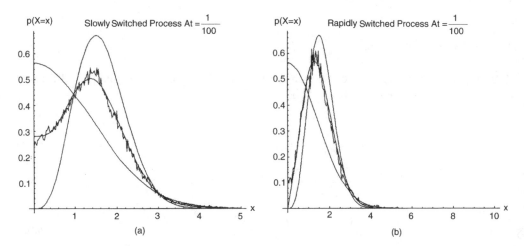

Figure 8.13 Results of the numerical simulation of SDE (8.55) with random structure: (a) slow fading; (b) fast fading.

8.3 A SIMULATOR OF RADAR SEA CLUTTER WITH A NON-RAYLEIGH ENVELOPE

Clutter, being an unwanted phenomenon which usually accompanies target detection, has been the subject of numerous investigations related to the performance of sea microwave radar [100–153]. In many real situations (low grazing angles, requirement of high resolution) the probability density function of the clutter envelope (EPDF) does not fit the Rayleigh law [100–153], i.e. it exhibits higher tails and a larger standard deviation–to–mean ratio. Here two models of a simulator of radar clutter, utilizing a second order SDE considered earlier in Chapter 7, and generating clutter with the K or Weibull distribution of the envelope, are considered. Such simulators can be used in the testing of radar target discrimination schemes ([117–153] and many others).

8.3.1 Modeling and Simulation of the K-Distributed Clutter

The K-distribution of the clutter envelope, which is now widely applied both to the description of microwave scattering by a rough ocean surface and to the synthesis of the target detection algorithms [101–108], provides a better agreement with experimental data than the Rayleigh one. In this case the envelope of the signal is distributed according to the following PDF

$$p_A(A) = \frac{b^{\nu+1}}{2^{\nu-1}\Gamma(\nu)} A^\nu K_{\nu-1}(bA), \qquad A \geq 0 \tag{8.70}$$

where $\Gamma(^\circ)$ is the Euler Gamma function [23], ν is a positive shape parameter, $K_\nu(^\circ)$ is the modified second kind Bessel function of order ν [23] and b is related to the average power σ^2 by the formula

$$b^2 = 2\nu/\sigma^2 \tag{8.71}$$

which generalizes the Rayleigh law in the case $\nu \to 0$.

A non-Rayleigh, and in particular K-distributed, correlated clutter may be presented as a narrow band stationary process with a symmetrical non-Gaussian distribution of instantaneous values [101–108]. In addition to having its power spectrum concentrated around its central frequency, the corresponding generating system has to be at least of the second order, and the clutter itself can be considered as a component of the two-dimensional Markovian process, i.e. as a solution, for example, of the following Duffing stochastic SDE

$$\ddot{x} + 2\alpha\dot{x} + F(x) = K\xi(t) \tag{8.72}$$

where $\xi(t)$ is the white Gaussian noise (WGN) of a unit one-sided power spectral density. To find the parameters of eq. (8.72) the following steps have to be taken.

1. Choose a non-linear function $F(x)$ which determines the desired distribution of the instantaneous values of the process $x(t)$.

2. Estimate the damping factor α and intensity K^2 of the WGN, which together define correlation properties of the process $x(t)$.

The relation between the stationary distribution $p_{x\,\text{st}}(x)$ of the solution of SDE (8.72) and the non-linear term $F(x)$ in its operator has been found to be (see Section 7.3.3, eq. (7.201))

$$p_{x\,\text{st}(x)} = C \exp\left[-\frac{4\alpha}{k^2}\int_{x_0}^{x} F(s)\,\mathrm{d}s\right] \tag{8.73}$$

where C is the normalization condition constant.

Function $F(x)$ in SDE (8.72), generally non-linear, can be determined by inverting eq. (8.73) (see eq. (7.202))

$$F(x) = -\frac{K^2}{4\alpha}\frac{\mathrm{d}}{\mathrm{d}x}\ln p_{x\,\text{st}(x)} \tag{8.74}$$

In order to use the synthesis equation (8.74) the exact form of the PDF $p_{x\,\text{st}}(x)$ of the process $x(t)$ must be evaluated from the known PDF of the envelope (8.70). Functions $p_A(A)$ and $p_{x\,\text{st}}(x)$ are related by the Blanc-Lapierre transformation (4.331) for narrow band processes

$$p_{x\,\text{st}}(x) = \frac{1}{2\pi}\int_{-\infty}^{\infty} \exp[i\,s\,x]\mathrm{d}s \int_{0}^{\infty} p_A(A)J_0(s\,A)\mathrm{d}A \tag{8.75}$$

This has been accomplished in Section 4.7.4.1, eq. (4.436)

$$p_{x\,\text{st}}(x) = \frac{2^{-\nu+\frac{1}{2}}b}{\Gamma(\nu)\sqrt{\pi}}(b|x|)^{\nu-\frac{1}{2}}K_{\nu-\frac{1}{2}}(b|x|) \tag{8.76}$$

To calculate the logarithmic derivative of the PDF (8.76) one can use the well known identity [23]

$$\frac{\mathrm{d}}{\mathrm{d}z}\left[\frac{K_\nu(z)}{z^\nu}\right] = -\frac{K_{\nu+1}(z)}{z^\nu} \tag{8.77}$$

which leads to

$$\frac{p'_{x\,\text{st}}(x)}{p_{x\,\text{st}}(x)} = -b\,\text{sgn}(x)\frac{K_{3/2-\nu}(b|x|)}{K_{1/2-\nu}(b|x|)} = -b\,\text{sgn}(x)\frac{K_{\nu-3/2}(b|x|)}{K_{\nu-1/2}(b|x|)} \tag{8.78}$$

As a result, the generating SDE of a narrow band random process with K-distributed envelope is given by

$$\ddot{x} + \alpha\dot{x} + \text{sgn}(x)\frac{K^2bK_{\nu-3/2}(b|x|)}{4\,\alpha K_{\nu-1/2}(b|x|)} = K\xi(t) \tag{8.79}$$

The next problem is to estimate the mean value of the central frequency of the narrow band process $x(t)$. For this purpose one can approximate the non-linear function $F(x)$ in eq. (8.79) by a polynomial series as in Section 7.3.3:

$$\ddot{x} + \alpha \dot{x} + \left(\sum_{i=0}^{\infty} a_{2i+1} x^{2i} \right) \omega_0 x = K \xi(t) \tag{8.80}$$

where $a_1 = 1$ and a_{2i+1} are the coefficients of $F(x)$'s power expansion. For this purpose one may use the following presentation of the modified Bessel function around zero [23]

$$K_\nu(x) \approx \frac{\Gamma(\nu) 2^{\nu-1}}{x^\nu} \left[1 + \frac{\Gamma(1-\nu)}{4\Gamma(2-\nu)} x^2 \right], \qquad x > 0 \tag{8.81}$$

Taking expansion (8.81) into account, the non-linear function $F(x)$ can be rewritten as

$$
\begin{aligned}
F(x) &= \frac{K^2 b^2 \Gamma(\nu - 3/2)}{8\, \alpha \Gamma(\nu - 1/2)} x \cdot \frac{\left[1 + \dfrac{\Gamma(-1/2 - \nu)}{4\Gamma(1/2 - \nu)} x^2 \right]}{\left[1 + \dfrac{\Gamma(1/2 - \nu)}{4\Gamma(3/2 - \nu)} x^2 \right]} \\[2mm]
&\approx \frac{K^2 b^2 \Gamma(\nu - 3/2)}{8\, \alpha \Gamma(\nu - 1/2)} x \left[1 + \left(\frac{\Gamma(-1/2 - \nu)}{\Gamma(1/2 - \nu)} - \frac{\Gamma(1/2 - \nu)}{\Gamma(3/2 - \nu)} \right) \frac{b^2 x^2}{4} \right]
\end{aligned}
\tag{8.82}
$$

Fig. 8.14 shows the satisfactory quality of such an approximation.

It is was shown in Section 4.7 that a narrow band process can be represented as

$$x(t) = A(t) \cos(\Psi(t)) = A(t) \cos[\omega_0 t + \varphi(t)] \tag{8.83}$$

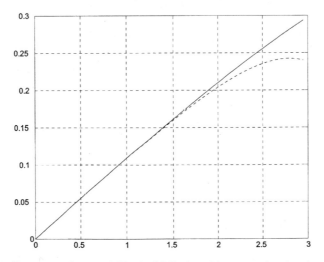

Figure 8.14 Non-linear term in eq. (8.79) (solid line) and its approximation in eq. (8.82) (dashed line).

where $A(t)$ and $\varphi(t)$ are the slowly varying envelope and phase, respectively, defined via the Hilbert transform $\hat{x}(t)$ of the process $x(t)$. In this case

$$A^2(t) = \hat{x}^2(t) + x^2(t)$$

$$\Psi(t) = \operatorname{atan} \frac{\hat{x}(t)}{x(t)}$$

(8.84)

Since for a narrow band process its Hilbert transform can be approximated by a derivative of the original process (see Section 4.7)

$$\dot{x} = -\omega_0 A(t) \sin \Psi(t)$$

(8.85)

an equivalent first order SDE can be obtained [154]

$$\frac{dA}{dt} = -2\alpha \sin \Psi \left[A \sin \Psi - \frac{1}{2\alpha\omega_0} \sum_{i=0}^{\infty} a_{2i+1} A \cos \Psi^{2i+1} \right] - \frac{\sin \Psi}{\omega_0} \xi(t)$$

(8.86)

$$\frac{dA}{dt} = -2\frac{\alpha}{A} \cos \Psi \left[A \sin \Psi - \frac{1}{2\alpha\omega_0} \sum_{i=0}^{\infty} a_{2i+1} A \cos \Psi^{2i+1} \right] - \frac{\cos \Psi}{A\omega_0} \xi(t)$$

(8.87)

which, in turn, can be simplified to

$$\frac{dA}{dt} = -2\alpha\theta_s + \frac{K^2}{4A} + \frac{K}{2\omega_0} \gamma_1(t)$$

(8.88)

$$\frac{d\Psi}{dt} = -2\alpha\theta_c + \frac{K^2}{4A\omega_0} + \frac{K}{2\omega_0} \gamma_1(t) = w(t)$$

(8.89)

where θ_s and θ_c are regular terms of eqs. (8.86) and (8.87), respectively [155], $\gamma_1(t)$, and $\gamma_1(t)$ are independent white Gaussian noises with spectral density $0.5 K^2$. Averaging both sides of eq. (8.89) over A, the mean angular frequency of the power spectrum of the process $x(t)$ can be obtained as

$$w = \omega_0 \left[1 + \sum_{i=1}^{\infty} \frac{a_{2i+1}(2i)!}{(i!)^2 2^i} \int_{-\infty}^{\infty} p_A(A) A^{2i+1} dA \right]$$

$$= \omega_0 \left[1 + \sum_{i=1}^{\infty} \frac{a_{2i+1}(2i)!}{(i!)^2 2^i} m_{2i+1} \right]$$

(8.90)

The initial moments of K-distribution are [101–103]

$$m_l = \int_0^{\infty} A^l p_A(A) dA = \frac{b^{\nu+1}}{2^{\nu-1}\Gamma(\nu)} \int_0^{\infty} A^{l+\nu} K_{\nu-1}(bA) dA = \left(\frac{2}{b}\right)_l \frac{\Gamma\left(\nu + \frac{l}{2}\right)\Gamma\left(1 + \frac{l}{2}\right)}{\Gamma(\nu)}$$

(8.91)

Using the latter expression and taking into account the fact that $b^2 = 2\nu/\sigma^2$, one can obtain the following expressions for the first three even moments (8.91)

$$m_2 = 2\sigma^2 \frac{\nu + 1}{\nu}$$

$$m_3 = \frac{6\sqrt{\pi}\,\Gamma(\nu + 3/2)}{b^3}\frac{}{\Gamma(\nu)} \tag{8.92}$$

$$m_4 = \frac{8\sigma^2(\nu + 2)(\nu + 1)}{\nu^2}$$

Approximation (8.82) now can be rewritten as

$$F(x) \approx \frac{K^2 b^2 \Gamma(\nu - 3/2)}{8\,\alpha\Gamma(\nu - 1/2)} x[1 + a_3 m_3 x^2] \tag{8.93}$$

where

$$a_3 = \left(\frac{\Gamma(-1/2 - \nu)}{\Gamma(1/2 - \nu)} - \frac{\Gamma(1/2 - \nu)}{\Gamma(3/2 - \nu)}\right)\frac{b^2}{4} \tag{8.94}$$

and

$$\omega_0^2 = \frac{K^2 b^2 \Gamma(\nu - 3/2)}{8\,\alpha\Gamma(\nu - 1/2)} \tag{8.95}$$

Taking into account only terms of order 3 in eq. (8.80), one can arrive at the following expression for the central frequency of the solution of SDE (8.80)

$$\omega_c = \omega_0(1 + a_3\,m_3) \tag{8.96}$$

The next step is to show that if the parameter ν approaches infinity, the generating SDE (8.79) tends to a linear second order equation which generates a narrow band Gaussian process with the corresponding Rayleigh PDF.

This can be accomplished by considering the following asymptotic expansion [23]

$$K_\nu(\nu z) \sim \sqrt{\frac{\pi}{2\nu}}\frac{\exp(-\nu\eta)}{(1 + z^2)^{1/4}} \tag{8.97}$$

where

$$\eta = \sqrt{1 + z^2} + \ln\frac{z}{1 + \sqrt{1 + z^2}} \tag{8.98}$$

Using eq. (8.97), one can obtain that

$$K_\nu(b|x|) = K_\nu(\nu z), z = \frac{b|x|}{\nu} \tag{8.99}$$

If x is fixed and $z \to 0$, eq. (8.98) can be simplified to

$$\eta \approx 1 + \ln \frac{z}{2} \tag{8.100}$$

thus producing a simple approximation for $K_\nu(\nu z)$ and $K_\nu(b|x|)$ in the following form

$$K_\nu(\nu z) \approx \sqrt{\frac{\pi}{2\nu}} \exp\left(-\nu \ln \frac{z}{2}\right) = \sqrt{\frac{\pi}{2\nu}} \exp(-0.307\,\nu) z^{-\nu} \tag{8.101}$$

$$K_\nu(b|x|) = \sqrt{\frac{\pi}{2\nu}} \exp(-0.307\,\nu) b^{-\nu)|x|^{-\nu}} \tag{8.102}$$

Finally, substituting expression (8.102) into eq. (8.78), one obtains that

$$\frac{p'_{x\,\mathrm{st}}(x)}{p_{x\,\mathrm{st}}(x)} = -b\,\mathrm{sgn}(x) \frac{K_{\nu-3/2}(b|x|)}{K_{\nu-1/2}(b|x|)}$$

$$\approx -b \frac{\sqrt{\pi/(2\nu - 3)} \exp(-0.307(\nu - 3/2)) b^{\frac{3}{2}-\nu}}{\sqrt{\pi/(2\nu - 1)} \exp(-0.307(\nu - 1/2)) b^{\frac{3}{2}-\nu}} \tag{8.103}$$

$$= -b^2 \frac{ex}{2} \sqrt{\frac{\nu - \frac{1}{2}}{\nu - \frac{3}{2}}} = -\omega_{0r}^2 x$$

The last expression corresponds to the linear function $F(x)$ in SDE (8.79)

$$\ddot{x} + \alpha \dot{x} + \omega_{0r}^2 x = K\xi(t) \tag{8.104}$$

which generates a narrow band Gaussian process with a central frequency ω_{0r} and with an envelope distributed by the Rayleigh law.

The next step is to tune the correlation function of the envelope. It is reasonable to assume that a small deviation of the PDF of the envelope does not significantly influence the correlation time, and to use the approximation of the K-distribution by the Nakagami law (8.4). Clearly, the most feasible approximation condition is the equality of the second and fourth moments of both distributions. In the case of the Nakagami PDF, they are equal to [6]

$$m_{N2} = \Omega \tag{8.105}$$

and

$$m_{4N} = \frac{m+1}{m} \Omega^2 \tag{8.106}$$

It follows from eq. (8.92) that

$$\Omega = 2\sigma^2 \frac{\nu + 1}{\nu} \tag{8.107}$$

$$\frac{m+1}{m} \Omega^2 = \frac{8\sigma^2(\nu + 2)(\nu + 1)}{\nu^2} \tag{8.108}$$

and, thus

$$m = \frac{\nu + 1}{\nu + 2} \tag{8.109}$$

The SDE which generates the Nakagami distribution is well known (7.41)

$$\dot{A} = \frac{K_N^2}{2}\left(\frac{2m-1}{A} - \frac{2mA}{\Omega}\right) + K_N\xi_N(t) \tag{8.110}$$

where WGN $\xi_N(t)$ has unit intensity and $K_N = 0.5\,K$. The correlation function of the SDE (8.110) solution may be written as (see eq. (7.43))

$$C_{AA}(\tau) = \frac{\Omega\,\Gamma_2\left(m+\frac{1}{2}\right)}{4\,m\,\pi\Gamma(m)}\sum_{i=1}^{\infty}\frac{\Omega\,\Gamma^2\left(i-\frac{1}{2}\right)}{\Gamma(m+1)}\exp\left(-\frac{2\,m\,k\,i}{\Omega}|\tau|\right) \tag{8.111}$$

Taking into account only the first term in eq. (8.92), one can obtain an approximation for the envelope covariance function and correlation interval, respectively

$$C_A(\tau) \approx \frac{\Omega\,\Gamma^2\left(m+\frac{1}{2}\right)}{16\,m\,\pi\,\Gamma(m)\Gamma(m+1)}\exp\left(-\frac{2\,m\,K_N^2}{\Omega}|\tau|\right) \tag{8.112}$$

$$\tau_{\text{corr}} = \frac{\Omega}{2\,K_N^2\,m} \tag{8.113}$$

The latter allows one to complete the set of equations required to design a simulator. If the parameters ω_c, τ_{corr}, ν and σ^2 of sea clutter are given, then the steps to generate the SDE model are the following:

1. Calculate the spectral density K_n of the excitation from eq. (8.113)

$$K_N = \sqrt{\frac{2\,\sigma^2(\nu+2)}{\tau_{\text{corr}}}} \tag{8.114}$$

2. Using eq. (8.96) calculate the frequency

$$\omega_0 = \frac{\omega_c}{1 + a_3\,m_3} \tag{8.115}$$

3. Calculation of the damping factor

$$\alpha = \frac{(\nu+2)\Gamma(\nu-3/2)}{2\,\omega_0^2\Gamma(\nu-1/2)} \tag{8.116}$$

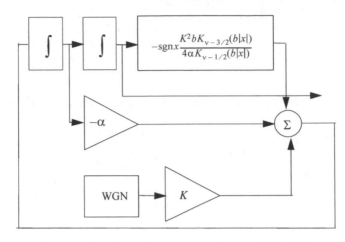

Figure 8.15   Form filter corresponding to SDE (8.79).

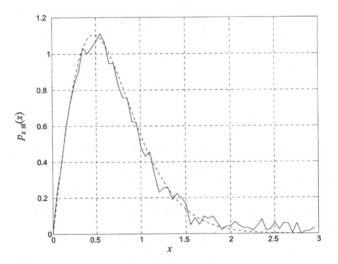

Figure 8.16   Desired (dashed line) and estimated (solid line) PDF obtained from SDE (8.79).

The form filter, corresponding to model (8.79), is shown in Fig. 8.15 and results of the numerical simulation are given in Fig. 8.16.

8.3.2   Modeling and Simulation of Weibull Clutter

A different kind of radar clutter is the Weibull distributed radar clutter [112,122–153]. In this case the PDF of the envelope has the following form

$$p_A(A) = \alpha \beta A^{\alpha-1} \exp[-\beta A^{\alpha}] \tag{8.117}$$

where α and β are related to the average power σ^2 of the quadrature components as

$$2\sigma^2 = \alpha^{-2/\beta}\Gamma\left(1 + \frac{2}{\beta}\right) \tag{8.118}$$

It is easy to see that, for the case of the Weibull distribution (8.118), eq. (7.244) can be rewritten in the form

$$\mathcal{F}_1(A) = \frac{\pi K^2 \alpha}{4\omega_0^2}[\beta A^\alpha - 1] \tag{8.119}$$

and, thus,

$$\frac{d}{dA}\mathcal{F}_1(A) = \frac{\pi K^2 \alpha^2 \beta}{4\omega_0^2}A^{\alpha-2} \tag{8.120}$$

$$\frac{d}{dA}\left[\frac{\frac{d}{dA}\mathcal{F}_1(A)}{A}\right] = \frac{\pi K^2 \alpha^2 (\alpha - 2)\beta}{4\omega_0^2}A^{\alpha-3} \tag{8.121}$$

The expression (8.120) has a zero limit when $A \to 0$ and $\alpha > 2$ (non-Rayleigh case) and is equal to $\pi K^2 \alpha^2 \beta / 4\omega_0^2$ in the Rayleigh case.

The integral on the left side of eq. (7.244) can easily be calculated by using eq. (8.121)

$$\int_0^x \frac{d}{dA}\left[\frac{\frac{d}{dA}\mathcal{F}_1(A)}{A}\right]\frac{dA}{\sqrt{x^2 - A^2}} = \frac{\pi K^2 \alpha^2 (\alpha - 2)\beta}{4\omega_0^2}\int_0^x \frac{A^{\alpha-3}dA}{\sqrt{x^2 - A^2}}$$

$$= \frac{\pi\sqrt{\pi}K^2\alpha^2(\alpha - 2)\beta}{4\omega_0^2}\frac{\Gamma\left(\dfrac{\alpha - 2}{2}\right)}{\Gamma\left(\dfrac{\alpha - 1}{2}\right)}x^{\alpha-3} \tag{8.122}$$

thus producing

$$f(x) = \frac{\pi\sqrt{\pi}K^2\alpha^2(\alpha - 2)\beta}{4\omega_0^2}\frac{\Gamma\left(\dfrac{\alpha - 2}{2}\right)}{\Gamma\left(\dfrac{\alpha - 1}{2}\right)}x^{\alpha-2} \tag{8.123}$$

and leading to a generating SDE of the van der Pol type (7.225)

$$\ddot{x} + \frac{\pi\sqrt{\pi}K^2\alpha^2(\alpha - 2)\beta}{4\omega_0^2}\frac{\Gamma\left(\dfrac{\alpha - 2}{2}\right)}{\Gamma\left(\dfrac{\alpha - 1}{2}\right)}x^{\alpha-2}\dot{x} + \omega_0^2 x = \omega_0^2\xi(t) \tag{8.124}$$

The next step is to find the correlation function of the solution $x(t)$. For this purpose one can rewrite eq. (8.124) in the Cauchy form

$$\dot{x} = y$$
$$\dot{y} = -\gamma x^{\alpha-2}y - \omega_0^2 x + \omega_0^2\xi(t) \tag{8.125}$$

where

$$\gamma = \frac{\pi\sqrt{\pi}K^2\alpha^2(\alpha-2)\beta}{4\omega_0^2}\frac{\Gamma\left(\dfrac{\alpha-2}{2}\right)}{\Gamma\left(\dfrac{\alpha-1}{2}\right)} \tag{8.126}$$

The drift coefficients K_{1x} and K_{1y} are thus defined as

$$\begin{aligned}
K_{1x}(x,y) &= y \\
K_{1y}(x,y) &= -\gamma x^{\alpha-2}y - \omega_0^2 x
\end{aligned} \tag{8.127}$$

The latter produces the following Gaussian approximation for the covariance functions [156,157]:

$$\frac{d}{d\tau}\begin{bmatrix} \langle x,x_\tau\rangle \\ \langle x,y_\tau\rangle \end{bmatrix} = \begin{bmatrix} \left\langle \dfrac{\partial}{\partial x}K_{1x}(x,y)\right\rangle & \left\langle \dfrac{\partial}{\partial y}K_{1x}(x,y)\right\rangle \\ \left\langle \dfrac{\partial}{\partial x}K_{1y}(x,y)\right\rangle & \left\langle \dfrac{\partial}{\partial y}K_{1y}(x,y)\right\rangle \end{bmatrix}\begin{bmatrix} \langle x,x_\tau\rangle \\ \langle x,y_\tau\rangle \end{bmatrix} \tag{8.128}$$

where $\langle^\circ\rangle$ denotes cumulant brackets (see section 4.4.2) and the initial conditions are

$$\begin{aligned}
\langle x,x_\tau\rangle|_{\tau=0} &= \sigma^2 \\
\langle x,y_\tau\rangle|_{\tau=0} &= 0
\end{aligned} \tag{8.129}$$

Using eq. (8.127) one can rewrite system of equation (8.128) in the following form

$$\frac{d}{d\tau}\begin{bmatrix} \langle x,x_\tau\rangle \\ \langle x,y_\tau\rangle \end{bmatrix} = \begin{bmatrix} 0 & 1 \\ -\gamma(\alpha-2)\langle x^{\alpha-3}y\rangle - \omega_0^2 & -\gamma\langle x^{\alpha-2}\rangle \end{bmatrix}\begin{bmatrix} \langle x,x_\tau\rangle \\ \langle x,y_\tau\rangle \end{bmatrix} \tag{8.130}$$

The solution $\langle x,x_\tau\rangle$ of eq. (8.130) is given by

$$\langle x,x_\tau\rangle = \frac{\sigma^2}{\lambda_2-\lambda_1}(\lambda_2 e^{\lambda_1\tau} - \lambda_1 e^{\lambda_2\tau}) \tag{8.131}$$

where λ_1 and λ_2 are the eigenvalues of eq. (8.130). Defining the correlation interval τ_{corr} as

$$\sigma^2\tau_{\text{corr}} = \int_0^\infty \langle x,x_\tau\rangle d\tau \tag{8.132}$$

one can obtain

$$\tau_{\text{corr}} = \frac{\lambda_1+\lambda_2}{\lambda_1\lambda_2} = -\frac{\gamma\langle x^{\alpha-2}\rangle}{\gamma\langle x^{\alpha-3}y\rangle - \omega_0^2} \tag{8.133}$$

We are now ready to present an algorithm for the synthesis of an SDE whose solution has the Weibull distributed envelope with parameters α, β, σ^2 related by formula (8.118) and correlation interval τ_{corr}. In fact, solving eq. (8.133) with respect to γ

$$\gamma = \frac{\pi\sqrt{\pi}K^2\alpha^2(\alpha-2)\beta}{4\,\omega_0^2}\frac{\Gamma\left(\dfrac{\alpha-2}{2}\right)}{\Gamma\left(\dfrac{\alpha-1}{2}\right)}\frac{\omega_0^2}{\langle x^{\alpha-2}\rangle - (\alpha-2)\langle x^{-3}y\rangle} \tag{8.134}$$

therefore

$$K^2 = \frac{\Gamma\left(\dfrac{\alpha-1}{2}\right)}{\Gamma\left(\dfrac{\alpha-2}{2}\right)}\frac{1}{\sqrt{\pi}\alpha^2(\alpha-2)\beta}\frac{8\,\omega_0^2}{\langle x^{\alpha-2}\rangle - (\alpha-2)\langle x^{\alpha-3}y\rangle} \tag{8.135}$$

Thus, all the parameters in eq. (8.124) are now known.

The form filter corresponding to the model (8.124) is given in Fig. 8.17 and the results of the numerical simulation are given in Fig. 8.18.

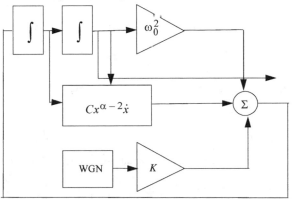

Figure 8.17 Form filter corresponding to SDE (8.124).

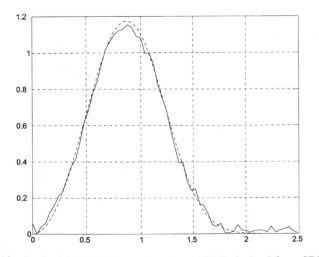

Figure 8.18 Desired (dashed line) and estimated PDF obtained from SDE (8.124).

8.4   MARKOV CHAIN MODELS IN COMMUNICATIONS

In contrast to continuous models of fading channels, considered above, this section deals with the attempt to model the error flow directly, i.e. focusing on the discrete channel itself. Such models are powerful tools in the investigation of coding and network performance [38–103,158–170]. Only a few topics are considered here. The reader is referred to [171] for more details and practical examples.

8.4.1   Two-State Markov Chain—Gilbert Model

Let a Markov chain with only two states, bad (B) and good (G) as shown in Fig. 8.19, be considered. This chain can be described by the following transitional matrix

$$\pi = \begin{bmatrix} 1-p & p \\ r & 1-r \end{bmatrix} \tag{8.136}$$

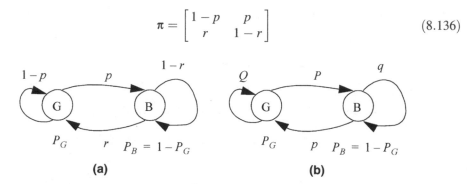

Figure 8.19   Two-state Markov chain: (a) Zorzi notation [62], (b) original Gilbert paper notation.

In the *good* state there are no errors, while in the *bad* state error appears with probability $1 - h$. We assume that the probability of error, P_{err}, and the parameter h are given. Since errors appear only in the *bad* state, one has

$$P_B = 1 - P_G = \frac{P_{err}}{1-h}, \qquad P_G = 1 - \frac{P_{err}}{1-h} \tag{8.137}$$

Furthermore, any two-state transitional matrix can be decomposed into the following form

$$T = (1-\lambda)\begin{bmatrix} P_G & P_B \\ P_G & P_B \end{bmatrix} + \lambda \begin{bmatrix} 1 & 0 \\ 0 & 1 \end{bmatrix} = \begin{bmatrix} (1-\lambda)P_G + \lambda & (1-\lambda)P_B \\ (1-\lambda)P_G & (1-\lambda)P_B + \lambda \end{bmatrix} \tag{8.138}$$

where parameter $0 \le \lambda \le 1$ reflects correlation properties of the Markov chain: $\lambda = 0$ corresponds to independent error occurrence, while $\lambda = 1$ results in completely deterministic error occurrence since the chain remains in only one of the states (see Table 8.3). Comparing eqs. (8.136) and (8.138) one obtains

$$r = (1-\lambda)P_G = (1-\lambda)\left(1 - \frac{P_{err}}{1-h}\right), \qquad p = (1-\lambda)P_B = (1-\lambda)\frac{P_{err}}{1-h} \tag{8.139}$$

Table 8.3 Correspondence between Markov chain parameters in different notations

Quantity	Zorzi [62]	Gilbert [31]	$P_{err} - \lambda$
Prob{G → G}	$1 - p$	Q	$1 - \dfrac{1 - \lambda}{1 - h} P_{err}$
Prob{G → B}	p	P	$\dfrac{1 - \lambda}{1 - h} P_{err}$
Prob{B → G}	r	p	$\dfrac{1 - \lambda}{1 - h}(1 - P_{err} - h)$
Prob{B → B}	$1 - r$	q	$\lambda + (1 - \lambda)\dfrac{P_{err}}{1 - h}$
Probability of a correct digit in B state	$h = 0$	h	h

If the error-free run statistics are used to estimate parameters of the Gilbert–Elliot model, one has to fit the experimental data $P(0^m|1)$ by a double exponential curve

$$P(0^m|1) \approx A J^m + (1 - A)L^m \tag{8.140}$$

In this case the parameters of the Gilbert model can be defined as [31]

$$h = \frac{LJ}{J - A(J - L)}, \quad P = \frac{(1 - L)(1 - J)}{1 - h}, \quad p = A(J - L) + (1 - J)\left(\frac{L - h}{1 - h}\right) \tag{8.141}$$

For a more general Elliot model, the same approximation (8.140) of the error-free run produces the following parameters of the model

$$\alpha = 1 - \frac{P_{err}}{JL}[A(J - L) - J] \tag{8.142}$$

$$\frac{P_{err} - 1 + \alpha JL - \sqrt{(P_{err} - 1 + \alpha JL)^2 - 4JL}}{2} \le h \le \frac{P_{err}}{\alpha - 1} \tag{8.143}$$

$$k = \frac{1 - h - P_{err}}{1 - \alpha h} \tag{8.144}$$

$$p = \frac{J + L - \dfrac{JL}{h} - h}{k - h} \tag{8.145}$$

$$P = 1 - \frac{JL}{hk} - p \tag{8.146}$$

8.4.2 Wang–Moayeri Model

One of the decisions which must be made by the simulator developer is to choose the proper order of Markov chain, N, and calculate elements of the transitional matrix. It was pointed out in [44] that the finite state models can produce unreasonable results if the model

$$R_k = p_k R$$

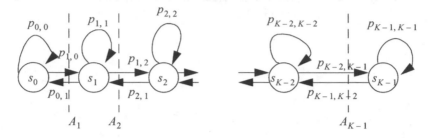

Figure 8.20   Wang–Moayeri model. Only transitions to a neighbouring state are allowed.

does not represent real physical characteristics of the continuous channel. One of the important features which must be represented is the so-called level crossing rate (LCR) of fading. Having this in mind, Wang and Moayeri have suggested a model [44] which is based on the assumption that transitions appear only between two neighbouring states and all other transitions are not allowed (see Fig. 8.20). Such an assumption is quite reasonable for the case of a very slow fading. Indeed, in this case the velocity

$$\nu_A(t) = \frac{\mathrm{d}}{\mathrm{d}t} A(t) \tag{8.147}$$

of the process is relatively small and one can assume that the signal level $A(t)$ is not able to change by an amount more than a quantization step

$$|A(t+T) - A(t)| \approx |\nu(t)|T < \min(\Delta A_i) \tag{8.148}$$

Of course this condition will be violated for a faster fading. However, for a large number of systems with high bit rate, the duration of bit, T, without interleaving can be accurately described by the Wang–Moayeri model, described below.

The range of magnitude variation can be divided into N regions by a set of thresholds A_k such that

$$0 = A_0 < A_1 < \cdots < A_K = \infty \tag{8.149}$$

The fading is said to be at state s_k if the magnitude of the signal resides inside interval $[A_k, A_{k+1})$. In each state, the probability of error e_k is assigned according to the modulation scheme used. The probability p_k of the channel being in state s_k can be calculated if the fading PDF $p_A(A)$ is known. The Wang–Moayeri model assumes that

$$p_k = \int_{A_k}^{A_{k+1}} p_A(A)\mathrm{d}A \tag{8.150}$$

Since only transitions into the neighbouring states are allowed, all transitional probabilities with indices differing by at least two are equal to zero:

$$p_{ij} = 0, \text{ for } |i - j| > 1, \quad 0 < i, \quad j < N - 1 \tag{8.151}$$

Now let one assume that the communication system considered transmits R symbols per second, i.e. the symbol rate is R. Then, on average, the number of symbols sent through the communication channel at state s_k is given by

$$R_k = p_k R \qquad (8.152)$$

A slow fading assumption implies that the level crossing rate $LCR(A_k)$ calculated at the level A_k is much smaller than R_k. In this case the transition probability $p_{k,k+1}$ can be approximated by the ratio of the upward level crossing rate $LCR(A_k)$ divided by an average symbol rate transmitted in the state s_k. Indeed, given an arbitrary time interval T_0, there will be approximately $LCR(A_k)T_0$ upward crossings of the level A_k, which corresponds to the channel state transitions from s_k to s_{k+1} (see Fig. 8.20). Thus

$$p_{k,k+1} = \frac{LCR(A_k)T_0}{R_k T_0} = \frac{LCR(A_k)}{R_k} = \frac{LCR(A_k)}{p_k R}, \quad 0 \le k \le N - 2 \qquad (8.153)$$

In the very same manner, transitional probabilities $P_{k,k-1}$ can be obtained as

$$p_{k,k-1} = \frac{LCR(A_{k-1})}{p_k R}, \quad 1 \le k \le N - 1 \qquad (8.154)$$

The rest of the probabilities can be found as follows

$$p_{0,0} = 1 - p_{0,1}$$
$$p_{k,k} = 1 - p_{k,k-1} - p_{k,k+1}, \quad 1 \le k \le N - 2 \qquad (8.155)$$
$$p_{K-1,K-1} = 1 - p_{K-2,K-1}$$

The next task is to verify that the suggested assignment of transitional probabilities is consistent with the stationary probabilities p_k of the channel states. In other words, one has to prove that

$$p_0 p_{0,0} + p_1 p_{1,0} = p_0$$
$$p_{k-1} p_{k-1,k} + p_k p_{k,k} + p_{k+1} p_{k+1,k} = p_k, 1 \le k \le N - 2 \qquad (8.156)$$
$$p_{k-2} p_{K-2,K-1} + p_{K-1} p_{K-1,K-1} = p_{K-1}$$

The first and the third equations can be easily verified. Indeed

$$p_0 p_{0,0} + p_1 p_{0,1} = p_0(1 - p_{0,1}) + p_1 p_{1,0} - p_0 - p_0 p_{0,1} + p_1 p_{1,0} = p_0$$
$$p_{K-2} p_{K-1,K-2} + p_{K-1} p_{K-1,K-1} = p_{K-2} p_{K-1,K-2} + p_{K-1}(1 - p_{K-2,K-1}) = p_{K-1}$$

$$(8.157)$$

since, according to eqs. (8.153) and (8.154),

$$p_0 p_{0,1} = p_1 p_{1,0} = \frac{LCR(A_1)}{R}, \quad p_{K-2} p_{K-1,K-2} = p_{K-1} p_{K-2,K-1} = \frac{LCR(A_{K-1})}{R} \qquad (8.158)$$

Finally, the second equation in eq. (8.156) can be rewritten as

$$p_{k-1}p_{k-1,k} + p_k(1 - p_{k,k-1} - p_{k,k+1}) + p_{k+1}p_{k+1,k} = p_k \tag{8.159}$$

or, equivalently

$$p_{k-1}p_{k,k-1} + p_{k+1}p_{k,k+1} = p_k p_{k,k-1} + p_k p_{k,k+1} \tag{8.160}$$

But both sides of eq. (8.156) are equal to $[LCR(A_k) + LCR(A_{k+1})]/R$, which proves that the designed transitional matrix has the required stationary probabilities of states.

A number of useful simulators can be based on the Wang–Moayeri model. The original paper [44] presents a model of a Rayleigh fading channel. In this case the instantaneous SNR $\gamma(t)$ is described by the exponential PDF

$$p_\gamma(\gamma) = \frac{1}{\bar{\gamma}} \exp\left(-\frac{\gamma}{\bar{\gamma}}\right) \tag{8.161}$$

Here $\bar{\gamma}$ represents the average SNR over the fading channel. If thresholds γ_k are chosen, then the probabilities of each state can be calculated through eq. (8.150) as

$$p_k = \int_{\gamma_k}^{\gamma_{k+1}} \frac{1}{\bar{\gamma}} \exp\left(-\frac{\gamma}{\bar{\gamma}}\right) d\gamma = \exp\left(-\frac{\gamma_k}{\bar{\gamma}}\right) - \exp\left(-\frac{\gamma_{k+1}}{\bar{\gamma}}\right) \tag{8.162}$$

Level crossing rates for such fading also can be easily calculated (see Section 4.6) to be

$$LCR(\gamma) = \sqrt{-\frac{\rho_0''}{2\pi}} \sqrt{\frac{\gamma}{\bar{\gamma}}} \exp\left(-\frac{\gamma}{\bar{\gamma}}\right) \tag{8.163}$$

This quantity depends on the exact shape of the normalized correlation function $\rho(\tau)$. For Clark's model one shall assume

$$\rho(\tau) = J_0^2(2\pi f_D \tau) \tag{8.164}$$

This results in

$$\rho_0'' = \frac{\partial^2}{\partial \tau^2} \rho(\tau) \Big|_{\tau=0}$$

$$= 8J_1^2(2\pi f_D \tau)\pi^2 f_D^2 - 8J_0(2\pi f_D \tau)\left[J_0(2\pi f_D \tau) - \frac{1}{2}\frac{J_1(2\pi f_D \tau)}{\pi f_D \tau}\right]\Big|_{\tau=0} = -4\pi^2 f_D^2 \tag{8.165}$$

and, thus, the level crossing rate can be calculated as

$$LCR(\gamma) = \sqrt{\frac{2\pi\gamma}{\bar{\gamma}}} f_D \exp\left(-\frac{\gamma}{\bar{\gamma}}\right) \tag{8.166}$$

The next question to be resolved is a proper choice of the threshold levels γ_k, and the number of states N. There are two conflicting requirements which must be taken into account. On the one hand, the range of the each state must be large enough to cover signal level variation during the bit (or block, or packet) duration. In this case it is valid to assume that the same state characterizes the channel. In addition, a wide range ensures that the following state of the channels does not differ greatly from the current state, thus justifying the assumption that only transitions to the neighbouring states are allowed. This requirement dictates that the number of states should be small and the probabilities of all states are approximately of the same order. On the other hand, a wide range of the signal levels inside a single state may result in a situation when two separate bits (blocks, or packets) experience two significantly different SNR levels, while formally belonging to the same state of the channel, thus resulting in significant discrepancy in the error rates. Thus a relatively large number of states is needed.

It is suggested in [61] that an appropriate partition is made based on average residence time $\bar{\tau}_k$ in each state. This quantity can be shown to be

$$\bar{\tau}_k = \frac{p_k}{LCR(\gamma_k) + LCR(\gamma_{k+1})} \tag{8.167}$$

The mean residence time can be normalized to a bit (block, packet) duration T_b as

$$c_k = \frac{\bar{\tau}_k}{T_b} \tag{8.168}$$

In the case of Clark's model, the expression (8.168) becomes

$$c_k = \frac{1}{f_m T_b} \frac{\exp\left(-\dfrac{\gamma_k}{\bar{\gamma}}\right) - \exp\left(-\dfrac{\gamma_{k+1}}{\bar{\gamma}}\right)}{\sqrt{\dfrac{2\pi\,\tau_k}{\bar{\gamma}}}\exp\left(-\dfrac{\gamma_k}{\bar{\gamma}}\right) + \sqrt{\dfrac{2\pi\,\gamma_{k+1}}{\bar{\gamma}}}\exp\left(-\dfrac{\gamma_{k+1}}{\bar{\gamma}}\right)} \tag{8.169}$$

If the threshold γ_k is known then the average residence time in each state c_k can be easily calculated. The inverse problem can also be formulated: given c_k, find a partition of the SNR, i.e. thresholds γ_k, such that the residence time is exactly c_k. In this case eq. (8.169) is an overdefined system since there will be K equations with only $K - 1$ unknown thresholds. One of the possibilities to circumvent this problem is to introduce an additional unknown scale factor c such that only the relative length $\bar{c}_k$ of the intervals is fixed

$$c_k = \bar{c}_k c \tag{8.170}$$

For example, the authors of [61] suggest that all $\bar{c}_k = 1$, i.e. the residence time is the same in all states. For a fixed Doppler spread $f_m T_b$ it is possible to vary the number of states K and investigate its effect on the average residence time. On the other hand, if the average residence time is confined to certain specific values, such a restriction allows one to determine the number of states of the Markov chain which satisfies this requirement. The results of numerical solution of eqs. (8.169) and (8.170) with constraints $\bar{c}_k = 1$ with $f_m T_b = 0.0338$ and $f_m T_b = 0.0169$ are shown in Fig. 8.21.

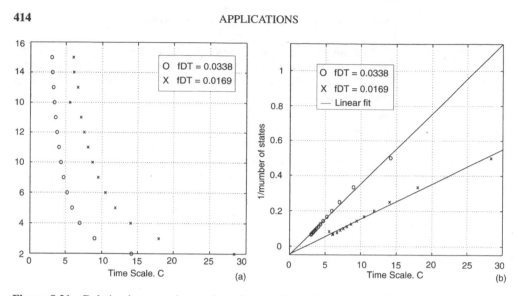

Figure 8.21 Relation between the number of states K and the average residence time c for a fixed fading velocity $f_m T_b$ as a parameter. The residence time in each state is the same. (a) Linear scale, (b) inverse scale. Choice of parameters corresponds to the carrier frequencies $f_0 = 900\,\text{MHz}$ and $1.9\,\text{GHz}$, vehicular velocity $v = 5\,\text{km/h}$, and the transmission rate is $1\,\text{Mb/s}$ or $100\,\text{kb/s}$.

It is interesting to observe that the number of states is approximately proportional to the inverse average residence time, which is emphasized by linear fit curves in Fig. 8.21(b). A large number of states used results in the fact that the chain changes from state to state more often, thus resulting in shorter residence time. This conclusion has been reached above. To obtain a similar value of the residence time c one requires a bigger number of states for a slower fading, i.e. for a smaller $f_m T_b$.

Once the number of states and corresponding thresholds are chosen, the transitional matrix is completely designed and the next step is to take into account the modulation technique used and the average SNR $\bar{\gamma}$ over the channel, since both of these factors contribute to the probability of error in each state.

The Wang–Moayeri model has proved to be very popular and useful. Babich and Lombardi [65] investigated a model for Rician fading described by the PDF

$$p_A(A) = \frac{A}{\sigma^2} \exp\left(-\frac{A^2 + A_s^2}{2\sigma^2}\right) I_0\left(\frac{A A_s}{\sigma^2}\right), \quad K = \frac{A_s^2}{2\sigma^2} \tag{8.171}$$

The Rice factor K reflects a ratio between the power of the line-of-sight component with magnitude A_s and the power of the diffusion component $\bar{\gamma} = \sigma^2$. Based on the general expression of the level crossing rate of the process with statistically independent derivative, the level crossing rate can be expressed as (see Section 4.6)

$$LCR(\gamma) = \sqrt{-\frac{\ddot{\rho}_0}{2\pi}} p_\gamma(\gamma) = f_D \sqrt{\frac{2\pi\gamma}{\bar{\gamma}}} \exp\left[-\left(K + \frac{\gamma}{\bar{\gamma}}\right)\right] I_0\left(2\sqrt{\frac{K\gamma}{\bar{\gamma}}}\right) \tag{8.172}$$

Generalization to a case of Nakagami fading can be performed in two ways. One approach was suggested in [172] and is based on an approximation of the Nakagami process with severity parameter $m > 1$ by a Rice process with the Rice factor K defined as [3,172]

$$K = \frac{\sqrt{m^2 - m}}{m - \sqrt{m^2 - m}} \tag{8.173}$$

thus resulting in the following expression

$$LCR(\gamma) = \sqrt{-\frac{\rho_0''}{2\pi} p_\gamma(\gamma)}$$

$$= f_D \sqrt{\frac{2\pi\gamma}{\bar{\gamma}}} \exp\left[-\left(\frac{\sqrt{m^2 - m}}{m - \sqrt{m^2 - m}} + \frac{\gamma}{\bar{\gamma}}\right)\right] I_0\left(2\sqrt{\frac{\sqrt{m^2 - m}}{m - \sqrt{m^2 - m}} \frac{\gamma}{\bar{\gamma}}}\right) \tag{8.174}$$

Another and more direct approach can be based on the fact that, for a certain class of Nakagami process, its derivative is indeed independent of the value of the process [3]. Thus, the general expression (see Section 4.6) can be used to produce

$$LCR(\gamma) = \sqrt{-\frac{\rho_0''}{2\pi} p_\gamma(\gamma)} = f_D \sqrt{\frac{2\pi\gamma}{\sigma^2} \frac{1}{\Gamma(m)} \left(\frac{m}{\bar{\gamma}}\right)^m \gamma^{m-1} \exp\left[-\frac{m\gamma}{\bar{\gamma}}\right]} \tag{8.175}$$

Of course, all three eqs. (8.172), (8.174) and (8.175) produce the same results for the case of Rayleigh fading, i.e. $K = 0$ and $m = 1$.

It is suggested by Babich and Lombardi [65] that the distribution corresponding to the Rice fading power be approximated by means of the Weibull distribution

$$p_\gamma(\gamma) = \frac{\gamma^{\lambda-1}}{\beta^\lambda} \exp\left[-\left(\frac{\gamma}{\beta}\right)^\lambda\right] \tag{8.176}$$

$$P_\gamma(\gamma) = \int_{-\infty}^{\lambda} p_\gamma(\gamma) d\gamma = 1 - \exp\left[-\left(\frac{\gamma}{\beta}\right)^\lambda\right] \tag{8.177}$$

Here the scale and the shape parameters β and λ are related to the parameters of the Rice distribution K, σ^2 and A_s^2 by the means of the following equalities, obtained by equalizing the first two moments of both distributions

$$\frac{(K+1)^2}{1 + 2K} = \frac{\Gamma\left(1 + \frac{2}{\lambda}\right)}{\Gamma^2\left(1 + \frac{1}{\lambda}\right)} - 1 \tag{8.178}$$

$$\beta = \frac{\sigma^2 + A_s^2}{\Gamma\left(1 + \frac{1}{\lambda}\right)}$$

Here $\Gamma(\bullet)$ is the Gamma function [23]. The advantage of using a Weibull PDF becomes apparent since a simple expression can be obtained for probabilities p_k of states of the Markov chain if a set of thresholds γ_k is chosen. The reverse procedure is also analytically treatable. Indeed, if one is given a set $\{p_k\}$ of the probabilities of the state of a Markov chain, then the corresponding thresholds can be found by noting that

$$\sum_{i=1}^{k} p_i = \text{Prob}\{\gamma \le \gamma_k\} = 1 - \exp\left[-\left(\frac{\gamma_k}{\beta}\right)^{\lambda}\right] = P(\gamma_k) \tag{8.179}$$

and thus

$$\gamma_k = \beta\left[\ln\frac{1}{1 - P(\gamma_k)}\right]^{1/\lambda} = \beta\left[-\ln\left(1 - \sum_{i=1}^{k} p_i\right)\right]^{1/\lambda} \tag{8.180}$$

Using thresholds obtained through eq. (8.180), the level crossing rates can be estimated as

$$LCR(\gamma_k) = \sqrt{-\frac{\rho_0''}{2\pi}p_A(A)} = \sqrt{-\frac{\rho_0''}{2\beta\pi}}\left[-\ln\left(1 - \sum_{i=1}^{k} p_i\right)\right]^{1-\frac{1}{2\lambda}}\left(1 - \sum_{i=1}^{k} p_i\right) \tag{8.181}$$

This can be further used in order to obtain simple analytical expressions for the elements of the transitional matrix of the Wang–Moayeri model.

Example: two-state model.[5]   Approximation (8.181) can be readily applied to the case of the Gilbert–Elliot model. In this case $p_1 = \varepsilon$ is the average probability of error in the channel and the transitional probabilities $p = p_{BG}$ from the "Bad" to "Good" state and $r = p_{GB}$ from the "Good" to the "Bad" state can now be approximated as [65]

$$r = \frac{T\sqrt{-\frac{\rho_0''}{2\beta\pi}[-\ln(1 - p_1)]^{1-\frac{1}{2\lambda}}(1 - p_1)}}{p_1} = T\sqrt{-\frac{\rho_0''}{2\beta\pi}\frac{1-\varepsilon}{\varepsilon}}[-\ln(1 - \varepsilon)]^{1-\frac{1}{2\lambda}} \tag{8.182}$$

$$p = 1 - \sqrt{-\frac{\rho_0''}{2\beta\pi}[-\ln(1 - \varepsilon)]^{1-\frac{1}{2\lambda}}} \tag{8.183}$$

Similar equations were derived independently in [25] for the case of Nakagami and log-normal distributions and Clark's correlation function:

$$r_N = \frac{4\,m^m R^{2m-1}\exp\left(-\frac{mR^2}{\Omega}\right)\sqrt{\left(-\frac{0.4439\,\Omega}{8\pi}\frac{\Omega}{m}\right)}\pi f_D NT}{\Gamma(m,0) - \Gamma\left(m,\frac{mb}{\Omega}\right)} \tag{8.184}$$

$$r_L = \frac{\sigma_c\exp\left[\frac{-(\ln R - m)^2}{2\sigma^2}\right]NT}{\frac{1}{2} + \frac{1}{2}\text{erf}\left(\frac{\ln(\sqrt{b}) - m}{\sqrt{2}\sigma}\right)} \tag{8.185}$$

---

[5] This example was developed in collaboration with Ms Vanja Subotic.

Both studies [65] and [25] compared the average block failure rate predicted by eqs. (8.183)–(8.185) with those obtained from numerical simulation using the Jakes simulator and other simulators. It can be seen that there is a good agreement between the numerical simulation and prediction based on the two-state Markov model (Figs. 8.22–8.24).

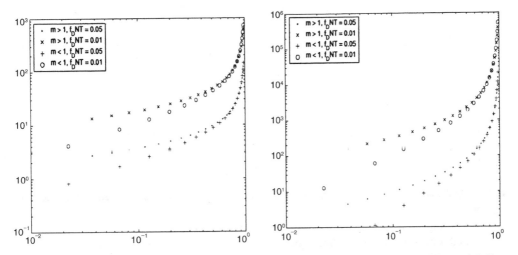

Figure 8.22   Average block error burst length vs. the probability of the block error. Nakagami fading mean (left) and variance (right).

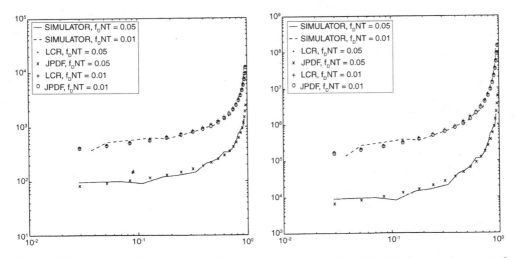

Figure 8.23   Average block error burst length vs. the probability of the block error. Log-normal fading mean values (left) and variance (right).

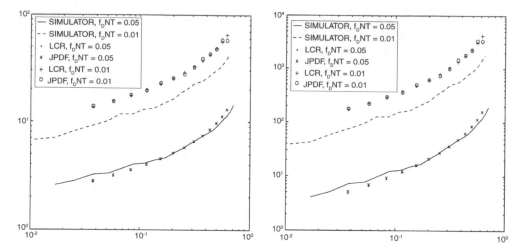

Figure 8.24 Average block error burst length vs. the probability of the block error. Two vehicles, Nakagami PDF: mean (left) and variance (right).

8.4.3 Independence of the Channel State Model on the Actual Fading Distribution

An interesting comment can be made about the fact that after a Markov chain model is associated with a continuous fading process, i.e. a set of the state probabilities p_k is assigned, it is impossible to restore the original distribution of the fading. Indeed, any two distributions can be discretized in such a way as to produce the same set of state probabilities of the corresponding Markov chain. This may be considered as a sort of a paradox. However, a proper model assigns a certain probability of error in each state[6] which reflects the average SNR in this state and the modulation–demodulation techniques. Those probabilities will be different for different types of fading, since the same probability of the states of a Markov chain would require different quantization thresholds, thus producing different values of the average SNR in different states. Furthermore, using a state splitting technique [171] a hidden Markov model can be converted to a Markov model. However, due to different error probability assignments, the results of such a conversion will be different for different types of fading.

8.4.4 A Rayleigh Channel with Diversity

Since the only characteristic required to obtain the Wang–Moayeri model is to be able to calculate the marginal PDF of the signal and its level crossing rate, it is possible to adopt the same technique for a case of diversity channels with a single output signal. At this stage it can be realized that there would be a significant number of states which are described either by a very small probability of appearance or very small probability of errors. Both such states could be excluded from further consideration to reduce the complexity of the model and speed up simulation.

---

[6]Thus leading to a Hidden Markov Model (HMM).

Following [91] one can consider a Rayleigh channel with N-branch pure selection combining with the following decision rule

$$R = \max(r_1, r_2, \ldots, r_N) \tag{8.186}$$

The simplest model assumes that the fading in different branches is independent and identically distributed. The cumulative distribution function of such a decision scheme is given by

$$FR(R) = \text{Prob}(r \leq R) = \prod_{n=1}^{N} \text{Prob}(r_i \leq R) = \prod_{n=1}^{N} \left(1 - \exp\left[-\frac{R^2}{2\,\sigma_n^2}\right]\right)$$

$$= \left(1 - \exp\left[-\frac{R^2}{2\sigma^2}\right]\right)^N \tag{8.187}$$

$$F_\gamma(\gamma) = \text{Prob}(SNR < \gamma) = \text{Prob}(r < \sqrt{\gamma}) = \left(1 - \exp\left[-\frac{\gamma}{\rho}\right]\right)^N, \quad \gamma = 2\sigma^2 \tag{8.188}$$

where σ_n^2 is the variance of the signal envelope in the n-th branch. Thus, stationary probabilities of the states of the approximating Markov chain can be obtained as

$$p_k = F_\gamma(\gamma_{k+1}) - F_\gamma(\gamma_k) = \left(1 - \exp\left[-\frac{\gamma_{k+1}}{\rho}\right]\right)^N - \left(1 - \exp\left[-\frac{\gamma_k}{\rho}\right]\right)^N \tag{8.189}$$

It is shown in [91] that the level crossing rate of the combined signal is given by

$$LCR_N(\gamma) = N \prod_{n=2}^{N} \int_{-\pi}^{\pi} p(\theta_{1n}) d\theta_{1n} \prod_{n=2}^{N} \int_{0}^{\sqrt{\gamma}} p(r_n) d r_n \int_{0}^{\infty} \dot{r}_1 p(\dot{r}_1|r_1 = \sqrt{\gamma}) d\dot{r}_1$$

$$= N \sqrt{\frac{2\pi\gamma}{\rho}} f_m \exp\left(-\frac{\gamma}{\rho}\right) \left[1 - \exp\left(-\frac{\gamma}{\rho}\right)\right]^{N-1} \tag{8.190}$$

$$LCR_N(\gamma_k) = \sqrt{\frac{2\pi\gamma_k}{\rho}} f_m \exp\left(-\frac{\gamma_k}{\rho}\right) \left[1 - \exp\left(-\frac{\gamma_k}{\rho}\right)\right]^{N-1} \tag{8.191}$$

Here θ_{1n} represents the phase of the signal in the n-th element of the antenna array. Combining eqs. (8.189) and (8.191) with eqs. (8.153) and (8.154) derived earlier for the elements of the transitional matrix T one can obtain the desired Markov chain.

8.4.5 Fading Channel Models

Swarts and Ferreira carried out a comparison study [81] of Gilbert, Elliot and Fritchman models as applied to GSM systems with the band-time product $BT = 0.3$, maximum Doppler shift $f_D = 40\,\text{Hz}$ and average SNR levels $\bar{\gamma} = 5$, 10, 15 and 20 dB. Parameters of the models can be estimated based on the error distribution statistics, obtained from the Jakes simulators and eqs. (8.141)–(8.146). Since only simple statistics were used, it appears that all three models perform almost identically.

Guan and Turner [47] suggested modelling the fading envelope by approximating it by a general N-state Markov chain of the first order. The interval of variation of the envelope $[0, \infty)$ is divided into N sub-intervals by a set of $N-1$ points τ_n, $1 \leq n \leq N-1$. It is suggested that elements of the transitional probability can be found from a joint probability density function $p_{AA}(A, A_\tau; \tau)$ of the envelope considered in two separate moments of time

$$p(A_k | A_l) = \frac{p(A_k, A_l)}{p(A_l)} = \frac{\int_{\tau_{k-1}}^{\tau_k} \int_{\tau_{l-1}}^{\tau_l} p_{AA}(A, A_\tau; \tau) \mathrm{d}A \, \mathrm{d}A_\tau}{\int_{\tau_{l-1}}^{\tau_l} p_A(A) \mathrm{d}A} \tag{8.192}$$

Such an assignment ensures that not only the stationary probabilities of each state approximate the distribution of the continuous random process but the second order statistics at two sequential moments of time are also well approximated. As an example of the application of their model the authors of [47] consider modeling of the Nakagami distributed random envelope

$$p(A_k, A_l) = \frac{4(AA_\tau)^m}{\Gamma(m)(1-\rho)\rho^{\frac{m-1}{2}}} \left(\frac{m}{P}\right)^{m+1} \exp\left[-\frac{m}{P}\frac{(A^2 + A_\tau^2)}{1-\rho}\right] I_{m-1}\left(\frac{2mAA_\tau\sqrt{\rho}}{P(1-\rho)}\right) \tag{8.193}$$

Here $P = E\{A^2\}$ is the average power of the envelope (proportional to the average SNR in the AWGN channel), $0.5 \leq m \leq \infty$ is the Nakagami parameter and ρ is the correlation coefficient between two sequential samples of the envelope separated by a duration of bit τ

$$\rho \approx J_0^2(2\pi f_D \tau) \tag{8.194}$$

While an analytical expression for integral (8.192) cannot be obtained in quadratures, a good approximation has recently been derived [3]. To validate the model, a comparison with the Wang–Moayeri approximation (see Section 8.4.2 for more details) has been conducted for the case of $m = 1$, $\tau = 10^{-5}$ s and $f_D = 10$, 100 Hz. A remarkable agreement between two models can be observed—most of the values differ by only the fifth decimal digit. The authors also consider an interleaved channel to find that the approximation of transitional probabilities calculated through the joint probability density (8.193) still produces good results, while the Wang–Moayeri approximation through the crossing rates becomes less accurate due to the fact that transitions to non-neighbouring states appear with greater probability. At the same time, larger time separation between samples, which is equivalent to larger Doppler spread, or a shorter correlation interval, also brings into consideration the non-Markov nature of Clark's model of fading. This clearly shows limitations of simple Markov chain models for relatively fast fading. The question of choosing a proper order of Markov chain is treated in [65].

Modeling of a slow fading is considered in [72]. In this paper, the authors obtain measurements of average signal intensity and its variation based on 1.9 GHz measurements, collected in Austin, TX. Measurements were taken by a moving car. This data was later used to obtain fluctuating slow fading by means of numerical simulation of a random driving path, with the car turning randomly at each intersection. The value of the signal along each simulated path was then retrieved from the measurement database, filtered and decimated to produce a signal with 1 Hz sampling rate. The slow fading signal trace, obtained through this process, was used as a reference signal for modeling.

An 80-state Markov chain was chosen to approximate real data. Threshold levels were set in such a way that all states had equal probability $p_i = 1/80$ and spanned a region between -20 dB and 50 dB. The authors explained that such a choice would allow them to avoid situations when the chain resided in a few states for a long time. Transitional probabilities were estimated from the experimental data by counting a number of transitions between any two states and normalizing it by a total number of data points. Thus the constructed model was used to generate a sample of a Markov process and compare its characteristics with the one obtained from the measurements. It was found that the cumulative distribution function was well approximated, since a sufficient number of states was selected. The authors also point out that the model is able to approximate the level crossing rate with reasonable accuracy. This can be explained by the fact that the model has been constructed based on the transitional probabilities, i.e. level crossing rates. However, the model fails to approximate the power spectral density of the signal. The authors attribute this discrepancy to the fact that the actual channel does not show a Markov property. However, no further studies of the memory of measurements were conducted.

It is clear that more accuracy in Markov models can be achieved by both increasing the number of states and increasing the memory of the Markov process.[7] Such an increase in complexity is not desired for fast and approximate simulation. Rao and Chen [79] suggest a mechanism to reduce the number of states by aggregating a number of states into new superstates. The authors use the lumpability property of a Markov chain to derive a two-state model of the fading process and compare performance of such a simplified model with those produced by higher order models; in particular, one error state Fritchman model[8] with four non-error states and different types of modulation schemes (FSK, DPSK, QPSK and 8-ary PSK), a hidden Markov model with six states (described in [173] and the Wang and Moayeri model[9] with eight states.

Many of the critics of Markov models point out that, for the correlation function, a simple Markov chain considered is a poor fit to realistic ones [170]. There is also ongoing discussion on applicability of modeling of fading as a Markov process [166–170,173]. While most of the researchers agree that the second order Markov models fit the canonical Jakes model of the envelope very well, there is no consensus on applicability of simple Markov models. In the view of the authors of the current book, such models are at least very helpful in discovering general trends and analysis performance of complicated codes.

8.4.6 Higher Order Models [170]

A number of methods allow one to extend Markov chain models to account for dependence between the future and the past of the process, given current observation. Such dependence leads to a non-Markov nature of the process under consideration.

Zorzi *et al.* suggest [170] that both the values of the random process and its derivative (velocity) are approximated by a Markov chain. If there are M levels m_i ($0 \leq i \leq M - 1$) of the random process $\xi(t)$ to be considered, and there are S levels $s_j (0 \leq j \leq S - 1)$ of the velocity of the random process $\dot{\xi}(t)$, then the Markov chain consists of $M \times S$ states obtained

---

[7]That is, to consider Markov processes of order more than 1. See the chapter for more details.
[8]See Section 12.1.4.
[9]See Section 8.4.2.

as all possible pairs (m_i, s_j). It is quite clear that the complexity of such a description grows very fast with the number of states. It is claimed in [170] that it is sufficient to choose only three levels of the velocity: $S = 3$, $s_0 = 0$, $s_1 = -s_2 = \varepsilon$. Here velocity levels reflect a well known fact that for many useful processes such as Rayleigh, Nakagami, etc. the velocity is a zero mean Gaussian process. A particular value of ε is chosen to compromise between the accuracy of approximation of the flat portion of the early time and the oscillatory portion in the late time of Clark's model. The authors proceed further to show an example of an ARQ system simulation that provides a significant improvement when compared to the model which approximates only the states of the process $\xi(t)$.

8.5 MARKOV CHAIN FOR DIFFERENT CONDITIONS OF THE CHANNEL

A number of applications use a finite state Markov chain to describe changing conditions of the communication channel [16]. It is characteristic for a wireless link to change if at least one of the mobile receivers changes position. For example, a pedestrian can move from a residential area with relatively mild obstructions by two-storey wooden houses and trees to either an open area with virtually no obstruction or to downtown areas with high rises completely shadowing the signal. Such dramatic changes can be modeled as separate states of the Markov chain with certain non-deterministic transitions.

One of the examples is considered in [56]. Here a land mobile satellite service (LMSS) channel is modeled as a three-state Markov chain. All states have contributions from weak scattered waves. In addition, the state A has a contribution from a strong, unattenuated line-of-sight ray; the state B is characterized by significant attenuation of the line-of-sight, and finally, the state C is described by fully blocked line-of-sight. As a result, fast fading statistics differ in each state. Probabilities of each state are p_A, p_B and p_C, respectively (Fig. 8.25). Of course

$$p_A + p_B + p_C = 1 \tag{8.195}$$

The three-state model presented in [56] is considered more general than the two-state models considered by Lutz [164] (only states B and C were considered, with a non-

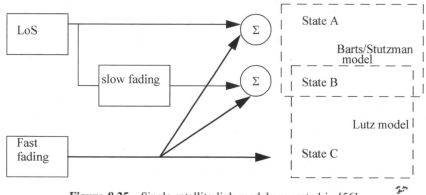

Figure 8.25   Single satellite link model suggested in [56].

Rayleigh model for the state *C*), and Barts and Stutzman [165] (only states *A* and *B* were considered).

In all states, contribution from the weak diffused scatterers is approximated by Rayleigh fading. In the state *A*, the Nakagami–Rice model is used. In the state *B*, attenuation of the direct ray is described by a log-normal distribution. Superimposed with Rayleigh fast fading, this results in the well known Loo model [163]. Fast fading in the state *C* is represented just by Rayleigh fading. According to this model, the probability density function of the received signal is a weighted sum of the density functions at each state, with weights equal to the probabilities of the states

$$p(r) = P_A p_A(r, P_{r,A}) + P_b p_B(r, m_L, \sigma_L, P_{r,B}) + P_C p_C(r, P_{r,C}) \tag{8.196}$$

Here

$$p_A(r, P_{r,A}) = \frac{2\,r}{P_{r,A}} \exp\left(-\frac{1+r^2}{P_{r,A}}\right) I_0\left(\frac{2\,r}{P_{r,A}}\right) \tag{8.197}$$

$$p_B(r, m_L, \sigma_L, P_{r,B}) = \frac{6.930\,r}{\sigma_L P_{r,B}} \int_0^\infty \frac{1}{z} \exp\left[-\frac{\{20\log(z) - m_L\}^2}{2\,\sigma_L^2} - \frac{r^2 + z^2}{P_{r,B}}\right] dz \tag{8.198}$$

$$p_C(r, P_{r,C}) = \frac{2\,r}{P_{r,C}} \exp\left(-\frac{r^2}{P_{r,C}}\right) \tag{8.199}$$

and $P_{r,A}, P_{r,B}$ and $P_{r,C}$ are the mean powers of the multipath components normalized to the direct ray power; m_l and σ_L are the mean and the variance of the signal fading in dB. In the original paper [56] these values were obtained from the measurements in different cities. Based on these measurements it is suggested that the following approximations are used for the state probabilities

$$P_A = 1 - \frac{(90 - \theta)^2}{\alpha}, \quad 10° < \theta < 90°, \alpha = \begin{cases} 7.0 \times 10^3 & \text{for urban areas} \\ 1.66 \times 10^4 & \text{for suburban areas} \end{cases} \tag{8.200}$$

A similar idea has been suggested in [77]. In this paper the authors suggest that a satellite communication channel can be described by three significantly different conditions, described by the Rayleigh, Rician and log-normal distributions of the signal's envelope. The transitions between states are described by a Markov chain of order 3. The main contribution of this paper is the fact that the model is implemented on a TMS320C30 microprocessor and can be expanded to model wide band channels by means of parallel computations. One of the drawbacks of this method is that the number of processors required is equal to the number of rays to be simulated, thus increasing the cost of such simulators [174–188].

REFERENCES

1. B. Evans, *Satellite Communication Systems*, London: IEEE, 1999.
2. M. Jakes, *Microwave Mobile Communications*, New York: Wiley, 1974.

3. M. Simon and M. Alouini, *Digital Communication over Fading Channels: A Unified Approach to Performance Analysis*, New York: John Wiley & Sons, 2000.

4. COST 207, Digital Land Mobile Radio Communications, *Final Report*, Office for Official Publications of the European Communities, Luxembourg, 1989.

5. H. Suzuki, A statistical model for urban radio propagation, *IEEE Trans. Communications*, **25**(7), 673–680, 1977.

6. M. Nakagami, The *m*-distribution—A general formula of intensity distribution of rapid fading, In *Statistical Methods in Radio Wave Propagation*, edited by W. C. Hoffman, New York: Pergamon, 1960, pp. 3–36.

7. C. Loo, A statistical model for a land mobile satellite link, *IEEE Trans. Vehicular Technology*, **34**(3), 122–127, 1985.

8. A. Aboudebra, K. Tanaka, T. Wakabayashi, S. Yamamoto and H. Wakana, Signal fading in land-mobile satellite communication systems: statistical characteristics of data measured in Japan using ETS-VI, *Proc. IEE Microwaves, Antennas and Propagation*, **146**(5), 349–354, 1999.

9. P. Taaghol and R. Tafazolli, Correlation model for shadow fading in land-mobile satellite systems, *IEE Electronics Letters*, **33**(15), 1287–1289, 1997.

10. H. Wakana, Propagation model for simulating shadowing and multipath fading in land-mobile satellite channel, *IEE Electronics Letters*, **33**(23), 1925–1926, 1997.

11. R. Akturan and W. Vogel, Elevation angle dependence of fading for satellite PCS in urban areas, *IEE Electronics Letters*, **31**(25), 2156–2157, 1995.

12. W. Vogel and J. Goldhirsh, Multipath fading at L band for low elevation angle, land mobile satellite scenarios, *IEEE Journal Selected Areas in Communications*, **13**(2), 197–204, 1995.

13. F. Vatalaro, G. Corazza, C. Caini and C. Ferrarelli, Analysis of LEO, MEO, and GEO global mobile satellite systems in the presence of interference and fading, *IEEE Journal Selected Areas in Communications*, **13**(2), 291–300, 1995.

14. Y. Chau and J. Sun, Diversity with distributed decisions combining for direct-sequence CDMA in a shadowed Rician-fading land-mobile satellite channel, *IEEE Trans. Vehicular Technology*, **45**(2), 237–247, 1996.

15. C. Amaya, and D. Rogers, Characteristics of rain fading on Ka-band satellite-earth links in a Pacific maritime climate, *IEEE Trans. Microwave Theory and Techniques*, **50**(1), 41–45, 2002.

16. P. Babalis and C. Capsalis, Impact of the combined slow and fast fading channel characteristics on the symbol error probability for multipath dispersionless channel characterized by a small number of dominant paths, *IEEE Trans. Communications*, **47**(5), 653–657, 1999.

17. A. Ismail and P. Watson, Characteristics of fading and fade countermeasures on a satellite-earth link operating in an equatorial climate, with reference to broadcast applications, *Proc. IEE Microwaves, Antennas and Propagation*, **147**(5), 369–373, 2000.

18. A. Verschoor, A. Kegel and J. Arnbak, Hardware fading simulator for a number of narrowband channels with controllable mutual correlation, *IEE Electronics Letters*, **24**(22), 1367–1369, 1988.

19. G. Arredondo, W. Chriss and E. Walker, A multipath fading simulator for mobile radio, *IEEE Trans. Communications*, **21**(11), 1325–1328, 1973.

20. E. Casas and C. Leung, A simple digital fading simulator for mobile radio, *IEEE Trans. Vehicular Technology*, **39**(3), 205–212, 1990.

21. M. Pop and N. Beaulieu, Limitations of sum-of-sinusoids fading channel simulators, *IEEE Trans. Communications*, **49**(4), 699–708, 2001.

22. P. Chengshan Xiao, Y. R. Zheng and N. C. Beaulieu, Second-order statistical properties of the WSS Jakes' fading channel simulator, *IEEE Trans. Communications*, **50**(6), 888–891, 2002.

23. M. Abramovitz and I. Stegun, *Handbook of Mathematical Functions*, New York: Dover, 1995.

24. K. Yip and T. Ng, A simulation model for Nakagami-*m* fading channels, $m < 1$, *IEEE Trans. Communications*, **48**(2), 214–221, 2000.

25. V. Subotic and S. Primak, Markov approximation of a fading communications channel, *Proc. 2002 European Microwave Week*, Milan, 2002.

26. V. Subotic and S. Primak, Novel simulator of a sub-Rayleigh fading channel, *IEE Electronics Letters*, 2003.

27. S. Primak, V. Lyandres, O. Kaufman and M. Kliger, On generation of correlated time series with a given PDF, *Signal Processing*, **72**(2), 61–68, 1999.

28. V. Kontorovich and V. Lyandres, Dynamic systems with random structure: an approach to the generation of nonstationary stochastic processes, *Journal of The Franklin Institute*, **336**(6), 939–954, 1999.

29. J. Weaver, V. Kontorovich and S. Primak, On analysis of systems with random structure and their applications to communications, *Proc. VTC-2003*, Fall, Orlando, Florida, 2003.

30. L. N. Kanal and A. Sastry, Models for channels with memory and their applications to error control, *Proc. of the IEEE*, **66**(6), 724–744, 1978.

31. E. Gilbert, Capacity of a burst-noise channel, *The Bell System Technical Journal*, 1253–1265, 1960.

32. E. L. Cohen and S. Berkovits, Exponential distributions in Markov chain models for communication channels, *Information and Control*, **13**, 134–139, 1968.

33. E. Elliot, Estimates of error rates for codes on burst-noise channels, *Bell Systems Technical Journal*, **42**, 1977–1997.

34. P. Trafton, H. Blank and N. McAllister, Data transmission network computer-to-computer study, *Proc. ACM/IEEE 2nd Symposium on Problems in the Optimization of Data Communication Systems, Palo Alto*, October, 1971, pp. 183–191.

35. B. Fritchman, A binary channel characterization using partitioned Markov chains, *IEEE Trans. on Information Theory*, **13**(2), 221–227, 1967.

36. I. Shji and T. Ozaki, Comparative study of estimation methods for continuous time stochastic processes, *Journal of Time Series Analysis*, **18**, 485–506, 1997.

37. W. Press, S. Teukolsky, W. Vetterling and B. Flannery, *Numerical Recipes in C*, Cambridge, England: Cambridge University Press, 2002.

38. J. P. Adoul, B. Fritchman and L. Kanal, A critical statistic for channel with memory, *IEEE Trans. Information Theory*, **18**(1), 133–141, 1972.

39. H. Blank and P. Trafton, ARQ analysis in channels with memory, *National Telecommunication Conference*, 1973, pp. 15B-1–15B-8.

40. J. Kemeny, Slowly spreading chains of the first kind, *Journal of Mathematical Analysis and Applications*, **15**, 295–310, 1966.

41. S. Tsai, Evaluation of burst error correcting codes using a simple partitioned Markov chain, *IEEE Trans. Communications*, **21**, 1031–1034, 1973.

42. P. McManamon, HF Markov chain models and measured error averages, *IEEE Trans. Communication Technologies*, **18**(3), 201–208, 1970.

43. R. McCullough, The binary regenerative channel, *Bell Systems Technical Journal*, **47**, 1713–1735, 1968.

44. H. Wang and N. Moayeri, Finite-state Markov channel—a useful model for radio communication channels, *IEEE Trans. Vehicular Technology*, **44**(1), 163–171, 1995.

45. I. Oppermann, B. White and B. Vucetic, A Markov model for wide-band fading channel simulation in micro-cellular systems, *IEICE Trans. Communications*, **E79-B**(9), 1215–1220, 1996.

46. S. Tsai and P. Schmied, Interleaving and error-burst distribution, *IEEE Trans. Communications*, **20**(3), 291–296, 1972.

47. Y. Guan and L. Turner, Generalized FSMC model for radio channels with correlated fading, *IEE Proc. Communications*, **146**(2), 133–137, 1999.

48. H. Kong and E. Shwedyk, Markov characterization of frequency selective Rayleigh fading channels, *IEEE Pacific Rim Conference on Communications, Computers and Signal Processing*, 1995, pp. 359–362.

49. H. Viswanathan, Capacity of Markov channels with receiver CSI and delayed feedback, *IEEE Trans. Information Theory*, **45**(2), 761–771, 1999.

50. J. Metzner, An interesting property of some infinite-state channels, *IEEE Trans. Information Theory*, **11**(2), 310–312, 1965.

51. M. Muntner and J. Wolf, Predicted performance of error-control techniques over real channels, *IEEE Trans. Information Theory*, **14**(5), 640–650, 1963.

52. F. Swarts and H. Ferreira, Markov characterization of channels with soft decision outputs, *IEEE Trans. Communications*, **41**(5), 678–682, 1993.

53. C. Tan and N. Beaulieu, On first-order Markov modeling for the Rayleigh fading channel, *IEEE Trans. Communications*, **48**(12), 2032–2040, 2000.

54. A. Haddad, S. Tsai, B. Goldberg and G. Ranieri, Interleaving and error-burst distribution, *IEEE Trans. Communications*, **23**(11), 1189–1197, 1975.

55. H. Wakana, Propagation model for simulating shadowing and multipath fading in land-mobile satellite channel, *Electronics Letters*, **33**(23), 1925–1926, 1997.

56. Y. Karasawa, K. Kimura and K. Minamisono, Analysis of availability improvement in LMSS by means of satellite diversity based on three-state propagation channel model, *IEEE Trans. Vehicular Technologies*, **46**(4), 1047–1056, 1997.

57. H. Wang and P.-C. Chang, On verifying the first-order Markovian assumption for a Rayleigh fading channel model, *IEEE Trans. Vehicular Technology*, **45**(2), 353–357, 1996.

58. A. Goldsmith and P. Varaya, Capacity, mutual information and coding for finite-state Markov channels, *IEEE Trans. Information Theory*, **42**(3), 868–886, 1996.

59. W. Turin, Throughput analysis of the go-back-*N* protocol in fading channels, *IEEE Journal Selected Areas in Communications*, **17**(5), 881–887, 1999.

60. S. Tsai, Markov characterization of the HF channel, *IEEE Trans. Communication Technology*, **17**(1), 24–32, 1969.

61. Q. Zhang and S. Kassam, Hybrid ARQ with selective combining for fading channels, *IEEE Journal Selected Areas in Communications*, **17**(5), 867–879, 1999.

62. M. Zorzi, R. Rao and L. Milstein, Error statistics in data transmission over fading channels, *IEEE Trans. Communications*, **46**(11), 1468–1477, 1998.

63. M. Zorzi, A. Chockalingam and R. Rao, Throughput analysis of TCP on channels with memory, *IEEE Trans. Vehicular Technology*, **49**(7), 1289–1300, 2000.

64. A. Chockalingam, W. Xu, M. Zorzi and B. Milstein, Throughput-delay analysis of a multichannel wireless access protocol in the presence of Rayleigh fading, *IEEE Trans. Vehicular Technology*, **49**(2), 661–671, 2000.

65. F. Babich and G. Lombardi, A Markov model for the mobile propagation channel, *IEEE Trans. Vehicular Technology*, **49**(1), 63–73, 2000.

66. F. Babich and O. Kelly, A simple digital fading simulator for mobile radio, *IEEE Trans. Vehicular Technology*, **39**(3), 205–212, 1990.

67. H. Kong and E. Shwedyk, Sequence detection and channel state estimation over finite state Markov channels, *IEEE Trans. Vehicular Technology*, **48**(3), 833–839, 1999.

68. Y. Kim and S. Li, Capturing important statistics of a fading/shadowing channel for network performance analysis, *IEEE Journal of Selected Areas in Communications*, **17**(5), 888–901, 1999.

69. H. Bischl and E. Lutz, Packet error rate in the non-interleaved Rayleigh channel, *IEEE Trans. Communications*, **43**(2–4), 1375–1382, 1995.

70. A. Chockalingam, M. Zorzi and L. Milstein, Performance of a wireless access protocol on correlated Rayleigh-fading channels with capture, *IEEE Trans. Communications*, **42**(3), 868–886, 1996.

71. M. Zorzi, Outage and error events in bursty channels, *IEEE Trans. Communications*, **46**, 349–356, 1998.

72. T. Su, H. Ling and W. Vogel, Markov modeling of slow fading in wireless mobile channels at 1.9 GHz, *IEEE Transactions on Antennas and Propagation*, **46**(6), 947–948, 1998.

73. R. Rao, Higher layer perspectives on modeling the wireless channel, *ITW*, 1998, Killarney, Ireland, pp. 137–138.

74. W. Turin and R. van Nobelen, Hidden Markov modeling of fading channels, *VTC 98*, 1234–1238.

75. A. Fulinski, Z. Grzywna, I. Mellor, Z. Siwy and P. N. R. Usherwood, Non-Markovian character of ionic current fluctuations in membrane channels, *Physical Review E*, **58**(1), 919–924, 1998.

76. H. Kong and A. Shwedyk, Sequence estimation for frequency non-selective channels using a hidden Markov model, *IEEE Vehicular Technology Conference*, 1251–1253, 1994.

77. J. An and M. Chen, Simulation techniques for mobile satellite channel, *Proc. MILCOM '94*, IEEE, 1994, pp. 163–167.

78. A. Abdi, Comments on "On verifying the first-order Markovian assumption for a Rayleigh fading channel model", *IEEE Trans. Vehicular Technology*, **48**(5), 1739, 1999.

79. A. Chen and R. Rao, On tractable wireless channel models, *IEEE International Symposium on Personal, Indoor and Mobile Radio Communications*, 825–830, 1998.

80. A. Beverly and K. Shanmugan, Hidden Markov models for burst errors in GSM and DECT channels, *GLOBECOM 1998*, **6**, 3692–3698, 1998.

81. J. Swarts and H. Ferreira, On the evaluation and application of Markov channel models in wireless communications, *Proc. VTC 1999—Fall*, 1999, pp. 117–121.

82. H. Steffan, Adaptive generative radio channel models, *IEEE Conference on Personal, Indoor and Mobile Radio Communications*, 1994, pp. 268–273.

83. F. Babich and G. Lombardi, On verifying a first-order Markovian model for the multi-threshold success/failure process for Rayleigh channel, *IEEE Conference on Personal, Indoor and Mobile Radio Communications*, PIMRC '97, 1997, pp. 12–16.

84. N. Nefedov, Generative Markov models for discrete channel modeling, *IEEE Conference on Personal, Indoor and Mobile Radio Communications*, PIMRC '97, 1997, pp. 7–11

85. Q. Cao and M. Gurkan, Markov model and delay analysis for group ALOHA systems, *Proc. VTC-1996*, **3**, 1996, pp. 1761–1765.

86. P. Soni and A. Chockalingam, Energy efficiency analysis of link layer backoff schemes on point-to-point Markov fading links, *Proc. PIMRC 2000*, **1**, 416–420, 2000.

87. D. Huang and J. Shi, TCP over packet radio, *Emerging Technologies Symposium: Broadband, Wireless Internet Access*, 2000.

88. M. Hueda, A Markov-based model for performance evaluation in multimedia CDMA wireless transmission, *Proc. IEEE-VTS Fall VTC 2000*, **2**, 2000, pp. 668–673.

89. M. Ouameur and D. Massicotte, A Markov chain and quadrature amplitude modulation fading based statistical discrete time model for multi-WSSUS multipath channel, *Canadian Conference on Electrical and Computer Engineering*, 2001, pp. 487–492.

90. B. Clerckx and D. Vanhoenacker-Janvier, Performance analysis of Rake receivers with maximum a posteriori channel state estimation over time varying frequency selective finite state Markov fading channels, *Proc. IEEE RAWCON 2001*, pp. 141–144.

91. J. Lu, K. Letaief, M. Liou and J. Chuang, Modeling correlated fading with diversity and its application to block codes performance estimation, *Proc. IEEE GLOBECOM 1998*, pp. 3657–3662.

92. D. Oosthuizen, H. Ferreira and F. Swarts, Markov models for block code error control systems on fading RF channels, *Proc. COMSIG 1989*, 1989, pp. 1–6.

93. T. Tao, J. Lu and J. Chuang, Hierarchical Markov model for burst error analysis in wireless communications, *Proc. VTC 2001 Spring*, **4**, 2843–2847, 2001.

94. D. Oosthuizen, H. Ferreira and F. Swarts, On renewal inner channels and block code error control super channels, *IEEE Trans. Communications*, **42**(9), 2645–2649, 1994.

95. C. Lee and Y. Su, Errors-and-erasures decoding of block codes in hidden Markov channels, *Proc. IEEE VTC 2001 Spring*, 1425–1429, 2001.

96. M. Hueda, On the equivalence of channel-state and block-error Markov models for the analysis of transmission over slow fading channels, *Proc. IEEE VTC 2001 Fall*, **2**, 1043–1047, 2001.

97. J. Hsieh, H. Lin and C. Yang, A two-level, multi-state Markov model for satellite propagation channels, *Proc. IEEE VTC 2001*, **4**, 3005–3009, 2001.

98. B. D. Carlson, and L.M. Goodman, Adaptive antenna system design for multi-element combining with transient interrupted channels, *IEEE Trans. Vehicular Technology* **51**(5), 799–807, 2002.

99. R. D'Agostino and M. Stephens, *Goodness-of-Fit Techniques*, New York: Marcel Dekker, 1986.

100. M. Skolnik, *A Review of Radar Sea Echo*, Technical report, NRL 2025, Naval Research Laboratory, Washington, 1969.

101. E. Jakeman and P. Pusey, A model for non-Rayleigh sea echo, *IEEE Trans. Antennas and Propagation*, **40**(4), 700–707, 1973.

102. E. Jakeman, Significance of K-distribution in scattering experiments, *Physical Review Letters*, **40**(9), 546–550, 1978.

103. E. Jakeman, On statistics of K-distributed noise, *Journal of Physics*, A: Mathematics and General, **13**(1), 31–48, 1980.

104. D. Kreithen and G. Hagan, Statistical analysis of K-band sea clutter, *Proc. Ocean-91*, Hadlay, Hawaii, 128–132, 1991.

105. T. Nohara and S. Haykin, East coast radar trials and the K-distribution, *IEE Proc. Radar and Signal Processing F*, **138**(2), 80–88, 1991.

106. F. Gini, M. Greco, M. Diani and L. Verrazzani, Performance analysis of two adaptive radar detectors against non-Gaussian real sea clutter data, *IEEE Trans. Aerospace and Electronic Systems*, **36**(4), 1429–1439, 2000.

107. G. Tsihrintzis and C. Nikias, Evaluation of fractional, lower-order statistics-based detection algorithms on real radar sea-clutter data, *IEE Proc. Radar, Sonar and Navigation*, **144**(1), 29–38, 1997.

108. S. Haykin, R. Bakker and B. Currie, Uncovering nonlinear dynamics—the case study of sea clutter, *Proc. of the IEEE*, **90**(5), 860–881, 2002.

109. C. Baker, K-distributed coherent sea clutter, *IEE Proc. Radar and Signal Processing*, **138**(2), 89–92, 1991.

110. T. Lamont-Smith, Translation to the normal distribution for radar clutter, *IEE Proceedings Radar, Sonar and Navigation*, **147**(1), 17–22, 2000.

111. A. Farina, F. Gini, M. Greco and L. Verrazzani, High resolution sea clutter data: statistical analysis of recorded live data, *IEE Proc. Radar, Sonar and Navigation*, **144**(3), 121–130, 1997.

112. T. Azzarelli, General class of non-Gaussian coherent clutter models, *IEE Proc. Radar, Sonar and Navigation*, **142**(2), 61–70, 1995.

113. R. Tough, C. Baker and J. Pink, Radar performance in a maritime environment: single hit detection in the presence of multipath fading and non-Rayleigh sea clutter, *IEE Proc. Radar and Signal Processing*, **137**(1), 33–40, 1990.

114. T. Hair, T. Lee and C. Baker, Statistical properties of multifrequency high range resolution sea reflections, *IEE Proc. F, Radar and Signal Processing*, **138**(2), 75–79, 1991.

115. M. Sletten, D. Trizna and J. Hansen, Ultrawide-band radar observations of multipath propagation over the sea surface, *IEEE Trans. Antennas and Propagation*, **44** (5), 646, 1996.

116. G. Corsini, A. Mossa and L. Verrazzani, Signal-to-noise ratio and autocorrelation function of the image intensity in coherent systems. Sub-Rayleigh and super-Rayleigh conditions, *IEEE Trans. Image Processing*, **5**(1), 132–141, 1996.

117. F. Gini and A. Farina, Vector subspace detection in compound-Gaussian clutter. Part I: survey and new results, *IEEE Trans. Aerospace and Electronic Systems*, **38**(4), 1295–1311, 2002.

118. J. Nakayama, Generation of stationary random signals with arbitrary probability distribution and exponential correlation, *IEICE Trans. Fundamentals*, **E77-A**(5), 917–921, 1994.

119. G. Fikhtengolts, *The Fundamentals of Mathematical Analysis*, edited by I. Sneddon, Oxford: Pergamon Press, 1965.

120. J. Proakis, *Digital Communications*, New York: McGraw-Hill, 1995.

121. B. Mandelbrot, *The Fractal Geometry of Nature*, New York: WH Freeman, 1983.

122. H. Chan, Radar sea clutter at low grazing angles, *IEE Proc. Radar and Signal Processing*, **137**(2), 102–112, 1990.

123. E. Conte, M. Longo, M. Lops and S. Ullo, Radar detection of signals with unknown parameters in K-distributed clutter, *IEE Proc. F: Radar and Signal Processing*, **138**(2), 131–138, 1991.

124. C. Pimentel and I. Blake, Enumeration of Markov chain and burst error statistics for finite state channel models, *IEEE Trans. Veh. Tech.*, **48**(2), 415–427, 1999.

125. C. Leung, Y. Kikumoto and S. Sorensen, The throughput efficiency of the go-back-*N* ARQ scheme under Markov and related error structures, *IEEE Trans. Communications*, **36**(2), 231–234, 1988.

126. H. Wang and P.-C. Chang, On verifying the first-order Markovian assumption for a Rayleigh fading channel model, *IEEE Trans. Vehicular Technology*, **45**(2), 353–357, 1996.

127. U. Krieger, B. Muller-Clostermann and M. Sczittnick, Modeling and analysis of communication systems based on computational methods for Markov chains, *IEEE J. Selected Areas in Communications*, **8**(9), 1630–1648, 1990.

128. A. Elwalid and D. Mitra, Effective bandwidth of general Markovian traffic source and admission control of high speed networks, *IEEE/ACM Trans. on Networking*, **1**(3), 329–343, 1993.

129. J. Timmer and S. Klein, Testing the Markov condition in ion channel recording, *Physical Review E*, **55**(3), 3306–3311, 1997.

130. M. Boguna, A. M. Berezhkovskii and G. Weiss, Residence time densities for non-Markovian systems. (I). The two-state systems, *Physica A*, **282**, 475–485, 2000.

131. S. Sivaprakasam and K. S. Shanmugaan, An equivalent Markov model for burst errors in digital channels, *IEEE Trans. Communications*, **43**, 1347–1355, 1995.

132. W. Turin, Fitting probabilistic automata via the EM algorithm, *Commun. Statist.-Stochastic Models*, **12**(3), 405–424, 1996.

133. W. Turin and M. Sondhi, Modeling error sources in digital channels, *IEEE Journal of Selected Areas in Communications*, **11**(3), 340–347, 1993.

134. H. Wang and P. Chang, Finite-state Markov channel—a useful model for radio communication channels, *IEEE Trans. Vehicular Technology*, **44**, 163–171, 1995.

135. M. Zorzi and R. R. Rao, On the statistics of block errors in bursty channels, *IEEE Trans. Communications*, **45**, 660–667, 1997.

136. H. Wang and N. Moayeri, Finite-state Markov channel—a useful model for radio communication channels, *IEEE Trans. Vehicular Technology*, **44**(1), 163–171, 1995.

137. M. Zorzi and R. Rao, Bounds on the throughput performance of ARQ selective-repeat protocol in Markov channels, *ICC'96*, Dallas, Texas, 1996.

138. Y. M. Zorzi, R. Rao and L. Milstein, A Markov model for block errors on fading channels, *Proc. PIMRC'96*, Taiwan, 1996, pp. 1–5.

139. A. Anastasopuolus and K. Chugg, An efficient method for simulation of frequency selective isotropic Rayleigh fading, *Proc. VTC*, 1997, pp. 2084–2088.

140. P. Bello, A generic channel simulator for wireless channels, *Proc. IEEE MILCOM*, 1997, pp. 1575–1579.

141. G. Burr, Wide-band channel modeling, using a spatial model, *Proc. IEEE 5-th Int. Symp. on Spread Spectrum Tech. and Applications*, 1998, pp. 255–257.

142. L.Clavier and M. Rachdi, Wide band 60 GHz indoor channel: characterization and statistical modeling, *Proc. IEEE VTC*, pp. 2090–2102.

143. S. Fechtel, A novel approach to modeling and efficient simulation of frequency selective radio channels, *IEEE Journal on Selected Areas in Communications*, **11**(3), 422–431, 1993.

144. J. Mastrangelo, J. Lemmon, L. Vogler, J. Hoffmeyer, L. Pratt and C. Behm, A new wideband high frequency channel simulation system, *IEEE Trans. on Communications*, **45**(1), 26–34, 1997.

145. R. Parra-Michel, V. Kontorovich and A. Orozco-Lugo, Modeling wideband channels using orthogonalizations, *IEICE Trans. on Electronics and communications*, **E85-C**(3), 544–551, 2002.

146. R. Parra-Michel, V. Kontorovich and A. Orozco-Lugo, Simulation of wideband channels with non-separable scattering functions, *Proc. of ICASSP-2002*, 2002, pp. 111.2829–111.2832.

147. K. Yip and T. Ng, Kahrunen–Loev expansion of WSSUS channel output and its applications to efficient simulation, *IEEE Journal on Selected Areas in Communications*, **15**(4), 640–646, 1997.

148. T. Zwick, D. Didascalou and W. Wysocki, A broadband statistical channel model for SDMA applications, *Proc. of the 5th Int.Symp.on Spread Spectrum Tech. and Applications*, **2**, 527–531, 1998.

149. A. Chockalingam, M. Zorzi, L. Milstein and P. Venkataram, Performance of a wireless access protocol on correlated Rayleigh-fading channels with capture, *IEEE Transactions on Communications*, **46**(5), 644–654, 1998.

150. E. Conte, M. Longo and M. Lops, Modeling and simulation of non-Rayleigh radar clutter, *IEE Proc. F: Radar and Signal Processing*, **138**(2), 121–130, 1991.

151. E. Conte, M. Longo and M. Lops, Performance analysis of CA-CFAR in the presence of compound Gaussian clutter, *IEE Electronics Letters*, **24**(13), 782–783, 1988.

152. M. Rangaswamy, D. Weiner and A. Ozturk, Non-Gaussian random vector identification using spherically invariant random processes, *IEEE Trans. Aerospace and Electronic Systems*, **29**(1), 111–124, 1993.

153. M. Rangaswamy, D. Weiner and A. Ozturk, Computer generation of correlated non-Gaussian radar clutter, *IEEE Trans. Aerospace and Electronic Systems*, **31**(1), 106–116, 1995.

154. R. L. Stratonovich, *Topics in the Theory of Random Noise,* New York: Gordon and Breach, 1967.

155. D. Klovsky, V. Kontorovich and S. Shirokov, *Models of Continuous Communications Channels Based on Stochastic Differential Equations*, Moscow: Radio i sviaz, 1984 (in Russian).

156. V. Pugachev and I. Sinitsin, *Stochastic Differential Systems: Analysis and Filtering*, New York: Wiley, 1987.

157. A. N. Malakhov, *Cumulant Analysis of Non-Gaussian Random Processes and Their Transformations*, Moscow: Sovetskoe Radio, 1978 (in Russian).

158. F. Babich and G. Lombardi, A measurement based Markov model for the indoor propagation channel, *Proc. IEEE VTC-1997*, 77–81, 1997.

159. W. Feller, *An Introduction to Probability Theory and Its Applications*, 3rd Edition, New York: John Wiley & Sons, 1968.

160. L. Sharf and R. Behrens, *First Course in Electrical and Computer Engineering: With MATLAB Programs and Experiments*, Reading, MA: Addison-Wesley, 1990.

161. P. Stoica and R. Moses, Introduction to spectral analysis, Upper Saddle River, NJ: Prentice Hall, 1997.

162. E. Parzen, *Stochastic Processes*, San Francisco: Holden-day, 1962.

163. C. Loo, Digital transmission through a land mobile satellite channel, *IEEE Trans. Communications*, **38**(5), 693–697, 1990.

164. E. Lutz, D. Cygan, M. Dippold, F. Dolainsky and W. Papke, The land mobile satellite communication channel—recording, statistics and channel model, *IEEE Trans. Vehicular Technology*, **40**, 375–386, 1991.

165. R. Barts and W. Stutzman, Modeling and simulation of mobile satellite propagation, *IEEE Trans. Antennas and Propagation*, **40**(4), 375–382, 1992.

166. J. Pierce, A Markov envelope process, *IRE Transactions on Information Theory*, **4**, 163–166, 1958.

167. C. Helmstrom and C. Isley, Two notes on a Markov envelope process, *IRE Transactions on Information Theory*, **5**, 139–140, 1959.

168. J. Pierce, Further comments on A Markov envelope process, *IRE Transactions on Information Theory*, **5**, 186–189, 1958.

169. A. Gray, On Gaussian noise envelopes, *IEEE Trans. Information Theory*, **16**(5), 522–528, 1970.

170. P. Bergamo, D. Maniezzo, A. Giovanardi, G. Mazzini and M. Zorzi, Improved Markov model for Rayleigh fading envelope, *Electronics Letters*, **38**(10), 477–478, 2002.

171. W. Turin, *Performance Analysis of Digital Transmission Systems*, New York: Computer Science Press, 1990.

172. M. Pham, *Performance Evaluation of LMDS Systems Supporting Mobile Users*, Ms. Eng. Sci. Thesis, UWO, 2001.

173. S. Sivarprakasam and K. Shanmugan, An equivalent Markov model for burst errors in digital channels, *IEEE Trans. Communications*, **43**(2–4), 1347–1355, 1995.

174. K. Watkins, *Discrete Event Simulation in C*, New York: McGraw-Hill, 1993.

175. T. Robertazzi, *Computer Networks and Systems*, New York: Springer, 2000.

176. A. Adas, Traffic models in broadband networks, *IEEE Communication Magazine*, **35**(7), 82–89, 1997.

177. A. Baiocchi, F. Cuomo and S. Bolognesi, IP QoS delivery in a broadband wireless local loop: MAC protocol definition and performance evaluation, *IEEE J. Selected Areas in Communications*, **18**(9), 1608–1622, 2000.

178. C. Ng, L. Yuan, W. Fu and L. Zang, Methodology for traffic modeling using two-state Markov-modulated Bernoulli process, *Computer Communications*, **99**(3), 1266–1273, 1999.

179. D. Goodman and S. Wei, Efficiency of packet reservation multiple access, *IEEE Transactions on Vehicular Technology*, **40**(1), 170–176, 1991.

180. D. Goodman, R. Valenzuela, K. Gayliard and B. Ramamurthi, Packet reservation multiple access for local wireless communication, *IEEE Transactions on Communications*, **37**(8), 885–890, 1989.

181. F. Cali, M. Conti and E. Gregori, Dynamic IEEE 802.11: design, modeling and performance evaluation, *Lecture Notes in Computer Science,* CNUCE, Institute of National Research Council Via Alfiery 1–56127, Pisa, Italy

182. F. Babich and G. Lombardi, A Markov model for the mobile propagation channel, *IEEE Transactions on Vehicular Technology*, **49**(1), 63–73, 2000.

183. G. Sater, IEEE 802.16.1mc-00/21r, Media Access Control Layer Proposal for the 802.16.1 Air Interface Specification, July 07, 2000, http://www.ieee802.org/16/mac/index.html.

184. G. L. Stuber, *Principles of Mobile Communication Second Edition*, Norwell: Kluwer Academic Publishers, 2001.

185. Hao Xu, T. S. Rappaport, R. J. Boyle and J. H. Schaffner, Measurements and models for 38-GHz point-to-multipoint radiowave propagation, *IEEE Journal Selected Areas in Communications*, **18**(9), 310–321, 2000.

186. S. Haykin, *Communication Systems*, New York: John Wiley & Sons, 1994.

187. J. Verhasselt and J. P. Martens, A fast and reliable rate of speech detector, *Proceedings ICSLP-96*, **4**, 2258–2261, 1996.

188. J. Yee and E. Weldon, Jr., Evaluation of the performance of error-correcting codes on a Gilbert channel, *IEEE Transactions on Communications*, **43**(8), 2316–2323, 1995.

Index

Analytical Signal 154
Average fade duration 145, 151

Birth process 223
Blanc-Lapierre transform 159, 348

Chapman equation 196, 204, 207, 214
Characteristic function 10, 15, 27, 63, 116,
    228, 254, 343, 367
Clutter
    K-distributed 397
    Weibull 404
Compound process 181
Conditional PDF 28, 62, 182, 287
Correlation coefficient 31, 74
Correlation interval 76–77, 133, 255, 290,
    322, 380
Covariance 31
    mutual 72
    of a modulated carrier 87
    of a periodic RP 85
    of a sum and product of RV 84
    of WGN 95
    properties 73
    relation to PSD 80, 83
Covariance function 67, 69, 71–72
    Jakes' 379
    of λ process 322
    of $\lambda - \omega$ 322
    of a Markov diffusion process 239
    of Gaussian Markov process 241
    of narrow band process 351
    of telegraph signal 200
    of the Wiener process 254
Covariance matrix 89, 181, 268
Cumulant
    brackets 48, 55
    definition 11, 68
    generating function 11
    linear transformation 135

non-linear transformation 52, 112
of a random vector 30
of a sum of random variables 38
relation to moments 16, 48, 68
Cumulant equations 49, 51, 118, 289

Death process 224
Diffusion 229, 235, 312, 330
Diffusion process 217, 260
Drift 215, 235, 312, 322, 330, 335

Envelope 154, 156, 170, 177, 328, 348, 380
Error flow 388

Fading 43, 44, 140, 149, 377
    Loo 378
    Nakagami 378, 382, 385, 391
    Rayleigh 41, 44, 76, 384
    Rice 41
    Suzuki 378
Fokker–Planck equation 217, 227, 393
    boundary conditions 231
    derivation 227
    generalized 287
    large intensity approximation 302, 310
    methods of solution 236
    small intensity approximation 297, 304
FPE, boundary conditions 219, 231, 288, 326
    absorbing 219, 232–233
    natural 232–233
    periodic 232–233
    reflecting 232

Gaussian process 71, 88, 115, 130, 146, 155, 166,
    240, 270, 298
Gaussian RV 32–34

Hilbert transform 154, 358, 400

Independence 29, 31, 44, 87, 89, 155, 282, 419

Stochastic Methods and Their Applications to Communications.
S. Primak, V. Kontorovich, V. Lyandres
© 2004 John Wiley & Sons, Ltd   ISBN: 0-470-84741-7

Joint Probability Density 26, 29, 32, 88, 146, 149, 239, 280

Kolmogorov equation 218, 227, 235
Kolmogorov–Feller equations 285–287, 369
  curtosis 18, 49, 120

Level crossing rate 140, 144, 149, 172, 410, 411
Log-characteristic function 10, 30, 77, 343

Markov chain 190
  ergodic 197
  Fritchman 419
  Gilbert–Elliot 408
  homogeneous 196
  models 408
  of order n 193
  simple 190, 287, 392
  stationary 196
  vector 193
  Wang–Moayeri 409
Markov process 212–213
  binary 209
  classification 189–190
  diffusion 215
  ergodic 214
  homogeneous 209
  with jumps 221
  with random structure 276
Markov sequence 190, 203
Master equation 217
Moment equations 51, 289
Moments 12, 64

Orthogonal polynomials 23, 115, 296, 345

PDF
  Cauchy 184
  Gaussian, *see* Gaussian Process
  K 173, 184, 368, 383, 397
  Nakagami 149, 177, 241, 328, 360, 378
  Poisson 9, 136
  Rayleigh 41, 262, 333, 361, 377
  Rice 41, 378
  Two-sided Laplace 184, 325
  uniform 10, 25, 41, 85
  Weibull 333, 397

  Yule–Farri 224
Pearson PDF 10, 65, 115, 290, 322
Phase 25, 39, 65, 81, 126, 154, 168, 327
Poisson PDF 9
Poisson process 136, 221, 285, 345, 369, 390
Process with jump 215, 217, 221, 285, 296, 335, 339
Process with random structure 275
  classification 279
  examples 290, 292, 298, 300, 309, 312, 313, 391
  Fokker–Planck equation 281, 288
  high intensity approximation 302, 310
  low intensity approximation 298, 304
  moments equation 288
  statistical description 280

Random telegraph process 120, 200

Schrödinger equation 244
SDE 245
  Ito form 251
  relation to Fokker–Planck equation 251, 266
  relation to Kolmogorov–Feller 369
  Stratonovich form 251
  vector 266
Skewness 18
Smoluchowsky equation 195, 227, 339
Spherical invariant process 65, 181
Stochastic integrals 246
  Ito 248
  rules of change of variable 250
  Stratonovich 249
Stratonovich 249, 252

Temporal symmetry 257
Tikhonov 358
Transitional matrix 196, 278, 394, 408, 412, 414
Transitional PDF 199, 280, 204

Vector random variable and process 26, 112, 182, 205, 252, 263, 268

white Gaussian noise 88, 95, 246, 250, 252, 321, 379, 393
Wiener–Khinchin theorem 80